My T. Thai • Panos M. Pardalos
Editors

Handbook of Optimization in Complex Networks

Theory and Applications

Springer

Editors
My T. Thai
Department of Computer and Information
Science and Engineering
University of Florida
Gainesville, FL 32611
USA
mythai@cise.ufl.edu

Panos M. Pardalos
Department of Industrial and Systems
Engineering
University of Florida
303 Weil Hall
Gainesville, FL 32611
USA
pardalos@ufl.edu

Laboratory of Algorithms
and Technologies for
Networks Analysis (LATNA)
National Research University
Higher School of Economics
20, Myasnitskaya st.
Moscow 101000
Russia

M.T. Thai is partially supported by National Science Foundation (NSF) and Department of Defense.

P.M. Pardalos is partially supported by LATNA Laboratory, National Research University Higher School of Economics, RF government grant, ag. 11.G34.31.0057.

ISSN 1931-6828
ISBN 978-1-4614-0753-9 e-ISBN 978-1-4614-0754-6
DOI 10.1007/978-1-4614-0754-6
Springer New York Dordrecht Heidelberg London

Library of Congress Control Number: 2011940819

Printed on acid-free paper

Springer is part of Springer Science+Business Media (www.springer.com)

Springer Optimization and Its Applications

VOLUME 57

Aims and Scope

Optimization has been expanding in all directions at an astonishing rate during the last few decades. New algorithmic and theoretical techniques have been developed, the diffusion into other disciplines has proceeded at a rapid pace, and our knowledge of all aspects of the field has grown even more profound. At the same time, one of the most striking trends in optimization is the constantly increasing emphasis on the interdisciplinary nature of the field. Optimization has been a basic tool in all areas of applied mathematics, engineering, medicine, economics, and other sciences.

The series *Springer Optimization and Its Applications* publishes undergraduate and graduate textbooks, monographs and state-of-the-art expository work that focus on algorithms for solving optimization problems and also study applications involving such problems. Some of the topics covered include nonlinear optimization (convex and nonconvex), network flow problems, stochastic optimization, optimal control, discrete optimization, multi-objective programming, description of software packages, approximation techniques and heuristic approaches.

For further volumes:
http://www.springer.com/series/7393

To our families!

Preface

The oldest, shortest words - "yes" and "no" - are those which require the most thought.

Pythagoras (Greek Philosopher 582 BC – 497 BC)

One common problem that spans several diverse applications is the management and derivation of knowledge from huge amounts of data, especially in scenarios involving human and social activities. In many practical situations, a real-life dataset can be represented as a large network (graph) – a structure that can be easily understood and visualized. Furthermore, special structures of graphs, when viewed in the context of a given application, provide insights into the internal structure and patterns of the data. Among the many examples of datasets that can be represented as graphs are the Web graph derived from the World Wide Web, the Call graph arising in telecommunications traffic data, and metabolic networks arising in biology. Of particular interest are social networks, in which vertices represent people or groups of people.

Although the concept of a network roots back to the ancient Greek philosopher Pythagoras in his theory of cosmos ($\kappa\acute{o}\sigma\mu o\varsigma$), the mathematical principles of networks were first developed in the last century. The first book in networks appeared in 1936 (D. König: Theory of Finite and Infinite Graphs). Since then, there has been a huge explosion of research regarding theoretical tools and algorithms in the analysis of networks.

One of the most exciting moments came at the dawn of the new Millennium, in 1999 with the discovery of new types of graphs, called complex networks. Examples of such well-known classes of complex networks are scale-free networks and small-world networks. These classes of networks are characterized by specific structural features such as the power-law vertex degree distribution (scale-free networks) and for the short path lengths, small diameter, and high clustering (small-world networks). Moreover, several other measures and features have been discovered, and are recently the focus of active research, that related to the structural properties of complex networks. A new area of complex networks has been rapidly developing, spanning several disciplines such as mathematics, physics, computer science, social science, biology, and telecommunications.

In our two volume handbook, an attempt was made to present a wide spectrum of recent developments with emphasis in both theory and applications on complex networks. The first volume focuses on basic theory and properties of complex networks, on their structure and dynamics, and optimization algorithmic approaches. The last part of the volume concentrates on some feature applications. The second volume, this volume, deals with the emerging issues on communication networks and social networks. It covers material on vulnerability and robustness of complex networks. The second part is dedicated to complex communication networks, discussing several critical problems such as traffic activity graph analysis, throughput optimization, and traffic optimization. The last part of this volume focuses on recent research topics on online social networks such as security and privacy, social aware solutions, and social based routing algorithms.

We would like to take this opportunity to thank all authors, the anonymous referees, and Springer for helping us to finalize this handbook. Our thanks also go to our students for their help during the processing of all contributions. We hope that this handbook will encourage research on the many intriguing open questions and applications on complex networks that still remain.

Gainesville, Florida

My T. Thai
Panos M. Pardalos

Contents

Contributors

Yaniv Altshuler Department of Information Systems Engineering and Deutsche Telekom Laboratories, Ben Gurion University of the Negev, Beer Sheva, Israel

MIT Media Laboratory, E15 Cambridge, MA, USA

Alex Arenas Departament d'Enginyeria Informàtica i Matemàtiques, Universitat Rovira i Virgili, Tarragona, Spain

Instituto de Biocomputación y Física de Sistemas Complejos (BIFI), Universidad de Zaragoza, Zaragoza, Spain

G.J. Baxter Departamento de Física, Universidade de Aveiro, Campus Universitário de Santiago, Aveiro, Portugal

Javier Borge-Holthoefer Departament d'Enginyeria Informàtica i Matemàtiques, Universitat Rovira i Virgili, Tarragona, Spain

Instituto de Biocomputación y Física de Sistemas Complejos (BIFI), Universidad de Zaragoza, Zaragoza, Spain

D. Brockmann Northwestern University, Evanston, IL, USA

R. Brune Max Planck Institute for Dynamics and Self-Organization, Göttingen, Germany

Ľuboš Buzna Faculty of Management Science and Informatics, Department of Transportation Networks, University of Žilina, Žilina, Slovak Republic

Guanrong Chen Department of Electronic Engineering, City University of Hong Kong, Hong Kong SAR, China

Juan Chen School of Mathematics and Statistics, Wuhan University, Wuhan, Hubei, China

Alexandre S. da Cunha Department of Computer Science, Federal University of Minas Gerais, Belo Horizonte, MG, Brazil

Shlomi Dolev Computer Science Department, Ben Gurion University of the Negev, Beer Sheva, Israel

S.N. Dorogovtsev A. F. Ioffe Physico-Technical Institute, Petersburg, Russia

Yuval Elovici Department of Information Systems Engineering and Deutsche Telekom Laboratories, Ben Gurion University of the Negev, Beer Sheva, Israel

Zheng Fang Department of Computer Science, University of Massachusetts, Lowell, MA, USA

L.M. Floría Departamento de Física de la Materia Condensada, Universidad de Zaragoza, Zaragoza, Spain

Institute for Biocomputation and Physics of Complex Systems, Zaragoza, Spain

Maria Fonoberova Aimdyn, Inc., Santa Barbara, CA, USA

Pavel Ghosh School of Computing, Informatics and Decision Systems Engineering, Arizona State University, Tempe, AZ, USA

Mark Goldberg Rensselear Polytechnic Institute, Troy, NY, USA

A.V. Goltsev A. F. Ioffe Physico-Technical Institute, Petersburg, Russia

Sergio Gómez Departament d'Enginyeria Informàtica i Matemàtiques, Universitat Rovira i Virgili, Tarragona, Spain

J. Gómez-Gardeñes Institute for Biocomputation and Physics of Complex Systems, Zaragoza, Spain

Weibo Gong Department of Electrical and Computer Engineering, University of Massachusetts, Amherst, MA, USA

D. Grady Northwestern University, Evanston, IL, USA

Jaroslav Janáček Faculty of Management Science and Informatics, Department of Transportation Networks, University of Žilina, Žilina, Slovak Republic

Ľudmila Jánošíková Faculty of Management Science and Informatics, Department of Transportation Networks, University of Žilina, Žilina, Slovak Republic

Stephen Kelley Oak Ridge National Laboratory, Bethel Valley Road, Oak Ridge, TN, USA

Benyuan Liu Department of Computer Science, University of Massachusetts, Lowell, MA, USA

Jun-an Lu School of Mathematics and Statistics, Wuhan University, Wuhan, Hubei, China

Malik Magdon-Ismail Rensselear Polytechnic Institute, Troy, NY, USA

Geraldo R. Mateus Department of Computer Science, Federal University of Minas Gerais, Belo Horizonte, MG, Brazil

Sandro Meloni Department of Informatics and Automation, University of Rome "Roma Tre", Rome, Italy

J.F.F. Mendes Departamento de Física, Universidade de Aveiro, Aveiro, Portugal

Konstantin Mertsalov Rensselear Polytechnic Institute, Troy, NY, USA

Marija Mitrović Department of Theoretical Physics, Jožef Stefan Institute, Ljubljana, Slovenia

Yamir Moreno Instituto de Biocomputación y Física de Sistemas Complejos (BIFI), Universidad de Zaragoza, Zaragoza, Spain

Dung T. Nguyen Department of Computer Information Science and Engineering, University of Florida, Gainesville, FL, USA

Khanh Nguyen Computer Science Department, University of Massachusets, Boston, MA, USA

J. Poncela Institute for Biocomputation and Physics of Complex Systems, Zaragoza, Spain

V.P. Roychowdhury Department of Electrical Engineering, University of California, Los Angeles, CA, USA

Arun Sen School of Computing, Informatics and Decision Systems Engineering, Arizona State University, Tempe, AZ, USA

Yilin Shen Department of Computer Information Science and Engineering, University of Florida, Gainesville, FL, USA

M.V. Simkin Department of Electrical Engineering, University of California, Los Angeles, CA, USA

Fernanda S.H. Souza Department of Computer Science, Federal University of Minas Gerais, Belo Horizonte, MG, Brazil

Ioannis Stavrakakis Department of Informatics and Telecommunications, National & Kapodistrian University of Athens, Ilissia, Athens, Greece

Bosiljka Tadić Department of Theoretical Physics Jožef Stefan Institute, Ljubljana, Slovenia

My T. Thai Department of Computer Information Science and Engineering, University of Florida, Gainesville, FL, USA

F. Theis Group leader CMB Professor for Mathematics in Systems Biology,

Institute of Bioinformatics and Systems Biology, Helmholtz Zentrum München, Ingolstädter Landstraße 1, 85764 Neuherberg, Germany

C. Thiemann Max Planck Institute for Dynamics and Self-Organization, Göttingen, Germany

Duc A. Tran Computer Science Department, University of Massachusetts, Boston, MA, USA

Al Wallace Rensselear Polytechnic Institute, Troy, NY, USA

Jie Wang Department of Computer Science, University of Massachusetts, Lowell, MA, USA

Choujun Zhan Department of Electronic Engineering, City University of Hong Kong, Hong Kong SAR, China

Part I
Basic Theory and Properties

Chapter 1
Optimization in Designing Complex Communication Networks

Fernanda S.H. Souza, Geraldo R. Mateus, and Alexandre Salles da Cunha

Abstract Complex networks are found in real world in different areas of science, such as technological, social and biological. These networks are many times characterized by a non-trivial topology, with connection patterns among their elements that are neither purely regular nor purely random. The interesting features presented by this class of networks may be useful in improving the overall efficiency of engineered networks as computer, communication and transportation ones. There is a conjecture indicating that such complex topologies normally appear as a result of optimization processes. Optimization techniques have been applied to design complex communication networks, showing that features such as small path length, high clustering coefficient and power-law degree distribution can be achieved through optimization processes. In this chapter, models and algorithms based on optimization techniques to generate complex network topologies are discussed. We review some models, heuristics as well as exact solution approaches based on Integer Programing methods to generate topologies owning complex features.

1.1 Introduction

The Network Science concept has its roots in graph theory dating back to the 1730s. However, it is evolving, since it reemerged in the late 1990s as a new science [1]. Even though a definitive description of its meaning is still open, there are many ways to define it, for sure. Network science involves the study of the theoretical foundations of network structure, dynamic behavior, and its application to many subfields [2–6]. Thus, the study of topics like structure, topology, emergence,

F.S.H. Souza (✉) • G.R. Mateus • A.S. da Cunha
Department of Computer Science, Federal University of Minas Gerais, Belo Horizonte, MG, Brazil
e-mail: fersouza@dcc.ufmg.br; mateus@dcc.ufmg.br; acunha@dcc.ufmg.br

M.T. Thai and P.M. Pardalos (eds.), *Handbook of Optimization in Complex Networks: Theory and Applications*, Springer Optimization and Its Applications 57,
DOI 10.1007/978-1-4614-0754-6_1,

dynamism, autonomy, and so on is certainly of great importance in the field. In particular, the origin of real world complex networks is constantly a topic of interest, as an attempt to clarify what kind of processes naturally is taking place on these networks. On the other hand, there is a conjecture whether such complex topologies normally appear as a result of some optimization processes. A challenge can be to identify these processes and properties and apply them to manage or to design the network from scratch.

Complex systems are found in real world in different areas of science, including the Internet, WWW, neural networks, friendship relationships, among others [4, 7–10]. All these networks are large scale networks and very different from traditional network problems explored by operations researchers. Also, they have an intense amount of activities and behaviors that cannot be fully explained. Since then, networks have been used to model and simulate complex interactions among elements of a system, providing support for a better understanding and analysis. Surely, it is important to emphasize that complex systems differ from complicated systems [6]. Large scale systems can be considered complicated, although their components and behaviors are well known. However, complex systems show diverse behaviors, not always known or predictable.

Complex networks can be defined as large scale networks with an intricate relationship among their components and many degrees of freedom in the possible actions of components [11]. In this context, the concept of complex is based on behaviors exhibited by the network that arise naturally and unplanned. On the other hand, a complicated network is also a large scale network where the components and the rules governing its functioning are known [6]. In this case, complex is associated with the difficulties to solve the problems using traditional approaches, including the computational complexity.

Optimization approaches have been applied successfully to solve engineering problems [12, 13]. Given the network structure with their components, interactions and constraints, the objective is to optimize a well known function resulting from that structure. This solution allows the user to control and design the network.The models consider costs, performance, resource, and design constraints. In complex networks, we have the inverse problem, or the reverse engineering, where the objective is to know how the observed structure supports a perceived function [11].

Complex networks are many times characterized by a non-trivial topology and present interesting features which may be useful in designing engineered networks. One of these features concerns the cheap price for sending information (or a packet, a commodity) through the network. Thereby, computer, communication, and transportation networks, just to name a few, could take advantage of being modeled to present specific complex features, to improve their overall efficiency. On the other hand, it is expected that the structure and function of a complex network can be interpreted from some optimization process. Consequently, the existing complex network can also be redesigned or restructured or a new network can be designed from scratch.

Regarding these considerations, the objective of this chapter is the investigation of how optimization strategies could be applied in the context of complex communication networks. We show that a given network can be tuned to satisfy a desired property or set of properties and to avoid others. This tuning can be guided by a set of patterns previously defined. This is possible by changing the structure level, i.e., modifying the physical topology of the network to get improvements in a function.

This chapter is organized as follows. Section 1.2 presents the main network metrics used in complex theory. These metrics are explored by the optimization models. In Sect. 1.3, different classical network structures and models are introduced; in particular, we present the regular, random, small world, and scale-free concepts of networks.

Section 1.4 presents optimization models and algorithms to create complex network topologies. Two mathematical formulations are proposed, the Arc-Flow and the Arc-Path. Both are Integer Linear Programs to treat the same combinatorial optimization problem. The first is based on network flows and is solved by a commercial software that is based on Branch-and-bound (BB). The Arc-Path Formulation is an implicit and stronger formulation, but demands specific methods as column generation and Branch-and-price [14, 15]. Variations of the proposed formulations are explored in the end of the section in order to capture diverse network features and properties.

Section 1.5 shows how complex networks can be reached through a heuristic approach. Some of the classical heuristic algorithms are discussed and some computational results are reported based on GRASP [16]. Comparisons between the heuristic solutions and the optimal solutions obtained with the mathematical formulations are presented to show the quality of the GRASP approach.

This chapter is concluded with final remarks in Sect. 1.6.

1.2 Measurements of Complex Networks

Complex systems have been modeled through network representation, making possible the analysis of topological features using informative measurements. A central issue in the study of complex systems is understanding the relationship between system structure and function. Network metrics are, therefore, of great importance while investigating network representation, characterization and behavior. This Section is devoted to the presentation of the key measurements of networks which will be discussed along the chapter.

Let an undirected graph $G = (V, E)$ where V is the set of vertices and E is the set of edges (also called links) connecting the nodes. The *degree* of a vertex $i \in V$ is the number of edges incident to vertex i and the *degree distribution* is the probability distribution of these degrees over the whole network. The *density* of a graph is the ratio between the number of edges and the upper bound on the number of edges.

A path connecting two vertices i, $j \in V$ is said to be minimal, if there is no other path connecting i to j with fewer links. Accordingly, the *average path length* of G is given by the average number of links in all shortest paths connecting all pairs of

Table 1.1 Network metrics

Metric	Formula
Number of vertices	$n = \|V\|$
Number of edges	$m = \sum_{i,j \in V: i<j} a_{ij}$ where $a_{ij} = \begin{cases} 1, & \text{if vertices } i \text{ and } j \text{ are connected} \\ 0, & \text{otherwise} \end{cases}$
Degree	$d_i = \sum_{j \in V} a_{ij}$
Degree distribution	$P(k) = \frac{n_k}{n}$ where n_k is the number of vertices with degree k
Density	$\rho = \frac{2m}{n(n-1)}$
Average path length	$L = \frac{1}{n(n-1)} \sum_{i,j \in V: i \neq j} d_{ij}$ where d_{ij} is the distance between vertices i and j
Diameter	$D = max\left\{d_{ij}\right\},\ \forall i,j \in V, i \neq j$
Clustering coefficient of a vertex	$C_i = \frac{2e_i}{r_i(r_i - 1)}$ where e_i is the number of edges between neighbors of i and r_i is the number of neighbors of vertex i
Clustering coefficient of a graph	$CC = \frac{1}{n} \sum_{i \in V} C_i$
Betweenness centrality	$B_i = \sum_{s,t \in N: s \neq t} \frac{\sigma(s,i,t)}{\sigma(s,t)},\ s \neq i, t \neq i$ where $\sigma(s,i,t)$ is the number of shortest paths between vertices s and t that pass through vertex i and $\sigma(s,t)$ is the total number of shortest paths between s and t
Global efficiency	$GE = \frac{1}{n(n-1)} \sum_{i,j \in V: i \neq j} \frac{1}{d_{ij}}$

vertices in V. The graph *diameter* is the maximum shortest path length between all pairs of vertices in V. The *clustering coefficient of a vertex i* is the ratio between the number of edges between neighbors of vertex i and the upper bound on the number of edges between them. For instance, given $i, j, k \in N$ and assuming that edges $(i, j), (i, k) \in E$, the clustering coefficient defines the probability that (j, k) also belongs to set E. The *clustering coefficient of a graph* is the average value of the clustering coefficients of all vertices in G. The *betweenness centrality* of a vertex i is associated with an importance measure, based on the number of paths between other pairs of vertices that include vertex i. The *global efficiency* of a network quantifies the efficiency in sending information between vertices, assuming that for a pair of vertices i and j it is proportional to the reciprocal of their distance. Table 1.1 summarizes the mathematical formulas for the main network metrics outlined above. See [17] for a complete review of measurements.

1.3 Network Models

Several network models were proposed and studied in an attempt to represent elements of a system and their relationships [1, 4–6]. In the following Sections, different network structures and the models used to generate them are introduced. Network models can be classified into static and dynamic. Three static and one dynamic network models are discussed.

1.3.1 Regular Networks

Regular networks are characterized by an associated regular graph structure where the connections between vertices follow a common pattern. They have some theoretical importance since other models can be derived by rewiring the regular networks. Early work on the design of complex networks [7, 18] contrasted the features of regular and random graphs. A regular network with vertices of degree k is called a k-regular network or regular network of degree k. Sparse k-regular networks are known to own high average path length and high clustering coefficient. As we will show in Sect. 1.3.3, k-regular networks are used on the design of small world networks.

Other examples of regular networks include rings, lattices, n-ary trees, stars and full or complete graphs [1]. Ring networks are a special case of a connected k-regular network, in which $k = 2$. In a lattice, the vertices are placed on a grid and connected to their immediate neighbors. A n-ary tree consists of a connected network without cycles, with a root vertex and each vertex which is not a leaf having at most n children. A star network is a special tree, where every vertex is connected to the root. In a full or complete network there is an edge between every pair of vertices. Figure 1.1 illustrates regular network structures. Figure 1.1a, in particular, depicts a 4-regular ring network.

1.3.2 Random Networks

Random networks have been studied since the 1950s, when they were independently defined by Erdös and Rényi [19] and Gilbert [20]. A random network is generated by a random process, in which a set of edges are added at random between pairs of vertices belonging to the network. The class of random networks contrasts directly to that of regular networks mainly in the structure aspect, being a useful baseline for comparison.

The main idea of the Gilbert [20] random network model is to add edges independently with probability p $(0 < p < 1)$ from the $n(n-1)/2$ potential edges of an undirected graph. Let $G(n, p)$ denote a graph G with n vertices and an associated

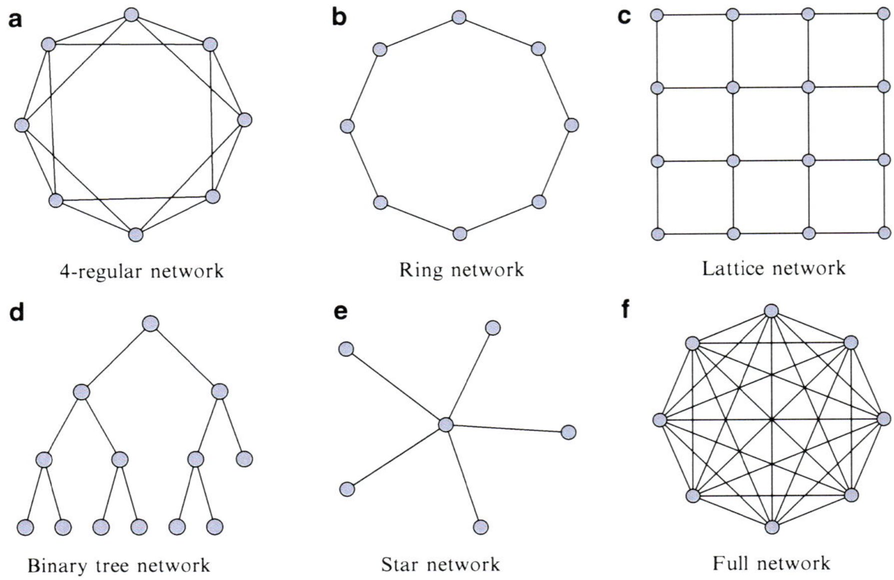

Fig. 1.1 Regular networks

probability p. One may note that the number of edges of a network added according to the $G(n,p)$ model is not known in advance. Moreover, the total number of possible graphs sums up to $2^{n(n-1)/2}$. On average, the resulting network ends up with $m = p[n(n-1)/2]$ and p also corresponds to the density of the network.

Another random network model, also known as ER model, was proposed by Erdös and Rényi [19]. Unlike Gilbert's procedure, the ER model is characterized by generating networks with a previously known fixed number m of edges. In the $G(n,m)$ model, equal probability is assigned to all graphs with exactly m edges. In other words, considering a stochastic process that starts with n vertices and no edges and at each step adds one new edge chosen uniformly from the set of missing edges, $G(n,m)$ represents a snapshot at a particular time (m) of this process.

The difference between ER and Gilbert models is that the ER model generates a network with a certain number of edges while the Gilbert model generates a network with a defined density. However, in both models the probability that a given vertex has degree k approaches a Poisson distribution for $n >> 1$, i.e., $P(k) = <k>^k e^{-k}/k!$, where $<k>$ is the average vertex degree. This means that random graphs tend to be homogeneous in vertex degree as the majority of the vertex degrees are close to the average value. The randomness attached to this class of networks induces properties based on two of the metrics presented before, small average path length and small clustering coefficient. Figure 1.2 shows an example of a random network (Fig. 1.2a) and its Poisson degree distribution curve (Fig. 1.2b).

For a long time, random networks were widely studied and used to model complex systems. Indeed, real world networks present an average path length

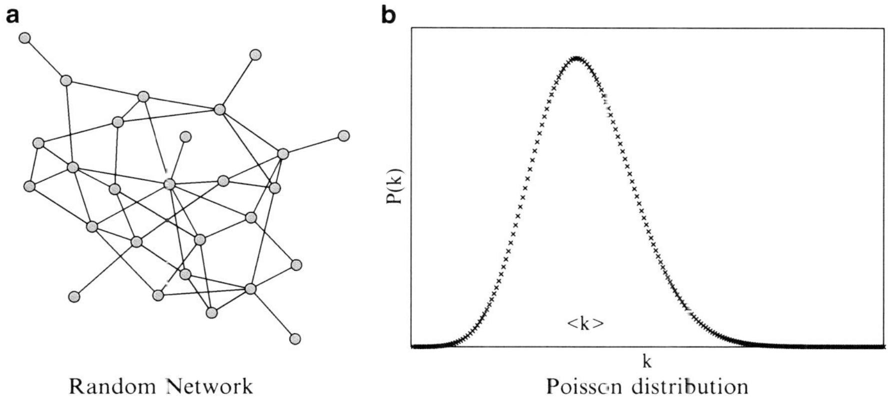

Fig. 1.2 Random network

close to the average path length of a random network with the same number of edges. However, in the last few years it was noticed that general properties of real world networks are quite different from random ones. For instance, the clustering coefficient of a network found in nature is remarkably larger than the clustering coefficient of a random network with the same number of vertices and edges. It seems that real networks present a kind of local interaction not observed in random ones. Furthermore, the typical degree distribution found in nature is significantly different from a Poisson distribution. Thus, new models were developed as Small World and Scale Free networks.

1.3.3 Small World Networks

The small world concept, first introduced from Milgram's experiment [21], showed that the "world is small" because a person can reach all other people in the world, directly or indirectly, through few intermediaries. In [7, 18], the authors formalized the small world concept and defined small world networks. The small world phenomenon is found in various networks like the Internet and the routing of messages in social environments [5, 22, 23]. Small world graphs are intriguing, because, among other reasons, they share characteristics of regular graphs (high clustering coefficient) and random graphs (small average path length).

In [7], a simple procedure to generate a small world network based on rewiring edges of the network was proposed by Watts and Strogatz (WS model). The WS model of a small world network is described as follows. Given a k-regular ring graph $G = (V,E)$ where each node is connected to its first k neighbors, rewiring edge $(i, j) \in E$ according to a probability p $(0 \leq p \leq 1)$, consists in randomly replacing one of its endpoints i or j by another vertex q. After all edges of E are attempted to be rewired, one at a time, multiple edges and loops are not allowed, another type

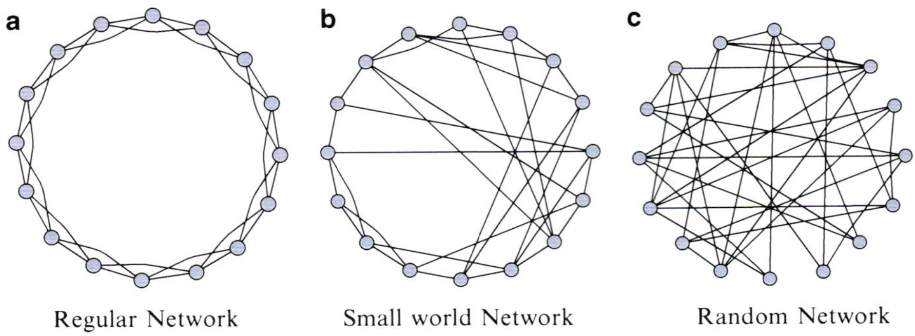

Fig. 1.3 WS model

of graph may emerge from this process. Depending on which probabilities are used, the graph obtained may exhibit small world features. An example of the rewiring procedure is presented in Fig. 1.3. Starting with the 4-regular graph ($p = 0$) depicted in Fig. 1.3a, two other types of graphs can be obtained. If a small value of p is used, a small world structure like that in Fig. 1.3b arises. On the other hand, if larger values of p are used, random graphs like that in Fig. 1.3c may appear.

A similar procedure grounded in a slight improvement of the method was proposed by Newman and Watts [18]. Instead of rewiring edges, the addition model starts with a k-regular graph $G = (V, E)$ and then adds new edges, according to probability p. As in the rewiring process, depending on the probability p, different graph structures may appear: small world graphs if p is small and random graphs if larger values of p are used.

This class of networks presents a high clustering coefficient for small values of p as the procedure starts with a regular graph, which has a high clustering coefficient value. However, the average path length falls significantly, since the random rewiring or addition of an edge works as a shortcut in the network, decreasing the distance among the vertices. This particular feature is extremely profitable in communication networks and indeed resides in real world networks such as social networks. As the global efficiency of the network (defined in Table 1.1) is based on the distance among its elements, the overall efficiency of small world networks is said to be improved comparing to regular networks.

1.3.4 Scale Free Networks

The dynamic behavior of real world systems leads to the emergence of another important class of networks, known as scale free [2–4, 8]. Static models formerly presented are not able to capture the constant growth of a large scale network or how to attach new vertices and to connect them to existing ones. Scale free networks, in contrast to random ER graphs that follow a Poisson distribution, are characterized by a power-law degree distribution, in which there is a small number of high-degreed vertices and a large number of low-degreed vertices. The few vertices with high degree are usually called *hubs*, resulting in a network with skewed degree distribution. A power-law distribution follows the form $P(k) \sim k^{-\gamma}$, where k is the degree ($1 < k < \infty$) and γ is an exponent ($2 < \gamma < 3$).

Barabási and Albert [24] introduced the idea of evolving networks and addressed the origin of this power-law degree distribution in many real networks in a pioneering work. As previously noticed, contrary to the idea that real networks could be represented by random networks, it was proven that many real networks obey a power-law degree distribution instead of a Poisson distribution. Examples include the WWW, the Internet, railroads, and protein–protein interaction networks [4], just to name a few.

The Barabási-Albert network, also called BA network, is generated through a constructive procedure known as *preferential attachment*. The preferential attachment is biased (not random), in which new vertices entering the network do not connect uniformly to existing nodes, but attach preferentially to vertices of a higher degree. This model better represents the evolving real world networks, creating hubs in an unequal addition of new components. It starts with a small number (m_0) of connected vertices and assume that every time step a new vertex is added, and $m \leq m_0$ edges are connecting the new vertex to m different vertices already present in the network.The preferential attachment is incorporated assuming that the probability Π_i that a new vertex will be connected to the existing vertex i depends on the degree k_i of that vertex, so that $\Pi_i = k_i / \sum_j k_j$. After t time steps, the model leads to a random network with $t + m_0$ vertices and m_t edges.

The main property of this class of networks is the extremely high hub degree. This also means that the betweenness values of these vertices tend to be high, since they can participate in many paths connecting vertices in the network. The average path length of this class grows as $\log(n)/\log(\log(n))$ and thus displays the small world property. A linear relationship between clustering coefficient versus number of edges is found in a scale free network. This means that the clustering coefficient increases linearly with density ($CC \sim O(\rho)$). At last, it was observed that random failures do not affect the usual operation of the network as the majority of vertices are those with a small degree and the likelihood that a hub be affected is almost minimum. On the other hand, if the vertices chosen to quit the network are specific, it may be turned into a set of isolated graphs easily. A scale free network is presented in Fig. 1.4a, while Fig. 1.4b shows a power-law degree distribution.

1.3.5 Summary of Network Models

Table 1.2 shows a summary of the features for each one of the network models presented.

A comparison between complex and engineered networks was introduced in the first section and optimization techniques can be applied in both contexts.

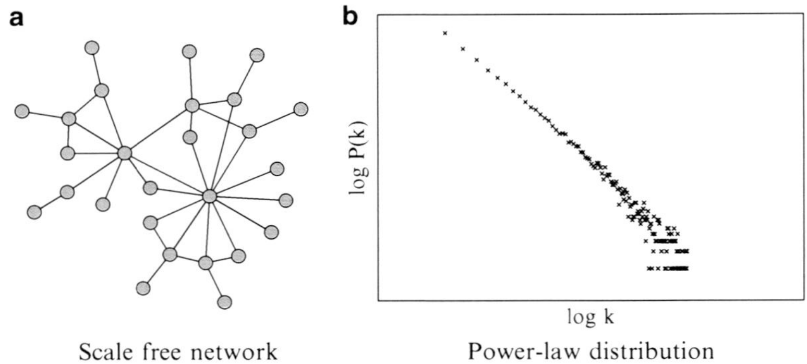

Fig. 1.4 Scale-free network

Table 1.2 Network models' features

Network model	Features
Regular networks	– High average path length – High clustering coefficient
Random networks	– Small average path length – Small clustering coefficient – Poisson degree distribution
Small world networks	– Small average path length – High clustering coefficient
Scale free networks	– Small average path length – High betweenness centrality – Power-law degree distribution (small number of high-degreed vertices and high number of low-degreed vertices)

To illustrate this idea, let us take a telecommunication network as motivation to establish a comparison. Although engineering metrics have been used for a long time as optimization criteria, it is possible to optimize complex metrics to achieve similar objectives. Thus, four widely known metrics in traffic engineering for telecommunication networks were chosen to establish our comparison: delay, load balancing, resilience, and vulnerability. Table 1.3 presents the relation between engineering and complex metrics.

1.4 Optimization Models for Complex Networks

As pointed out in Sect. 1.3, early work in complex systems has focused mainly on how complex networks can be obtained by means of stochastic algorithms. In contrast, this section is dedicated to introduce optimization models representing networks that, if solved to optimality by exact solution algorithms, allow such

Table 1.3 Engineering metrics × Complex metrics

Telecommunication metric	Complex metric
Delay	– Small average path length – High betweenness centrality
Load balancing	– High average path length – Small betweenness centrality
Resilience	– High average path length – High clustering coefficient – Poisson degree distribution
Vulnerability	– Small average path length – High betweenness centrality – Power-law degree distribution

networks to exhibit complex features. Our goal is to show that some complex features (such as small path length, high clustering coefficient and power-law degree distribution), often desirable for engineered networks outside the complex network domain, may arise as a result of deterministic optimization processes and algorithms.

In order to attain the goals we have set, we start with a network given by a directed graph $D = (V, A)$ having weights and capacities assigned to its arcs and state the *core* optimization problem we deal with in the section, the Optimal Topology Design Problem (OTDP). We formulate OTDP as two different Integer Programs (IPs) and describe algorithms for solving each of them. As we detail the models and algorithms, we review some basic aspects on Integer Programing. Advantages of one program (and associated algorithm) over the other are also highlighted. We close the section indicating how small modifications in the core optimization problem allow us to obtain engineered networks where other complex features arise.

1.4.1 *The Optimal Topology Design Problem*

The OTDP for complex networks was introduced by [25, 26] and is defined as follows. Given a complete directed graph $D = (V, A)$ with set of vertices V and arcs A, costs $\{c_{ij} = c_{ji} \geq 0 : \forall i, j \in V, j \neq i\}$ assigned to the arcs of A, and a wire budget B, OTDP consists in defining a subset S of arcs of A such that $\sum_{(i,j)\in S} c_{ij} \leq B$, the subgraph (V, S) of D is connected and exhibits complex network features. By minimizing the average path length among all pairs of vertices under a limited budget, complex features may arise as we show in the sequence.

An IP to model an optimization problem like OTDP can be defined in many different ways, depending on our choices to select decision variables, to state the constraint set and the objective function (the function we wish to minimize). Usually, there is a close connection between the way the model is formulated and the algorithms we devise to solve it.

Two Integer Programing Formulations for OTDP were proposed in [26]. In the first one, named Arc-Flow Formulation, connectivity between each pair of vertices is enforced through network flows [27]. In the second one, the Arc-Path Formulation, connectivity is guaranteed by imposing that one path connecting every pair of vertices must be available in the subgraph of D implied by the arcs selected in a solution. In the next two sections, we discuss the two formulations. Each one leads to a different exact algorithm.

1.4.1.1 Arc-Flow Formulation and BB Algorithm

Let us assume that, given $D = (V,A)$, A_j^- and A_j^+ respectively denote the set of arcs arriving and leaving $j \in V$. To model OTDP, we make use of the following sets of decision variables:

- $\{x_{ij}^{st} \in \mathbb{R}_+ : (i,j) \in A, \forall s,t \in V, s \neq t\}$. Variable x_{ij}^{st} indicates the amount of flow of a commodity that is sent from vertex s to t, that passes through arc (i,j).
- $\{m_{ij} \in \{0,1\} : (i,j) \in A\}$. Variable m_{ij} takes value 1 if arc (i,j) is included in the solution we are aiming for (0, otherwise).

OTDP can now be stated as the following IP

$$w = \min \sum_{s\in V} \sum_{t\in V} \sum_{(i,j)\in A} x_{ij}^{st} \tag{1.1}$$

s.t.

$$\sum_{j\in A_s^+} x_{sj}^{st} = 1 \; \forall s,t \in V, s \neq t, \tag{1.2}$$

$$\sum_{i\in A_t^-} x_{it}^{st} = 1 \; \forall s,t \in V, s \neq t, \tag{1.3}$$

$$\sum_{i\in A_j^-} x_{ij}^{st} - \sum_{k\in A_j^+} x_{jk}^{st} = 0 \; \forall s,t,j \in V, s \neq t, s \neq j, t \neq j, \tag{1.4}$$

$$\sum_{(i,j)\in A: i<j} c_{ij} m_{ij} \leq B, \tag{1.5}$$

$$x_{ij}^{st} - m_{ij} \leq 0 \; \forall s,t \in V, \forall (i,j) \in A, \tag{1.6}$$

$$m_{ij} - m_{ji} = 0 \; \forall (i,j) \in A, \tag{1.7}$$

$$0 \leq x_{ij} \leq 1 \; \forall (i,j) \in A, \tag{1.8}$$

$$m_{ij} \in \{0,1\} \; \forall (i,j) \in A. \tag{1.9}$$

Constraints (1.2)–(1.4) impose flow balance conditions for each pair of vertices s,t. Note that (1.2) ensures that one unity of a commodity will be sent from s

to t, for every possible pair of distinct vertices $s,t \in V$, while (1.3) guarantees that every commodity sent by s to t will arrive its destination vertex. Constraints (1.4) guarantee the flow conservation of a commodity associated with the pair s,t in vertices that are neither the origin nor the destination of the flow. Inequalities (1.6) couple flow and binary variables, imposing that an arc cannot be used to send flow if it is not included in the solution. Constraints (1.7) impose that whenever arc (i,j) is selected to be in a solution, so is (j,i). The knapsack inequality (1.5) assures that the selected arcs do not violate the budget. Note that whenever arcs (i,j) and (j,i) are included in the solution, the cost c_{ij} is considered only once in the left hand side of (1.5). Finally, objective function (1.1) aims to minimize the average path length among all pairs of vertices. Note that in model (1.1)–(1.9), variables x_{ij}^{st} were not imposed to assume binary values. However, due to constraints (1.6) and (1.9), whenever $\{m_{ij} : (i,j) \in A\}$ variables assume integer values, $\{x_{ij}^{st} : (i,j) \in A, s,t \in V, s \neq t\}$ variables also do.

One approach to solve IPs like (1.1)–(1.9) is to provide valid lower $\underline{w}$ and upper $\overline{w}$ bounds on the optimal objective function w, such that $\overline{w} = \underline{w} = w$. Usually, this task is conducted by an algorithm that provides a sequence of non-decreasing lower bounds (through relaxations or from duality) and non-increasing upper bounds (through heuristics, for example) that, in a given moment, converge to the same (optimal) value. One widely known algorithm to solve IPs like (1.1)–(1.9) that explores such idea is the BB method (see Chap. 7 in [28] for further details). Depending on how the lower bounds are calculated, different types of BB algorithms arise. In the sequence, we illustrate a Linear Programing (LP) based BB algorithm tailored for OTDP model (1.1)–(1.9).

For the particular case considered here, BB calculates lower bounds on w through LP relaxations of model (1.1)–(1.9). A LP relaxation of model (1.1)–(1.9) is given by the LP problem obtained when constraints (1.9) are replaced by their continuous version: $\{0 \leq m_{ij} \leq 1 : (i,j) \in A\}$. Note that when the integrality on m_{ij} variables is relaxed, the set of feasible solutions of the (relaxed) Linear Program contains all the solutions of the IP it derived from. Consequently, the optimal value of the (relaxed) Linear Program is a valid lower bound on w. Usually, in LP based BB algorithms, the LP relaxations of the IP are computed through the Simplex Algorithm [29] proposed by George Dantzig.

Assume now that the optimal solution to the LP relaxation of (1.1)–(1.9) is given by vector $(\underline{x},\underline{m})$ and that the optimal objective function to the LP relaxation is $\underline{w}$. If $(\underline{x},\underline{m})$ has only $0-1$ entries, constraints (1.9) are satisfied and, thus, $(\underline{x},\underline{m})$ is an optimal solution to the original IP as well. Note also that if $(\underline{x},\underline{m})$ has only $0-1$ entries, $\underline{w}$ provides, at the same time, valid lower and upper bounds for w.

Let us now assume that $(\underline{x},\underline{m})$ has at least one entry (say, variable $m_{pq} : \underline{m}_{pq} \notin \{0,1\}$ for a given $(p,q) \in A$) that is not binary and that a valid upper bound on w, $\overline{w}$, is available (if not, set $\overline{w} = \infty$). If $\underline{w} < \overline{w}$, to proceed with the resolution, BB must resort to some kind of enumeration of the solution space. This is accomplished by following the *divide and conquer* paradigm. Two new IPs are created, with the same objective function (1.1). In one of them, one variable for which the corresponding entry in $(\underline{x},\underline{m})$ is not integer is imposed to be at its (integer) lower bound. In the other, the same variable is fixed to its (integer) upper bound. More precisely, assuming that m_{pq}: $\underline{m}_{pq} \notin \{0,1\}$ is the *branching variable*, we create two new IPs. The constraint set of the first is (1.2)–(1.9) and $m_{pq} = 0$, while the constraint set of the second is (1.2)–(1.9) and $m_{pq} = 1$. Note that if the two new IPs are solved to optimality, we actually solve the original one. This is true since the whole domain of the original program is covered by the union of the domains of the two programs that derived from it. Therefore, by following the approach outlined above, the problem of solving the original IP was replaced by the problem of solving two new IPs, each one defined on a smaller domain. Let us call the IPs that were created as a consequence of branching as IP subproblems or simply subproblems. Both are stored in a list of subproblems to be investigated. After implementing branching, BB iteratively picks one subproblem from the list and tries to solve it, using the LP relaxation procedure outlined above. BB finishes when the list of subproblems is empty and, therefore, the original IP was solved.

When a subproblem in the list is picked and its LP relaxation is computed, the following cases may occur:

- The LP relaxation gives a lower bound $\underline{w} : \underline{w} < \overline{w}$. We proceed as explained above, picking a variable to branch on and creating two new subproblems that are added to the list.
- The LP relaxation is unfeasible. In this case, the IP subproblem is unfeasible as well and the subproblem is said to be *pruned by unfeasibility*. In this case, BB just picks another subproblem from the list and the search goes on.
- The LP lower bound implied by the LP relaxation is equal or greater than the best known valid upper bound $\overline{w}$ for the original IP. In this case, it does not make sense to branch and to create two new subproblems from the current one, since any feasible solution to them costs at least $\overline{w}$. The subproblem is thus said to be *pruned by bounds*.
- The solution to the LP relaxation is integer feasible. As explained above, the LP relaxation solution solves the subproblem and no branching is needed. In this case, the subproblem is said to be *pruned by optimality*. If the implied upper bound improves on the best known upper bound, the latter is updated.

Putting in another way, BB is a divide and conquer algorithm that partitions the original solution space into a number of smaller subsets and then solves the resulting smaller IPs in a recursive fashion. Usually, the algorithm starts with an upper bound $\overline{w} = \infty$. As new subproblems are solved and their LP relaxations turn out to be integer feasible, the best upper bounds are updated. At the end, when the list of subproblems is empty, the best lower and upper bounds match.

It is common to associate the set of subproblems in the list with a tree, called the BB search tree. Each *node* in the tree connects to its parent (the subproblem it derived from) and to its children (the subproblems that were originated from it).

Therefore, we interchangeably use the terms IP subproblem or tree node to refer to one subproblem in the list. The original IP, in this sense, is called the root node of the search tree.

The practical performance of BB algorithms, i.e., the rate in which the sequence of the best lower and upper bounds converge to the optimal solution, depends heavily on the following aspects:

- The quality of the lower bounds implied by the LP relaxations of the subproblems. The stronger the lower bounds (i.e., the greater the $\underline{w}$ figures), the more likely BB becomes to prune subproblems by bounds. As a consequence, fewer subproblems are created.
- The computing time needed by the Simplex Algorithm to solve each LP relaxation. Usually, the larger the number of constraints and variables in the Linear Program, the larger the computing time needed to solve the LP relaxations.
- How fast valid upper bounds derived from integer feasible solutions are found through the search. In order to prune nodes by bounds, BB must have a valid upper bound on w.
- How variables on which to branch are chosen (the *branching policy*).
- How subproblems are picked from the list (the *node selection policy*). In order to choose a node to explore from the list, many approaches may be considered, such as the breadth first, the depth first and the best bound. In the breadth first, the nodes of the tree are explored in the same order in which they were created. Depth first selection policy explores the last node created, going deeper in the BB tree. The best bound or best first chooses the node having the lowest value of the LP relaxation among all BB nodes.

Several commercial solvers for IPs that operate under the LP based BB framework are available. Such optimization packages include an implementation of the Simplex Algorithm, as well other procedures to manage the BB tree (to implement branching, node selection, etc.).

For solving OTDP by a LP based BB algorithm, we have chosen the state-of-the-art commercial solver CPLEX [30]. The advantage of the LP based BB approach we just described for OTDP is that, after model (1.1)–(1.9) is stated and loaded into an optimization package, little additional programing effort is needed, once one has in hands an Integer Programing solver like CPLEX.

Formulation (1.1)–(1.9), however, has $O(n^4)$ variables and constraints. Therefore, even with the help of a highly sophisticated optimization package like CPLEX, only OTDP instances of limited size are expected be solved to proven optimality by LP based BB algorithms that rely on this formulation. Therefore, the drawback of the approach we have just described is that, in practice, only limited size instances of OTDP can be actually solved in a reasonable amount of time.

In order to overcome this difficulty, in the following, we present a reformulation for OTDP that, despite having exponentially many variables, can lead to a specialized BB procedure, named Branch-and-price (BP) algorithm [14, 15]. Roughly speaking, the procedure consists in applying the Delayed Column Generation method [31] to derive lower bounds to be used in the search.

Another motivation for using the Delayed Column Generation method is to provide stronger LP bounds through the BB search. In doing so, fewer BB subproblems should be solved. In the sequence, we briefly explain how the technique works, for the particular case of OTDP. For additional details regarding Delayed Column Generation algorithms and Branch-and-price implementations, see [14, 31–33].

1.4.1.2 Arc-Path Formulation, Delayed Column Generation and Branch-and-Price Algorithm

Given a pair of vertices $s,t \in V, s \neq t$, assume that P^{st} denotes the set of all simple directed paths connecting s to t in D. The main idea of the Arc-Path Formulation to enforce connectivity is to impose that, for every s,t, out of all paths in P^{st}, exactly one that uses only the arcs included in the solution will be used to compute the average paths we aim to minimize.

Accordingly, in addition to binary variables m_{ij} defined previously, our second model uses binary variables $\{\lambda_p^{st} : s,t \in V, s \neq t, \forall p \in P^{st}\}$ (taking value 1 if path $p \in P^{st}$ is selected, 0, otherwise) to choose the paths. Assume that $a_{ij}^p \in \{0,1\}$ denotes a binary parameter that indicates whether arc (i,j) belongs ($a_{ij}^p = 1$) or not ($a_{ij}^p = 0$) to path p. The Arc-Path Formulation for OTDP is given by the IP:

$$w = \min \sum_{s \in V} \sum_{t \in V} \sum_{p \in P^{st}} |\lambda_p^{st}| \lambda_p^{st} \tag{1.10}$$

s.t.

$$\sum_{p \in P^{st}} \lambda_p^{st} = 1 \quad \forall s,t \in V, s \neq t, \tag{1.11}$$

$$\sum_{(i,j) \in A} c_{ij} m_{ij} \leq B, \; i < j, \tag{1.12}$$

$$\sum_{p \in P^{st}} a_{ij}^p \lambda_p^{st} - m_{ij} \leq 0 \quad \forall s,t \in N, \forall (i,j) \in A, \tag{1.13}$$

$$\lambda_p^{st} \in \{0,1\} \quad \forall s,t \in V, \forall p \in P^{st}, \tag{1.14}$$

$$m_{ij} \in \{0,1\} \quad \forall (i,j) \in A, \tag{1.15}$$

where $|\lambda_p^{st}|$ stands for the number of arcs in path p. As before, the objective function (1.10) minimizes the average path length among all pairs of vertices. Convexity constraints (1.11) ensure that exactly one path connecting every pair of distinct vertices s,t will be selected. Inequalities (1.13) assure that if at least one selected path crosses arc (i,j), this arc should be included in the solution.

Formulation (1.10)–(1.15) involves exponentially many λ_p^{st} variables, one for each possible path connecting each pair s,t of vertices in D. Therefore, if we plan to use formulation (1.10)–(1.15) to provide LP lower bounds in a BB algorithm, we have to find a way to deal with the large number of columns (variables) in the model. It should be clear that, even for very small instances (for small values of $|V|$), the memory requirements to load the LP Relaxation of (1.10)–(1.15) into a computer are huge. Therefore, to derive lower bounds from (1.10)–(1.15), we must deal with the excessive number of columns in our formulation *implicitly*.

The exact algorithm we plan to implement to solve model (1.10)–(1.15) is a LP based BB method, where the LP relaxations of the subproblems are solved through the Delayed Column Generation approach [31]. As its name suggests, the technique for solving the LP problems (relaxations) does not consider all columns at once. Instead, it starts with a restricted set of columns, solves a Linear Program defined over that restricted set of columns (usually this Linear Program is called the Restricted LP Master or simply Restricted Master Program) and adds new columns *on-the-fly*, only when needed. A new Restricted LP Master, enlarged with new columns, is then solved. The procedure iterates until no further columns need to be added. At this point, the Linear Program has been solved. Typically, the total number of columns in the last Restricted Master Program is a tiny fraction of the total number of columns.

Let us now describe, more precisely, how the Delayed Column Generation proceeds to solve the LP Relaxation implied by (1.10)–(1.15). Firstly, recall that the LP relaxation of (1.10)–(1.15) is the Linear Program obtained by replacing constraints (1.14) and (1.15) by their continuous counterparts. Assume that dual variables $\pi^{st} \in \mathbb{R}, \gamma \leq 0$ and $\beta_{ij}^{st} \leq 0$ are associated with constraints (1.11), (1.12) and (1.13), respectively, in the LP Relaxation of (1.10)–(1.15). The LP Dual associated wtih the LP relaxation of (1.10)–(1.15) is given by:

$$\max \sum_{s\in V}\sum_{t\in V} \pi^{st} + B\gamma \tag{1.16}$$

s.t.

$$\pi^{st} + \sum_{(i,j)\in A} a_{ij}^{p}\beta_{ij}^{st} \leq |\lambda_p^{st}| \quad \forall s,t \in V, s \neq t, \forall p \in P^{st} \tag{1.17}$$

$$c_{ij}\gamma - \sum_{(s,t)\in V} \beta_{ij}^{st} \leq 0 \quad \forall (i,j) \in A, \tag{1.18}$$

$$\pi \in \mathbb{R}, \tag{1.19}$$

$$\gamma \leq 0, \tag{1.20}$$

$$\beta \leq 0. \tag{1.21}$$

Let us now consider what happens when, instead of including in the LP relaxation of (1.10)–(1.15) all the columns associated with the λ_p^{st} variables, we formulate and solve another Linear Program, related to it, that involve a small subset of these columns. More precisely, let us assume that we have formulated the LP

Relaxation of (1.10)–(1.15), in terms of a restricted set of λ_p^{st} variables, defined only for those paths included in restricted sets $C^{st} \subset P^{st}, \forall s,t \in V, s \neq t$. Of course, for our procedure to work, it is mandatory that each C^{st} has very few paths, i.e., $|C^{st}| << |P^{st}|$. Then assume that, given the sets $C^{st}, \forall s,t$, the associated Restricted Master Program (RMP) reads:

$$\min \sum_{s \in V} \sum_{t \in V} \sum_{p \in C^{st}} |\lambda_p^{st}| \lambda_p^{st} \tag{1.22}$$

s.t.

$$\sum_{p \in C^{st}} \lambda_p^{st} = 1 \quad \forall s,t \in V, s \neq t, \tag{1.23}$$

$$\sum_{(i,j) \in A} c_{ij} m_{ij} \leq B, \ i < j, \tag{1.24}$$

$$\sum_{p \in C^{st}} a_{ij}^p \lambda_p^{st} - m_{ij} \leq 0 \quad \forall s,t \in N, \forall (i,j) \in A, \tag{1.25}$$

$$\lambda_p^{st} \geq 0 \quad \forall s,t \in V, \forall p \in C^{st}, \tag{1.26}$$

$$m_{ij} \geq 0 \quad \forall (i,j) \in A. \tag{1.27}$$

Note that the Linear Program (1.22)–(1.27) is exactly the LP Relaxation associated with (1.10)–(1.15), if sets P^{st} are replaced by C^{st} for all $s,t \in V, s \neq t$.

Consider now that (1.22)–(1.27) has one basic feasible solution $\hat{\lambda}, \hat{m}$. Let $\hat{\pi}, \hat{\gamma}$ and $\hat{\beta}$ be the corresponding optimal dual solutions. LP Duality theory states that, if for all pairs s,t, no path $p \in P^{st} \setminus C^{st}$ violates the dual constraints

$$\hat{\pi}^{st} + \sum_{(i,j) \in A} a_{ij}^p \hat{\beta}_{ij}^{st} \leq |\lambda_p^{st}|, \tag{1.28}$$

then, $\hat{\lambda}, \hat{m}$ solves the LP relaxation of (1.10)–(1.15) and the corresponding optimal LP objective function value gives a valid lower bound $\underline{w}$ on w. Otherwise, for a given pair s,t there must be a path in $P^{st} \setminus C^{st}$ that violates (1.28) that must be included in C^{st} and a new RMP must be formulated and re-optimized.

The problem of finding a path $p \in P^{st} \setminus C^{st}$ such that the corresponding dual constraint (1.28) is violated or to provide a certificate that there is no violated dual constraint is called the *pricing problem*. If after solving the pricing problems for all pairs of distinct vertices s,t, no constraint (1.28) is found to be violated, the Column Generation procedure stops and the LP relaxation of (1.10)–(1.15) is computed. Otherwise, if a solution (path) p to one of the s,t pricing problems violates the corresponding dual constraint, a new column (variable λ_p^{st}) associated with the path $p \in P^{st} \setminus C^{st}$ that violates the constraint is included in the new Restricted Master Program. The new RMP, enlarged with the sets of paths associated with violated constraints (1.28), is re-optimized (through the Simplex Algorithm). The Column

Generation procedure goes on until no further dual constraints are found to be violated.

Let us now discuss how, for a given pair $s,t \in V$, the associated pricing problem is solved. Firstly, recall that $|\lambda_p^{st}|$ denotes the number of arcs in path p. Define the reduced cost of a path $p \in P^{st}$ as $\overline{c}_p^{st} := -\hat{\pi}^{st} + \sum_{(i,j)\in p}(1-\hat{\beta}_{ij}^{st})$. In the previous definition, we allow ourselves the abuse of notation $(i,j) \in p$ to mean the arcs that are included in path p. Note that constraint (1.28) associated with path $p \in P^{st}$ is violated if and only if $\overline{c}_p^{st} < 0$. Instead of checking for the reduced cost of each possible path $p \in P^{st} \setminus C^{st}$, the pricing problem is formulated as an optimization problem as well. It involves finding a constrained shortest path connecting s to t in digraph D, under arcs costs given by $\{1-\hat{\beta}_{ij}^{st} : (i,j) \in A\}$ (note that these costs are non negative since $\hat{\beta}_{ij}^{st} \leq 0$). The paths are constrained to use a budget that does not exceed B. If the sum of the optimal constrained path length plus $-\hat{\pi}^{st}$ is negative, the corresponding path has negative reduced cost, i.e., a violated constraint (1.28) has been found. For details on algorithms for the resolution of the Resource Constrained Shortest Path Problems see [34, 35]. For the particular case considered here, for each pair s,t the associated pricing problem can be solved in $O(Bn)$ time.

As we pointed out, when no more columns with negative reduced costs are found, the LP relaxation of the Arc-Path formulation has been computed. If the LP solution is integer, it solves the original problem (1.10)–(1.15). Otherwise, being fractional, we must resort to branching.

It should be clear that applying a traditional BB algorithm for the RMP obtained at the end of the column generation process does not guarantee that an optimal (nor even feasible) solution to OTDP will be found. In contrast to that, we must embed the whole Delayed Column Generation procedure in the BB framework, leading to what is called a Branch-and-price algorithm, where new columns are likely to be generated at each node in the enumeration tree.

One key issue in the implementation of BP algorithms is how branching is performed [36, 37]. Since at each node of the enumeration tree the pricing subproblems are called repeatedly and solving them accounts for much of the computing time in BP algorithms, ideally, branching rules should not destroy their structure (in the OTDP case, the constrained shortest path structure).

To illustrate, assume that λ_p^{st} is fractional and BP implements branching on variable dichotomy (by imposing $\lambda_p^{st} = 1$ and $\lambda_p^{st} = 0$ on the two child nodes). While the first branching decision can be easily accommodated, the latter cannot. Note that for the first branch ($\lambda_p^{st} = 1$), it is sufficient to remove from the associated RMP all the other columns associated with paths that connect s,t and not to solve the subproblem defined by the pair s,t. However, the $\lambda_p^{st} = 0$ branch cannot be enforced by solving a simple constrained shortest path problem. This is true since there is no guarantee that variable λ_{st}^p will not be regenerated again and again. Although this issue could be tackled by finding the next cheapest shortest path, as we go deeper in the enumeration tree and many other decision variables have been fixed to 0, the complexity of finding such next shortest paths increases and solving the pricing subproblems become more and more inefficient.

In order to go around the difficulty associated with the $\lambda_{st}^p = 0$ branch, we implemented a branching rule proposed in [38]. The idea to deal with the $\lambda_p^{st} = 0$ case is to create several child nodes, one for each arc in path p. In order to guarantee that the path will not be regenerated, we create $|\lambda_p^{st}|$ subproblems and, in each of these, we impose that $m_{ij} = 0$ for an arc (i, j) in path p. Indeed, in each of these branches, we are avoiding that path p (and any other path that includes (i, j)) to be priced once again. The main advantage of this additional branching rule is that it does not affect the structure of the pricing problem. It only requires the elimination of the corresponding arc from the input graph, when solving the constrained shortest path problem between s and t.

In our BP implementation, the selection of which variable to branch on is based on the fractional λ_{st}^p variable associated with the maximal integer unfeasibility (farthest from integrality). This means that among all paths of all s,t pairs, the variable closest to $\frac{1}{2}$ is to be selected. A best-first strategy for selecting nodes from the list is also implemented.

In contrast to solving OTDP through the LP BB algorithm described in Sect. 1.4.1.2, the implementation of a BP algorithm like the one described here imposes additional challenges. For example, we can not rely on CPLEX's pre-implemented procedures for managing the search tree. We only use the CPLEX LP solver in our own code. All other procedures, being problem specific, have to be implemented by ourselves. On the positive side, BP algorithms typically scale much better than BB methods based on network flow formulations. Consequently, BP allows us to solve larger instances of OTDP.

1.4.1.3 Computational Results

In this section, we report computational results for OTDP, obtained with both formulations/algorithms discussed previously. Our aim is to illustrate the main advantages of the Branch-and-price method based on the Arc-Path formulation for the problem.

Different network sizes $n \in \{30, 45, 60\}$ were tested, each one with eight budget values B in the interval $[bmin, bmax]$. It was observed that the small world phenomenon is achieved for small values of probability p (adding or rewiring links) in the stochastic model presented in Sect. 1.3.3. Following the same idea, the sample budget values grow in a logarithmic scale, to better achieve the desired topologies, since they tend to be sparse. Parameter $bmin$ corresponds to the cost of a minimum spanning tree of D (regardless of arc orientations) while $bmax = \sum_{(i,j)\in A: i<j} c_{ij}$. We consider that all arcs have unitary costs ($c_{ij} = 1, \;\; \forall (i, j) \in A$).

All computational results reported in this section were conducted with a Intel Xeon Core 2 Quad machine, with 2GHz and 8Gb of RAM memory, running under Linux operating system. CPLEX release 10.2 was used for both algorithms.

Since computing the LP relaxations of the Arc-Flow formulation are very time consuming, for that model, we only report associated LP lower bounds, for the root node. For the Arc-Path formulation, on the other hand, we report results associated to the LP relaxation as well as results given by the full Branch-and-price method.

In Table 1.4, we present results for both formulations. The first two columns in the table indicate the network size n and the budget value B. In the next two columns, we report GAP (in % figures), the LP duality gap implied by the Arc-Flow formulation and t_{LP}, the time (in seconds) taken to evaluate the LP relaxation bound. In the next three columns, similar entries are given for the Arc-Path Formulation: the implied LP duality gap (GAP), the time t_{LP} (also in seconds) taken by the Delayed Column Generation, at the root node, to evaluate the bound, and, the number of columns ($\#col$) generated at the root node. In the last three columns, we provide more detailed results for the Branch-and-price search tree. They are the optimal solution values (under headings opt), the total time t_{BP} (in seconds) spent in the search, the total number of columns ($\#col$) priced out, and finally, the total number of nodes ($\#nod$) investigated in the tree.

Results in Table 1.4 indicate that BP was able to solve all instances for each network size. As one can observe, the smaller the budget B, the larger the CPU time to solve the instances to proven optimality. These results confirm our expectations that harder OTDP instances have small values of B. Sparse graphs tend to be harder since each single edge represents a high contribution minimizing the average path length and where it should be placed may be tough. Although the LP bounds given by both formulations are the same for these instances, the time taken to compute the bounds by the Delayed Column Generation algorithm is just a tiny fraction of the times needed to compute them by the Arc-Flow Formulation. Since the number of nodes explored in the BP tree is very small, the total computing time needed by BP is also very small compared to the computing time needed to evaluate the LP bound given by the Arc-Flow Formulation.

1.4.2 Exploring Objective Functions and Constraints

As mentioned before, depending on the desired complex network features and properties (network structure, function, and the different metrics), different criteria may be explored in the objective function or even in the constraints of an optimization problem.

In what follows, minor modifications into the core optimization problem, OTDP, are introduced. These modifications account for different objective functions and constraints in order to capture other network features into our mathematical models. For simplicity, such functions and constraints are formulated to replace or to be appended in the Arc-Flow Formulation. Equivalent counterparts could be presented in terms of the variables of the Arc-Path Formulation. The optimization problems discussed next, thus, correspond to variants of OTDP, in which some complex metrics and network models are explored.

Generally, the formation of hubs (high-degreed vertices) is associated with a sort of vulnerability to external attacks and does not prioritize the resilience [39, 40].

Table 1.4 Budget model

		Arc-flow formulation		Arc-path formulation						
		Linear relaxation		Linear relaxation			Branch-and-price			
n	B	GAP (%)	t_{LP}(s)	GAP (%)	t_{LP}(s)	$\#col$	opt	t_{BP}(s)	$\#col$	$\#nod$
30	29	0	820.67	0	18.28	2,452	841	19.62	2,452	1
	34	0	1,333.22	0	5.83	1,472	836	6.47	1,472	1
	38	0	605.08	0	7.75	1,594	832	8.52	1,594	1
	48	0	455.02	0	18.71	1,691	822	23.99	1,691	4
	70	0	856.65	0	0.81	841	800	1.25	841	1
	117	0	216.16	0	1.01	841	753	1.47	841	1
	218	0	19.32	0	1	841	652	1.47	841	1
	435	0	1.77	0	0.91	841	435	1.33	841	1
45	44	*	3,600	0	57.76	3,513	1,936	62.32	3,513	1
	54	*	3,600	0	55.38	3,489	1,926	59.03	3,489	1
	65	*	3,600	0	354.1	4,600	1,915	358.88	4,600	1
	88	*	3,600	0	6.59	1,936	1,892	8.71	1,936	1
	139	*	3,600	0	6.65	1,936	1,841	8.75	1,936	1
	248	*	3,600	0	6.58	1,936	1,732	8.74	1,936	1
	484	0	221.17	0	6.57	1,936	1,496	8.74	1,936	1
	990	0	11.54	0	5.74	1,936	990	7.87	1,936	1
60	59	*	3,600	0	839.31	9,982	3,481	885.56	9,982	1
	77	*	3,600	0	460.35	6,931	3,463	476.52	6,931	1
	96	*	3,600	0	47.36	5,244	3,444	125.07	5,358	4
	139	*	3,600	0	23.87	3,481	3,401	30.78	3,481	1
	231	*	3,600	0	24	3,481	3,309	30.89	3,481	1
	428	*	3,600	0	24.1	3,481	3,112	31.03	3,481	1
	854	*	3,600	0	23.93	3,481	2,686	30.78	3,481	1
	1,770	0	44.83	0	20.77	3,481	1,770	27.59	3,481	1

* Not available, since the time limit of 3,600 s was exceeded

A practical way to avoid them is to impose constraints limiting vertex degrees from above. Expected topologies tend to show a regular structure as the vertex degrees will be close to the average value, a regular structure in its characteristic of high average path length and a small betweenness for all nodes. Compared to ODTP model (1.10)–(1.15), the IP

$$\min \ \omega \tag{1.29}$$

s.t.

$$\sum_{j \in V} m_{ij} \leq \omega, \quad \forall i \in V, \tag{1.30}$$

$$\sum_{(i,j) \in A} c_{ij} m_{ij} = B, \quad i < j, \tag{1.31}$$

$$(1.2)\text{–}(1.4), (1.6)\text{–}(1.9)$$

makes use of a new decision variable ω that means the maximum degree of a node in a solution, in addition to x_{ij}^{st} and m_{ij} variables that keep their previously defined meaning. Note that constraints (1.30) assure that the degree of each vertex is no more than ω, the variable that it is minimized through objective function (1.29). Another important difference between the two models is that inequality (1.12) is now turned into equality form, constraint (1.31). In this case, different topologies can be designed by varying the budget value that needs to be completely spent adding B arcs of unitary cost to the network.

Applying the complex metric to minimize the maximal degree of each vertex in the telecommunication context, implies to improve the load balance and the resilience aspects and consequently to reduce the vulnerability.

A third formulation can be proposed with the objective of decreasing the maximum distance (in number of hops) between pairs of vertices in the network and, therefore, leading to a network with smaller average length. Following this idea, the created network is characterized by the small world phenomenon and scale free properties since it can generate a subset of high-degreed vertices with high betweenness and a subset of low-degreed vertices. To attain this goal, we formulated the next model:

$$\min \ \chi \tag{1.32}$$

s.t.

$$\sum_{(i,j) \in A} x_{ij}^{st} \leq \chi, \quad \forall s,t \tag{1.33}$$

$$(1.2)\text{–}(1.4),(1.6)\text{–}(1.9), (1.31).$$

The model uses variable χ that represents the maximum number of arcs in any path connecting two vertices of the network. Constraints (1.33) assure that only paths with less than χ arcs can be used in a solution. The model can be seen as a telecommunication problem where the objective is to minimize delay, a well known quality of service (QoS) metric in the context of routing problems.

The objective functions of both problems use number of arcs to define ω and χ. Therefore, a mono-objective problem can also be obtained considering the sum of both functions. In this case, it is possible to balance the minimization of the maximum path length of the network and the maximum degree of the vertices in the same model:

$$\min \ \omega + \chi \tag{1.34}$$

s.t.

$$\sum_{j \in V} m_{ij} \leq \omega, \quad \forall i \in V, \tag{1.35}$$

$$\sum_{(i,j) \in A} x_{ij}^{st} \leq \chi, \quad \forall s, t \tag{1.36}$$

$$(1.2)\text{–}(1.4), (1.6)\text{–}(1.9), (1.31).$$

Note that the objective function (1.34) seeks a trade off between average path length and vulnerability to external attacks, since it minimizes the sum of ω and χ. All the complex characteristics emphasized for both models are competing for the optimal structure. For large and sparse networks, the χ variable has a great influence. On the other hand, for small and dense networks the variable ω will predominate in the network structure.

In the next model, capacities $\{F_{ij} \geq 0 : (i, j) \in A\}$ are assigned to the use of each arc in D. The value F_{ij} limits from above the number of paths that make use of arc (i, j), in order to connect two vertices in a solution. Assume that a variable α is associated with the maximum occupation of any arc in the network. The occupation of an arc means the fraction of its capacity that is used in a solution.

$$\min \ \alpha \tag{1.37}$$

s.t.

$$\sum_{s \in V} \sum_{t \in V} x_{ij}^{st} \leq \alpha F_{ij}, \quad \forall (i, j) \in A, \tag{1.38}$$

$$0 \leq \alpha \leq 1, \tag{1.39}$$

$$(1.2)\text{–}(1.9),$$

The model attempts to balance the flow distribution in the whole network minimizing the maximum occupation of an arc. Note that constraints (1.38) guarantee that the occupation of arc (i,j) is no more than α times its capacity F_{ij}. Solutions to this model tend to follow a regular structure, where the average path length cannot be too small since capacity constraints on the use of the arcs are now imposed and the betweenness is small for majority of the vertices.

Another result is that the associated telecommunication network is likely to show more load balancing and resilience because the balanced flow distribution allows smaller losses in case of failure.

It may be the case that, under a very tight budget constraint, not all connections between pairs of vertices of V can be established. In such cases, it may be interesting to have a model that maximizes the number of pairs of vertices that actually have a path connecting them. The next model represents this case. The model uses $\{\theta^{st} : \forall s,t \in V, s \neq t\}$ variables to indicate whether or not there will be a directed path connecting s to t in the solution. The model reads:

$$\min \sum_{s \in V} \sum_{t \in V} (1 - \theta^{st}) \tag{1.40}$$

s.t.

$$\sum_{j \in A_s^+} x_{sj}^{st} = \theta^{st} \quad \forall s,t \in V, s \neq t, \tag{1.41}$$

$$\sum_{i \in A_t^-} x_{it}^{st} = \theta^{st} \quad \forall s,t \in V, s \neq t, \tag{1.42}$$

$$\theta \geq 0, \tag{1.43}$$

$$(1.4)\text{–}(1.9).$$

Note that the right hand side of the convexity constraints (1.41) and (1.42) may take a zero value. When that happens, no path to the corresponding pair of vertices will be available in the solution. The objective function (1.40) thus maximizes the number of pairs of vertices that can be connected. None of the complex metrics is being considered in this case, but in the context of telecommunications the objective is to maximize the number of requests accepted.

1.4.2.1 Results and Discussion

Let us now illustrate how different network features arise when OTDP and its variations introduced in the last section are solved for different values of the budget B.

Firstly, let us evaluate the impact of the choice of the budget B on the optimal solution to the core ODTP, given by IP (1.10)–(1.15). In Fig. 1.5, we present the optimal network topologies found for 8 different values of B in the range

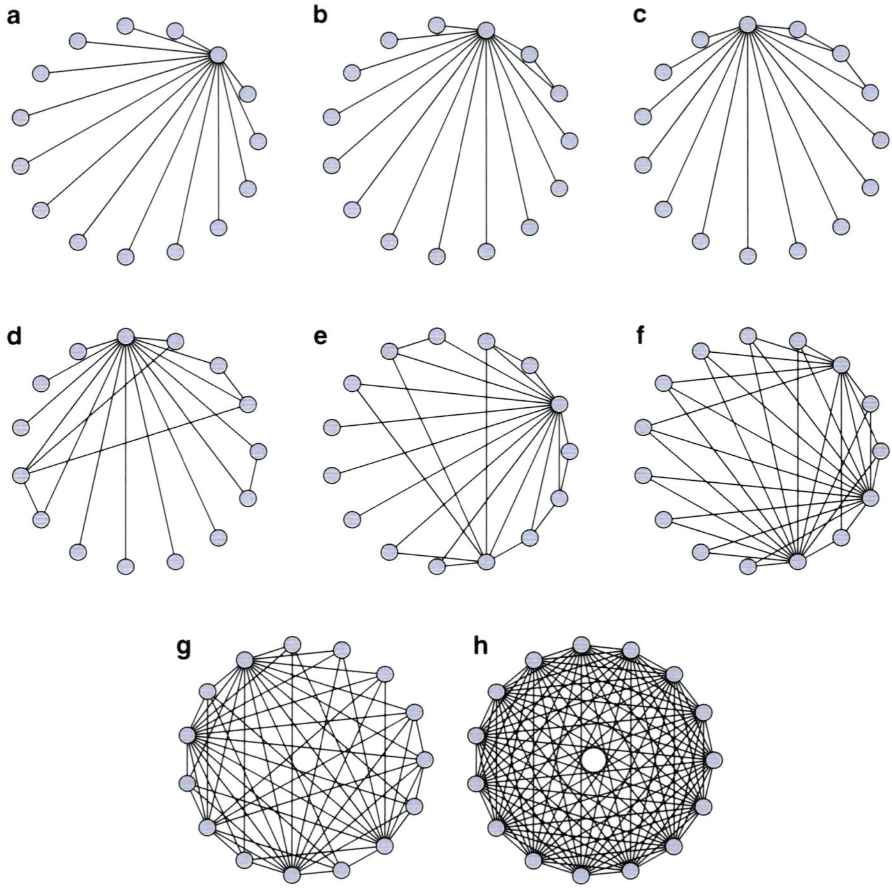

Fig. 1.5 Optimal solutions for eight budget values, minimizing objective function (1.10), $n = 15$

$[bmin, bmax]$ for $n = 15$.[1] Recall that, in this case, we are minimizing objective function (1.10). The upper-left-most figure accounts for the smallest value of B in our test bed, while the lower-right-most figure accounts for the largest value of B considered in our study. Note that smaller budgets lead to a simple spanning tree topology (Fig. 1.5a) or the union of a spanning tree and just few more arcs (Fig. 1.5b, c). Increasing the budget, topologies showing the small world phenomenon arise, where the average path length is small, the clustering coefficient

[1]The BP algorithm outlined previously is capable of solving OTDP instances with up to 60 vertices in less than one desktop computer hour. However, we depict instances with only 15 vertices since, as n and B grow, the higher number of crossing arcs does not allow one to identify the features we plan to discuss.

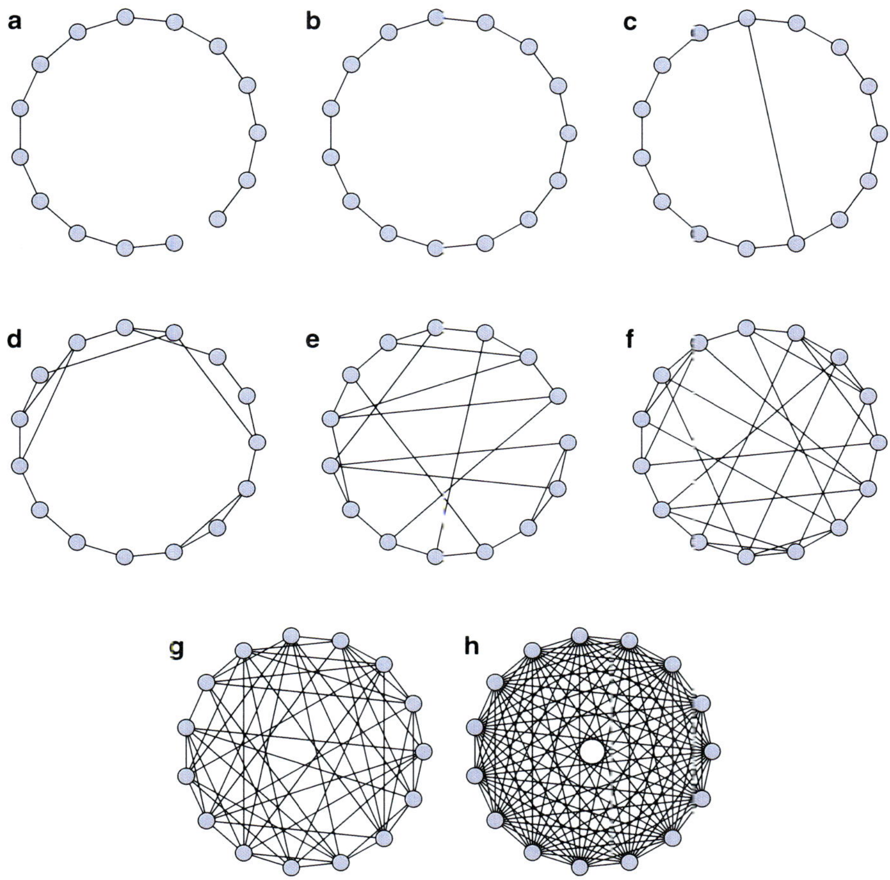

Fig. 1.6 Optimal solutions for eight budget values, minimizing objective function (1.29), $n = 15$

is high and graph densities are low (see Fig. 1.5d–f). For such intermediate values of B, we also identify the formation of hubs, associated with a power-law degree distribution. On the other hand, budget values near *bmax* induce dense graphs (Fig. 1.5 g) and finally a complete graph in Fig. 1.5h. In this sense, the budget value B plays here a similar role played by the probability of addition or rewiring arcs in random procedures for generating network topologies. The main feature of this set of topologies is the extremely small average path length. As mentioned before, the formation of hubs is crucial to reduce the average path length in the whole network.

Figure 1.6 presents network topologies found when optimizing objective (1.29), i.e., when the maximum vertex degree is minimized. As before, the upper-left-most figure accounts for the smallest value of B. The smallest budget value leads to a chain topology (Fig. 1.6a), with very high average path length. As the budget increases a bit, we obtain a ring (Fig. 1.6b) and a ring with few extra arcs, (Fig. 1.6c),

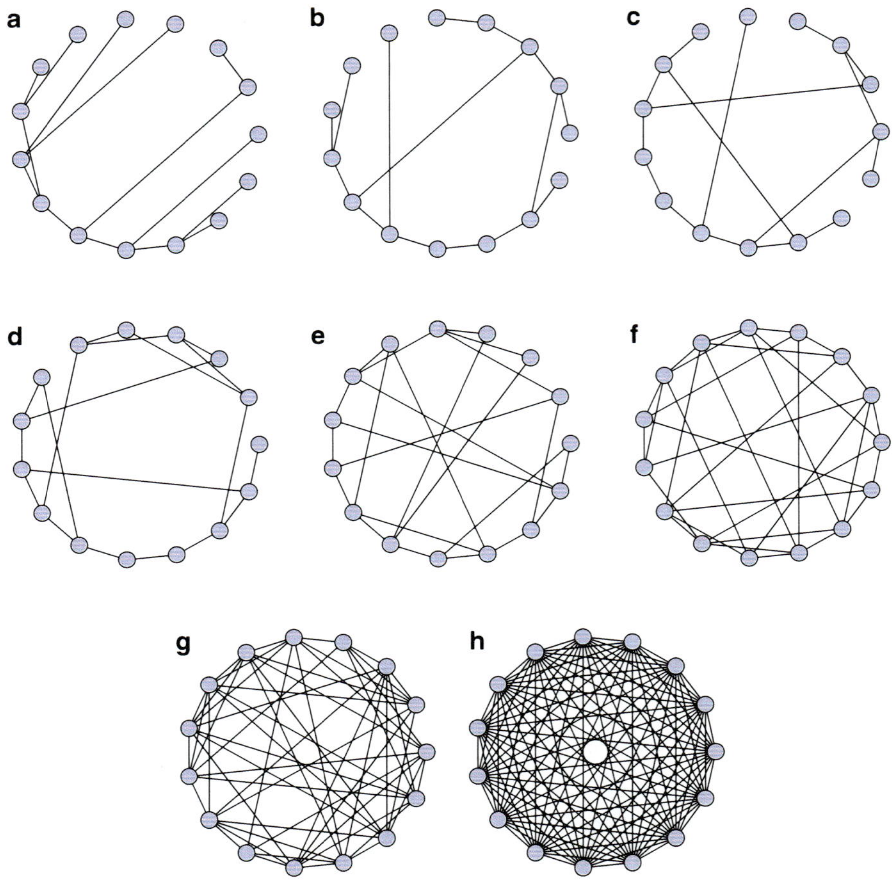

Fig. 1.7 Optimal solutions for eight budget values, minimizing objective function (1.34), $n = 15$

for which the solution still presents high average path length. Increasing the budget value a bit further, average length starts to decrease, improving the communication efficiency of the network (see Fig. 1.6d–f). We clearly observe that hub formation is indeed avoided. Budget values near *bmax* also induce dense graphs, but with fewer hubs (Fig. 1.6g) and, finally, a complete graph in Fig. 1.6h. Topologies indicated in Fig. 1.6 directly contrast to the previous set. The average path length is much higher due to the regular structure. However, random failures are less dangerous in this case.

Lying between minimizing the average path length and the maximum vertex degree, objective (1.34) leads to intermediate topologies, if compared to the previous solutions. Figure 1.7 illustrates the case where objective function (1.34) is minimized. Once again, by varying the budget values, different topologies arise. Figure 1.7a–c are characterized by fragments of a regular structure together with some additional arcs connecting opposite vertices, that contribute to decrease the average path length. As the budget increases, we observe that the degrees increase as well, but not to the point of generating hubs (see Fig. 1.7d–f). As expected, budget values near $bmax$ induce dense graphs (Fig. 1.7g, h). By the way the two metrics were combined into the objective function, the interesting aspect about solutions in Fig. 1.7 is that they share features like regularity and hub formation. Topologies like these may be quite interesting in engineered networks where one seeks for a certain efficiency in communication and a level of robustness against possible failures.

1.5 Heuristic Approaches

High computing times involved in LP based BB algorithms (and in BP as well) preclude their use to solve huge network optimization problems. In such cases, we must resort to heuristics. Differently from exact approaches, heuristics are concerned in seeking good solutions, not necessarily the optimal one. They have been applied in the solution of several practical problems. The progress in that approach gave rise to a set of metaheuristics [41, 42]. Metaheuristics are based on stochastic selection and iteratively seek for an improved candidate solution regarding a given measure of quality. Despite not guaranteeing that an optimal solution will be found, quite often they are capable of providing near-optimal solutions if properly implemented. In this section, we present related work in which metaheuristics were applied in order to create complex network topologies. In the following we present three models that have proven their merit for generating networks with complex features and share the same principle: desired topologies are generated by means of randomized algorithms. It is important to emphasize that all models were developed independently and although each approach aimed to create complex topologies, each one formulated a different starting optimization problem in order to attain that goal.

1.5.1 Small World Optimization Algorithm

In order to achieve an efficient communication system in which the information among entities should be exchanged as fast as possible, one aims to minimize the average path length, taking into account that it is wasteful to wire everything to everything else. An optimization model based on simulated annealing [43] is proposed for that purpose in [44]. The idea is to investigate whether the emergence of small world topologies could arise as a tradeoff between maximal connectivity (small path length) and minimal wiring (physical distance as a goal criterion).

The input k-regular graph is composed by vertices symmetrically placed along a ring, similar to the WS model. The size of the graph, as well as the total number of edges is fixed. Given a weighting factor $\mu \in [0, 1]$, the procedure minimizes the function $O = \mu L + (1-\mu)W$, where L and W, respectively, denote the normalized average path length among all pairs of vertices and the normalized wiring cost (which corresponds to the Euclidean distance between pairs of vertices). The characteristic path length L is normalized by $L(0)$ (which is the path length in the corresponding k-regular network); W (which is a physical distance measure) is normalized by the total wiring cost that results when the edges at each vertex are the longest possible, namely, when each vertex is connected to its diametrically opposite vertex, and to the vertices surrounding it.

By weighting two goals into the objective function the optimization of either one or the other will result in two extremes. At $\mu = 0$, when the optimization lies on minimizing the cost of wiring edges, a regular network emerges with a uniform degree and a high average path length ($L \sim n$). On the other hand, at $\mu = 1$, when only the average path length is minimized, the resulting network is random ($L \sim \log n$). At intermediate values of μ, the emergence of hub vertices is observed, due to the contribution of L to the objective function. Moreover, due to the contribution of W, hubs tend to be formed by connections to the closest vertices. Also, in order to reduce the path length, hubs may appear connected. Thus, opposite to the WS model, in which the average path length decreases by the presence of long range shortcuts, the reduction here is due to a small fraction of significant hub vertices.

Analyzing metrics as average path length, clustering coefficient and wiring cost and comparing them to the metrics from WS model, the following became apparent: (1) in both models L shows a sharp drop related to the small word behavior, such that the drop caused by hub formation is much sharper; (2) the drop observed in CC in the WS model is not valid for the optimized network, since hub formation keeps the CC at values higher than those for regular networks; and (3) the minimal wiring objective makes a clearly difference between the two models, as for larger values of μ the amount of wiring in the WS model is much greater than in the optimization model.

The authors in [44] conclude that the optimized networks are more clustered than corresponding regular networks, and have a smaller average degree of separation than their corresponding random graphs. Besides, small world topologies that arise from optimization consumes less wiring than their WS counterparts, being useful when wiring is expensive.

1.5.2 Scale Free Optimization Algorithm

An evolutionary algorithm for optimized network design is presented in [45], combining into the objective function, the minimization of the graph density and the average path length. These objectives include two relevant aspects of network performance: the cost of physical links and the communication speed among entities. Observing that most complex networks are extremely sparse and exhibit the so-called small world phenomenon, a minimization procedure based on these two criteria was expected to lead to small world and hub formation features.

The proposed procedure consists in optimizing an energy function defined as $E(\phi) = \phi d + (1-\phi)\rho$, where $0 \leq \phi, d, \rho \leq 1$. ϕ is a parameter controlling the linear combination of d (average path length) and ρ (density), which are normalized accordingly. The minimization of $E(\phi)$ involves the simultaneous minimization of distance and number of links (which is associated with cost).

The algorithm in [45] works with discrete time intervals. Starting at time $t = 0$, the network is set up with a density $\rho(0)$ following a Poisson distribution of degrees (connectedness is enforced). At time $t > 0$, the graph is modified by randomly changing the state of some pairs of vertices. For instance, with probability ν, each a_{ij} can switch from 0 to 1 or vice-versa. The new adjacency matrix is accepted if $E(\phi, t+1) < E(\phi, t)$. Otherwise, a different set of changes is tested. The algorithm stops when the modifications applied are not accepted a given number of times in sequence. This number is a parameter of the algorithm, set up by the designers.

Depending on how density and path length are weighted into the objective function, four main types of networks can be found: (a) exponential networks, (b) scale free networks, (c) star networks, and (d) dense networks. Analyzing some basic properties such as density, clustering coefficient and path length as a function of ϕ along with another measure defined as degree entropy, the authors identified four different phases, separated by three sharp transitions at $\phi_1^* \approx 0.25, \phi_2^* \approx 0.80$ and $\phi_3^* \approx 0.95$. Examining the degree distributions achieved by the procedure, it is possible to note that the transitions are easily explained since from (a) to (b) hub formation emerges, from (b) to (c) a hub competition leads to a central vertex and finally a dense graph (d) results when a progressive increase in the average degree of non-central vertices occurs and a sudden loss of the central vertex.

The results in [45] suggest that preferential attachment networks (scale free) might emerge at the boundary between random attachment networks (a) and forced attachment (all vertices linked to a central vertex) networks (c). Exponential like networks appear when the path length is minimized under high density weight. When linking cost substantially decreases, the reduction of vertex–vertex distance is enforced heading to a complete graph for high values of ϕ.

1.5.3 Small World Topologies Using GRASP

The problem of generating a small world topology treated in [46] consists in turning a regular graph into a small world graph with the principle of adding new edges to it (such as the addition model presented in Sect. 1.3.3), minimizing its average path length. Instead of the probability p used in the stochastic model from the literature, the authors make use of an additional parameter B called budget, which defines the number of shortcuts (edges) that may be included in the graph. Therefore, the input parameters are the original regular graph and the budget value.

A GRASP approach is adopted for solving the studied problem. GRASP (Greedy Randomized Adaptive Search Procedure) is an iterative method for solving

Table 1.5 Path length improvement

	$n = 15$			$n = 30$	
B	GRASP (%)	LP BB (%)	B	GRASP (%)	LP BB (%)
1	3.2	3.2	3	8.1	8.5
4	8	9	25	7.2	7.7
5	1.1	2.2	38	2.9	3.5
6	4.4	7.1	64	0	0.6
10	6.9	6.9	107	0	0.6
19	1.2	1.2	148	0	0
60	0	0	240	0	0

optimization problems proposed by Feo and Resende [16], composed by two phases: a construction phase, in which a solution is built from scratch; and a refinement phase (local search), in which a local optima solution is reached. The best solution found during the GRASP iterations is returned as the result of the algorithm.

In the particular application considered here, at each iteration of the construction phase, a new solution is created by adding to the graph as many edges that can be fitted with the budget. A higher priority is given to edges offering best benefit (low cost and high impact in reducing the average shortest path length of the graph). However, a portion of randomness is also included in order to avoid a purely greedy behavior that allows the procedure to be applied repeatedly, in a multi-start scheme. The refinement phase, which is applied only to the best solution found in the construction phase, works as follows. For each edge in the current solution, we attempt to replace it by an edge that if included in the solution does not violate the budget and decreases the average shortest path length. Note that the evaluation of the benefit of the movement (the operation of replacing one edge by another) is very time consuming. That is the reason why in the proposed implementation of GRASP, the local search is applied only to the best solution, and not to all solutions provided by GRASP at the end of its first phase.

In Table 1.5, we show how the network obtained by the application of GRASP compares to the optimal one, in terms of their average path length. Optimal networks were obtained by means of the BB algorithm based on the Arc-Flow Formulation (named here LP BB) proposed in Sect. 1.4.1.1 with an additional constraint set that forces every edge initially included in the regular graph to be included in the final solution as well. In doing so, the quality of the solution provided by GRASP can be compared to the optimal one.

For each value of B in our test bed, we report in Table 1.5, how much the solutions provided by GRASP and LP BB improve the average path length of the solution provided by the stochastic model from Sect. 1.3.3. For example, for $B = 25$, the average shortest path length of the optimal solution is 7.7% inferior than the average shortest path length of the solution obtained by the stochastic model.

As one could expect, graphs generated by the optimization approaches present better values of the average path length (indicated by the improvement values) compared to the stochastic method, except when the budget value B is very high (in such cases, all approaches provided solutions with identical average path length). This means that using the same number of additional links the optimization approaches are capable of finding small world networks with a smaller average path length, increasing the efficiency of the whole network.

1.6 Final Remarks

In this chapter, we presented different alternatives to design complex communication networks. Besides the stochastic methods from the literature, we focused on how optimization techniques both based on exact solution methods as well as in heuristics may be applied to generate complex communication networks. The main difference between both approaches concerns in the scalability provided by them. The Branch-and-price method is able to solve problems to optimality while heuristics can find feasible approximate solutions, with no guarantee of optimality.

Irrespective of how the optimization models are solved (through heuristics or exact methods), it can be observed that all optimization techniques have shown that features such as small path length, high clustering coefficient and power-law degree distribution can be achieved. It has been shown that the optimization of different criteria in the objective function and constraints leads to diverse complex network topologies.

Acknowledgements This work is supported by Conselho Nacional de Desenvolvimento Científico e Tecnológico (CNPq) under grants 302276/2009-2 and 477863/2010-8 and by Fundação de Amparo à Pesquisa do Estado de Minas Gerais (FAPEMIG) under grants 14016*1.

References

1. T. G. Lewis. *Network Science: Theory and Applications*. Wiley Publishing Hoboken, NJ, USA, 2009.
2. A-L. Barabási, R. Albert, and H. Jeong. Scale-free characteristics of random networks: the topology of the world-wide web. *Physica A: Statistical Mechanics and its Applications*, 281(1-4):69–77, June 2000.
3. R. Albert and A-L. Barabási. Statistical mechanics of complex networks. *Reviews of Modern Physics*, 74(1):47–97, January 2002.
4. M. E. J. Newman. The structure and function of complex networks. *SIAM Review*, 45:167–256, 2003.
5. D. J. Watts. *Six Degrees: The Science of a Connected Age*. W. W. Norton & Company, New York, NY, USA, 2003.
6. H.P. Thadakamalla, S. R. T. Kumara, and R. Albert. *Complexity and Large-Scale Networks*, volume 4 of *Operations Research Series*, chapter 11. CRC Press, Taylor and Francis Group, Pennsylvania State University, University Park, USA, 2008.
7. D. J. Watts and S. H. Strogatz. Collective dynamics of 'small-world' networks. *Nature*, 393(6684):440–442, June 1998.

8. M. Faloutsos, P. Faloutsos, and C. Faloutsos. On power-law relationships of the internet topology. In *SIGCOMM '99: Proceedings of the conference on Applications, technologies, architectures, and protocols for computer communication*, pages 251–262, New York, NY, USA, 1999. ACM.
9. H. P. Thadakamalla, U. N. Raghavan, S. Kumara, and R. Albert. Survivability of multiagent-based supply networks: A topological perspective. *IEEE Intelligent Systems*, 19:24–31, 2004.
10. R. Albert, H. Jeong, and A-L. Barabási. The diameter of the world wide web. *Nature*, 401:130, 1999.
11. David L. Alderson. Catching the 'network science' bug: Insight and opportunity for the operations researcher. *Operations Research*, 56(5):1047–1065, September-October 2008.
12. M. G.C. Resende and P. M. Pardalos, editors. *Handbook of Optimization in Telecommunications*. Springer, New York, NY, USA, 2005.
13. John W. Chinneck, Bjarni Kristjansson, and Matthew J. Saltzman. *Operations Research and Cyber-Infrastructure*. Springer Publishing Company, Incorporated, 1 edition, New York, NY, USA, 2009.
14. C. Barnhart, E.L. Johnson, G.L. Nemhauser, M.W.P. Savelsbergh, and P.H. Vance. Branch-and-price: column generation for solving huge integer programs. *Operations Research*, 48: 318–326, 1998.
15. C. Barnhart, C. A. Hane, and P. H. Vance. Using branch-and-price-and-cut to solve origin-destination integer multicommodity flow problems. *Operations Research*, 48:318–326, March 2000.
16. Thomas A. Feo and Mauricio G.C. Resende. Greedy randomized adaptive search procedures. *Journal of Global Optimization*, 6:109–133, 1995.
17. L. F. Costa, F. A. Rodrigues, G. Travieso, and P. R. Villas Boas. Characterization of complex networks: A survey of measurements. *Advances In Physics*, 56:167–242, 2007.
18. M. E. J. Newman and D. J. Watts. Scaling and percolation in the small-world network model. *Physical Review E*, 60(6):7332–7342, December 1999.
19. P. Erdös and A. Rényi. On random graphs i. *Publicationes Mathematicae Debrecen*, 6: 290–297, 1959.
20. E.N. Gilbert. Random graphs. *Annals of Mathematical Statistics*, 30:1141–1144, 1959.
21. S. Milgram. The small world problem. *Psychology Today*, 2:60–67, 1967.
22. J. M. Kleinberg. Navigation in a small world. *Nature*, 406(6798):845–845, August 2000.
23. D. J. Watts, P. S. Dodds, and M. E. J. Newman. Identity and search in social networks. *Science*, 296:1302, 2002.
24. A.-L. Barabasi and R. Albert. Emergence of scaling in random networks. *Science*, 286(5439):509–512, 1999.
25. F. S. H. Souza, A. S. Cunha, and G. R Mateus. Optimal topology design of complex networks. In *INFOCOM'09: Proceedings of the 28th IEEE International Conference on Computer Communications Workshops*, pages 278–283, Piscataway, NJ, USA, 2009. IEEE Press.
26. F. S. H. Souza, A. S. Cunha, and G. R Mateus. On the design of complex networks through a branch-and-price algorithm. In *IEEE GLOBECOM Workshop on Complex and Communication Networks*, 2010.
27. R. K. Ahuja, T. L. Magnanti, and J. B. Orlin. *Network Flows: Theory, Algorithms and Applications*. Prentice Hall, USA, 1993.
28. L. Wolsey. *Integer Programming*. John Wiley and Sons, New York, NY, USA, 1998.
29. G.B. Dantzig. *Linear Programming and Extensions*. Princeton University Press, Princeton, NJ, USA, 1963.
30. IBM ILOG. CPLEX Optimizer. http://www-01.ibm.com/software/integration/optimization/cplex-optimizer, [Online; accessed 08-Oct-2010].
31. L. S. Lasdon. *Optimization Theory for Large Scale Systems*. McMillan Company, New York, USA, 1970.
32. J. Desrosiers and W.E. Lübbecke. A primer in column generation. In G. Desaulniers, J. Desrosiers, and M.M. Solomon, editors, *Column Generation*, pages 1–32. Springer US, New York, NY, USA, 2005.

33. D. Huisman, R. Jans, M. Peeters, and A. P.M. Wagelmans. Combining column generation and lagrangian relaxation. In G. Desaulniers, J. Desrosiers, and M.M. Solomon, editors, *Column Generation*, pages 247–270. Springer US, New York, NY, USA, 2005.
34. D. Feillet, P. Dejax, M. Gendreau, and C. Gueguen. An exact algorithm for the elementary shortest path problem with resource constraints: Application to some vehicle routing problems. *Networks*, 44(3):216–229, 2004.
35. S. Irnich and G. Desaulniers. Shortest path problems with resource constraints. In G. Desaulniers, J. Desrosiers, and M.M. Solomon, editors, *Column Generation*, pages 33–65. Springer US, New York, NY, USA, 2005.
36. F. Vanderbeck. Branching in branch-and-price: a generic scheme. *Mathematical Programming*, Springer Berlin/Heidelberg, Berlin, Germany, pages 1–46, 2010. 10.1007/s10107-009-0334-1.
37. T. Koch, A. Martin, and T. Achterberg. Branching rules revisited. *Operations Research Letters*, 33:42–54, 2004.
38. M. Parker and J. Ryan. A column generation algorithm for bandwidth packing. *Telecommunications Systems*, 2:185–195, 1994.
39. R. Albert, H. Jeong, and A-L. Barabási. The internet's achilles' heel: Error and attack tolerance of complex networks. *Nature*, 406:378–382, 2000.
40. R. Albert, I. Albert, and G.L. Nakarado. Structural vulnerability of the north american power grid. *Physical Review E*, 69(2), 2004.
41. T.F. Gonzalez, editor. *Handbook of Approximation Algorithms and Metaheuristics (Chapman & Hall/CRC Computer & Information Science Series)*. Chapman and Hall/CRC, Boca Raton, FL, USA, 1 edition, May 2007.
42. E-G. Talbi. *Metaheuristics: From Design to Implementation*. Wiley Publishing, Hoboken, NJ, USA, 2009.
43. S. Kirkpatrick, C. D. Gelatt, and M. P. Vecchi. Optimization by simulated annealing. *Science*, 220(4598):671–680, 1983.
44. N. Mathias and V. Gopal. Small worlds: How and why. *Physics Review E*, 63(2):021117, January 2001.
45. R. F. i Cancho and R. V. Sole. Optimization in complex networks. Working Papers 01-11-068, Santa Fe Institute, November 2001.
46. F. S. H. Souza, D.L. Guidoni, and A.A.F. Loureiro. Uma abordagem baseada em grasp para geração de topologias small world. In *XL Simpósio Brasileiro de Pesquisa Operacional*, SOBRAPO, João Pessoa, Brazil, 2008.

Chapter 2
Fitness-Based Generative Models for Power-Law Networks

Khanh Nguyen and Duc A. Tran

Abstract Many real-world complex networks exhibit a power-law degree distribution. A dominant concept traditionally believed to underlie the emergence of this phenomenon is the mechanism of preferential attachment which originally states that in a growing network a node with higher degree is more likely to be connected by joining nodes. However, a line of research towards a naturally comprehensible explanation for the formation of power-law networks has argued that degree is not the only key factor influencing the network growth. Instead, it is conjectured that each node has a "fitness" representing its propensity to attract links. The concept of fitness is more general than degree; the former may be some factor that is not degree, or may be degree in combination with other factors. This chapter presents a discussion of existing models for generating power-law networks, that belong to this approach.

2.1 Introduction

The last decade has seen much interest in studying complex real-world networks and attempting to find theoretical models that elucidate their structure. Although empirical networks have been studied for some time, a surge in activity is often seen as having started with Watts and Strogatz's paper on "small-world networks" [23]. More recently, the major focus of research has moved from small-world networks to "scale-free" networks, which are characterized by having power-law degree

K. Nguyen • D.A. Tran (✉)
Computer Science Department, University of Massachusets, Boston, MA, USA
e-mail: knguyen@cs.umb.edu; duc@cs.umb.edu

M.T. Thai and P.M. Pardalos (eds.), *Handbook of Optimization in Complex Networks: Theory and Applications*, Springer Optimization and Its Applications 57,
DOI 10.1007/978-1-4614-0754-6_2, © Springer Science+Business Media LLC 2012

Table 2.1 Real-world networks: power-law exponent (λ)

Case	λ
Mathematics coauthorship	2.2
Film actor collaborations	2.1–2.3
WWW	2.1
Internet backbone	2.15–2.2
Protein interactions	2–2.4
ER graph	NA
BA graph	3

distributions [3]; that is, if $p(k)$ is the fraction of nodes in the network having degree k (i.e., having k connections to other nodes), then (for suitably large k)

$$p(k) = ck^{-\lambda}, \tag{2.1}$$

where $c = (\lambda - 1)m^{\lambda-1}$ is a normalization factor and m is the minimum degree in the network. This distribution is observed in many real-world networks, including the WWW [1], the Internet [12], metabolic networks [15], protein networks [14], co-authorship networks [21], and sexual contact networks [19]. In these networks, there are a few nodes with high degree and many other nodes with small degree, a property not found in standard Erdós–Rényi (ER) random graphs [9].

The near ubiquity of heavy-tailed degree distributions such as the power-law (2.1) for real-world complex networks, together with the inadequacy of the ER random graphs as a theoretical model for such networks, brings into sharp relief the fundamental problem of obtaining a satisfactory theoretical explanation for how heavy-tailed degree distribution can naturally arise in complex networks.

A dominant concept traditionally believed to underlie the emergence of the power-law phenomenon is the mechanism of preferential attachment, proposed by Barabási and Albert [3]: the higher degree a node has, the more likely it is to be connected by new nodes. This model, hereafter referred to as the BA model, leads to a growing random network which simulations and analytic arguments show has a power-law degree distribution with exponent $\lambda = 3$. Despite its elegance and simplicity, a deficiency of this mechanism is due to its fixed power-law exponent. As real-world networks exhibit a wide range of exponents, typically between 2 and 3 (see Table 2.1 for examples), the BA model may only explain a small subset of complex networks.

Consequently, other mechanisms have been proposed. Some, e.g., [8], are merely formulaic without a natural interpretation. Others use different connectivity information, not merely degree, of each node to influence the formation of a network, such as the mechanism in [17]. Still, a universally accepted explanation that works for not just one network but also others remains to be found. For example, if we use the BA-based models to explain the sexual contact network studied in [19], which is known to be power-law, a new individual will prefer to have sexual contact with those individuals who already have a large number of sexual contacts, while the explanation according to [17] will infer that a new individual will have sexual contact with some existing partners of a randomly chosen individual. These explanations seem bizarre for human sexual behavior.

In searching for a more natural explanation for the formation of power-law networks in the real world, there is a line of research, [4–6, 13, 16, 22], founded based on the conjecture that in many complex networks each node will have associated to it a "fitness" representing the propensity of the node to attract links. Using the fitness concept to explain the sexual contact network above, we could say that it is the fitness of an individual that attracts other individuals; an individual wants to have sexual contact with another individual because of the latter's fitness, not the latter's connectivity. The fact that a node has many contacts may just be a consequence of its high fitness. The key challenge in the design of fitness-based models is how fitness is defined; for example, what is fitness? what is it made of? what is its influence? Fitness may be just degree, or something not, or a combination of many factors, explicit or implicit. Fundamental differences in the approach to addressing these questions is discussed in the remaining sections of this chapter.

It is important to note that, as argued in [18], there are a rich variety of "emergent" topological signatures beyond mere power-law degree distributions that are also present in real-world complex networks. To date, there has been no perfect network generative model satisfying all signatures. This chapter is focused only on the power-law degree property of complex networks.

2.2 Early Network Models

One of the earliest theoretical models of a complex network was proposed and studied in detail by Erdós and Rényi [9–11] in a famous series of papers in the 1950s and 1960s. The ER random graph model consists of n nodes (or vertices) joined by links (or edges), where each possible edge between two vertices is present independently with probability p and absent with probability $1-p$. The probability $p(k)$ that a node has exactly degree k is given by the binomial distribution

$$p(k) = \binom{n-1}{k} p^k (1-p)^{n-k-1}. \tag{2.2}$$

In the limit when $n \gg kz$, where $z = (n-1)p$ is the mean degree, the degree distribution becomes the Poisson distribution

$$p(k) = \frac{z^k e^{-z}}{k!}. \tag{2.3}$$

The Poisson distribution is strongly peaked about the mean z, and has a tail that decays very rapidly as $1/k!$. This rapid decay is completely different from the heavy-tailed power-law nature of the tail of the degree distribution that is observed in many real-world complex networks.

Most widely known among the theoretical models for complex networks with a degree distribution similar to real-world networks is the BA model [3], which works as follows. Starting from a small number (n_0) of nodes (which could, for example, be chosen to form a random graph), at every time-step we add a new node with $m \leq n_0$ edges linking the new node to the m distinct nodes already present in the network such that the probability Π_i that the new node will connect to node i is proportional to the degree k_i of that node; that is,

$$\Pi_i = \frac{k_i}{\sum_j k_j}. \tag{2.4}$$

Consider a node i that joins the network at time i. Denote by $k_i(t)$ the degree of this node at time t. When the network is sufficiently large, $k_i(t)$ can be represented as a continuous function of time t, which grows at the following rate:

$$\frac{\partial k_i}{\partial t} = \frac{mk_i}{k_1 + k_2 + \cdots + k_n}, \tag{2.5}$$

where $n = n_0 + t$ is the network size at time t, and k_j's are the degrees of the current nodes. For large n, we can ignore the value of n_0 and since m new links are added after each time step (resulting in $2m$ degree increase), the sum of all nodes' degrees, $k_1 + k_2 + \cdots + k_n$, can be approximated by $2mt$. Thus, (2.5) can be rewritten approximately as

$$\frac{\partial k_i}{\partial t} = \frac{mk_i}{2mt} = \frac{k_i}{2t}.$$

Hence, $k_i(t)$ must be of the form $k_i(t) = c\sqrt{t}$ for some constant c. When node i just joins the network, that is, $t = i$, its degree is m and so we have $m = c\sqrt{i}$ or $c = m/\sqrt{i}$. Consequently, the degree of node i as a function of time t is

$$k_i(t) = m\sqrt{t/i}.$$

Given k, the probability that $k_i(t)$ is less than k is

$$Pr[k_i(t) < k] = Pr\left[m\sqrt{t/i} < k\right] = Pr\left[i > tm^2/k^2\right].$$

Because node i can equally likely be any node among $(n_0 + t)$ current nodes (consisting of n_0 initial nodes and t nodes added by time t), we have

$$Pr[i > i_0] = 1 - \frac{i_0}{n_0 + t}, \tag{2.6}$$

for any given i_0. Consequently,

$$Pr[i > tm^2/k^2] = 1 - \frac{tm^2/k^2}{n_0 + t} = 1 - \frac{m^2/k^2}{n_0/t + 1} \approx 1 - m^2/k^2$$

(when t is large).

The probability density function (pdf) of node i's degree is

$$P(k) = \frac{\partial Pr[k_i(t) < k]}{\partial k} = 2m^2/k^3.$$

Thus, BA results in a power-law degree distribution with $\lambda = 3$ (independent of m, which only changes the mean degree of the network).

A different generative model for power-law networks is the copy model studied by Kumar et al. [17], in which the new node chooses an existing node at random and copies a fraction of the links of the existing node. This model assumes that the new node knows who the chosen existing node is connected to. This assumption is reasonable for web graphs for which the model is originally proposed. However, the assumption is not always reasonable in other circumstances, as is clear from the case of sexual contact networks discussed in Sect. 2.1.

2.3 Combination of Degree and Fitness

An early fitness-based model for constructing power-law networks was proposed in [5]. This model assumes that the evolution of a network is driven by two factors associated with each node, its node degree and its fitness. These factors jointly determine the rate at which new links are added to the node. While node degree represents an ability to attract links that is increasing over time, node fitness represents something attractive about the node, that is constant. For example, in the WWW network, there can be two web pages published at the same time (thus, same degree) but later one might be much more popular than the other; this might be because of something intrinsic about one page, e.g., its content, that makes it more attractive than the other page.

In the proposed model, each node i has a fitness Φ_i which is chosen according to some distribution $\rho(\Phi_i)$. The network construction algorithm generalizes the BA algorithm as follows:

- Parameters
  - n_0: the size of the initial network which can be any graph.
  - $m \leq n_0$: the number of nodes a new node connects to when it joins the network.
  - n: number of nodes in the final network.
- Procedure
  1. Initially, start with the initial network of n_0 nodes, each assigned a random fitness according to distribution ρ.

2. Each time a new node is added; stop after the nth node is added.
   - Assign a random fitness to the new node according to distribution ρ
   - Add m edges linking the new node to m distinct existing nodes such that the probability Π_i for connecting to an existing node i is taken to be proportional to its fitness Φ_i:

$$\Pi_i = \frac{k_i \Phi_i}{\sum_j k_j \Phi_j}. \tag{2.7}$$

Suppose that i is the time node i joins the network. When the network is sufficiently large, the degree of node i over time, k_i, can be represented as a continuous function of time t. The value of this function increases over time as follows:

$$\frac{\partial k_i}{\partial t} = m \left(\frac{k_i \Phi_i}{\sum_j k_j \Phi_j}. \right) \tag{2.8}$$

It can be shown that $k_i(t)$ follows a power law

$$k_i(t) = m \left(\frac{t}{i} \right)^{\Phi_i / A},$$

with A given by

$$1 = \int_0^{\Phi_{\max}} \mathrm{d}x \rho(x) \frac{1}{\frac{A}{x} - 1}.$$

(Here, $\Phi_{\max}$ defines the maximum fitness.)

The degree of a node therefore grows faster if it has a larger fitness. This allows a node with a higher fitness to enter the network late but still become more popular than nodes that have stayed in the network for a much longer period.

If every node has the same fitness, this model is identical to the orignal BA model, resulting in power-law networks of exponent $\lambda = 3$. In the case that fitness is chosen uniformly in the interval [0, 1], we have $A = 1.255$ and the fraction of nodes in the network having degree k follows a generalized power-law with an inverse logarithmic correction:

$$p(k) \propto \frac{k^{-2.255}}{\ln k}. \tag{2.9}$$

This estimation has been confirmed by numerical simulations [5].

2.4 Fitness-Based Model with Deletion

The BA model cannot explain the heavy-tail degree distribution in complex networks where there are many node deletions [20]. On the other hand, the deletion rate can be significant for certain networks, such as the WWW. In the study of [16], based on the result of year-long web crawls to track the creation and deletion of web pages on the Internet, it was found that at least 0.76 pages were removed for every new web page created. Towards an approach that accommodates deletion, a model was proposed in [16] which extends the fitness-based model of [5] discussed in the previous section by allowing a probability for deleting a node at each time step. Specifically, the network is constructed as follow:

- Parameters
  - n_0: the size of the initial network which can be any graph.
  - $m \leq n_0$: the number of nodes a new node connects to when it joins the network.
  - n: number of nodes in the final network.
- Procedure
  1. Initially, start with the initial network of n_0 nodes, each assigned a random fitness according to distribution ρ.
  2. Each time a new node is added; stop after the nth node is added.
     - Assign a random fitness to the new node according to distribution ρ.
     - Add m edges linking the new node to m distinct existing nodes such that the probability Π_i for connecting to an existing node i is taken to be proportional to its fitness Φ_i:

$$\Pi_i = \frac{k_i \Phi_i}{\sum_j k_j \Phi_j}. \tag{2.10}$$

     - With probablity c, a random node is removed, along with all its edges.

It can be shown that the evolution of the degree of node i over time follows a power law

$$k_i(t) = m \left(\frac{t}{i}\right)^{\beta(\Phi_i)},$$

where the growth exponent β is a function of fitness and deletion rate

$$\beta(\Phi_i) = \frac{\Phi_i}{A} - \frac{c}{1-c},$$

with A given by

$$1 = \int_0^{\Phi_{\max}} \mathrm{d}x \rho(x) \frac{1}{\frac{A}{x}\frac{1+c}{1-c} - 1}.$$

If fitness is identical for every node, this model is exactly the same as the BA model applied to networks with deletion, resulting in a power-law exponent $\lambda = 1 + 2/(1-c)$ for small c. As c is closer to 1 (i.e., a high deletion rate), the degree distribution diverges rapidly from power-law. This explains why the BA model is not appropriate for networks with deletion.

In the case that fitness follows a truncated exponential distribution, which has been shown to empirically characterize the fitness distribution of web pages defined as its degree growth rate, the result is that the power-law exponent is not affected by the deletion rate and stablizes around two. In other words, the network as it is growing remains power-law, regardless of the rate of node deletion.

The models discussed in this section and the previous section combine fitness with degree to influence the growth of a network. A fundamentally different model is presented in the next section, which uses fitness as the only driver for the network growth.

2.5 Preferential Attachment Using Fitness Only

In a citation network such as [21] the different nodes (i.e., papers) will have different propensities to attract links (i.e., citations). The various factors that contribute to the likelihood of a paper being cited could include the prominence of the author(s), the importance of the journal in which it is published, the apparent scientific merit of the work, the timeliness of the ideas contained in the paper, etc. Moreover, it is plausible that the overall quantity that determines the propensity of a paper to be cited depends essentially multiplicatively on such various factors. The multiplicative nature is likely in this case since if one or two of the factors happen to be very small then the overall likelihood of a paper being cited is often also small, even when other factors are not small; e.g., an unknown author and an obscure journal were enough to bury a fundamentally important scientific paper.

The lognormal fitness attachment (LNFA) model, proposed by Ghadge et al. in [13], was motivated by the observation above. In this model, the fitness Φ_i representing the property of each node i to attract links is formed multiplicatively from a number of factors $\{\phi_1, \phi_2, ..., \phi_L\}$ as follows:

$$\Phi_i = \prod_{l=1}^{L} \phi_l, \tag{2.11}$$

where each factor ϕ_l is represented as a real non-negative value. Since there may be many factors contributing to the a node's attractiveness, explicit or implicit, we assume that the number of factors ϕ_i is reasonably large and that they are statistically independent. The fitness Φ_i will therefore be lognormally distributed, irrespective of the manner in which the individual factors are distributed. Indeed, we have

$$\ln \Phi_i = \sum_{l=1}^{L} \ln \phi_l, \tag{2.12}$$

and the Central Limit Theorem implies that this sum will converge to a normal distribution. Therefore, $\ln \Phi_i$ will be normally distributed. Since a random variable X has a lognormal distribution if the random variable $Y = \ln X$ has a normal distribution, Φ_i will be lognormally distributed. The density function of the normal distribution is

$$f(y) = \frac{1}{\sqrt{2\pi}\sigma} \mathrm{e}^{-(y-\mu)^2/(2\sigma^2)}, \tag{2.13}$$

where μ is the mean and σ is the standard derivation (i.e., σ^2 is the variance). The range of the normal distribution is $y \in (-\infty, \infty)$. It follows from the logarithmic relation $Y = \ln X$ that the density function of the lognormal distribution is given by

$$f(x) = \frac{1}{\sqrt{2\pi}\sigma x} \mathrm{e}^{-(\ln x-\mu)^2/2\sigma^2}. \tag{2.14}$$

It is conventional to say that the lognormal distribution has parameters μ and σ when the associated normal distribution has mean μ and standard deviation σ. The range of the lognormal distribution is $x \in (0, \infty)$. The lognormal distribution is skewed with mean $\mathrm{e}^{\mu+\sigma^2/2}$ and variance $(\mathrm{e}^{\sigma^2} - 1)\mathrm{e}^{2\mu+\sigma^2}$.

Thus, the basic hypothesis that each of the nodes has associated to it a fitness of the form (2.11) entails that under quite general conditions this fitness will be lognormally distributed. In the LNFA model, without loss of generality, one can assume that $\mu = 0$; hence, the fitness distribution is characterized by only a single parameter σ. This lognormal distribution has mean $\mathrm{e}^{\sigma^2/2}$ and variance $(\mathrm{e}^{\sigma^2} - 1)\mathrm{e}^{\sigma^2}$; examples are shown in Fig. 2.1.

The network construction algorithm works as follows:

- Parameters
  - σ: parameter for the lognormal fitness distribution.
  - n_0: the size of the initial network which can be any graph.
  - $m \leq n_0$: the number of nodes a new node connects to when it joins the network.
  - n: number of nodes in the final network.

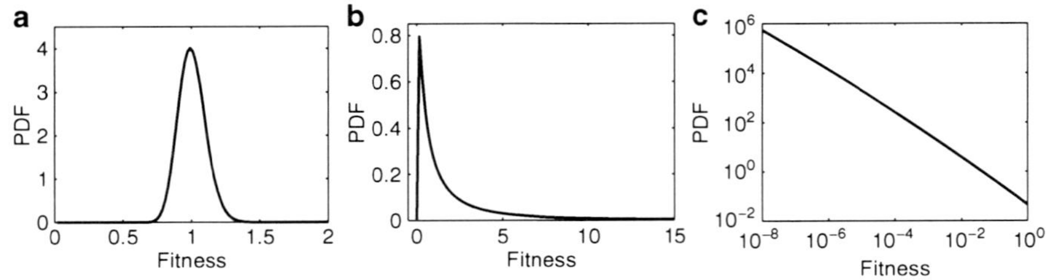

Fig. 2.1 Lognormal fitness distribution for three representative values of σ. The extreme cases are when σ is small (0.1) or large (9). In most cases, the distribution will have the shape similar to the case $\sigma = 1.5$. Figure 2.5 is plotted in the log-log scale to emphasize that all the nodes except a few have fitness zero (or extremely close) while the rest (very few) have high fitness

- Procedure
  1. Initially, start with the initial network of n_0 nodes, each assigned a random fitness according to the lognormal distribution.
  2. Each time a new node is added; stop after the nth node is added.
     - Assign a random fitness to the new node according to the lognormal distribution.
     - Add m edges linking the new node to m distinct existing nodes such that the probability Π_i for connecting to an existing node i is taken to be proportional to its fitness Φ_i:

$$\Pi_i = \frac{\Phi_i}{\sum_j \Phi_j}. \tag{2.15}$$

LNFA is almost identical to the BA model, the difference being that fitness information is used in place of degree information. Although this difference seems to be minor, it makes a fundamental shift in how the network is formed. To make this point, recall that in the BA protocol, the degree of a new node at the time it joins the network is small (m) and so it takes this node a long time before it may become a preferential choice for future new nodes to attach to. In LNFA, the new node may have a large fitness at the time it joins the network, making itself a preferential choice immediately. This is naturally reasonable because the attractiveness of a node may not result from how many nodes it is connected to; it may instead result from the "inner self" factors such as the personality of a person in a friendship network and his or her age.

As demonstrated in Fig. 2.2, LNFA can be used to generate a large spectrum of networks as seen in the real world and it is possible to do so by varying the

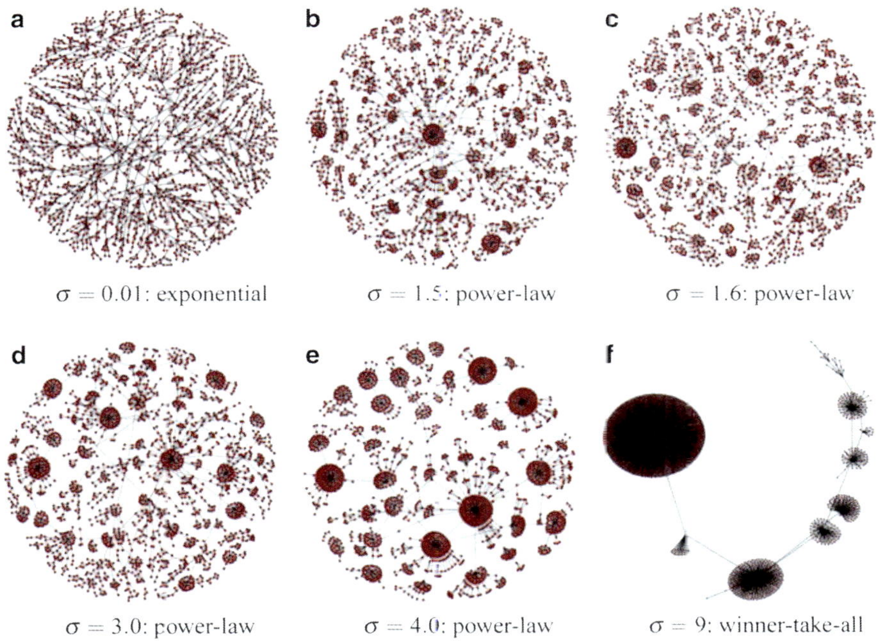

Fig. 2.2 2,000-node networks resulting from increasing σ. Transitions from exponential to power-law to winner-take-all graphs are observed

parameter σ. Consider two extreme cases of this parameter. In the first case, if σ is zero, nodes have exact same fitnesses and so, basing on Formula 2.15, each time a new node joins the network it chooses an existing node as neighbor with equal chance. This construction is simply the random graph model of [7] which yields a network with exponential degree distribution. On the other extreme, if σ is increased to reach a certain threshold, few nodes will stand out having very large fitnesses while all the other nodes will have very low fitnesses (zero or near zero; see Fig. 2.5). Consequently, an extremely high number of connections will be made to just a single node, resulting in a monopolistic network; this "winner-take-all" degree pattern is also observed in the real world [2]. Between these two extreme cases (exponential and monopolistic) we find a spectrum of power-law networks.

The transition between the regimes of degree distributions is not discontinuous. It is observed that when σ reaches 1 power-law degree distributions emerge and remain until σ is beyond 4. After that ($\sigma > 4$), the network becomes less power-law and more of a winner-take-all network. This is illustrated in Fig. 2.3 which shows the exponent λ as a function of σ. This figure plots the exponent values for 20 different networks of size ranging from 10,000 nodes to 20,000 nodes, that are constructed with the same σ value (ranging from 1 to 4.5). Here exponent values are not shown for the case $\sigma < 1$ because the corresponding resultant networks are not power-law; they are exponential random graphs. Seen from the figure, as σ increases from

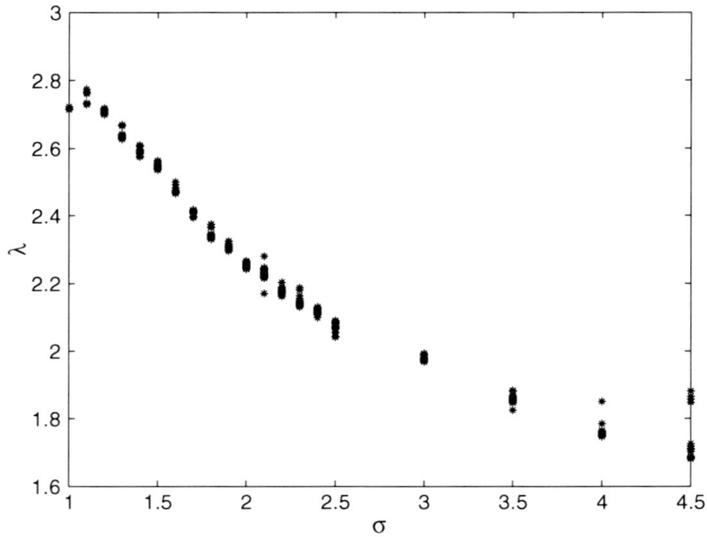

Fig. 2.3 Power-law exponent λ as a function of parameter σ. For each value of σ, 20 networks with size ranging from 10,000 to 200,000 nodes were considered. Each sample ("dot") in the plot shows the exponent for each network (for each value of σ)

1 to 3.5, there is remarkable consistency among all the 20 networks considered: they are all power-law with nearly identical exponent (it is noted that in the figure the increase in fluctuations seen at larger σ ($\sigma \geq 4$) is a consequence of the cross-over from power-law to a more monopolistic degree distribution). For example, when $\sigma = 2$, all networks have the same exponent 2.25. This implies that with LNFA the power-law degree distribution will emerge quickly as the network grows (in our observation, networks with 10,000 nodes already show their scale-freeness). It is also important to see that when the power-law pattern is observed, the exponent ranges between 1.8 and 2.8, which is close to the range [2, 3] typically observed in the real world. The exponent is monotonically decreasing as a function of σ.

2.6 Fitness-Based Model with Mutual Benefit

It is argued that in many cases of interest the power-law degree behavior is neither related to dynamical properties nor to preferential attachment. Also, the concept that the likelihood of a new node to link to an existing node depends solely on the latter's fitness might only apply to certain networks. The model proposed in [6] is suited for complex networks where it is the mutual benefit that makes two nodes link to each other. This model puts the emphasis on fitness itself without using the preferential attachment rule. Further, it is a static model building a network by growing links instead of growing nodes.

Specifically, the network construction algorithm starts with a set of n isolated nodes, where n is the size of the network to be built. Similar to the models discussed in the previous sections, each node i has a fitness Φ_i drawn from some distribution ρ. Then, for every pair of nodes, i and j, a link is drawn with a probability $f(\Phi_i, \Phi_j)$ which is some joint function of Φ_i and Φ_j. This model can be considered a generalization of the ER random graph model [9]. Rather than using an identical link probability for every pair of nodes as in the ER model, here two nodes are linked with a likelihood depending jointly on their fitnesses. A general expression for $p(k)$ can be easily derived. Indeed, the mean degree of a node of fitness Φ is

$$k(\Phi) = n \int_0^\infty f(\Phi, x)\rho(x)\,\mathrm{d}x = nF(\Phi). \tag{2.16}$$

Assuming F to be a monotonous function, and for large enough n, the following form can be obtained for $p(k)$:

$$p(k) = \rho\left[F^{-1}\left(\frac{k}{n}\right)\right]\frac{\mathrm{d}}{\mathrm{d}k}F^{-1}\left(\frac{k}{n}\right). \tag{2.17}$$

Thus, one can choose an appropriate formula for ρ and f to achieve a given distribution for $p(k)$. It is shown that a power law for $p(k)$ will emerge if fitness follows a power law and the linking probability for two nodes is proportional to the product of their fitnesses. For example, one can choose

$$f(\Phi_i, \Phi_j) = (\Phi_i \Phi_j)/\Phi_{\max}^2$$

and

$$\rho(\Phi) \propto \Phi^{-\beta},$$

(corresponding to a Zipf's behavior with Zipf coefficient $\alpha = 1/(\beta - 1)$).

In the case that fitness does not follow a power law, it is possible to find a linking function that will result in a power-law degree distribution. For example, considering an exponential fitness distribution, $\rho(\Phi) \propto \mathrm{e}^{-\Phi}$ (representing a Poisson distribution), one can choose

$$f(\Phi_i, \Phi_j) = \theta[\Phi_i + \Phi_j - z(n)]$$

where θ is the usual Heaviside step function and $z(n)$ is some threshold, meaning that two nodes are neighbors only if the sum of their fitness values is larger than the threshold $z(n)$. Using these rules, the degree distribution has a power law with exponent $\lambda = 2$. This is interesting because it shows that power-law networks can emerge even if fitness is not power-law. The same behavior also emerges if a more generic form of the linking function is used,

$$f(\Phi_i, \Phi_j) = \theta[\Phi_i^c + \Phi_j^c - z^c(n)],$$

(where c is an integer number); however, the power-law degree distribution has logarithmic corrections in some cases.

Closely related to the above model is the work of [22] which assumes the same concept that the linking probability is a joint function of the fitnesses of the end nodes. In this related work, it is concluded that for any given fitness distribution $\rho(\Phi)$ there exists a function $g(\Phi)$ such that the network generated by $\rho(\Phi)$ and $f(\Phi_i, \Phi_j) = g(\Phi_i)g(\Phi_j)$ is power-law with an arbitrary real exponent. Mutual-benefit based linking can also be combined with preferential attachment to drive the growth of a power-law network as shown in the work of [4].

2.7 Summary

This chapter has provided a review of existing models aimed at constructing power-law complex networks, that are inspired by the idea that there is some intrinsic fitness associated with a node to drive its evolution in the network. This fitness might be causal to why a node has a high degree or low, or might be an independent factor which together with the node's degree affect the node's ability to compete for links. The models discussed differ in how they interpret fitness and its influence on growing the network. It is suggested in [5, 16] that both degree and fitness jointly determine the growth rate of node degree. This may apply to complex networks such as a social network where a person's attractiveness is a combination of both his or her experience (represented by node degree) and talent (represented by fitness), or the WWW network where a web page is popular because its long time staying online (represented by node degree) and quality of its content (represented by fitness). A different approach is proposed in [6,13,22] where all the attractiveness factors associated with a node can be combined into a single factor (fitness). While the models in [6, 22] are motivated by static networks where two nodes require mutual benefit in order to make a connection, the lognormal fitness model in [13] is suitable to explain growing networks where a node wants to be a neighbor of another solely because of the latter's fitness, regardless of the former's. Although none of these models is one-size-fits-all, they do represent a vast population of complex networks. The current models, however, assume that fitness is an intrinsic factor that does not change over time. In practice, there are cases of networks where the overall attractiveness of a node might increase for one period of time and decrease for another. For the future work, it is thus an interesting research problem to explore fitness models that allow fitness to have its own evolution. The future research should also pay great attention to (in)validating theoretical models with the data collected from real-world networks.

Acknowledgements The authors would like to thank UMass Boston colleagues, Shilpa Ghadge, Bala Sundaram, and Timothy Killingback, for valuable discussions regarding the lognormal fitness model presented in this chapter.

References

1. Albert, R., Jeong, H., Barabási, A.L.: Diameter of the world-wide-web. Nature **400**, 107–110 (1999)
2. Barabási, A.: The physics of the web. Physics World **14-7**, 33–38 (2001)
3. Barabási, A.L., Albert, R.: Emergence of scaling in random networks. Science **286**, 509–512 (1999)
4. Bedogne, C., Rodgers, G.J.: Complex growing networks with intrinsic vertex fitness. Phys. Rev. E **74**, 046,115 (2006)
5. Bianconi, G., Barabsi, A.L.: Competition and multiscaling in evolving networks. Europhysics Letters **54**, 436–442 (2001). DOI 10.1209/epl/i2001-00260-6. URL http://dx.doi.org/10.1209/epl/i2001-00260-6
6. Caldarelli, G., Capocci, A., Rios, P.D.L., Munoz, M.: Scale-free networks from varying vertex intrinsic fitness. Phys. Rev. Lett. **89-25**, 258,702 (2002)
7. Callaway, D.S., Hopcroft, J.E., Kleinberg, J.M., Newman, M.E.J., Strogatz, S.H.: Are randomly grown graphs really random? Physical Review E **64**, 041,902 (2001). URL http://www.citebase.org/abstract?id=oai:arXiv.org:cond-mat/0104546
8. Dangalchev, C.: Generation models for scale-free networks. Physica A: Statistical Mechanics and its Applications **338**, 659–671 (2004)
9. Erdós, P., Rényi, A.: On random graphs. Publicationes Mathematicae **6**, 290–297 (1959)
10. Erdós, P., Rényi, A.: On the evolution of random graphs. Publ. of Math. Inst. of the Hungarian Acad. of Sci. **5**, 17–61 (1960)
11. Erdós, P., Rényi, A.: On the strength of connectedness of a random graph. Acta Mathematia Scientia Hungary **12**, 261–267 (1961)
12. Faloutros, M., P. Faloutros, Faloutros, C.: On power-law relationship of the internet topology. ACM SIGCOM 99, Comp. Comm. Rev. **29**, 251–260 (1999)
13. Ghadge, S., Killingback, T., Sundaram, B., Tran, D.A.: A statistical construction of power-law networks. International Journal on Parallel, Emergent, and Distributed Systems **25**, 223–235 (2010)
14. Jeong, H., Mason, S., Oltvai, R., Barabási, A.: Lethality and centrality in protein networks. Nature **411**, 41 (2001)
15. Jeong, H., Tombor, B., Albert, R., Oltvai, N., Barabási, A.: The large-scale organization of metabolic networks. Nature **607**, 651 (2000)
16. Kong, J.S., Sarshar, N., Roychowdhury, V.P.: Experience versus talent shapes the structure of the web. PNAS **105**(37), 13724–13729 (2008)
17. Kumar, R., Raghavan, P., Rajagopalan, S., Sivakumar, D., Tomkins, A., Upfal, E.: Stochastic models for the web graph. In: FOCS '00: Proceedings of the 41st Annual Symposium on Foundations of Computer Science, p. 57. IEEE Computer Society, Washington, DC, USA (2000)
18. Li, L., Alderson, D., Doyle, J., Willinger, W.: Towards a Theory of Scale-Free Graphs: Definition, Properties, and Implications. In Proceedings of Internet Mathematics (2005)
19. Liljeros, F., Edling, C.R., Amaral, L.A.N., Stanley, H.E., Aberg, Y.: The web of human sexual contacts. Nature **411**, 907 (2001)
20. Moore, C., Ghoshal, G., Newman, M.E.J.: Exact solutions for models of evolving networks with addition and deletion of nodes. Phys. Rev. E **74**(3), 036,121 (2006). DOI 10.1103/PhysRevE.74.036121
21. Neuman, M.: The structure of scientific collaboration networks. Proc. Nat. Acad. Sci. USA **98**, 404–409 (2001)
22. Vito, D., Caldarelli, G., Butta, P.: Vertex intrinsic fitness: How to produce arbitrary scale-free networks. Phys. Rev. E **70**, 056,126 (2004)
23. Watts, D., Strogatz, S.: Collective dynamics of "small-world" networks. Nature **393**, 440–442 (1998)

Chapter 3
Double Pareto Lognormal Distributions in Complex Networks

Zheng Fang, Jie Wang, Benyuan Liu, and Weibo Gong

Abstract This article elaborates the mathematical concept of double Pareto lognormal distribution and provides an overview of complex networks and natural phenomena that exhibit double Pareto lognormal distributions. These include the number of friends in social networks, the number of downloads on the Internet, Internet file sizes, stock market returns, wealth in human societies human settlement sizes, oil field reserves, and areas burnt from forest wildfire.

3.1 Introduction

Power-law distributions have been found in a good number of complex networks and natural phenomena of significant scientific interests. For example, population size distribution of cities, wealth distributions, intensities of earthquakes, and sizes of particles are all thought to follow the power law, and they cannot be correctly characterized by median or average values. For instance, the average population of all cities and towns is not a useful concept for most purposes because a significant fraction of the total population lives in big cities (e.g., New York, Los Angeles, and Chicago), which is substantially larger than those of many other cities by several orders of magnitude. Studies on complex networks such as the World Wide Web and online social networks also reveal that certain attributes of interests exhibit power law behaviors.

Z. Fang (✉) • J. Wang • B. Liu
Department of Computer Science, University of Massachusetts, Lowell, MA 01854, USA
e-mail: zfang@cs.uml.edu; wang@cs.uml.edu; bliu@cs.uml.edu

W. Gong
Department of Electrical and Computer Engineering, University of Massachusetts, Amherst, MA 01003, USA
e-mail: gong@ecs umass.edu

M.T. Thai and P.M. Pardalos (eds.), *Handbook of Optimization in Complex Networks: Theory and Applications*, Springer Optimization and Its Applications 57,
DOI 10.1007/978-1-4614-0754-6_3, © Springer Science+Business Media, LLC 2012

Pareto distribution, named after the Italian economist Vilfredo Pareto, is a commonly used canonical power law distribution. Pareto originally used this distribution to describe the allocation of wealth among individuals, for he observed that it depicts rather accurately the phenomenon that a larger portion of the wealth of any society is owned by a smaller percentage of the people in that society. Pareto also used it to describe the income distribution. This idea is sometimes referred to as the Pareto principle or the 80–20 rule, which says that 20% of the population controls 80% of the wealth [40]. Another popular power law distribution is the Zipfian distribution, also known as Zipf's law, which is widely cited in linguistic research of natural languages.

Pareto and Zipfian distributions only exhibit a single tail. It has come to people's attention in recent years that certain power-law phenomena, if observed more closely, actually exhibit two tails: a lower tail and an upper tail. While the two tails would share a similar shape with lognormal distributions, how to correctly characterize such phenomena remains a challenge, which has stimulated many debates among researchers. It is true that the lognormal model could better fit the bodies of empirical distributions, but it does not seem to fit well with the power law behavior in the tails.

The concept of double Pareto lognormal distributions has been shown useful in modeling distributions of various complex networks and natural phenomena that consist of a lognormal body and Pareto tails, including computer networks, social networks, economics, finance, geography, geology, and physical sciences [36, 42, 43]. In this article, we will provide and elaborate mathematical background on the concept of double Pareto lognormal distribution, demonstrate how it would fit into empirical data, and present possible explanations of the causes for the power law behaviors.

The rest of this article is organized as follows. In Sect. 3.2 we will provide a brief overview of the classical power law and lognormal distributions. We will then introduce the double Pareto lognormal distribution and several models that generate it. In Sect. 3.3 we will present examples from diverse areas that fit the double Pareto (lognormal) distributions. These examples include wealth distributions in human societies, stock market returns, Internet file sizes, the number of friends distributions in social networks, downloads distributions on the Internet, human settlement sizes, oil field reserves, and areas burnt from forest wildfire. Section 3.4 concludes the article.

3.2 Generation Model of Double Pareto Distribution

We start this section by presenting the mathematical background for the power law distribution and lognormal distribution, and then derive the double Pareto lognormal distribution as an exponential mixture of lognormal distributions. We will also introduce several stochastic models of differential equations that converge to the double Pareto distribution under different sets of scenarios.

3.2.1 Power Law Distributions

Let X be a random variable. The complementary cumulative distribution function (ccdf) of X is defined by

$$\overline{F}_X(x) = Pr(X \geq x). \tag{3.1}$$

A non-negative random variable X is said to follow a power law distribution if its ccdf satisfies

$$\overline{F}_X(x) \sim x^{-\alpha}, \tag{3.2}$$

for some constant $\alpha > 0$. Here, $f(x) \sim g(x)$ denotes $\lim_{x\to\infty} f(x)/g(x) = c$ for some constant $c > 0$. The plot of the density of such random variable X has a long tail, which is referred to as *power tail* (a.k.a. *heavy tail*).

The Pareto distribution is one of the canonical power law distributions with the following ccdf on random variable X:

$$\overline{F}_X(x) = \begin{cases} \left(\frac{x}{x_m}\right)^{-\alpha}, & x \geq x_m \\ 1, & x < x_m, \end{cases} \tag{3.3}$$

for some $\alpha > 0$ and $x_m > 0$. Note that the Pareto distribution has the density function $f(x) = \alpha x_m^{\alpha} x^{-\alpha-1}$ when $x \geq x_m$. The ith moment of the density function can be represented as

$$E(x^i) = \int_{x_m}^{\infty} x^i f(x)\mathrm{d}x = \alpha x_m^{\alpha} \int_{x_m}^{\infty} x^{-\alpha-1+i}\mathrm{d}x = \alpha x_m^{\alpha} \left. \frac{x^{-\alpha+i}}{i-\alpha} \right|_{x_m}^{\infty}. \tag{3.4}$$

Note that when $\alpha < 1$ the first moment does not exist and when $\alpha < 2$ the second moment does not exist. Therefore, if α falls in the range of $(0,1]$, the function has an infinite mean. If α is in $(1,2]$, the function has a finite mean but infinite variance. Only when $\alpha > 2$ will the function have both finite mean and finite variance.

Another popular power law distribution is the Zipfian distribution (a.k.a. Zipf's law), which is widely used in linguistic studies of natural languages. Zipf's law states that frequency of occurrences of an event, as a function of the ranking of frequencies, is a power law function with the exponent α close to unity, where a higher frequency has a smaller rank. For example, in the English language the word "the" has the highest frequency, and so its rank is equal to 1. The word "of" has the second highest frequency and so its rank is equal to 2. Zipf's law predicts that in a population of N elements, the frequency of elements that is ranked k is determined by the following formula:

$$f(k) = \frac{k^{-\alpha}}{\sum_{n=1}^{N} n^{-\alpha}}, \tag{3.5}$$

where α is the value of the exponent characterizing the distribution, typically close to 1.

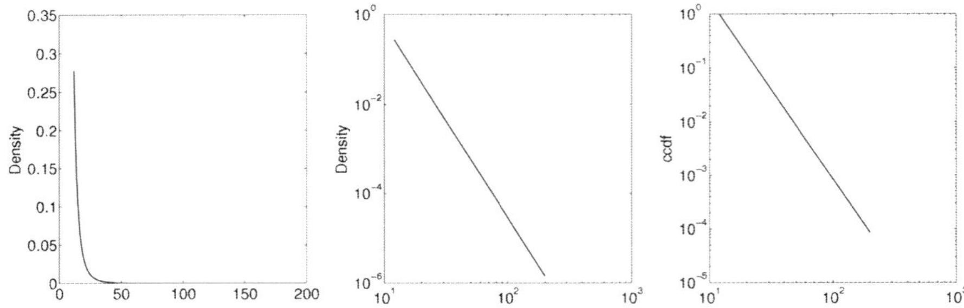

Fig. 3.1 A Pareto density in log-log plot and ccdf

It is customary to plot the power law distribution by taking logarithm on both sides, referred to as log-log plot, which produces a straight line for the asymptotic behavior of the ccdf. This is the basis for testing power-law behaviors. The same is true for the power law density function, which might be easier to work with mathematically under certain circumstances. For example, for the Pareto distribution, the logarithm of the density function is linear with the following form:

$$\ln f(x) = -(\alpha+1)\ln x + \alpha \ln x_m + \ln \alpha, \tag{3.6}$$

Plotting ccdf in the logarithmic scale also emphasizes the tail region, providing a good visual when fitting empirical data into a power law model. Figure 3.1 shows the log–log plot of a Pareto density function and the ccdf.

3.2.2 *Lognormal Distributions*

A positive random variable X is said to be lognormally distributed with parameters (μ, σ^2) if the random variable $Y = \ln X$ is normally distributed with mean μ and variance σ^2. The density function for a lognormal distribution is determined by

$$f(x) = \frac{1}{\sqrt{2\pi}\sigma x} \mathrm{e}^{-\frac{(\ln x-\mu)^2}{2\sigma^2}}. \tag{3.7}$$

The lognormal distribution, in contrast to the normal distribution, is skewed, with median e^{μ}, mean $\mathrm{e}^{\mu+\frac{1}{2}\sigma^2}$, and mode $\mathrm{e}^{\mu-\sigma^2}$. Although the lognormal distribution has finite moments and the Pareto distribution has infinite moments, their plot shapes are extremely similar in that a large portion of the body of the density function and the ccdf can appear linear. To be more specific, take logarithm on a lognormal distribution density function, we have

$$\ln f(x) = -\ln x - \ln \sqrt{2\pi}\sigma - \frac{(\ln x - \mu)^2}{2\sigma^2}. \tag{3.8}$$

When σ is sufficiently large, the value of $\ln f(x)$ is barely affected by the quadratic term for a large range of x values. Therefore, the logarithm of the density function

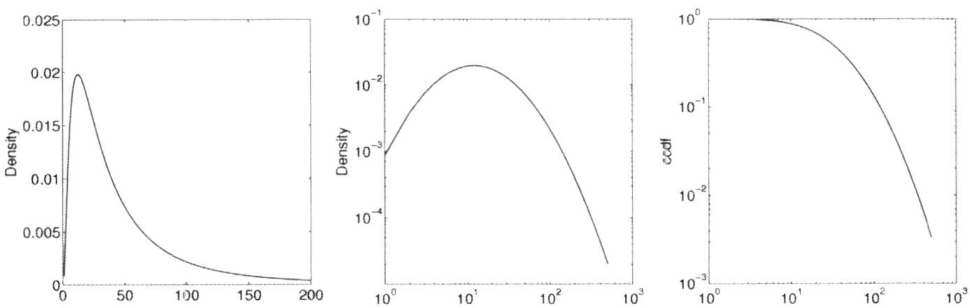

Fig. 3.2 A lognormal density in log-log plot and ccdf

will appear as a straight line for a large range of x values. The same holds true for the ccdf, which is shown in Fig. 3.2.

While the normal distribution can be thought of as an additive accumulation of a number of independent random variables, a variable could be thought of as lognormal if it is the multiplicative product of a number of independent, positive random variables, for

$$\ln Y = \ln X_1 + \ln X_2 + \cdots + \ln X_n = \ln(X_1 X_2 \cdots X_n). \tag{3.9}$$

For example, a long-term discount factor in a financial market can be derived from the product of short-term discount factors. In wireless communications, for another example, the attenuation caused by shadowing or slow fading from random objects is often assumed to be lognormally distributed.

This phenomenon may be better illustrated using the following example. Suppose we start with an initial state X_0. At each step, the state may change with a certain percentage (e.g., changes of price, size, and volume) denoted by an independent random variable C_i. Thus, each state X_t can be described as

$$X_t = X_{t-1} C_t = X_0 \prod_{i=1}^{t} C_i. \tag{3.10}$$

Taking logarithm on both sides we have

$$\ln X_t = \ln(X_{t-1} C_t) = \ln X_0 + \sum_{i=1}^{t} \ln C_i. \tag{3.11}$$

Since each C_i is independent from each other, applying the Central Limit Theorem to the summation term yields a convergence to a normal distribution for sufficiently large t. Hence, random variable X_t is well approximated by a lognormal distribution. Lognormal distributions are natural distributions for modeling growth of population, growth of wealth, growth of organisms, and any process where the underlying change rate is a random factor over a time step independent of the current size.

3.2.3 Double Pareto Distribution, A Mixture of Lognormals

A double Pareto distribution can be generated by mixing a number of lognormal distributions. This can be done based on a model defined in Sect. 3.2.2 that yields a lognormal distribution.

Suppose in $X_t = X_{t-1}C_t$ we have $X_0 = c$ for a constant $c > 0$, and C_t is a lognormally distributed random variable with parameters (μ, σ^2). We may view the subscript of X as a moment in time, then at time $t = T$, the random variable X_T is also lognormally distributed with parameters $(\mu T, \sigma^2 T)$. We may also view the time t itself as a continuous random variable and let the process run for some random time θ and obtain a random variable that comes from a mixture of lognormal distributions with parameters $(\mu\theta, \sigma^2\theta)$. Specifically, if the process stops at a constant rate λ, the time variable is exponentially distributed with density

$$f_t(\theta) = \lambda \mathrm{e}^{-\lambda\theta}, \quad \theta \geq 0.$$

Reed et al. [44] show that this mixture of an exponentially distributed number of lognormal distributions exhibits power law behaviors for both the upper tail and the lower tail regions, and name it double Pareto distribution. The resulting density function for this mixture is

$$\begin{aligned} f(x) &= \int_{\theta=0}^{\infty} \lambda \mathrm{e}^{-\lambda\theta} \frac{1}{\sqrt{2\pi\theta}\sigma x} \mathrm{e}^{-\frac{(\ln x - \theta\mu)^2}{2\theta\sigma^2}} \mathrm{d}\theta \\ &= \frac{2\lambda \mathrm{e}^{\mu \ln x/\sigma^2}}{\sqrt{2\pi} x\sigma} \int_{u=0}^{\infty} \mathrm{e}^{-\left(\lambda + \frac{\mu^2}{2\sigma^2}\right)u^2} \mathrm{e}^{-\frac{\ln^2 x}{2\sigma^2 u^2}} \mathrm{d}u \\ &= \frac{\lambda}{\sigma\sqrt{(\mu/\sigma)^2 + 2\lambda}} \begin{cases} x^{-1+\mu/\sigma^2+\sqrt{(\mu/\sigma)^2+2\lambda}/\sigma}, & 0 < x \leq 1 \\ x^{-1+\mu/\sigma^2+\sqrt{(\mu/\sigma)^2+2\lambda}/\sigma}, & x \geq 1. \end{cases} \end{aligned}$$

Reed [44] suggests the following simpler form of double Pareto distribution density function:

$$f(x) = \frac{\alpha\beta}{\alpha+\beta} \begin{cases} x^{\beta-1}, & 0 < x \leq 1 \\ x^{-\alpha-1}, & x \geq 1 \end{cases} \tag{3.12}$$

where α and $-\beta$ $(\alpha > 0, \beta > 0)$ are the two roots of the following quadratic equation

$$\frac{\sigma^2}{2} z^2 + \left(\mu - \frac{\sigma^2}{2}\right) z - \lambda = 0.$$

Instead of using exponential distribution, Mitzenmacher [35] shows that using a geometric distribution to randomly sample lognormal distributions can lead to a distribution that is extremely similar to a double Pareto distribution. This approach assumes that the random process stops with a probability p. That is, the process stops at time k with probability $p(1-p)^{k-1}$. Using this discrete geometric mixture we can obtain the following distribution density:

$$f(x) = \sum_{k=1}^{\infty} p(1-p)^{k-1}\left(\frac{1}{\sqrt{2\pi k}x\sigma}\mathrm{e}^{-\frac{(\ln x - k\mu)^2}{2k\sigma^2}}\right). \tag{3.13}$$

This summation can be nicely approximated using the following integral when the absolute value of $\ln x$ is sufficiently large.

$$f(x) \approx \int_{k=1}^{\infty} \frac{p}{\sqrt{2\pi k}x\sigma(1-p)}\mathrm{e}^{k\ln(1-p)-\frac{(\ln x - k\mu)^2}{2k\sigma^2}}\mathrm{d}k. \tag{3.14}$$

The geometric approach produces essentially the same power-tail behavior as the exponential mixture does, but technically the geometric mixing of lognormal distributions only yields an approximation to the double Pareto distribution according to Reed's definition. Mitzenmacher [35] showed the following theorem.

Theorem 3.1. *There exist positive constants $\alpha, \beta, c_1, c_2, c_3, c_4, m$ and ε such that the density function in equation 3.13 satisfies $c_1 x^{\beta-1} \le f(x) \le c_2 x^{\beta-1}$ for $x < \varepsilon$, and $c_3 x^{-\alpha-1} \le f(x) \le c_4 x^{-\alpha-1}$ for $x > m$. (Constants c_i may depend on p, μ and σ but not on x.)*

It follows from this theorem that at the tails, the density function, the cumulative distribution function (cdf) and the ccdf of the geometric mixture are each bounded by two power law distributions that differ only by a constant factor [35]. Thus, the geometric mixture produces a valid and practical approximation to the double Pareto distribution for it is the tail behavior that characterizes the power law distribution.

To the extent of double Pareto distribution, Reed [44] also suggests a more generalized form called double Pareto lognormal distribution, by removing the constraint that the initial state X_0 must be a constant. Assume that the initial state X_0 also follows a lognormal distribution with parameters (ν, τ^2). Mixing with the exponential-time distribution we can show that the distribution of random variable $Y = \ln X$ can be represented as the sum of two independent random variables, where one variable is normally distributed and the other is double exponentially distributed. It follows that the density function $f_X(x)$ can be obtained from the density function $f_Y(y)$ using $f_X(x) = f_Y(\ln x)/x$, which in turn can be found by convolving a normal density and a double exponential density. The final density function of X is

$$\begin{aligned} f_X(x) &= \frac{\alpha\beta}{\alpha+\beta}(A+B) \\ A &= x^{-\alpha-1}\mathrm{e}^{\alpha\nu+\alpha^2\tau^2/2}\Phi\left(\frac{\ln x - \nu - \alpha\tau^2}{\tau}\right) \\ B &= x^{\beta-1}\mathrm{e}^{-\beta\nu+\beta^2\tau^2/2}\Phi^{c}\left(\frac{\ln x - \nu + \beta\tau^2}{\tau}\right), \end{aligned} \tag{3.15}$$

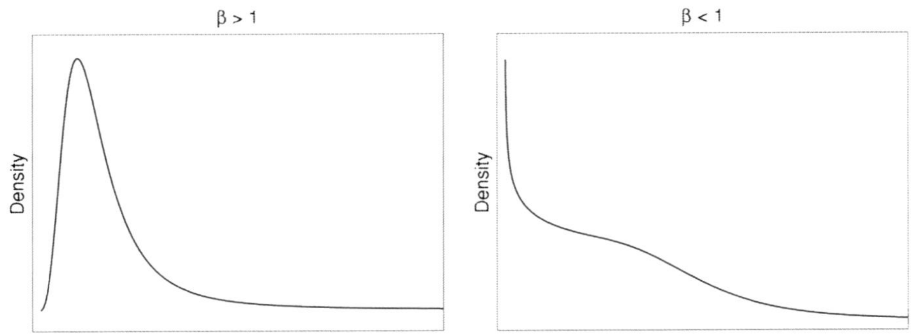

Fig. 3.3 The double Pareto lognormal distributions have completely different shapes with $\beta > 1$ and $\beta < 1$

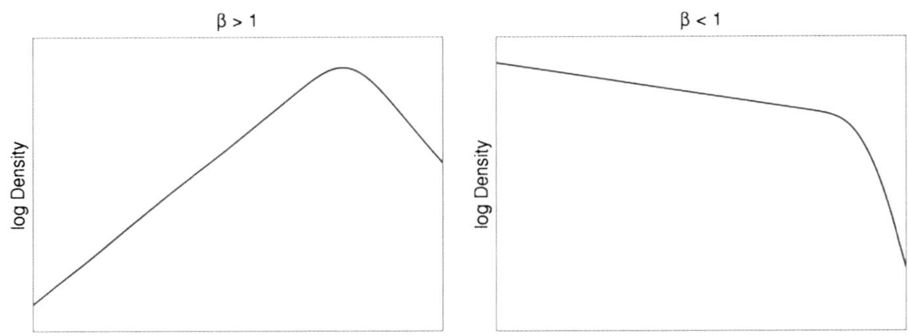

Fig. 3.4 The log–log plot consisting of two straight line segments

where Φ is the cdf of the standard normal distribution and Φ^c the compliment of Φ, that is, Φ^c is the ccdf of the standard normal distribution. Figure 3.3 shows the density of the double Pareto lognormal distributions with $\beta > 1$ and $\beta < 1$ (Note that changing the value of α does not alter the general shape of the distribution), and Fig. 3.4 shows the two straight line segments when log–log plot the double Pareto lognormal distribution.

The double Pareto lognormal distribution provides a reasonable model to describe a random process that allows random variables to start from different initial values, as long as they obey the same lognormal distribution. Releasing the initial value constraint would make the model more useful in empirical studies. For example, in surveying personal wealth accumulations, it would be more reasonable to assume that different people begin their asset accumulation with different starting salaries. Thus, the double Pareto lognormal distribution might allow closer matches with empirical data distributions.

The double Pareto distribution exhibits power law behavior at both upper and lower tails. That is, both the ccdf and cdf each has a linear tail on a log–log plot. This is an important characteristic, which is often used to test if a distribution

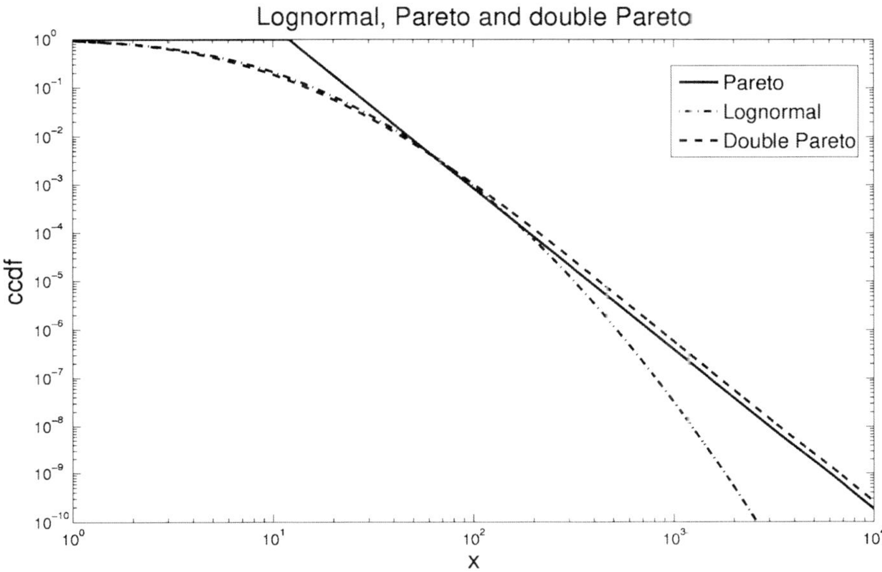

Fig. 3.5 Tail comparisons of lognormal, Pareto, and double Pareto

has a double Pareto distribution. That is, check whether the ccdf and the cdf of a distribution each have a linear tail on a log-log plot.

The double Pareto distribution falls nicely between the lognormal distribution and the Pareto distribution. The Pareto distribution and the double Pareto distribution both are power law distributions, but they have the following distinction: The log–log plot of the density function of the Pareto distribution is a single straight line, and the log–log plot of the density function of the double Pareto distibution consists of two straight line segments that meet at a transition point. This is similar to the lognormal distribution, which has a transition point around its median. Hence, an appropriate double Pareto distribution can closely match the body of a lognormal distribution and the tail of the Pareto distribution. Figure 3.5 shows the ccdfs of lognormal distribution, Pareto, and double Pareto distributions in the log–log plot. We can see that the double Pareto distribution matches well with the lognormal distribution in the body, and matches well with the Pareto distribution in the tail.

3.2.4 Power-Law Through Stochastic Differential Equations

The lognormal distribution plays an essential role in generating a double Pareto (lognormal) distribution. However, how a lognormal distribution is generated is an interesting topic in its own right. More specifically, one would like to know what kind of processes that model natural phenomena will result in a lognormal distribution.

A stochastic process, a.k.a. a random process, describes the probability distribution of possible realities of how the process might evolve over time [48]. A stochastic differential equation (SDE) is a differential equation in which one or more of the terms is a stochastic process, thus resulting in a solution that is itself a stochastic process. SDEs are used to model diverse phenomena such as personal income figures, human settlement sizes, fluctuating stock prices, or physical systems subject to thermal fluctuations. Typically, SDEs incorporate white noise that can be thought of as the derivative of the Brownian motion (or the Wiener process). However, it should be mentioned that other types of random fluctuations are also possible, including jump processes and Poisson counters.

The first SDE we would like to discuss is a Geometric Brownian Motion (GBM) process. Reed [42–44, 47] uses this model to explain the double Pareto lognormal distribution. A random variable X is said to follow GBM if its behavior over time is governed by the following differential equation

$$\mathrm{d}X = (\mu \mathrm{d}t + \sigma \mathrm{d}B)X, \tag{3.16}$$

where $\mathrm{d}B$ is the increment of a standard Brownian motion (a.k.a. the white noise). For a GBM the proportional increment of X in time $\mathrm{d}t$ comprises a systematic component $\mu\mathrm{d}t$, which is a steady contribution to X, and a random component $\sigma\mathrm{d}B$, which is fluctuated over time. Thus, the GBM can be seen to be a stochastic version of simple exponential growth.

Assuming a constant initial state X_0. Applying Itō's lemma [26] on this SDE produces the following equation at time $t = T$:

$$\ln X_T = \ln X_0 + \left(\mu - \frac{1}{2}\sigma^2\right)T + \sigma B_T. \tag{3.17}$$

Since B_T is normally distributed with parameters $(0, T)$, it is evident that the random variable $\ln X_T$ is also normally distributed with mean

$$\ln X_0 + \left(\mu - \frac{1}{2}\sigma^2\right)T,$$

and variance $\sigma^2 T$, which means that the random variable X_T is lognormally distributed with the same set of parameters. Since in a random process the stopping time (or a starting time) T for each individual instance may be different, it is a random variable. If it is exponentially distributed, then it can be shown that, as discussed in Sect. 3.2.3, the mixture of states follows a double Pareto distribution. If we further assume that the initial state X_0 follows a lognormal distribution, the distribution of X becomes a double Pareto lognormal distribution.

In addition to the GBM model, Jiang et al. [28] propose several other generalized forms of first order SDEs that also exhibit power law behaviors. In particular, they consider a different scenario involving the steady state density associated with an

SDE. The SDE describes a scenario where the quantity of interest decays to zero exponentially, that is,

$$dX = -\alpha X,$$

but is incremented by a fixed amount of σ at a random moment of time. Under the assumption that the random variable of time is exponentially distributed, they show that for a range of parameter values the steady state distribution of X exhibits a lower-tail power law, and through a simple transformation $Y = X^{-1}$ the distribution can be converted into an upper-tail power law. In their model, the random fluctuation component is implemented by a Poisson counter.

Poisson counter driven SDEs are studied in Brockett [6], one typical SDE with Poisson counter is

$$\mathrm{d}X = -\alpha X \mathrm{d}t + \sigma \mathrm{d}N, \tag{3.18}$$

where $\alpha > 0, \sigma > 0$, and N is a Poisson process of intensity λ.

Jiang et al. [28] showed that, through the transformation that converts the lower-tail power law into an upper-tail power law, the resulting distribution converts to a Pareto distribution as $t \to \infty$.

It is also possible to let Brownian motion and Poisson counter co-exist in the SDE. Adding a Brownian motion component, the SDE becomes

$$\mathrm{d}X_t = (\mu \mathrm{d}t + \sigma \mathrm{d}B_t)X_t + (x_0 - X_{t-})\mathrm{d}N_t, \tag{3.19}$$

which is a GBM with Poisson jumps that always reset the motion to a fixed state x_0. This SDE is similar to the one analyzed by Reed [42, 44], and the result is also similar, for Reed showed that as $t \to \infty$ the steady-state density distribution of X_t converts to a double Pareto distribution.

Jiang et al. [28] also studied the power-law behavior near a critical point and derived an SDE comprises of two independent bi-directional Poisson counters to demonstrate this behavior. They showed that a discontinuity at this critical point occurs in a surprising way, which might be of interest in statistical physics.

Early studies have all indicated that random multiplication with exponentially distributed stopping time will lead to power law behaviors. This result may serve as a guideline in explaining real-world phenomena that obey the power law.

3.3 Power-Law Behaviors in Complex Networks and Natural Phenomena

The most effective way to evaluate the accuracy a power-law model is to use it to explain the cause of certain power-law phenomenon and validate the prediction using empirical data. The goodness of fit is a typical measure of a theoretical model.

A good model should make this an intuitive process. In this section, we will present the possible explanations to various real-world complex networks and natural phenomena that exhibit power-law behaviors, and demonstrate the goodness-of-fit figures with empirical data.

3.3.1 Income

It is well known that the distribution of a population's income and wealth follows the power law. The original observation was made in 1907 by Pareto. He noticed that 80% of the wealth in Italy was owned by 20% of the population [40]. He then surveyed a number of various types of countries and found to his surprise that a similar income distribution applied. For over a century, most of the studies have been focusing on the distribution itself as well as the impacts imposed by Pareto's principle. Not much has been done to explain the underlying reasons that cause this phenomenon. Reed's GBM model for the double Pareto lognormal distribution based was the first-known model that provides an intuitive explanation for the income distribution.

The distribution of incomes over a population is the same as the probability distribution of the income of an individual randomly selected from that population. Thus, a stochastic model for the generation of the income of such an individual can be used to explain the observed distribution of incomes in a population or in random samples. For an individual, the more income he or she currently receives, the more income he or she will accumulate in the next time interval based on an expected rate of increase (e.g., interests of deposits in saving accounts and annual raises of salaries). Similarly, finance uncertainties (e.g., good or bad investment, market depression, marriage or divorce) will directly affect his or her income as well. This argument suggests that the income behavior over time could be modeled as a GBM model, which was discussed in Sect. 3.2.4 and is presented here for convenience:

$$\mathrm{d}X = (\mu \mathrm{d}t + \sigma \mathrm{d}B)X,$$

where μ represents the instantaneous rate of increase on a riskless asset, σ the volatility of the income, and $\mathrm{d}B$ the infinitesimal change in a Brownian motion over the next instant of time (a.k.a. white noise). Assume that individual's income follows GBM process X with initial state X_0 being a constant, then for a randomly selected individual from the group of people with the same working time T, his or her income is lognormally distributed.

If an individual is randomly selected from the entire workforce, the time T that he or she has been in the workforce will be a random variable. To find the distribution of time T, we consider a simple case when the workforce or population is growing at a constant rate ν. We assume that all individuals will eventually merge into the workforce at certain time. A simple analysis gives us the following equalities:

$$\begin{aligned} F_T(t) &= 1 - Pr(T \geq t) \\ &= 1 - \frac{N_{T-t}}{(1+\nu)^t N_{T-t}} \\ &= 1 - (1+\nu)^{-t}, \\ f_T(t) &= F_T'(t) \\ &= \ln(1+\nu)(1+\nu)^{-t} \\ &= \lambda \mathrm{e}^{-\lambda t}. \end{aligned}$$

In this case, the time T has an exponential distribution with a probability density function $f_T(t) = \lambda \mathrm{e}^{-\lambda t}$. Therefore, the income distribution from the entire workforce will be an exponential mixture of lognormal distributions, leading to a double Pareto distribution. Now let us consider the initial state X_0. It would be more realistic to assume that individuals' initial incomes will also vary and evolve over time, which can be described by another GBM. In this case, the income distribution is changed to a double Pareto lognormal distribution [44].

Figure 3.6 demonstrates the double Pareto lognormal distribution fitted to empirical income data (originally fitted by Reed [47]): the United States household income of 1997 and the Canadian personal earnings of 1996, respectively. The data fits in the theoretical curves quite well, not only in the upper tail region, but also in the lower tail region.

3.3.2 *Stock Market Returns*

Stock market returns is another example that fits the double Pareto distribution and can potentially be explained using a GBM model. During the last several decades researchers have investigated the statistical distribution of returns and have concluded that returns are "fat tailed," which is a key power law characteristic. Several studies have been devoted studying to the stochastic behaviors of the stock price changes [21,29,31,32,39]. Bachelier in his 1900 dissertation [4] suggested that stock prices in successive periods follow a random process that is best described by a Wiener process (i.e., a Brownian motion).

This suggests that we could treat the return rate of each transaction as an independent random variable. A simple model can be described as $A_T = A_0 \times X_0 \times \cdots \times X_T$ where A_T represents the current stock assets, A_0 the initial stock investment, and X_0 through X_T the average return rate for a certain time period. Thus, at a certain time the total return could be seen as a sequence of multiplication of such independent random variables with initial values, which is analogous to a long-term

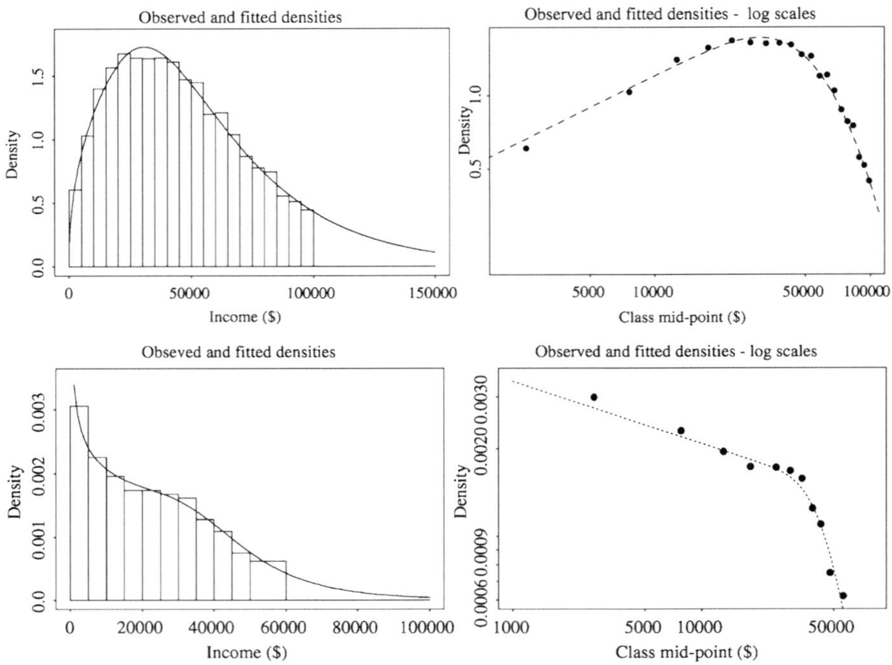

Fig. 3.6 The double Pareto lognormal distribution fitted to the US household income (1997) data and the Canadian personal earnings (1996) data [47]

price discount, except that it should not always be a discount. If we assume the stock is a good pick and has a steady increased rate with small fluctuation, it would be better to apply the GBM model of $\mathrm{d}X = (\mu \mathrm{d}t + \sigma \mathrm{d}B)X$, where μ is the expected return increase rate and $\sigma \mathrm{d}B$ the volatility of the return over time. The stochastic multiplicative process yields a lognormally distributed variable representing the stock returns at a fixed time. If we consider a random killing process with a constant killing rate, by killing here it means to cash out the stocks and settle, and the returns is observed at the killing time, then the time span a process is kept alive is exponentially distributed. Thus, if we observe the population of stock returns, the distribution will be the mixture of lognormal distributions with exponentially distributed time span, which is a double Pareto distribution. The initial wealth can be viewed as any wealth naturally accumulated at certain time, which is lognormally distributed as the result of multiplicative process. Hence, the distribution of stock returns can be extended to the double Pareto lognormal distribution.

Figure 3.7 shows a good fit to the returns of IBM's ordinary stocks from Jan 1, 1999 to Sept 18, 2003, originally plotted by Reed [44]. The figure at the left-hand side shows a density histogram and the fitted double Pareto lognormal density; the figure at the right-hand side shows the fitted density in logarithmic scale.

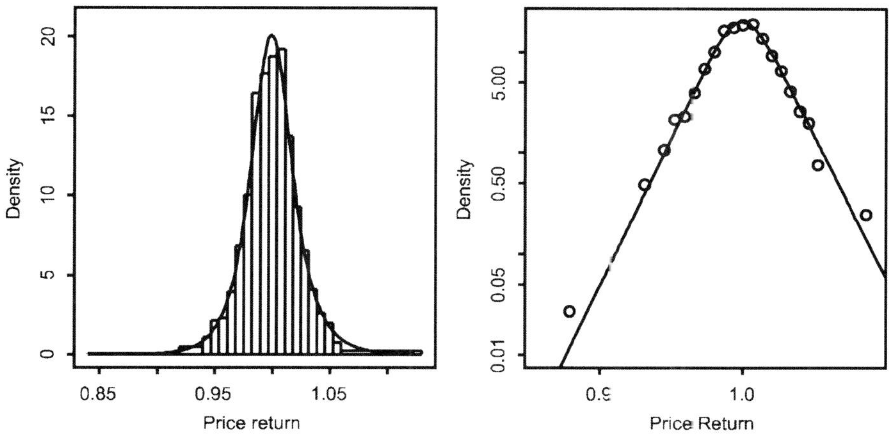

Fig. 3.7 The double Pareto lognormal distribution fitted to stock market returns [44]

Fig. 3.8 ccdf of file sizes statistics from University of Calgary Web server. Compare with a lognormal distribution and a Pareto distribution [11]. The empirical data has the lognormal body and Pareto tail, which indicates a double Pareto distribution

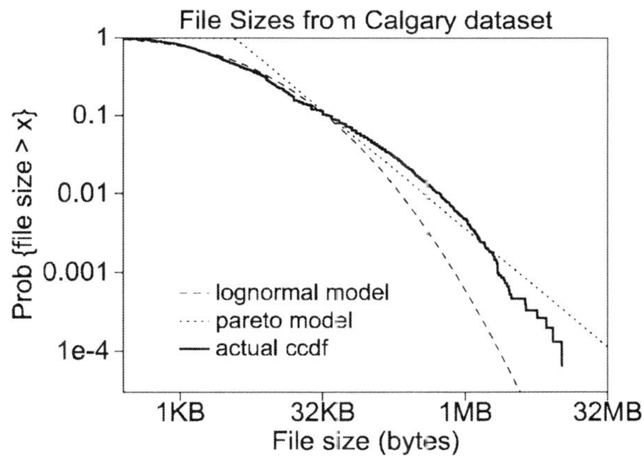

3.3.3 *Internet File Sizes*

It has been observed that the file size distribution over the internet seem to follow a double Pareto distribution [11, 14, 41]. Figure 3.8 shows a file size statistics from a Web server at the University of Calgary, from traces collected by Arlitt and Williamson [3]. There have been attempts to explain the file size distribution. Among them, Downey's multiplicative file size model [12] and Mitzenmacher's recursive forest file model [35] have received much attention. The latter can be viewed as an improvement over Downey's model by introducing a dynamic insertion and deleting process.

Downey's model [12] is based on the following observation: Users tend to create new files from old files by copying, editing, or filtering in some way, and the size of the new file will differ from that of the old file by a multiplicative factor f from a given distribution D. That is, the system begins with a single root file and repeatedly performs the following actions: randomly choose a file and create a new file from it with size equal to fs, where s is the size of the chosen file.

The assumption behind this model is that creating a new file from a template file through copying, editing, or filtering will cause the size of the new file to differ from that of the old file by a factor from a given distribution D. For any file in this system, the history of the creation can always be traced back to the original root file. Thus, the size of the file can be viewed as the result of a random multiplicative process. Downey, therefore, suggests that the entire file size distribution is lognormal. Downey's model, however, does not address the situations of insertion and deletion, and the result of empirical fitting on these two operations is not satisfiable.

To overcome this problem, Mitzenmacher [35] introduces a recursive forest file model by modifying Downey's model to include dynamic insertion and deleting process. In this model, the system begins with a collection of one or more files, whose sizes are drawn from a distribution D_1. New files are generated repeatedly as follows:

1. With probability γ, add a new file with size chosen from a given distribution D_1.
2. With probability η, select a file uniformly at random, delete this file.
3. With probability $1-\gamma-\eta$, select a file S with size s uniformly at random, choose a multiplicative factor f from a given distribution D_2, and create a new file S' with size fs.

Mitzenmacher reasons [35] that the file size density in this model converges to a double Pareto distribution if D_2 is a lognormal distribution D_2, and a double Pareto lognormal distribution if D_1 is also a lognormal distribution. This model explains why file size distribution may appear to have a lognormal body and a Pareto tail, making it more appealing compared to other attempted explanations. It also provides a flexible framework that may be extend to handle additional operations in the file system.

3.3.4 Friends in MySpace

The observation that complex networks often exhibit power-law behaviors has attracted much attention in recent years [8, 11, 14, 19]. Since Huberman and Adamic [23] suggested in 1999 that the exponential growth of the World Wide Web network could explain its power-law degree distribution, many studies have attempted to migrate this idea to explain other complex networks [5, 13, 15, 34]. The rise of online social networks has generated overwhelmingly huge amount of data, making it possible to carry out quantitative analysis on human social behaviors in a large scale.

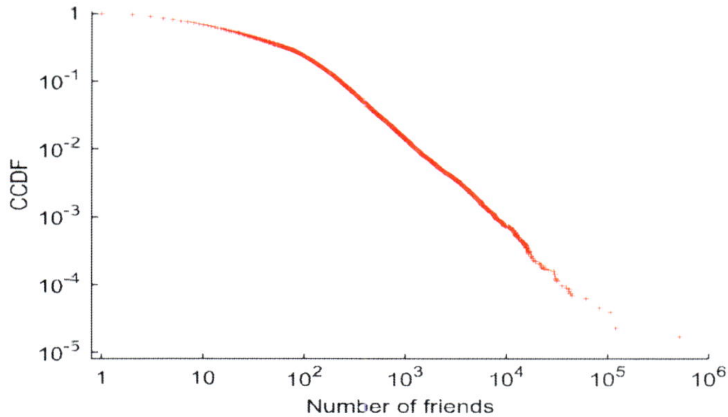

Fig. 3.9 Empirical ccdf of the number of friends in MySpace, plotted in the log–log scale [49]

Ribeiro et al. [49] recently investigated MySpace and showed that the distribution of the number of friends follows a double Pareto like distribution, which is shown in Fig. 3.9.

Making friends in a virtual world is much easier, for the click of "add as friend" button really does not need any social skill. Thus, the meaning of friends in an online social network may be weak, which may simply mean "somehow related." However, this observation still suggests a reasonable assumption that for a user with a large number of friends, most of the friends are added through passive referrals, assuming that all requests of making friends are automatically approved. In addition, if we assume that every user has a different referral probability to different people and the referral is always accepted, the growth of the number of friends can be modeled roughly as a multiplicative process $X_t = P_t X_{t-1}$, where each P_t can be computed as the average referral probability of the current group of friends. According to the analysis presented in Sect. 3.2.2, for a fixed period of time the number of friends follows a lognormally distribution.

Ribeiro et al. [49] also observe that the time span of MySpace accounts is distributed exponentially. This scenario might be explained by assuming a fixed increase rate of MySpace accounts. Combining a exponentially distributed time span with the lognormally distributed number of friends at fixed time, the overall friend distribution for all MySpace accounts converges to a double Pareto distribution.

3.3.5 Downloads from SourceForge

SourceForge is a Web-based source code repository that provides a centralized location for software developers to control and manage open source software development. As of February 2009, the SourceForge repository hosts more than 230,000 projects and has more than 2 million registered users.

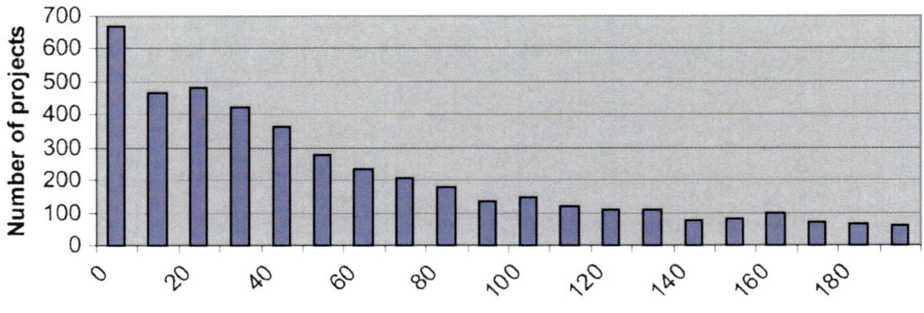

Fig. 3.10 The distribution of downloads in 30 days for active projects [24]

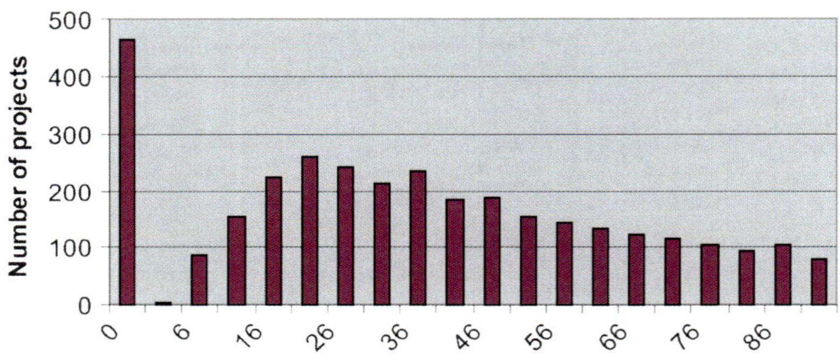

Fig. 3.11 A closer look at the distribution of downloads [24]

Hunt and Johnson [24] use SourceForce as a downloading platform and study the statistics of downloads. The data on the number of downloads was collected from all projects listed on the most active project list on October 22, 2001 for 30 days, which was partially shown in Figs. 3.10 and 3.11. The distribution obtained is heavily skewed which exhibits a significant sign of power tail. Taking a closer look at the few download region, it is evident that the download distribution also exhibits a lower tail, which implies that the double Pareto model could be able to provide an explanation to both of these tails.

It is reasonable to assume that during these 30 days every download for a single project comes from different users for the following reasons: People would not download the same file over and over again unless the download is not successful or the file has being updated, and the 30 days time span is considered to be a relatively short cycle that a file only has a small chance of being altered. Popular projects would attract more user downloads and people tend to broadcast satisfying user experience. Thus, the popular downloads are more likely to be introduced to new users and become even more popular, which would in turn increase the download counts. This process is similar to the growth of the number of MySpace friends.

The same model can be applied to conclude that the number of downloads of projects by the same age group would exhibit a lognormal distribution.

No public data is available concerning the distribution of ages for all the projects. If we assume that the total number of projects increase approximately at a fixed rate, then at time $T+1$ the number of projects would be $N_{T+1} = (1+\nu)N_T$. From the analysis in Sect. 3.3.1 it follows that the cumulative distribution of time T is $F_T(t) = 1-(1+\nu)^{-t}$ and the frequency distribution is $f_T(t) = F_T'(t) = \ln(1+\nu)(1+\nu)^{-t} = \lambda e^{-\lambda t}$ by replacing $\ln(1+\nu)$ with λ. Therefore, the ages of projects would follow an exponential distribution. Thus, the distribution of number of downloads over all active projects is a double Pareto distribution, which exhibits both the lower tail and the upper tail.

3.3.6 *Sizes of Human Settlements*

Auerbach (1913) was the first to discover that the distribution of city sizes can be well approximated by a power-law distribution. That is, if we rank the cities based on population, then the size of the largest city is twice as that of the second largest city, three times as that of the third largest city, and so on. This distribution was believed to follow Zipf's law, a.k.a. the rank-size distribution. A number of studies have since contributed evidence and provided support to this statement until recently [7, 18, 27, 38]. In a wide spread article [16, 17], Eeckhout points out that the old evidence is problematic since the early studies only dealt with truncated samples and only focused on large cities, and the Zipf's law does not hold when taking all settlements of a country into consideration. The Zipf's law is still valid for the tail behavior since the tail mostly consists of large cities. When adding more cities of smaller sizes to the distribution, it gradually shows a lognormal body. Eeckhout [16, 17] then model the growth of settlements using the pure form of Gibrat's law (a.k.a Gibrat's rule of proportionate growth) and generate lognormal size distribution. Gibrat's law states that the size and growth rate are independent. However, the lognormal distribution does not fit well with the power-law tail.

The double Pareto lognormal seems more appropriate since it comprises a lognormal body and power law tails. Reed [45] suggests a GBM model, similar to the one that models personal incomes, for obtaining the settlement size distribution. Individual human settlements grow in many different ways. At the macro level a GBM process $dX = (\mu dt + \sigma dB)X$ can be used to model the size growth by assuming a steady systematic growing rate μ and a random component σdB. The steady growing rate reflects the average growth rate over all settlements and times, and the random component reflects the variability of the growth rate. The time when a city is founded varies from settlement to settlement. If we assume in the time interval $(t, t+dt)$ any existing settlement can form a new satellite settlement with probability λdt, the creation of settlements is a Yule process [52], which was first proposed as a model for the creation of new biological species. Under Yule process, the expected number of settlements is $e^{\lambda t}$ after t time since the first settlement. That

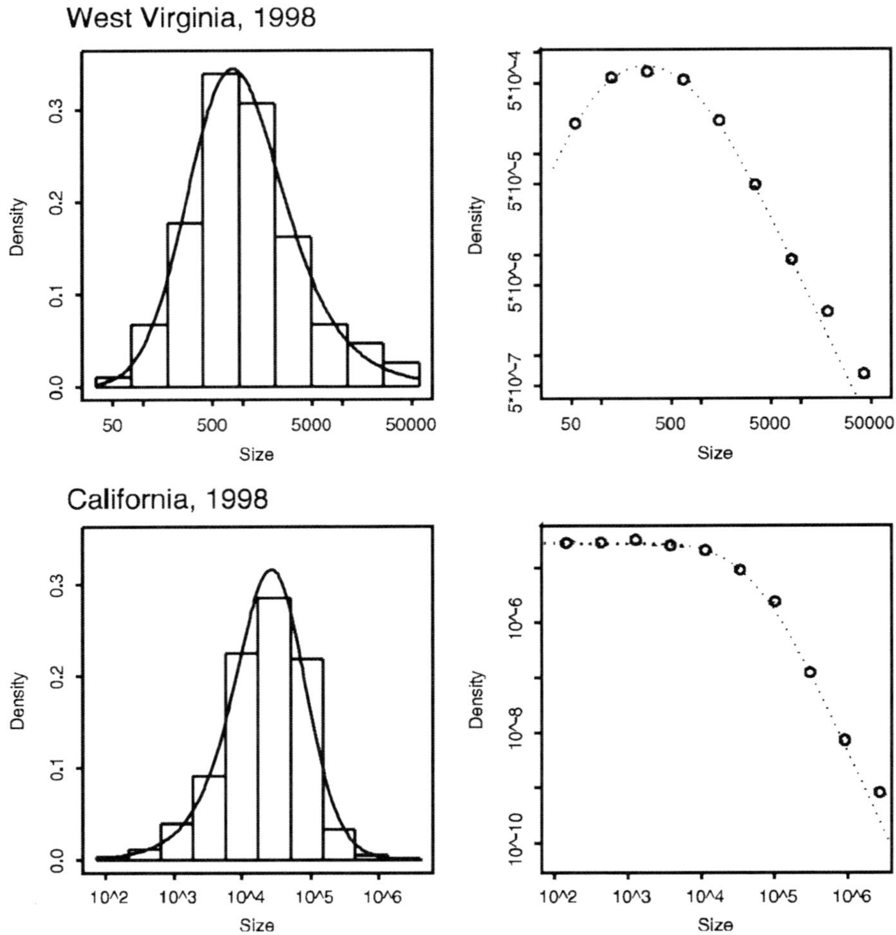

Fig. 3.12 The double Pareto lognormal distribution fitted to empirical city size data [45]

is, the number of settlements is growing at rate λ. Therefore, the existing time for all settlements is exponentially distributed. It is straightforward to conclude that under GBM and Yule processes, the overall settlements size distribution will is a double Pareto distribution. If we further assume a lognormal initial settlement size, the result will converge to the double Pareto lognormal distribution.

Figure 3.12 shows empirical and fitted double Pareto lognormal distribution originally plotted by Reed [45] on the West Virginia data and California data obtained in 1998. The left-hand panel demonstrates the empirical density histogram in logarithmic size scale and fitted theoretical curve. The right-hand panel shows the fitted double Pareto lognormal distribution density in the log–log scale. It is evident that the double Pareto lognormal distribution provides a nice fit to the data in each region.

3.3.7 Volumes of Oil Field Reserves

The distribution of oil field sizes (i.e., the volumes of oil field reserves) has been the subject of much study for decades. Allais [1] was the first to propose to use a lognormal distribution for mineral resources and Kaufman [30] used this distribution for a population of oil or gas field in a petroleum basin. After that, the Pareto distribution has also been commonly used [20, 22, 33, 37] since the petroleum exploration practice has indicated that the probability of discovering large oil pools is low while the probability of discovering medium and small-sized pools is high and the Pareto distribution is consistent with this type of structure. However, distinguishing lognormal and Pareto features that are both shown in the oil field size distribution may be difficult, which suggests that the double Pareto lognormal distribution might be a better choice.

An oil field can be thought of as a percolation cluster. The percolation theory provides a useful model of connectivity and dynamics in complex geometries (see [50] for a comprehensive introduction). The typical problem of percolation is to consider a lattice of $n \times n$ sites, each of which is either occupied or unoccupied with a certain probability. Clusters are formed when neighboring sites are occupied. The objective is to understand the relationship among groups of clusters. If we think of an oil field as a percolation cluster, the growth of oil field sizes can then be considered as a stochastic process of merging adjacent regions. Thus, the growth of an oil field size can be assumed to be proportional to its current size, for larger oil fields would have more possible regions to merge, which implies that the size distribution follows a lognormal distribution for a fixed percolating time.

The initial percolation cluster was formed by burying a huge amount of organically rich materials (e.g., plants and animal bodies), which could be caused by geologic hazards such as earthquake, landslides, and mudflows. Such extreme event occurs with small probability and percolation clusters are formed randomly with a small constant rate. Thus, the total percolating time span for a cluster is distributed exponentially. If we also assume a lognormal distribution on the initial burying amount, the overall size distribution of oil field would be a double Pareto lognormal distribution.

Figure 3.13 shows empirical and fitted double Pareto lognormal distribution by Reed [44] for the volume of 634 oil fields in the West Siberian basin. The left-hand figure demonstrates the empirical density histogram in logarithmic volume scale and fitted theoretical curve. The right-hand figure shows the fitted double Pareto lognormal distribution density in the log–log scale.

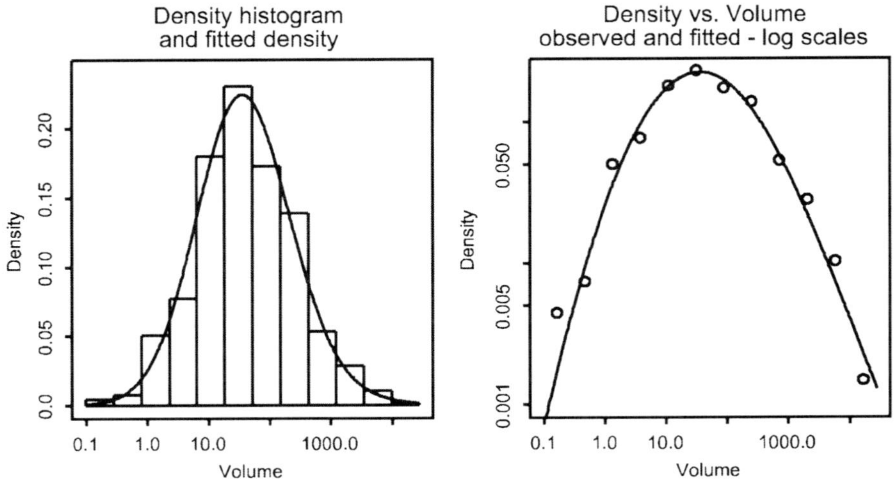

Fig. 3.13 The double Pareto lognormal distribution fitted to sizes of oil fields [44]

3.3.8 Areas Burned from Forest Wildfire

A number of researches have examined the distribution of burned areas from forest fires [2, 9, 10, 25, 46, 51] and claimed that it exhibits power-law behaviors, as shown in Fig. 3.14. Studying the fire size distribution would be useful to help construct a wildfire spreading and distinguishing model. Finding a good model that could precisely describe the process of a forest fire is still an active and open topic in ecological science. We only provide a sketch model to demonstrate a forest fire process and omit the details.

The start of a forest fire could just be a random lit up. The growth of a fire is a very complicated affair, depending on water, topology, temperature, humidity, plant types (i.e., fuel types), and many other factors. Percolation theory models forest fire spreads as a multiplicative stochastic process, which is similar to many biological entities that grow and die in a monotonic and stochastic fashion. That is, the size of burning area grows proportional to its current burning size as time spans. Therefore, the sizes of burnt area follows approximately a lognormal distribution for a fixed burning duration.

The causes of the forest fire extinguishment varies under different situations. For a small fire, it could extinguish because of lack of fuel in surrounding areas (burned up or changes of materials). Another possible cause of extinguishment could come from intervention of human. However, people involved in fire fighting commonly believe that suppression cannot put out very large fires. The effect of suppression is more likely to reduce the spread (e.g., by putting separation lines around the back and sides of fires) rather than to actually extinguish the fires [53].

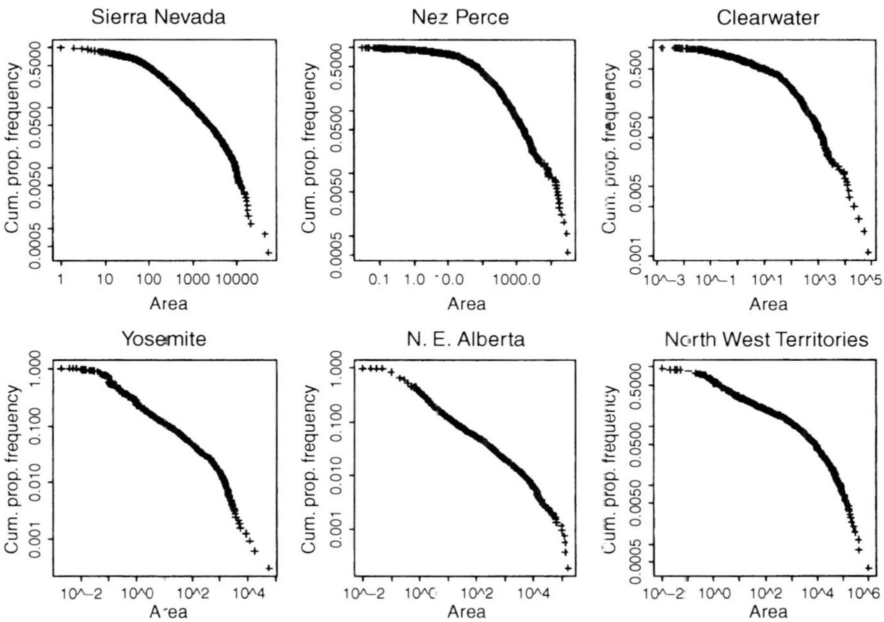

Fig. 3.14 A log–log plot of cdf for the size distribution of areas burned for the six dataset [46], measured in hectares

The only effective cause of extinguishment is precipitation. A few days of heavy rainfall could put off any size of forest fire. A weaker precipitation might only slow down the spread, which could be viewed as similar to that of human intervention. If we assume an equal chance that fires can be put off by natural causes, the time a forest fire lasts would follow an exponential distribution. Mixing the lognormal distribution of burned area size and the exponential time span will lead to a double Pareto distribution.

3.4 Conclusion and Future Directions

In this article, we presented an overview of recent significant research results in the studies of power law that occur in complex networks and natural phenomena, explored a two-tailed power-law model called the double Pareto (lognormal) model and presented a number of real-world examples that can be explained using this model. The diversity of these examples shows the robustness of the double Pareto (lognormal) model.

The good fit of double Pareto (lognormal) distribution, however, comes with costs. The physical meanings of parameters α and β become less intuitive comparing with those in the Pareto and lognormal distribution, and making a good estimation of the two parameters for the double Pareto distribution (four parameters for the double Pareto lognormal distribution) requires more mathematic skills. While we apply double Pareto (lognormal) distribution to empirical data for a close fit, seeking the underneath mechanism that makes various phenomena follow this distribution is also crucial. Although "random multiplication with exponentially distributed stopping time" provides a plausible explanation to the double Pareto (lognormal) distribution, it is not universal and should not be the only explanation. Is "exponentially distributed time" necessary to form the double Pareto (lognormal) distribution? If not, what are the alternatives? Questions like these are vital and warrant careful studies. The GBM model that essentially inherits from the random multiplication process and is adopted by numerous cases has its own limitation, especially in the examples of oil field reserves and forest fire, where a simple GBM seems not sufficiently sophisticated to precisely capture the nature of those phenomena. Other good models are still yet to be found.

Nevertheless, we still expect that this model can be further applied to other areas that are yet to be explored, including the knowledge networks. We would like to know whether the linkage of human knowledge also exhibits the power law. If we could manage to divide the knowledge networks into subject domains, would the different bodies of domain knowledge share the structural similarity? If so, can we even control the knowledge accumulation process? These questions seem interesting, for they may help discover new knowledge. The world is expecting continuous excitement from new findings on power law research.

Acknowledgements The author Z. Fang is supported in part by the NSF under grant CCF-0830314. J. Wang is supported in part by the NSF under grants CCF-0830314, CNS-0958477, and CNS-1018422. B. Liu is supported in part by the NSF under grants CNS-0721626, CNS-0953620, and CNS-1018303. W. Gong is supported in part by NSF under grant EFRI-0735974 and by Army Research Office under Contract W911NF-08-1-0233.

References

1. Allais, M.: Methods of appraising economic prospects of mining exploration over large territories. Management Science, Vol.3, 284–357 (1957)
2. Alvarado, E., Sandberg D., Pickford S.: Modeling large forest res as extreme events. Northwest Science Vol. 72, 66–75 (1998)
3. Arlitt, M.F., Williamson, C.L.: Web server workload characterization: the search for invariants. In Proceedings of the 1996 ACM SIGMETRICS international conference on measurement and modeling of computer systems, Philadelphia, Pennsylvania, United States, ACM, New York, NY, USA, 126–137 (1996)
4. Bachelier, L.: Théorie de la spculation, Annales Scientifiques de lÉcole Normale Suprieure, Vol. 3, No.17, 21–86 (1900)
5. Barabási, A.L., Jeong, H., Néda, Z., Ravasz, E., Schubert, A., Vicsek, T.: Evolution of the social network of scientific collaborations. Physica A: Statistical Mechanics and its Applications, Vol. 311, No. 3–4, 590-614 (2002)
6. Brockett, R.W.: Talk at Kyoto University, Kyoto, Japan (2007)
7. Chen, Y., Zhou, Y.: Multi-fractal measures of city-size distributions based on the three-parameter Zipf model. Chaos, Solitons & Fractals, Vol. 22, No. 4, 793–805 (2004)

8. Crovella, M.E., Taqqu, M.S., Bestavros, A.: Heavy-tailed probability distribution in the World Wide Web. A practical guide to heavy tails, Birkhauser Boston Inc., Cambridge, MA, USA, 3–25 (1998)
9. Cumming, S.G.: A Parametric models of the fire-size distribution. Forest Research Vol. 31, No. 8, 1297–1303 (2001)
10. Cui, W., Perera A.H.: What do we know about forest fire size distribution, and why is this knowledge useful for forest management? International Journal of Wildland Fire, CSIRO Publishing, Collingwood, Victoria, Australia, Vol. 17, 234–244 (2008)
11. Downey, A.B.: Evidence for long-tailed distributions in the internet. In Proceedings of the 1st ACM SIGCOMM Workshop on Internet Measurement (IMW '01), CSIRO Publishing, Collingwood, Victoria, Australia, 229–241 (2001)
12. Downey, A.B. The structural cause of file size distributions. In Proceedings of the 2001 ACM SIGMETRICS international conference on measurement and modeling of computer systems, Cambridge, Massachusetts, United States, ACM, New York, NY, USA, 328–329 (2001)
13. Dorogovtsev, S.N., Mendes, J.F.F.: Language as an evolving word web. Proc. R. Soc. Lond. B 22, Vol. 268, No. 1485, 2603–2606 (2001)
14. Downey, A.B.: Lognormal and Pareto distributions in the Internet. Comput. Commun. 28, 7, 790–801 (2005)
15. Ebel, H., Mielsch, L., Bornholdt, S.: Scale-free topology of e-mail networks. Phys. Rev. E Vol. 66, No. 3 (2002)
16. Eeckhout, J.: Gibrat's law for (all) cities. The American Economic Review, Vol. 94, No. 5, 1429–1451 (2004)
17. Eeckhout, J.: Gibrat's law for (all) cities: reply. The American Economic Review, Vol. 99, No. 4, 1676–1683 (2009)
18. Giesen, K., Zimmermann, A., Suedekum, J.: The size distribution across all cities – double Pareto lognormal strikes. Journal of Urban Economics, Vol. 68, 129–137 (2010)
19. Gong, W., Liu, Y., Misra, V., Towsley, D.: Self-similarity and long range dependence on the internet: a second look at the evidence, origins and implications. Comput. Netw. 48, 3, 377–399 (2005)
20. Greenman, J.V., Fryer, M.J.: Hydrocarbon Field Size Distributions: A Case Study in Mixed Integer Nonlinear Programming. The Journal of the Operational Research Society Vol. 47, No. 12, 1433–1442 (1996)
21. Gu, G., Chen, W., Zhou, W.: Empirical distribution of Chinese stock returns at different microscopic timescales. Physica A, 387, 495–502 (2008)
22. Houghton, J.C.: Use of the truncated shifted Pareto distribution in assessing size distribution of oil and gas fields. Mathematical Geology, Vol. 20, No. 8 (1988)
23. Huberman, B., Adamic, L.: Growth dynamics of the World Wide Web. Nature, page 130–130 (1999)
24. Hunt, F., Johnson, P.: On the Pareto distribution of SourceForge projects. In C. Gacek and B. Arief (eds.), Proc. Open Source Software Development Workshop, pps. 122–129, University of Newcastle, UK (2002)
25. Holmes, T.P., Huggett, R.J., Westerling, A.L.: Statistical Analysis of Large Wildfires. Forestry Sciences, Vol. 79, II, 59–77 (2008)
26. Itō, K.: On stochastic differential equations. Memoirs, American Mathematical Society 4, 1–51 (1951)
27. Jiang, B., Jia, T.: Zipf's law for all the natural cities in the United States: A geospatial perspective. Preprint, http://arxiv.org/abs/1006.0814
28. Jiang, B., Brockett, R., Gong, W., Towsley, D.: Stochastic differential equations for power law behaviors. Submitted to Journal of Applied Probability, Applied Probability Trust (2010)
29. Jondeau, E., Rocklinger, M.: The tail behavior of stock returns: Emerging versus mature markets. Les Cahiers de Recherche, HEC Paris, 668 (1999)
30. Kaufman, G.M.: Statistical decision and related techniques in oil and gas exploration. Prentice Hall, Englewood Cliffs (1963)

31. Klass, O.S., Biham, O., Levy, M., Malcai, O., Solomon, S.: The Forbes 400 and the Pareto wealth distribution. Economics Letters, Vol. 90, 290–295 (2006)
32. Levy, M.: Market efficiency, the Pareto wealth distribution, and the Levy distribution of stock returns. Economy as an Evolving Complex System III, Oxford University Press (2006)
33. Liu, X., Jin, Z., Chen, S., Liu, L.: Generalized Pareto distribution model and its application to hydrocarbon resource structure prediction of the Huanghua depression. Petroleum Science, Vol. 3, No. 2, 22–27 (2006)
34. Milo, R., Shen-Orr, S., Itzkovitz, S., Kashtan, N., Chklovskii, D., Alon, U.: Network Motifs: Simple building blocks of complex networks. Science, Vol. 298, No. 5594, 824–827 (2002)
35. Mitzenmacher, M.: Dynamic models for file sizes and double pareto distributions. Internet Mathematics, Vol. 1, No. 3, 305–333 (2004)
36. Mitzenmacher, M.: A history of and new directions for power law research. Invite talk at University at Buffalo (2008)
37. Michel, B.: Oil production: A probabilistic model of the Hubbert curve. Applied Stochastic Models in Business and Industry, Vol. 27, No. 4, 434–449, John Wiley & Sons, Ltd. (2011)
38. Nishiyama, Y., Osada, S., Morimune, K.: Estimation and testing for rank size rule regression under pareto distribution. In Proceedings of the International Environmental Modelling and Software Society iEMSs, University of Osnabrck, Germany, (2004)
39. Onour, I.A.: Extreme risk and fat-tails distribution model: Empirical analysis. Journal of Money, Investment and Banking ISSN 1450-288X Issue 13, EuroJournals Publishing, Inc. (2010) http://www.eurojournals.com/jmib_13_03.pdf
40. Pareto, V.: Un applicazione di teorie sociologiche, published in Revista Italiana di sociologia, 1901, p. 402–456. (1971. Manual of political economy. Translated by Ann S. Schwier. Edited by Ann S. Schwier and Alfred N. Page. New York: A.M. Kelley)
41. Park, K., Kim, G., Crovella, M.: On the relationship between file sizes, transport protocols, and self-similar network traffic. In Proceedings of the 1996 International Conference on Network Protocols (ICNP '96), IEEE Computer Society, Washington, DC, USA, (1996)
42. Reed, W.: The Pareto, Zipf and other power laws. Economics Letters, Vol. 74, No. 1, 15–19 (2001)
43. Reed, W., Hughes, B.D.: From gene families and genera to incomes and internet file sizes: Why power laws are so common nature. Physical Review E, Vol. 66, No. 6 (2002)
44. Reed, W., Jorgensen, M.: The double Pareto-lognormal distribution - A new parametric model for size distributions. Commun. in Statistics – Theory and Methods, Vol. 33, No. 8 (2004)
45. Reed, W.: On the rank-size distribution for human settlements. Journal of Regional Science Vol. 42, No. 1, 1–17 (2002)
46. Reed, W., McKelvey, K.S.: Power-law behaviour and parametric models for the size-distribution of forest fires. Ecological Modelling, Vol. 150, 239–254 (2002)
47. Reed, W.: A parametric model for income and other size distributions and some extensions. International Journal of Statistics, Vol. LXIV, No. 1, 93–106 (2006)
48. Ross, S.M.: Stochastic Processes, Second Edition, Wiley. ISBN 9780471120629, (1995)
49. Ribeiro, B., Gauvin, W., Liu, B., Towsley, D.:On MySpace Account Spans and Double Pareto-Like Distribution of Friends. Second International Workshop on Network Science for Communication Networks (NetSciCom), pp. 1–6 (2010)
50. Stauffer, D., Aharony, A.: Introduction to percolation theory, Second Edition. London: Taylor and Francis (1992)
51. Schoenberg, F.P., Peng R., Woods J.: On the distribution of wildfire sizes. Environmetrics, Vol. 14, No. 6, 583–592 (2003)
52. Yule, G.U.: A Mathematical Theory of Evolution, based on the Conclusions of Dr. J. C. Willis, F.R.S.. Philosophical Transactions of the Royal Society of London, Ser. B 213: 21–87 (1925)
53. Incident Operations Standards Working Team (2010), Incident Response Pocket Guide, National Wildfire Coordinating Group (NWCG), pp. i–101

Chapter 4
Laplacian Spectra and Synchronization Processes on Complex Networks

Juan Chen, Jun-an Lu, Choujun Zhan, and Guanrong Chen

Abstract The spectrum of the Laplacian matrix of a network contains a great deal of information about the network structure and plays a fundamental role in the dynamical behavior of the network. This chapter is to explore and analyze the Laplacian eigenvalue distributions of several typical network models, to study the network dynamics towards synchronization at a mesoscale level of description, and to report the finding of a relation between the spectral information of the Laplacian matrix and the dynamics in the network synchronization process. First, an example of adding long-distance edges is given to show that the network synchronizability may not be directly inferred from statistical properties of the network. Then, the Laplacian eigenvalues of several representative complex networks are shown to possess very different properties, and yet they also share some common features meanwhile. Further, the correlation between the Laplacian spectrum and the node-degree sequence of a network is investigated, revealing that scale-free networks have the highest correlation values, followed by random networks and then by small-world networks. To that end, a simple local prediction–correction algorithm is presented for approximating the eigenvalue λ_{i+1} from λ_i, $i = 1, 2, \cdots, N$, where N is the network size. Finally, it is shown that the processes of synchronization and generalized synchronization (GS) display different patterns, depending intrinsically on the topological structures of the networks. It is found that in the process of synchronization (or GS), roughly speaking, synchronization (or GS) first starts from a small part of hub nodes and then spreads to the other nodes with smaller degrees.

J. Chen • Jun-an Lu
School of Mathematics and Statistics, Wuhan University, Wuhan, Hubei 430072, China
e-mail: juanchen1220@yahoo.com.cn; jalu@whu.edu.cn

C. Zhan • G. Chen (✉)
Department of Electronic Engineering, City University of Hong Kong, Hong Kong SAR, China
e-mail: zchoujun2@student.cityu.edu.hk; eegchen@cityu.edu.hk

M.T. Thai and P.M. Pardalos (eds.), *Handbook of Optimization in Complex Networks: Theory and Applications*, Springer Optimization and Its Applications 57,
DOI 10.1007/978-1-4614-0754-6_4,

It is also demonstrated that, for community networks, a typical synchronization process generally starts from partial synchronization through cluster synchronization to evolve to global complete synchronization.

4.1 Introduction

In recent years, the theory of complex networks has attracted wide attention and become an area of great interest [1–5], for its advances in the understanding of many natural and social systems. One subject in the studies of complex networks that has received a great deal of attention is the topological characterization of various networks. Indeed, there has been considerable interest in investigating how the statistical properties of a network, such as the degree distribution, average distance, clustering coefficient, betweenness, and so on, are related to the dynamical processes taking place on the network [6–14]. However, it was shown in [15] that some statistical network properties are not sufficient to determine various complex dynamical patterns. It is found that unfortunately statistical properties may actually infer completely opposite conclusions about some large-scale complex networks sometimes.

From a graph-theoretic perspective, the spectrum (i.e., the set of eigenvalues) of the Laplacian matrix of a network contains tremendous information about the underlying network, which provides useful insights into the intrinsic structural features of the network [16]. A prototype example is the synchronizability of a network [17–20], which is crucially determined by the ratio of the smallest nonzero eigenvalue to the largest one of the corresponding Laplacian matrix. Also, the eigenvectors of the Laplacian matrix are known to be useful for detecting the community structure of a network [21].

Synchronization, as an emerging phenomenon of a population of dynamically interacting units, has fascinated humans since the ancient times. Synchronization phenomena and processes are ubiquitous in nature and play a vital role within various contexts in biology, chemistry, ecology, sociology, technology, and even visual arts. To date, the problem of how the structural properties of a network influence the performance and stability of the fully synchronized states of the network have been extensively investigated and discussed, both numerically and theoretically. Regarding partial synchrony, however, research results obtained thus far are much less and unmature [22–32]. It is well known today that synchronization analysis, synchronization processes and topological scales are all crucially determined by the whole eigenvalues spectrum of the Laplacian matrix of the network [15, 24, 25]. As such, a careful investigation on the eigenvalues spectrum of a complex network is of great significance especially regarding the evolution of the dynamical behaviors of the network.

The present chapter studies the Laplacian spectra of complex networks and their effects on the synchronization processes over the networks. Specifically, it is to further explore and analyze the Laplacian eigenvalue distributions of several typical network models, to study the network dynamics towards synchronization at a mesoscale level of description, and to report a connection between the spectra of the Laplacian matrix and the dynamical process through the emergence of network synchronization.

Section 4.2 introduces the notion of eigenvalues spectrum of the Laplacian matrix of a network, and some well-known estimations of the eigenvalues. As reported in [15], some networks with the same statistical properties may have very different synchronization characteristics. Here, in addition, an example is given to show that adding long-distance edges can sensitively affect the average distance, a typical statistical property, while the smallest nonzero eigenvalue remains essentially unchanged.

Section 4.3 introduces and analyzes the spectral distributions of regular, random, small-world, scale-free, and community networks. The main finding is that the Laplacian eigenvalues of these four types of complex networks have very different properties in general and yet they also share some common features meanwhile. The spectral distributions of regular, random, and small-world networks are homogeneous, whereas that of the scale-free networks is quite heterogeneous. Furthermore, for random and small-world networks, the smallest nonzero eigenvalue depends approximately linearly on the connection probability adopted in network generation. There exist some big gaps between consecutive eigenvalues in community networks, while for a network with k prominent communities there are $k-1$ eigenvalues near zero. In various realizations of the same type of network topology of the same size, the variations of their Laplacian eigenvalues are quite small; and this is roughly the same for all different types of network models studied.

In Sect. 4.4, it is revealed that the distributions of the Laplacian eigenvalues are very similar to the distributions of the node-degree sequence. It is found that the correlation between the spectrum and the node-degree sequence of a scale-free network is the highest, followed by random networks and then by small-world networks. Meanwhile, a simple local prediction-correction algorithm is designed to determine the eigenvalue $\lambda_i + 1$ from λ_i, $i = 1, 2, \ldots, N$ where N is the size of the network. It is also demonstrated that the eigenvalue curves are quite different from their corresponding node-degree distributions in some regions with small indexes, although they are quite similar in other regions.

Section 4.5 studies the dynamics towards synchronization in complex networks at the mesoscale level of description. The dynamical processes towards synchronization show different patterns depending intrinsically on the network topological structures. It is found that the processes of synchronization and generalized synchronization (GS) display different patterns, depending intrinsically on the topological structures of the networks. It is also found that in the process of synchronization (or GS), roughly speaking, synchronization (or GS) first starts from a small part of hub nodes and then spreads to the other nodes with smaller degrees. Finally, it is demonstrated that, for community networks, a typical synchronization process generally starts from partial synchronization through cluster synchronization to global complete synchronization.

4.2 The Laplacian Matrix and Its Eigenvalues

Consider a network G of N nodes, with edges linking certain pairs of nodes together. Suppose that there are no self-loops and no multiple edges between the same pair of nodes. The Laplacian matrix L of the network is defined by

$$L_{uv} = \begin{cases} d_v & \text{if } u = v, \\ -1 & \text{if } u \text{ and } v \text{ are adjacent}, \\ 0 & \text{otherwise.} \end{cases}$$

where d_v denotes the degree of node v. The adjacency matrix $A = (a_{uv})$ is defined by $a_{uv} = 1$ if u is adjacent to v or 0 others. Thus, the Laplacian matrix L is defined to be $L = D - A$, where $D = (d_{ij})$ is the diagonal matrix of all node degrees, that is, with d_{ii} equals the degree of node i, $i = 1, 2, \ldots, N$. Only undirected networks are considered here, for which all the Laplacian matrices are symmetric and positive semi-definite, with nonnegative real eigenvalues arranged as $\lambda_1 \leq \lambda_2 \leq \cdots \leq \lambda_N$.

Since the Laplacian matrix, described as above, has zero row-sums (hence, zero column-sums), the smallest eigenvalue $\lambda_1 = 0$ with the corresponding eigenvector $(1, 1, \ldots, 1)^{\mathrm{T}}$; in particular, λ_2 is nonzero if and only if the network is connected, and furthermore the number of connected components is equal to the multiplicity of the 0 eigenvalue.

4.2.1 *Basic Properties of Laplacian Eigenvalues*

Let $d_{\min}$ and $d_{\max}$ denote the smallest and the largest degree of a network, respectively, and use the index i to order the degrees of its nodes: $d_{\min} = d_1 \leq d_2 \leq \cdots \leq d_N = d_{\max}$. The following estimates are well-known [33]:

$$\lambda_2 \leq \frac{N}{N-1} d_{\min} \leq \frac{N}{N-1} d_{\max} \leq \lambda_N \leq 2 d_{\max}\,. \tag{4.1}$$

Similarly, using the average degree d_{avg}, it was shown in [34] that

$$d_{\text{avg}} < \lambda_N(A) \leq N, \tag{4.2}$$

where $\lambda_N(A)$ is the largest eigenvalue of the adjacency matrix A and, moreover, $\lambda_N(A) = N$ if and only if the complementary graph of G is disconnected [35]. However, the above estimations are generally conservative.

Fiedler [36] further established the following bounds relative to the node connectivity and the edge connectivity:

$$\lambda_2 \leq \upsilon(G) \leq e(G),$$

where $\upsilon(G)$ is the node connectivity of G, namely the minimal number of nodes whose removal together with the adjacent edges would result in losing the connectivity of G, while $e(G)$ is the edge connectivity, defined as the minimal number of edges whose removal would result in losing the connectivity of G.

Fiedler [36] also established a lower bound relative to the edge connectivity or to the largest degree, as follows:

$$\lambda_2 \geq 2e(G)(1-\cos(\pi/N)),$$
$$\lambda_2 \geq 2[\cos(\pi/N)-\cos(2\pi/N)]-2\cos(\pi/N)(1-\cos(\pi/N))d_{\max}.$$

The second lower bound is better if and only if $2e(G) > d_{\max}$. Relatively, one also has $2(1-\cos(\pi/N)) = \lambda_2(P_N)$, where $\lambda_2(P_N)$ the second smallest eigenvalue of the adjacent matrix corresponding to P_N, a path through N nodes.

Anderson and Morley [37] showed that

$$\lambda_N \leq \max\{d_u+d_v : \ u \text{ and } v \text{ are adjacent}\},$$

The equality holds if and only if G is a semi-regular bipartite graph.

Merris [38] showed that

$$\lambda_N \leq \max\{d_v+m_v\},$$

where m_v is the average degree of all the neighbors of node v.

Rojo, Soto, and Rojo [39] gave another upper bound on λ_N, as follows:

$$\lambda_N \leq \max\{d_u+d_v-|N_u \cap N_v| : \ u \neq v\},$$

where $|N_u \cap N_v|$ denotes the number of the common neighbors of nodes u and v.

Li and Pan [40] proved that

$$\lambda_{N-1} \geq d_{N-1},$$

where the equality holds for a complete bipartite graph or a tree with a degree sequence $\{N/2, N/2, 1, \ldots, 1\}$. Prior to this, it was known [41] that

$$\lambda_N \geq d_{\max}+1,$$

where the equality holds if and only if $d_{\max} = N-1$.

Last but not least, Duan, Liu, and Chen [42] reported some useful estimation bounds for some Laplacian eigenvalues and the eigenvalue ratio about complex network synchronizability, with respect to subgraphs, complementary and product graphs.

4.2.2 Statistical Properties Versus Laplacian Spectra

In [13], it was shown that the synchronizability of a large class of networks is determined by the eigenvalue ratio λ_2/λ_N, of which it was shown that the main dependence is on the smallest nonzero eigenvalue λ_2 [43, 44]. Since in most real networks, $d_{\min} = 1$, it follows from (4.1) that

$$\lambda_2 \leq N/(N-1),$$

which, of course, is a rather conservative estimation.

From (4.1), one immediately gets that

$$\frac{\lambda_2}{\lambda_N} \leq \frac{d_{\min}}{d_{\max}} \leq 1. \tag{4.3}$$

Therefore, generally speaking, the closer to 1 the ratio is, the better the network synchronizability will be. For example, that scale-free networks have $d_{\min}/d_{\max} \ll 1$; therefore, as has been well experienced, they have relatively poorer synchronizability in comparison to most other types of network structures in general.

It should be noted, however, that (4.3) does not imply that a more homogeneous degree distribution always means a better synchronizability. It was shown in [15] that the effect of small structural changes on synchronizability may not average out within a large-scale network, therefore the synchronizability may not be described by the statistical network properties. Moreover, some examples were presented in [15, 16] to show that networks with the same degree distribution can have very different synchronizability characteristics.

According to [15], furthermore, an upper bound of λ_2 can be estimated by

$$\lambda_2 \leq 2\frac{|\partial S|}{|S|}, \tag{4.4}$$

where S is any subset of nodes satisfying $0 < |S| \leq N/2$, $|S|$ is the total number of nodes in S, and $|\partial S|$ is the number of common edges between S and its complementary graph. It was also reported in [15] that the estimation (4.4) plays a key role in understanding why the statistical properties of a network may fail to determine λ_2. An important observation is that the above bound on λ_2 is determined by the properties of some subgraph S but not in general by the network itself. In particular, S can be very small comparing to the whole network, so in this case the statistical properties of the network need not be reflected by S, therefore the former may not be very crucial for bounding the value of λ_2.

Here, an example is given to further show that adding long-distance edges to a network can sensitively affect the average distance, but they have very little effect

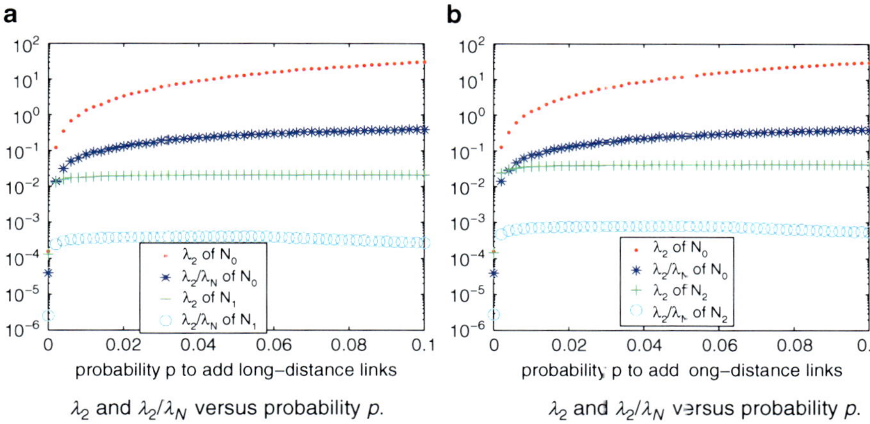

Fig. 4.1 Synchronizability comparison: (**a**) between network N_0, composing of a small-world model H of 500 nodes, and network N_1, composing of a small-world model H of 500 nodes and a fully connected model S of 50 nodes, with only one edge connecting H and S together; (**b**) between network N_0 and network N_2, composing of a small-world model H of 500 nodes and a fully connected model S of 50 nodes, with two edges connecting H and S together

on the smallest nonzero eigenvalue. To do so, consider a community network N_1 composing of a huge part H and a small part S, where H is a small-world model of 500 nodes and S is a fully connected model of 50 nodes, which is connected to H by only one edge. For comparison, define another network, N_0, composing of H only, without the small part S. Then, add some long-distance edges into H with probability p, so as to shorten the average network distance. Under this framework, the synchronizability of the networks N_0 and N_1 is analyzed numerically, with results as shown in Fig. 4.1a.

From Fig. 4.1a, one can observe the following: (1) Without the small part S, if p is increased, then λ_2 and λ_2/λ_N both rise up, implying that increasing p yields a better synchronizability of the network. (2) With the small part S, when p is large enough, λ_2 first increases and then saturates, while the largest eigenvalue λ_N continues to increase. Therefore, the ratio λ_2/λ_N eventually decays, thereby reducing the synchronizability. In other words, the network does not become more synchronizable, even though the probability p with long-distance connections increases, that is, the average network distance becomes smaller. This means that one cannot use the statistical properties to measure the synchronizability of networks in general, which is consistent with the observations reported in [15]. (3) It is easier for a network to synchronize in the case without the small community S than the case with S.

Furthermore, when there exists a small part S, as p is increased, the numerical result shows that the final value of the eigenvalue λ_2 remains to be about 0.021 (see Fig. 4.1a). In terms of (4.4), the theoretical value is $\lambda_2 \leq 2/50 = 0.04$, which is not of significant difference from the numerical result. In addition, when there exist two edges between H and S, the final value of λ_2 remains to be about 0.042 (see Fig. 4.1b), implying that the value of λ_2 is positively proportional to the number of edges between the two parts, at least for the present simple cases with one or two connections. In fact, this conclusion may also be deduced from (4.4).

4.3 Spectral Properties of Several Typical Networks

The spectrum of a network is the set of eigenvalues of the network's Laplacian matrix. While there are strict mathematical formulas to describe the spectra of some very regular networks, much less is known about the spectra of many real-world networks for their complex and irregular topological structures. A critical limitation in addressing the spectra of these real-world networks is the lack of theoretical tools for analysis; therefore, numerical simulation becomes the only way for investigation today.

For all the numerical results reported below, each value is obtained through averaging 50 simulation runs.

4.3.1 Regular Networks

Some very regular cases are easy to analyze.

For a fully connected network, the Laplacian matrix is

$$L = \begin{pmatrix} N-1 & -1 & \cdots & -1 \\ -1 & N-1 & \cdots & -1 \\ \ddots & \ddots & \ddots & \ddots \\ -1 & -1 & \cdots & N-1 \end{pmatrix},$$

with eigenvalues

$$\lambda_1 = 0,\ \lambda_2 = \cdots = \lambda_N = N. \tag{4.5}$$

For a star-shaped network, the Laplacian matrix is

$$L = \begin{pmatrix} N-1 & -1 & -1 & \cdots & -1 \\ -1 & 1 & 0 & \cdots & 0 \\ \vdots & & & \ddots & \\ -1 & 0 & \cdots & 0 & 1 \end{pmatrix},$$

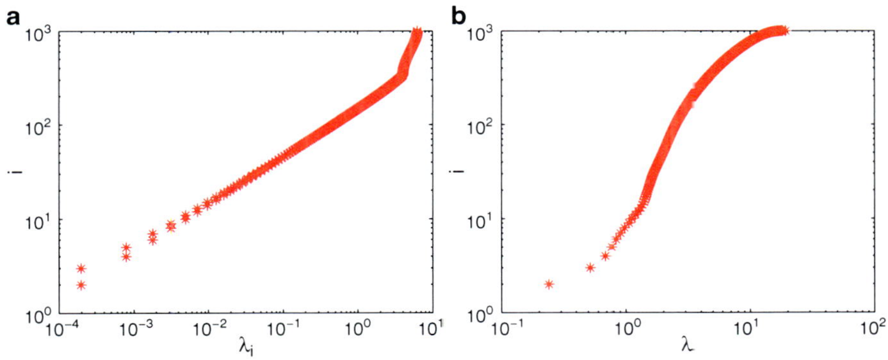

Fig. 4.2 Rank index i versus eigenvalues of the Laplacian matrices L: (**a**) $2K$-ring network with $N = 1000$ and $K = 2$; (**b**) random network with $N = 1000$ and $p = 0.007$

with eigenvalues

$$\lambda_1 = 0,\ \lambda_2 = \cdots = \lambda_{N-1} = 1,\ \lambda_N = N. \tag{4.6}$$

In a $2K$-ring network with degree sequence $\{2K, 2K, \ldots, 2K\}$, each node is connected to its $2K$ nearest neighbors, thereby forming a ring-shape of graph. Its Laplacian matrix is a circulant matrix:

$$L = \begin{pmatrix} 2K & -1 & \cdots & -1 & 0 & \cdots & 0 & -1 & \cdots & -1 \\ -1 & 2K & -1 & \cdots & -1 & & & & & \\ \ddots & \ddots & \ddots & \ddots & & & & & & \\ -1 & \cdots & -1 & 0 & \cdots & 0 & -1 & \cdots & -1 & 2K \end{pmatrix},$$

with eigenvalues 0 and $2K - 2\sum_{l=1}^{K} \cos\frac{2\pi il}{N} = 4\sum_{l=1}^{K} \sin^2\frac{\pi il}{N}$, $i = 1, 2, \ldots, N-1$.

To show some spectral properties of the $2K$-ring networks, consider the simple case of $K = 2$ as an example, and compute the eigenvalues of its Laplacian matrix for size $N = 1{,}000$. It can be seen from Fig. 4.2a that the nonzero eigenvalues come out equally in pairs, except the eigenvalue 4. Moreover, the smallest nonzero eigenvalue is $\lambda_2 = 0.00019739$ and the largest one is $\lambda_N = 6.25$. The eigenvalue ratio λ_2/λ_N is so small, implying that this network is very difficult to synchronize.

4.3.2 *Random Networks*

In contrary to the completely regular networks, a completely random network can be described by a random graph. One classical model of random graphs was first defined and then studied extensively by Erdös and Rényi [45]. An ER random

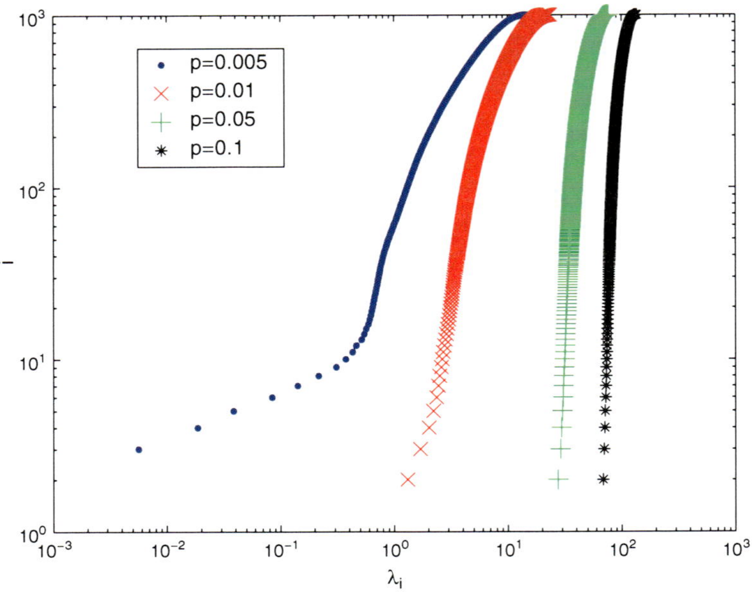

Fig. 4.3 Rank index i versus eigenvalues of the Laplacian matrices L, for random networks of size $N = 1{,}000$ with different connection probabilities

network has all edges established at random, with probability p, between each possible pair of nodes in the network. Thus, a random graph of N nodes with probability p will have $pN(N-1)/2$ edges statistically.

For example, consider such a network with $N = 1{,}000$ nodes. When $p = 0.007 > p_c$, where $p_c = (1+\varepsilon) ln N/N \approx 0.0069$, the random network will almost surely be connected. The eigenvalues are uniformly distributed in the interval $[0, 20)$, as can be seen from Fig. 4.2b. The spectrum has a short span and is quite homogeneous. It is found that the smallest nonzero eigenvalue $\lambda_2 = 0.3712$ and the largest one is $\lambda_N = 19.0554$, giving the ratio $\lambda_2/\lambda_N = 0.0195$.

Figure 4.3 displays the spectra of random networks, which are changing with the connection probability p. One can observe that as p increases, the spectral width decreases quickly, meanwhile the smallest nonzero eigenvalue λ_2 and the largest one λ_N both increase. The main reason is that the increased p not only reduces the total number of isolated subgraphs, leading to increase of λ_2, but also raises the largest degree $d_{\max}$, indicating the increase of λ_N according to (4.1).

In the case of $p = 0.005$, the network has very sparse edges, and there exist some disconnected subgraphs, since $p < p_c$. Hence, $\lambda_2 = 0$, consistent with the numerical result. As a result, such random networks generated with $p = 0.005$ are impossible to synchronize in general. When $p = 0.01$, the smallest nonzero eigenvalue and the eigenvalue ratio are changed to $\lambda_2 = 1.4512$ and $\lambda_2/\lambda_N = 0.0610$, respectively. As the probability is further increased to $p = 0.1$, one has $\lambda_2 = 68.0447$ and the ratio

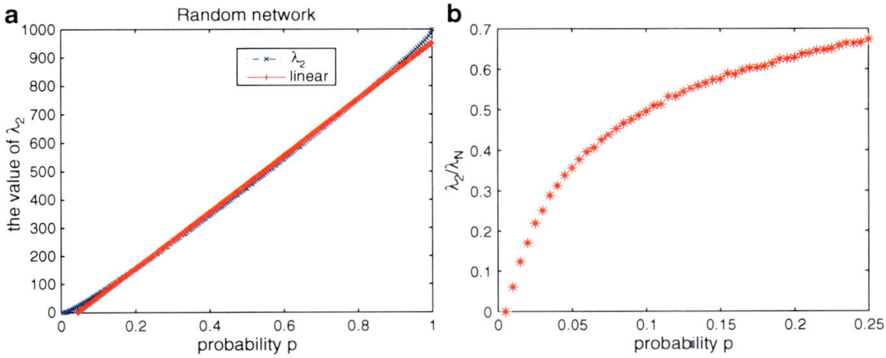

The second eigenvalue λ_2 versus the connection probability p.

The eigenvalue ration λ_2/λ_N versus the connection probability p.

Fig. 4.4 λ_2 and λ_2/λ_N of random networks of size $N = 1{,}000$ for different probabilities

becomes $\lambda_2/\lambda_N = 0.4979$. When the probability p is very close to 1, the model becomes very much like a fully connected network, and so the eigenvalue ratio λ_2/λ_N tends to 1. Therefore, from a statistical point of view, the synchronizability is enhanced gradually as the connection probability p is increased, as can be clearly seen from Fig. 4.4.

Interestingly, for random networks, a prominent approximately linear dependence between λ_2 and the connection probability p can be observed from Fig. 4.4a. Thus, $\lambda_2(p)$ is used to denote the dependence of λ_2 on the probability p. For a fixed size N, if $\lambda_2(p_1)$ and $\lambda_2(p_2)$ are computed from two suitable numbers of p_1 and p_2 ($p_1 \neq p_2$), then one can estimate the smallest nonzero eigenvalue $\lambda_2^*(p)$ for any p by using the following formula:

$$\lambda_2^*(p) \approx \frac{\lambda_2(p_1) - \lambda_2(p_2)}{p_1 - p_2}(p - p_1) + \lambda_2(p_1).$$

Since $p > p_c$, these random networks will almost surely be connected; therefore, it is reasonable to choose both $p_1, p_2 > p_c$. Here, take $p_1 = 0.1$ and $p_2 = 0.25$ to compute the corresponding λ_2^* of other probabilities p, resulting the plots shown in Fig. 4.5. It can be observed that the relative error $|\lambda_2 - \lambda_2^*|/\lambda_2$ is quite big when $p < 0.05$, while it is small when $p > 0.1$, and these estimations are almost exact.

4.3.3 *Small-World Networks*

Small-world networks are neither completely regular nor completely random; one representative model is the NW small-world network, proposed by Newman and Watts [46]. In this model, some edges are added to an initial $2K$-ring at random, with a probability $p \in [0,1]$.

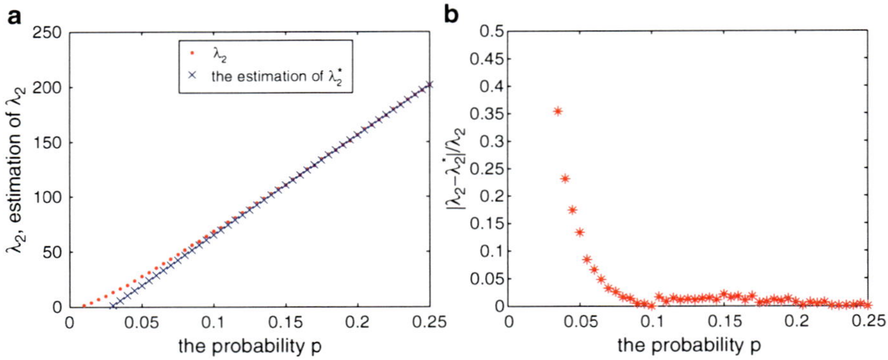

λ_2 and the estimated λ_2^* versus the connection probability p.

The relative error $|\lambda_2 - \lambda_2^*|/\lambda_2$ versus the connection probability p.

Fig. 4.5 The estimated λ_2^* and the relative error $|\lambda_2 - \lambda_2^*|/\lambda_2$ in random networks of size $N =$ 1,000 for different connection probabilities

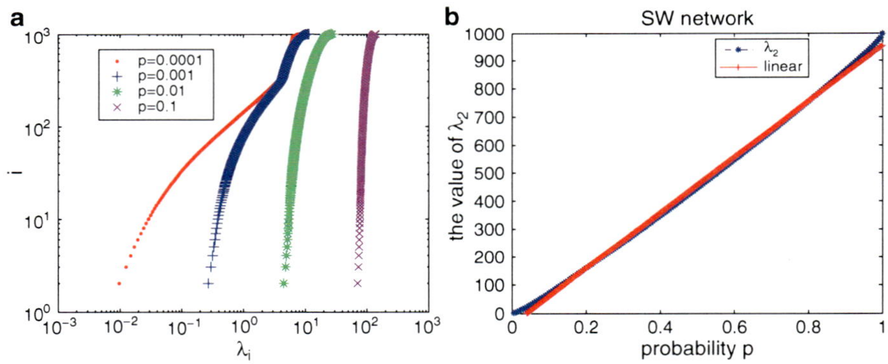

Fig. 4.6 Small-world networks of size $N = 1{,}000$: (**a**) Rank index i versus eigenvalues of Laplacian matrices L; (**b**) the smallest nonzero eigenvalue λ_2 versus the connection probability p

With $p = 0$, the network is the initial $2K$-ring network. As p is increased, the smallest nonzero eigenvalue λ_2 and the largest one λ_N both will increase (see Fig. 4.6a). However, λ_2 grows much faster than λ_N, resulting in an increased eigenvalue ratio λ_2/λ_N. It implies that the synchronizability of the small-world network is improved as the connection probability is increased.

For $0 < p < 1$, it can be seen from Figs. 4.3 and 4.6a that, similarly to random networks, the spectral range of a small-world network becomes narrower when the connection probability p is increased.

With $p = 1$, the small-world model becomes a fully connected graph, so the spectrum of this case is similar to the fully connected network discussed before.

Thus, it can be concluded that the spectrum of an NW small-world network, which is quite homogeneous, transits from the spectrum of a $2K$-ring network to that of a fully connected network, as the connection probability p is increased.

For the NW small-world network model, a similar phenomenon of a prominent approximately linear dependence of λ_2 on the connection probability p can be observed, as seen from Fig. 4.6b. By linear fitting, the estimation of λ_2 for an NW small-world network of $N = 1{,}000$ can be obtained, as $\lambda^*_{2\text{nw}}(p) = 992.48 * p - 38.831$. It should be noted that these values are related to several parameters such as the probability p, the ring constant K, and the size N of the network.

4.3.4 Scale-Free Networks

Scale-free networks are characterized by the power-law form of their degree distributions, which can be generated with the Barabási–Albert's preferential attachment algorithm [47]. Staring from an initial set of m_0 fully connected nodes, one node is added along with m new edges, at every time step. Nodes in the existing network with higher degrees have higher probabilities, proportional to their degrees, to be connected by the new node through a new edge, where multiple connections are prohibited. This algorithm yields a typical BA scale-free network, which is growing until the process stops.

Figure 4.7a shows some spectral properties of a scale-free network of size $N = 1{,}000$. It can be seen that the eigenvalues are distributed in a very heterogeneous way. Part of the eigenvalues are located in the interval $[0,20]$, while some larger ones are located far away from this interval. The smallest nonzero eigenvalue is $\lambda_2 = 0.5391$ and the largest one is $\lambda_N = 81.6367$, resulting in the ratio $\lambda_2/\lambda_N = 0.0066$. It can also be observed that a large difference exists between λ_2 and λ_N, implying that the spectral distribution range of the scale-free network is wide, distinctive from regular, random and small-world networks.

Figure 4.7b compares the spectra between NW small-world and BA scale-free networks. It can be seen that the span of the eigenvalues distribution in the small-world network is smaller than that of the scale-free network. For the latter, most of its eigenvalues are comparatively concentrated, and meantime there are some very large eigenvalues.

In summary, the spectral properties of different networks are clearly different. Thus, one can tell which category a network belongs to, by simply looking at its spectral properties. The spectra of regular, random, and small-world networks are homogeneous, whereas those of scale-free networks are heterogeneous. Furthermore, for random and small-world networks, the synchronizability is increased as the connection probability p is increased. In particular, the smallest nonzero eigenvalue is almost linearly dependent on the connected probability for both ER random and NW small-world networks.

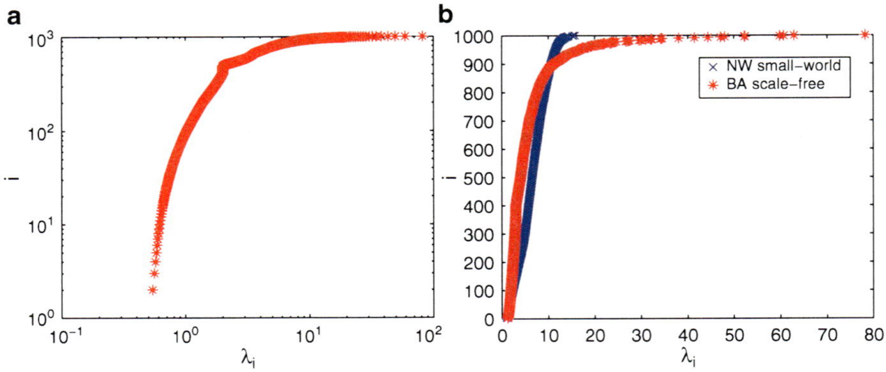

Fig. 4.7 Rank index i versus Eigenvalues of the Laplacian matrices L: (**a**) scale-free networks with $N = 1{,}000$, $m_0 = 5$ and $m = 2$; (**b**) scale-free networks with $N = 1{,}000$, $m_0 = 5$ and $m = 3$, and small-world networks with $N = 1{,}000$ and $p = 0.0025$

4.3.5 Community Networks

Community structure means densely connected groups of nodes with sparse connections between them. This section discusses the spectra of networks with prominent community structures, called community networks.

First, consider the spectra of networks composing of two communities, each of which is (1) a fully connected graph (Fig. 4.8a); (2) a random network (Fig. 4.8b); (3) a small-world network (Fig. 4.8c); and (4) a scale-free network (Fig. 4.8d). There are some random edges between two communities. It can be seen from Fig. 4.8 that there is a big gap between λ_2 and λ_3. Moreover, as the number of edges between two communities is increased, λ_2 is increased and the gap between λ_2 and λ_3 decreases, indicating that the community structure becomes blurred. It can also be seen that the other eigenvalues basically remain unchanged, reflecting the robustness of the spectral properties of the corresponding subgraphs.

Next, consider some networks composing of three communities: (1) each community is a small-world network (Fig. 4.9a); (2) each community is a scale-free network (Fig. 4.9b). There exist several random edges between every two communities. It can be seen that there is a big gap between λ_3 and λ_4, and the increased number of random edges among the three communities led to the increased values of λ_2 and λ_3. Furthermore, λ_2 and λ_3 are increased much faster than the other eigenvalues; therefore, the difference between λ_3 and λ_4 is reduced, and the community structure becomes blurred.

Remark 4.1. From Figs. 4.8c (4.9a), namely, and 4.8d (4.9b), one can find that the distributions of the eigenvalues in Figs. 4.8c and 4.9a are more homogeneous than those shown in Figs. 4.8d and 4.9b. This is probably due to the heterogeneity of the BA scale-free subnetworks and the homogeneity of the NW small-world

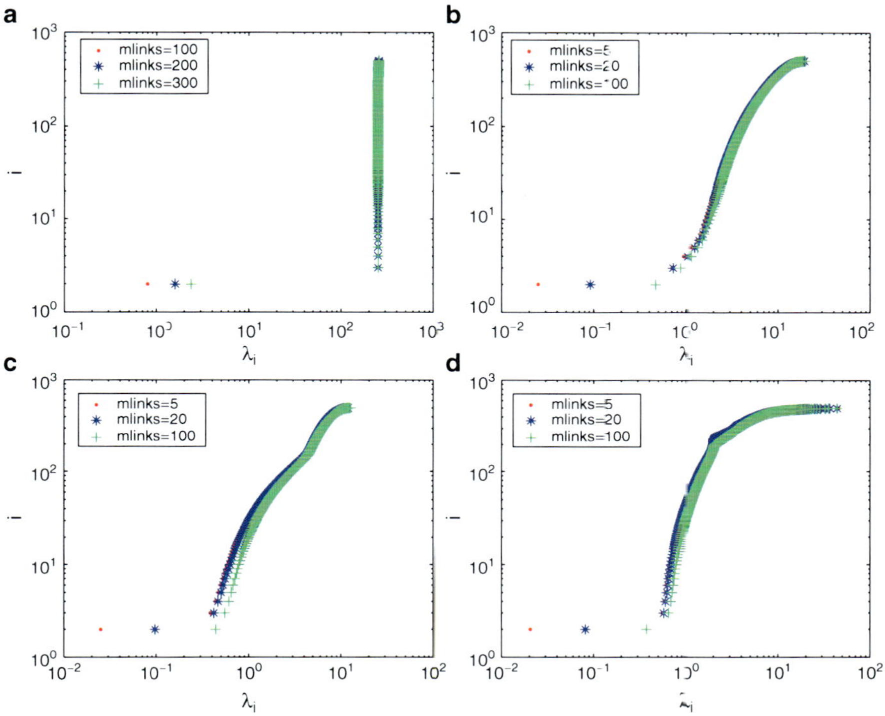

Fig. 4.8 Rank index i versus eigenvalues of the Laplacian matrix L: (**a**) a network composing of two fully connected subgraphs of size $N = 250$, with random edges 100, 200 and 300 between two communities, respectively; (**b**) a network composing of two random subgraphs of size $N = 250$, with $p = 0.03$, having 5, 20 and 100 random edges between two communities, respectively; (**c**) a network composing of two small-world subgraphs of size $N = 250$, with $p = 0.005$, having 5, 20 and 100 random edges between two communities, respectively; (**d**) a network composing of two scale-free subgraphs of size $N = 250$, with $m_0 = 5$ and $m = 2$, having 5, 20 and 100 random edges between two communities, respectively

subnetworks, since there exists a positive correlation between the spectrum and the degree sequence, as will be further discussed later in Sect. 4.4. Thus, the more diverse the degree distribution, the more heterogeneous the eigenvalues of the BA scale-free subnetworks.

In summary, the spectra of community networks bring to light many aspects of the network topologies: (1) the number of zero eigenvalues equals the number of isolated communities; (2) a gap between eigenvalues indicates the existence of a community structure; (3) for a network with k prominent communities, there are $k-1$ eigenvalues near zero, the gap between λ_k and λ_{k+1} is larger than the difference between any other consecutive eigenvalues, and larger eigenvalues in the last part of the sequence reflect some major properties of communities; and (4) the increase of the number of edges between communities can result in the increase of the value of λ_2, which also means a better synchronizability.

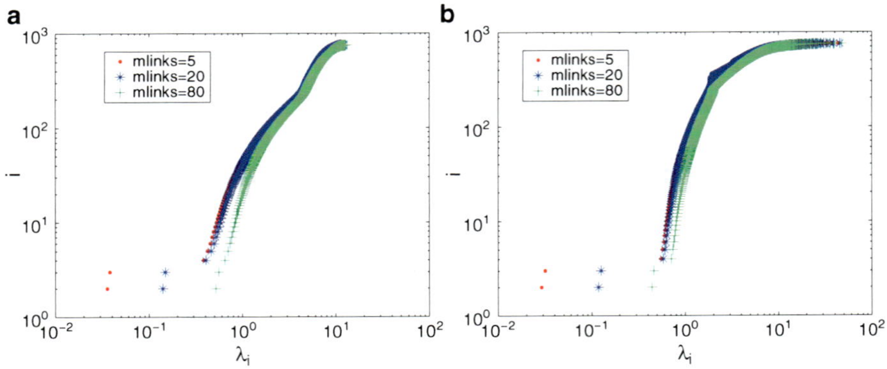

Fig. 4.9 Rank index i versus eigenvalues of the Laplacian matrix L: (**a**) a network composing of three small-world subgraphs of size $N = 250$, with $p = 0.005$, 5, 20, and 80, respectively, having random edges between two communities; (**b**) a network composing of three scale-free subgraphs of size $N = 250$, with $m_0 = 5$ and $m = 2$, 5, 20, and 80, respectively, having random edges between two communities

4.4 Relation Between the Spectrum and the Degree Sequence

Real-world networks are usually very large in size; thus, it is generally difficult and time-consuming to compute their spectra. However, the degree sequence is easy to obtain. So, if one can find the relation between a spectrum and a degree sequence, it will be reasonable to make use of the degree sequence for estimating the spectrum of a real network. The question is, then, whether or not they have any easily described and computable relations?

4.4.1 Theoretical Analysis

In order to identify some intrinsic relations between the Laplacian eigenvalues and the degree sequence of a network, the following lemma is needed.

Lemma 4.1 (Wielanelt-Hoffman Theorem [48]). *Suppose $C = A + B$, where $A, B, C \in R^{N\times N}$ are symmetric matrices, and let the eigenvalue sets $\lambda(B)$ and $\lambda(C)$ be arranged in the non-decreasing order. Then,*

$$\sum_{i=1}^{N} |\lambda_i(C) - \lambda_i(B)|^2 \leq ||A||_F^2 ,$$

where $||A||_F = \left(\sum_{ij} |a_{ij}|^2\right)^{1/2}$ denotes the Frobenius norm of matrix A.

Theorem 4.1 ([49]). *Let G be a graph of N nodes with Laplacian matrix L, and arrange the node-degree set and the eigenvalue set in the vector form, as $d = (d_1, d_2, \ldots, d_N)^{\mathrm{T}}$ and $\lambda(L) = (\lambda_1, \lambda_2, \ldots, \lambda_N)^{\mathrm{T}}$, respectively, both in the non-decreasing order. Then,*

$$\delta = \frac{||\lambda(L) - d||_2}{||d||_2} \leq \frac{\sqrt{||d||_1}}{||d||_2} \leq \sqrt{\frac{N}{||d||_1}}.$$

Remark 4.2. Theorem 4.1 shows that for the Laplacian matrix L, the difference between Laplacian eigenvalues $\lambda(L)$ and node-degrees d is bounded by $||d||_1/||d||_2^2$. Thus, for a large-scale complex network with a huge number of edges, $||d||_1/||d||_2^2$ typically have a small value, implying that the distribution of the Laplacian eigenvalues is indeed similar to the distribution of the node degrees. On the other hand, however, for a large-scale network with sparse connections, the value of $N/\left(\sum_{i=1}^N d_i\right)$ is usually small, implying that the inequality is quite conservative.

Remark 4.3. It is reported in [49] that the differences between Laplacian eigenvalues $\lambda(L)$ and node-degrees d are very small for random, small-world and scale-free networks, according to extensive numerical simulations. In the next subsection, it will be shown that the eigenvalues distributions are very closely related to the node-degree sequences, which is consistent with the results of [49].

Theorem 4.2 ([49]). *Let λ_j be the eigenvalues of the Laplacian matrix L of a graph with N nodes, and d_i be the degree of node i, $i = 1, 2, \cdots, N$. Then, in every interval $[d_i - \sqrt{d_i}, d_i + \sqrt{d_i}]$, there is at least one eigenvalue $\lambda^* \in \{\lambda_j | j = 1, 2, \ldots, N\}$ of L, that is,*

$$(d_i - \sqrt{d_i}) \leq \lambda^* \leq (d_i + \sqrt{d_i}), \quad i = 1, 2, \cdots, N.$$

Remark 4.4. In Theorem 4.2, some intervals $[d_i - \sqrt{d_i}, d_i + \sqrt{d_i}]$ may overlap, and some eigenvalues of L may not fall into any of such intervals at all.

Remark 4.5. By the Gerschgorin Theorem, the eigenvalues of L satisfy that

$$|\lambda - l_{ii}| \leq \sum_{j=1, j\neq i}^{N} |l_{ij}|.$$

Since $l_{ii} = d_i$, and L has zero row-sum, one has $|\lambda - d_i| \leq d_i$, namely,

$$0 \leq \lambda \leq 2d_i, \quad i = 1, 2, \ldots, N.$$

Thus, one can see that the result given by Theorem 4.2 is less conservative than this estimation given by the Gerschgorin Theorem.

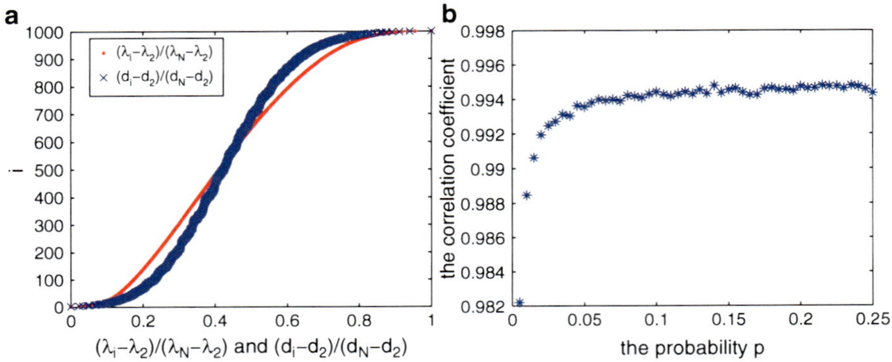

Fig. 4.10 ER random networks of size $N = 1{,}000$: (**a**) rank index i versus the relative spectrum and relative degree sequence, when $p = 0.0258$; (**b**) correlation coefficient between the spectrum and degree sequence versus the probability p

4.4.2 Numerical Results

To visualize the relation between the eigenvalues distribution and the degree sequence, extensive numerical simulations were performed and analyzed.

For convenience of analysis, define the relative spectrum

$$R_{\rm e}(i) := (\lambda_i - \lambda_2)/(\lambda_N - \lambda_2),\ i = 2,3,\ldots,N,$$

and the relative degree sequence

$$R_{\rm d}(i) := (d_i - d_2)/(d_N - d_2),\ i = 1,2,\ldots,N.$$

Numerical results illustrated in Figs. 4.10a, 4.11a and 4.12a show the relations between the relative spectrum $R_{\rm e}$ and the relative degree sequence $R_{\rm d}$. The total numbers of edges, in the random networks with $p = 0.0258$, the small-world networks with $p = 0.0218$, and the scale-free networks with $m_0 = m = 13$, are all equal to 12909 ± 10.

It can be easily seen from Fig. 4.10a that the spectrum and the degree sequence of a random network seems to be correlated, but how much the two are correlated is unclear, so the correlation coefficient between the spectrum and the degree sequence is calculated. When the connection probability $p = 0.0258$, from Fig. 4.10b, one can see that the correlation coefficient is 0.9925. That is, the spectrum and degree sequence are quite correlated. And, the correlation coefficient is increased, though slowly, as p is increased.

Figure 4.11a shows the correlation between the spectrum and degree sequence of a small-world network of 1,000 node with the connection probability $p = 0.0218$. Calculation yields the correlation coefficient 0.9921 in this case, implying that the spectrum and degree sequence are comparatively correlated. Fig. 4.11b shows that

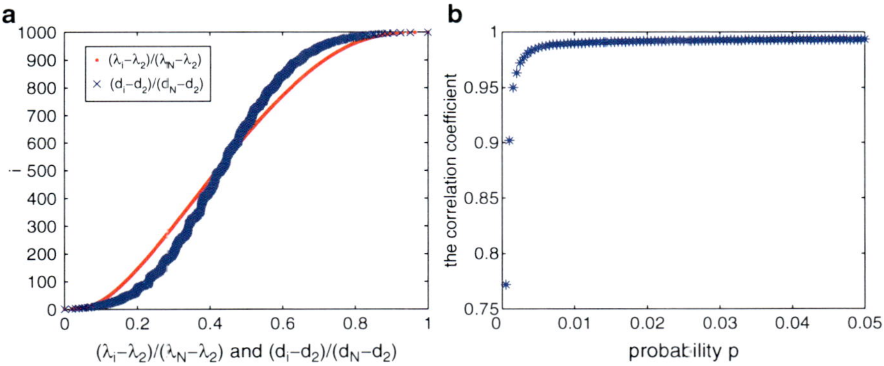

Fig. 4.11 NW small-world networks of size $N = 1{,}000$: (**a**) rank index i versus the relative spectrum and relative degree sequence, when $p = 0.0218$; (**b**) correlation coefficient between the spectrum and degree sequence versus the probability p

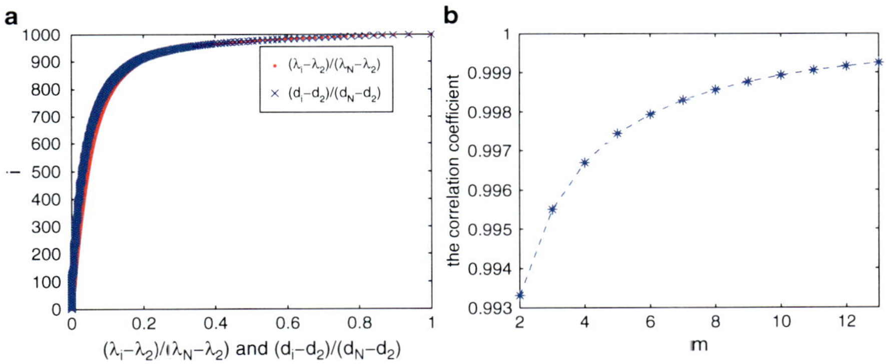

Fig. 4.12 BA scale-free networks of size $N = 1{,}000$: (**a**) rank index i versus the relative spectrum and relative degree sequence of scale-free networks of size $N = 1{,}000$, when $m_0 = m = 13$; (**b**) correlation coefficient between the spectrum and degree sequence versus m

the correlation coefficient ascends to a certain degree, and then converges to a constant value of 0.9935 as the probability p is increased further.

From Fig. 4.12, one can see that the spectrum and degree sequence of a BA scale-free network are closely correlated. Indeed, when $m_0 = m = 13$, the correlation coefficient equals 0.9992. Figure 4.12b shows the increasing correlation coefficient with the increasing m.

Moreover, it can also be observed that the correlation between the spectrum and degree sequence of a scale-free network is the highest, followed by random networks and then by small-world networks. Therefore, in the following, a scale-free network with $m_0 = m = 13$ is used as an example to estimate all the eigenvalues in terms of λ_2, λ_N, and the degree sequence.

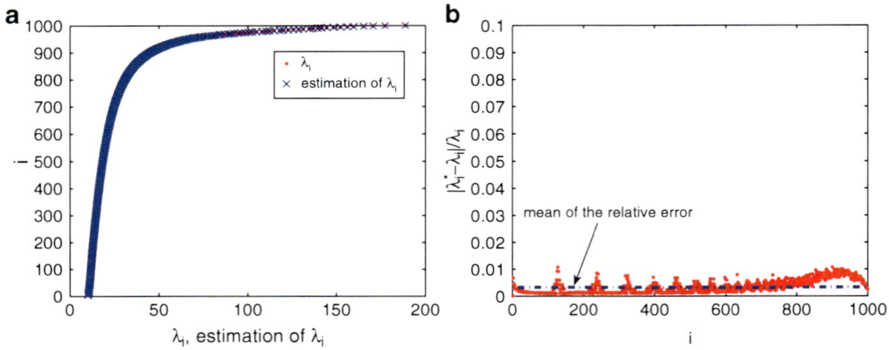

Fig. 4.13 Scale-free networks of size $N = 1{,}000$: (**a**) λ_i and the estimation λ_i^* versus the rank index i; (**b**) λ_i and the estimation λ_i^* versus the rank index i

It is known that if the correlation coefficient between two random variables X, Y is close to 1, then it means that the probability of the linear relation $Y = aX + b$ is close to 1. But the parameters a and b are usually unknown, so it is difficult to compute the whole spectrum of eigenvalues (the so-called global method) from the degree sequence, λ_2 and λ_N. The following is an efficient local algorithm for computing approximations of λ_{i+1} from λ_i, $i = 1, 2, ..., N$.

- Initial conditions: $\bar{\lambda}_1 = \lambda_1^* = 0$.
- Step 1 (Prediction). Compute $\bar{\lambda}_i$ in terms of the degree sequence, λ_2, and λ_N, as follows:

$$\bar{\lambda}_i = \frac{d_i - d_2}{d_N - d_2}(\lambda_N - \lambda_2) + \lambda_2.$$

- Step 2 (Correction). Compute the approximation λ_{i+1}^* from λ_i, iteratively, by

$$\lambda_{i+1}^* = \lambda_i + (\bar{\lambda}_{i+1} - \bar{\lambda}_i), \ i = 1, 2, \ldots, N-1.$$

Figure 4.13a shows a comparison between the approximated λ_i^* and the exact λ_i, $i = 2, 3, \ldots, N$. It is clear that the estimations are very accurate. From Fig. 4.13b, one can see that the relative error $(\lambda_i - \lambda_i^*)/\lambda_i$ is very small, with an average of 0.3263%, demonstrating that the local algorithm for estimating λ_{i+1} from λ_i is indeed highly effective.

4.5 Synchronization Processes on Complex Networks

Complex networks in various physical systems can be described at different scales. This section discusses networks at the "mesoscale," as defined in [25], addressing subgraphs rather than at the "microscale" which addresses individual nodes or at the "macroscale" which addresses the network as a whole. The mesoscale is an intermediate scale examining substructures such as motifs, cliques, cores, loops, and communities. In particular, the community detection problem concerning the determination of mesoscopic structures that have functional, relational or even social dynamics and impacts is an important and yet also challenging subject for investigation in the field of complex networks.

This section focuses only on synchronization processes on complex networks at the mesoscale level of description. Synchronization is a generic feature of networked dynamical systems such as cells and oscillators. Previous studies have discussed the onset of synchronization and the impact of structural properties on network synchronizability. The interest here is the regions outside the onset of phase synchronization and the role of network topology in the synchronization processes.

4.5.1 Paths to Synchronization on Complex Networks

One of the most popular models for coupled oscillators is the Kuramoto model [50]:

$$\frac{\mathrm{d}\theta_i}{\mathrm{d}t} = \omega_i + \lambda \sum_{j=1}^{N} A_{ij} \sin(\theta_j - \theta_i), \; i = 1,2,\ldots,N. \tag{4.7}$$

where ω_i represents the natural frequency of the ith oscillator and λ is the coupling constant.

This model can be studied in terms of an order parameter, r, that measures the extent of phase synchronization in the network, defined by

$$r\mathrm{e}^{\mathrm{i}\Psi} = \frac{1}{N} \sum_{j=1}^{N} \mathrm{e}^{\mathrm{i}\theta_j},$$

where Ψ represents the average phase of the network. The parameter $0 \leq r \leq 1$ displays a second-order phase transition in the coupling strength, with $r = 0$ being the value of the incoherent solution and $r = 1$ the value of the complete phase synchronization.

In [26], a new parameter, r_{edge}, is defined, as

$$r_{\mathrm{edge}} = \frac{1}{2N_l} \sum_i \sum_{j \in \Gamma_i} \left| \lim_{\Delta t \to \infty} \int_{t_\mathrm{r}}^{t_\mathrm{r}+\Delta t} \mathrm{e}^{\mathrm{i}[\theta_i(t) - \theta_j(t)]} \mathrm{d}t \right|$$

where Γ_i is the set of neighbors of node i and N_l is the total number of edges. This parameter represents the fraction of edges with which the network achieves phase synchronization, averaging over a large enough time interval Δt after the network is relaxed at a large time instant t_r.

In [26], the dynamics of (4.7) for ER and scale-free (SF) networks were studied, with respect to both global and local synchronization. It shows that the evolution

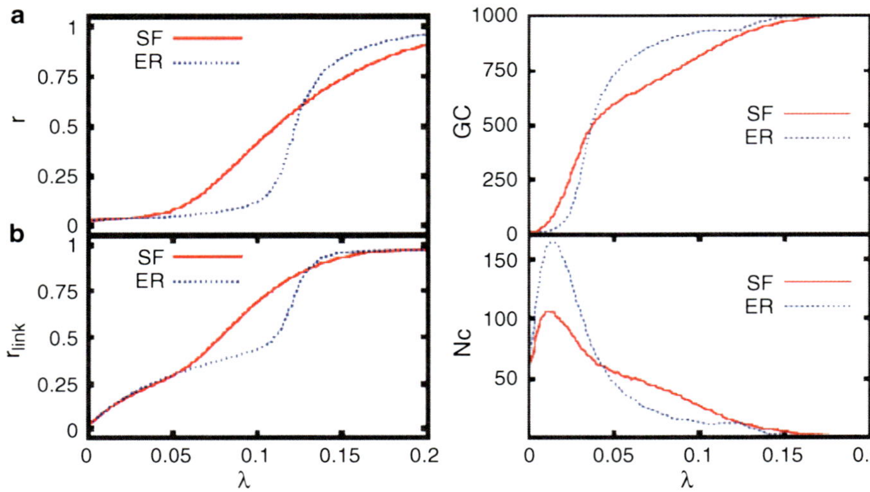

Fig. 4.14 *Left*: evolution of the order parameter r, and the fraction of synchronized edges $r_{link}(= r_{\mathrm{edge}})$ as a function of λ; *Right*: size of the largest synchronized connected component (GC) and the number of synchronized connected components (N_C), as a function of λ, for the different topologies considered [26]

of the order parameter r, as λ increases, can capture the global coherence of synchronization in the network, and that r_{edge} can be used to measure the local formation of synchronization patterns thereby revealing how global synchronization is achieved.

Synchronization processes in ER and SF networks were studied numerically in [26]. It can be seen from Fig. 4.14 that the global coherence of the synchronized state, represented by r and implying the onset of synchronization, first occurs for the SF network. If λ is further increased, there is a value of r for the ER network curve to cross over the SF network curve. From this value up, in λ, the ER network remains slightly better in synchrony than the SF network. The behavior of r_{edge} provides some additional information about the processes between ER and SF networks. Interestingly, the nonzero values of r_{edge} for very small λ indicate that some local synchronization patterns has occurred even in the regime of global incoherence ($r \approx 0$). Right at the onset of synchronization for the SF network, its r_{edge} value grows faster than the ER network. This implies that for the SF network the locally synchronized structures rise at a faster rate than the ER network. Finally, when λ is further increased, for the ER network, the growth in its synchronization patterns increases drastically up to those obtained from the SF network, and even higher.

From the evolution of the number of synchronized clusters and the size of the largest generated clusters (GC) shown in Fig. 4.14, the emergence of clusters of synchronized pairs of oscillators (edges) in the networks shows that for small λ values, still in the incoherent range with $r \approx 0$, both ER and SF networks have developed a largest cluster of synchronized pairs of oscillators involving about 50%

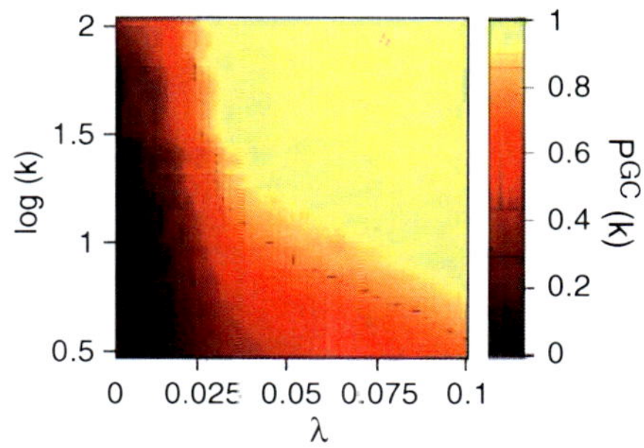

Fig. 4.15 Correlation between the likelihood that a node belongs to the GC of pairs of synchronized oscillators and its degree k as a function of the coupling strength λ in the SF network [26]

of the network nodes, with an equal number of smaller synchronization clusters. From this point of view, in the SF network the GC grows up but the number of smaller clusters goes down, whereas for the ER network the growth exploits. These results indicate that although SF networks present more coherence in terms of both r and r_{edge}, the mesoscopic evolution of the synchronization patterns is slower than the ER networks, which are far more locally synchronizable than heterogeneous networks in general (see [26]).

In [26], it argued that the above observed differences in the local behavior are resulted from the growth of the GC. It is shown in [26] that for ER networks pairs of oscillators synchronize to merge and form many different clusters and then form a GC when the coupling strength is increased. Many small clusters join together to produce a giant component consisting of synchronized pairs, the size of which is almost the same as the whole network, as soon as the global coherent state is achieved. However, this is far from the case for SF networks, where the GC is formed from a core consisting of about a half of the nodes in the network, and then new pairs of oscillators are incorporated into the GC one by one, as the coupling strength is increased.

In Fig. 4.15, it can be seen that for the SF network, the probability that a node of degree k belongs to the GC is an increasing function of k for every fixed λ, hence the more connected a node is, the more likely it takes part in the cluster of synchronized entries, as reported in [26]. Therefore, one can conclude that synchronization starts from the node with the largest degree and then spreads to the rest nodes in the network in this scenario.

Remark 4.6. The observed phenomena may be understood from the master-stability-function point of view. In this setting, the dynamics of a network of N coupled oscillators is described by

$$\dot{x}_i(t) = f(x_i(t)) - \varepsilon \sum_{j=1}^{N} L_{ij} \Gamma x_j, \quad i = 1, 2, \ldots, N, \tag{4.8}$$

where x_i is the state of oscillator i and $f(x_i(t))$ governs its dynamics, Γ is the inner coupling matrix, and $L = (L_{ij})$ is the Laplacian matrix of the network. Denote the completely synchronizing state of (4.8) as $\{x_j(t) = s(t), \forall j \,|\, \dot{s}(t) = f(s(t))\}$. A small perturbation on $s(t)$ yields the following linear variational equations:

$$\dot{\delta x}_i(t) = Df(s(t))\delta x_i - \varepsilon \sum_{j=1}^{N} L_{ij}\Gamma(\delta x_j), \; i = 1,2,\ldots,N. \tag{4.9}$$

Further, (4.9) can be diagonalized into N decoupled blocks of the form

$$\dot{\eta}_i = [Df(s(t)) - \varepsilon\lambda_i\Gamma]\eta_i, \; i = 1,2,\ldots,N. \tag{4.10}$$

From (4.10), the speed of node i converging to the synchronization manifold is mainly determined by λ_i, with $f(\cdot)$ given. It has been shown in the above that the distribution of the Laplacian eigenvalues of a network is strongly related to the node-degree distribution; therefore, one can conclude that synchronization typically starts from the node with the largest degree.

4.5.2 Paths to Generalized Synchronization on Complex Networks

In [32], the processes of generalized synchronization (GS) of complex networks was studied, mainly on NW small-world networks and BA scale-free networks, using the adaptive coupling strategy in the following form:

$$\dot{x}_i = f_i(x_i(t)) - c_i(t) \sum_{j=1}^{N} a_{ij}\Gamma(x_i(t) - x_j(t)), \qquad i = 1,2,\ldots,N, \tag{4.11}$$

where x_i is the state of node i, $f_i(\cdot)$ is a continuous vector function, $c_i(t) > 0$ is a time-varying coupling strength to be designed using only the neighborhood information of node i, Γ is the inner coupling matrix, and $A = (a_{ij})$ is the adjacency matrix of the network.

Construct an auxiliary network of the form

$$\dot{x}'_i = f_i(x'_i) - c_i(t) \sum_{j=1}^{N} a_{ij}\Gamma(x'_i - x_j), \qquad i = 1,2,\ldots,N, \tag{4.12}$$

denote $e_i = x'_i - x_i$, and apply the adaptive controller

$$\dot{c}_i(t) = \gamma_i \sum_{j=1}^{N} a_{ij}{e_i}^T\Gamma e_i, \qquad i = 1,2,\ldots,N, \tag{4.13}$$

where γ_i are positive constants.

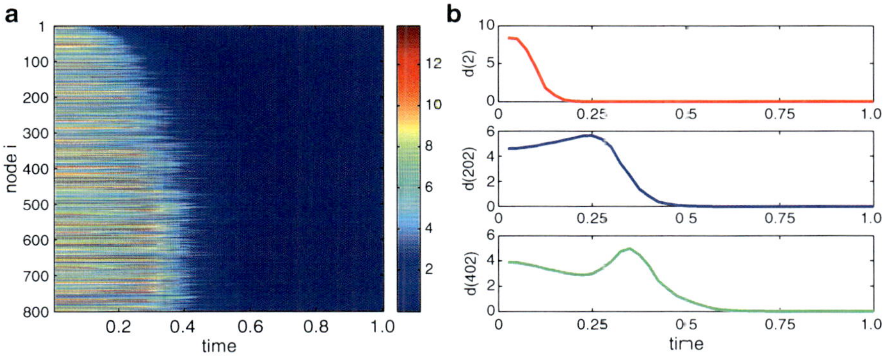

$E_i(t)$ in the two-dimensional parameter space (t,i), depicting the development of GS.

The evolution process of $E_i(t)$, for different nodes $i = 2,202,402$.

Fig. 4.16 Visualization of the evolutionary process to reach global GS via adaptive coupling strength regulation on a BA network. Total number of nodes $N = 800$ and total number of edges is 2388 ± 2 [32]

To describe the process of GS on network (4.11), define the following error signals at each time step:

$$E_i(t) = ||x_i'(t) - x_i(t)||_2, \qquad i = 1,2,\ldots,N, \tag{4.14}$$

and label the nodes according to their decreasing degree ordering $d(1) \geq d(2) \geq \cdots \geq d(N)$. Thus, $i = 1$ denotes the node with the largest degree, $i = 2$ the node with the second largest degree, and so on.

Numerical results illustrated by Figs. 4.16 and 4.17 show the processes towards global GS. Figure 4.16a shows the synchronization process of a BA network with $m = 3$; Fig. 4.16b displays the evolution process of $E_i(t)$, with $i = 2$, 202 and 402, respectively, for $t \in [0, 1.0]$. Figure 4.17a shows the synchronous process of a NW network, with connection probability $p = 0.0025$; Fig. 4.17b displays the evolution process of $E_i(t)$, with $i = 2$, 202, and 402, respectively. for $t \in [0, 1.0]$. The total numbers of edges in the BA network and the NW network are both equal to 2388 ± 2.

Clearly, not all nodes can achieve a common GS state simultaneously. Also obviously, there is a transition process from non-GS to global GS. It can be seen from Fig. 4.16a that, as time evolves, GS starts from the nodes with the largest degree and then spreads to the rest of the network. This can also be verified by (4.11)–(4.12), from which linearization results in

$$\dot{e}_i = \left[Df_i(x_i) - c_i(t) \sum_{j=1}^{N} a_{ij} \Gamma \right] e_i, \tag{4.15}$$

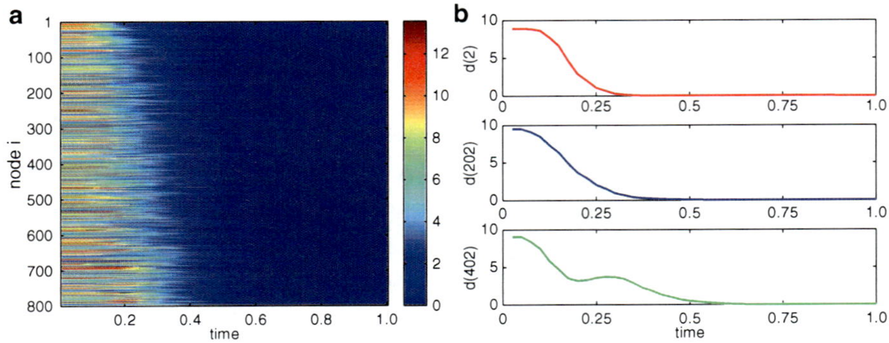

$E_i(t)$ in the two-dimensional parameter space (t,i), depicting the development of GS.

The evolution process of $E_i(t)$, for three different nodes, $i = 2,202,402$.

Fig. 4.17 Visualization of the process to global GS via adaptive coupling strength regulation of a NW network. The total number of nodes $N = 800$ and the total number of edges is 2388 ± 2 [32]

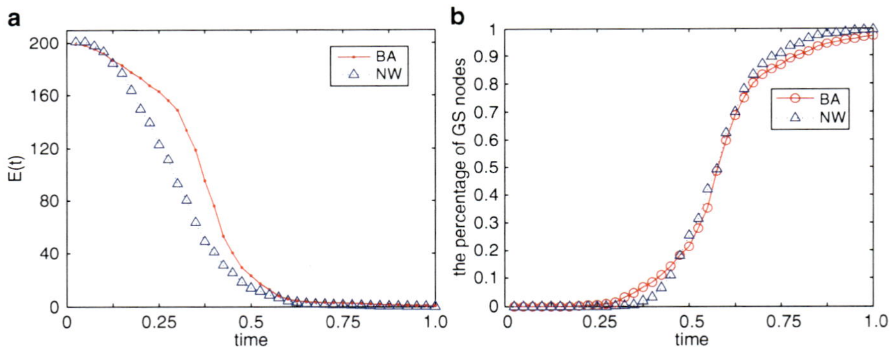

Fig. 4.18 A comparition of BA network and NW network. (**a**) shows the evolution of the total errors $E(t)$; (**b**) shows the percentage of nodes which have achieved GS in the adaptive evolution process [32]

where $Df_i(x_i)$ denotes the Jacobian matrix of $f_i(x)$ at x_i. In (4.15), $\sum_{j=1}^{N} a_{ij}$ is the degree of node i; hence, the larger a node's degree is, the faster its synchronization error converges, as can be seen in Figs. 4.16b and 4.17b.

Next, define $E(t) = \sum_{i=1}^{N} E_i(t)$.

Figure 4.18a shows the evolution of $E(t)$ for a BA scale-free network and a NW small-world network, while Fig. 4.18b shows the percentage of synchronized nodes in the GS process.

From Fig. 4.18, one can see that the BA network is easier to reach GS in the early stage, but harder in the later stage, in the process. This can be explained by the fact that the NW network has a comparative homogeneous node-degree distribution while the BA network has some hubs with extremely large degrees. In fact, in the beginning of the process, the existence of these hubs accelerates the GS speed on the BA network. But, at the later stage, the speed to achieve GS is dominated by the large number of nodes with smaller degrees. Thus, for the BA network, although the hubs achieve GS rather quickly, the distant nodes are much more difficult to achieve GS. So, as a result on the whole large network, it is easier for the NW network to achieve global GS than the BA network with the same numbers of nodes and edges.

4.5.3 Paths to Synchronization on Community Networks

In this subsection, synchronization processes on community networks are investigated.

Again, consider the network of N coupled identical oscillators, described by (4.7) with $\omega_i = \omega$, $\forall i$. The objective is to achieve $\theta_i \rightarrow \theta$, $\forall i$ as $t \rightarrow \infty$. This problem was studied in [24, 25], where the following concepts were defined to characterize the dynamic time scales:

- The average of the correlations between pairs of oscillators

$$\rho_{ij}(t) = \langle \cos(\theta_i(t) - \theta_j(t)) \rangle. \tag{4.16}$$

  where the brackets stand for the average over initial random phases;
- A connectively matrix with a given threshold T based on the above average correlations between pairs of oscillators

$$\mathscr{D}_T(t)_{ij} = \begin{cases} 1 & \text{if } \rho_{ij}(t) > T, \\ 0 & \text{if } \rho_{ij}(t) < T. \end{cases} \tag{4.17}$$

  For large enough T, the evolution of this matrix unravels the process of nodes merging into groups or communities.

In this part, a community network is composed of a huge part H and a small part S, where H is a NW small-world model of 500 nodes with connection probability $p = 0.01$, and S is a fully connected model of 50 nodes and is connected to H via only one edge. Through numerical simulations, some relation between the dynamic time scales and the Laplacian spectrum are obtained, as shown in Fig. 4.19.

In Fig. 4.19a, one can see the evolution of the oscillators, and find a path to the final global complete synchronization on a community network. When time $t < 0.02$, the whole network is in the state of "non-synchronization." As time goes on, the nodes in the small community S begin to synchronize, while those in the huge community H do not. This situation is referred to as "partial synchronization," which is determined by the community topological structure, where nodes inside the communities are first to synchronize. The small community S is fully connected, so is easier to synchronize than the huge one H. When $t > 0.5$, the nodes in

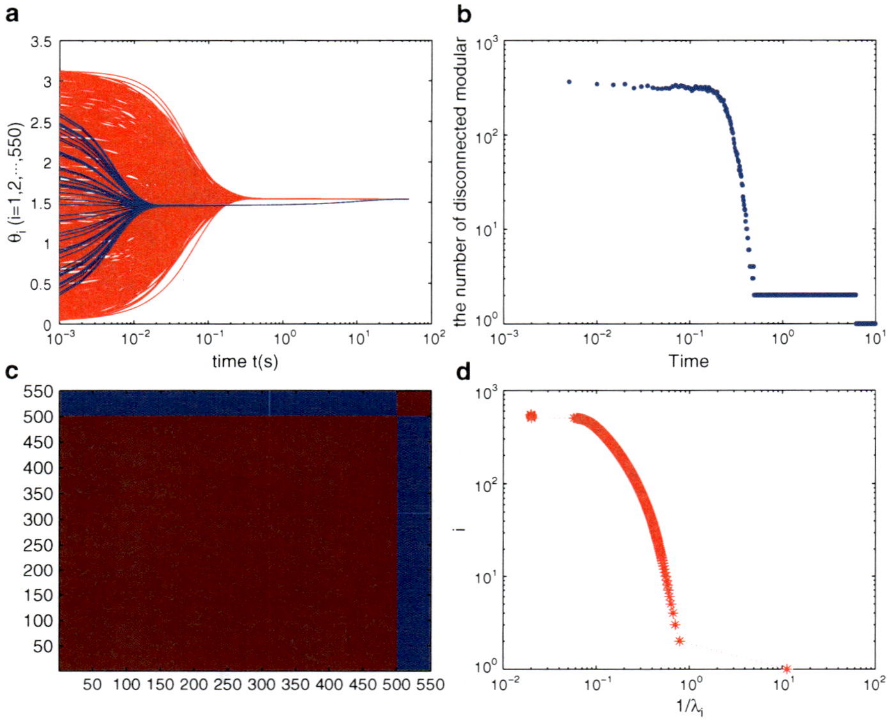

Fig. 4.19 Network with two communities: one is a huge part H of a small-world model of 500 nodes with $p = 0.01$; another is a small part S of a fully connected model of 50 nodes; with only one edge linking the two communities. (**a**) The time evolution of oscillators θ_i $(i = 1, 2, \cdots, N)$. Red line: nodes in H; Blue line: nodes in S; (**b**) number of disconnected components as a function of time t; (**c**) average of the correlation between pairs of oscillators. The colors are graded from blue (0) to dark red (1); (**d**) the inverse of the corresponding eigenvalues of the Laplacian matrix of L versus the rank index i

the huge community H also achieve synchronization, but the synchronous state of H is different from that of S. At this time, the synchronous state of the whole network achieves the so-called "cluster synchronization." Obviously, this regime is a particular transition to the global complete synchronization for the community network. A side benefit is that one can easily detect the community structure of a network during the stage of cluster synchronization. Finally, when the time is long enough, with $t > 30$ here, all oscillators in the whole network are entrained to the "global complete synchronization" state.

According to [24,25], the number of zero eigenvalues of $\mathscr{D}_T(t)$ in (4.17) indicates the number of connected components of the dynamical (synchronized) network. Figure 4.19b plots the number of disconnected components as a function of time. At the beginning, all nodes are uncorrelated, so there are N disconnected sets. As time goes on, some nodes become synchronized to each other and then merge into groups until a single synchronized component is formed after a long enough time. One can observe two plateau regions here, which indicate the relative stability of the dynamics at a given time scale. Notice that the plateau of 300 communities is shorter than the plateau of 2 communities, namely, the synchronization of 2 groups is much more stable than 300 groups, indicating that the 2-group community has better coherence in topological structure.

From Figs. 4.19b and d, one can see that there is a link between the stability of these regions and the spectrum of the Laplacian matrix. The huge gaps between eigenvalues indicates the existence of a relatively stable community structure. It can be observed that three groups of eigenvalues are separated by gaps. Each gap separates two communities, with 550, 300, or 2 groups of nodes. The synchronization dynamics and the spectrum of the Laplacian matrix, which reflects the topological structure of the network, show a surprising similarity.

Finally, Fig. 4.19c presents $\rho_{ij}(t)$ at a fixed time instant, $t = 2s$. At this instant, the whole network is at the stage of cluster synchronization, and it is easy to identify the separation of two communities. Therefore, the network is very close to a state in which two communities are synchronized individually, with different synchronous states from each other, once again proving the side benefit of cluster synchronization in community identification.

Similar synchronization processes can be observed for other communities, as shown in Fig. 4.20. Here, Fig. 4.20a displays the evolution of oscillators in a network consisting of two fully connected communities. There exists a clear transition to global complete synchronization. Figure 4.20b also verifies that the community structure of the two communities is very stable. In Fig. 4.20c, one can see that the network with three communities has a trend moving towards cluster synchronization. However, this transition is not as obvious as that of the community network shown in Fig. 4.20a. This is because the network in Fig. 4.20c has an inapparent community structure, where the first community is a small-world model of 100 nodes with $p = 0.01$, which has very sparse edges. Figure 4.20d shows an even fuzzier community structure of three communities, which is consistent with Fig. 4.20c.

Based on the above-described analysis, one can draw the following conclusions. (1) For community networks, there exists a general path to achieve global complete synchronization as time goes on: non-synchronization $\rightarrow$ partial synchronization $\rightarrow$ cluster synchronization $\rightarrow$ global complete synchronization. (2) Synchronization processes can be used to identify topological scales, that is, communities at different time scales. (3) Synchronization dynamics have a strong relation with the spectrum of the Laplacian matrix, which reflects the topological structure of the network.

Finally, it is remarked that this chapter only discusses complex networks of identical nodes. As to networks of non-identical nodes, spectral analysis becomes much more complicated, even for cluster synchronization (see, e.g., [51,52]), which is beyond the scope of the present study.

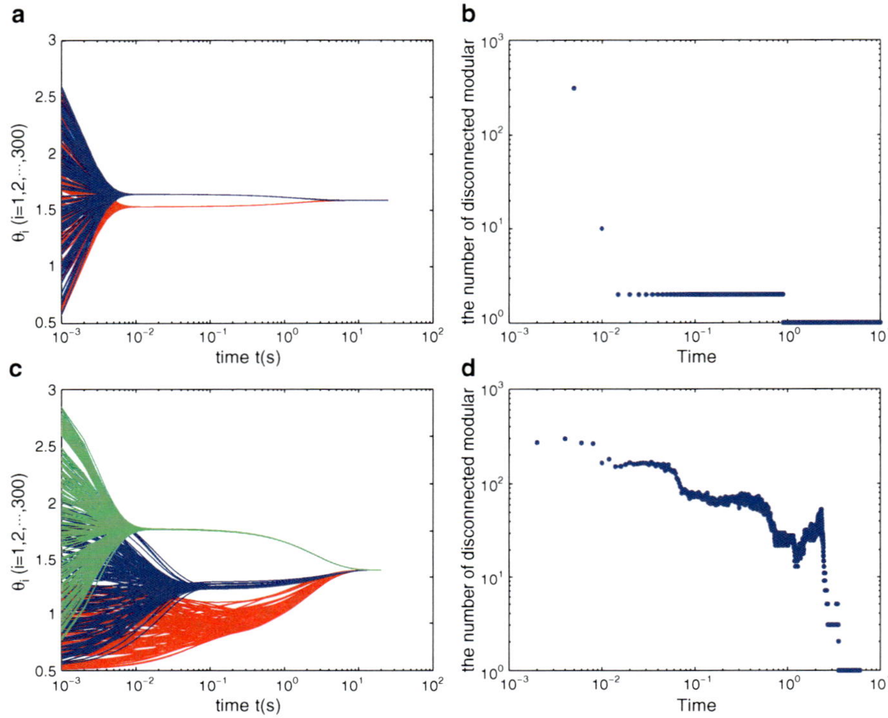

Fig. 4.20 *Top*: Network with 2 communities; each is a fully connected model of 150 nodes. There are 10 edges between the two communities. (**a**) The time evolution of oscillators θ_i $(i = 1, 2, \cdots, N)$. Red line: nodes in one community; Blue line: nodes in the other community. (**b**) Number of disconnected components as a function of time t. *Bottom*: Network with 3 communities: the first one is a small-world model of 100 nodes, with $p = 0.01$; the second is a small-world model of 150 nodes, with $p = 0.1$; the third is a small-world model of 200 nodes, with $p = 0.5$. There are 5 edges between every pair of communities. (**c**) The time evolution of oscillators θ_i $(i = 1, 2, \cdots, N)$. *Red line*: nodes in the first community; *Blue line*: nodes in the second community; *Green line*: nodes in the third community. (**d**) Number of disconnected components as a function of time t

4.6 Conclusions

The Laplacian spectra of several typical complex networks, particularly community networks, have been analyzed mainly from a numerical simulation approach. It is found that four representative complex networks have completely different spectra, where for ER random and NW small-world networks, the smallest nonzero eigenvalue λ_2 depends approximately linearly on the connection probability p. For community networks, the number of eigenvalues near zero reflects the number of communities identifiable from the network. In particular, for random, small-world, and scale-free networks, their spectra are positively correlated to their degree sequences. To find an approximated eigenvalue λ_{i+1} from λ_i, a local prediction-correction algorithm has been proposed, which is shown to be very effective. Furthermore, paths to complete synchronization and generalized synchronization of different networks have been investigated, concluding that the synchronization processes are different with respect to different topological structures, and that nodes with the largest degree firstly achieve synchronization (and generalized synchronization) and then synchronous dynamics spread out to the rest nodes in the network. It has also been found that there is a general path towards global complete synchronization: non-synchronization $\to$ partial synchronization $\to$ cluster synchronization $\to$ global complete synchronization. Finally, it has been revealed that the gaps existing in a Laplacian spectrum are largely dependent on the stability of the communities of the networks at different time scales. All these new findings should provide useful insights to a better understanding of complex network synchronization.

Acknowledgements This work is supported in part by the Chinese National Natural Science Foundation (Grant Nos. 11172215, 60804039 and 60974081), in part by the National Basic Research 973 Program of China under Grant No. 2007CB310805, and in part by the Hong Kong Research Grants Council (Grants NSFC-HK N-CityU107/07 and CRF1117/10E).

References

1. Albert, R., Barabási, A.-L.: Statistical mechanics of complex networks. Reviews of Modern Physics **74**, 47–92 (2002).
2. Dorogovtsev, S.N., Mendes, J.F.F.: Evolution of Networks. Advances in physics **51**, 1079–1187 (2002).
3. Newman, M.E.J.: The structure and function of complex networks. SIAM Review **45**, 167–256 (2003).
4. Boccaletti, S., Latora, V., Moreno, Y., Chavez, M., Hwanga, D.-U.: Complex networks: Structure and dynamics. Physics Reports **424**, 175–308 (2006).
5. Newman, M.E.J.: Networks: An Introduction. Oxford University Press, UK (2010).
6. Albert, R., Jeong, H., Barabási, A.-L.: Error and attack tolerance of complex networks. Nature (London) **406**, 378–382 (2000).
7. Callaway, D.S., Newman, M.E.J., Strogatz, S.H., Watts, D.J.: Network robustness and fragility: Percolation on random graphs. Phys. Rev. Lett. **85**, 5468–5471 (2000).
8. Pastor-Satorras, R., Vespignani, A.: Epidemic spreading in scale-free networks. Phys. Rev. Lett. **86**, 3200–3203 (2001).
9. Moreno, Y., Pastor-Satorras, R., Vespignani, A.: Epidemic outbreaks in complex heterogeneous networks. Eur. Phys. J. B **26**, 521–529 (2002).
10. Nishikawa, T., Motter, A.E., Lai, Y.-C., Hoppensteadt, F.C.: Heterogeneity in oscillator networks: Are smaller worlds easier to synchronize? Phys. Rev. Lett. **91**, 014101 (2003).
11. Sorrentino, F., di Bernardo, M., Garofalo, F.: Synchronizability and synchronization dynamics of complex networks with degree-degree mixing. Int. J. Bifurcation Chaos, **17**, 2419–2434 (2007).
12. Hong, H., Kim, B.J., Choi, M.Y., Park, H.: Factors that predict better synchronizability on complex networks. Phys. Rev. E **65**, 067105 (2002).
13. Barahona, M., Pecora, L.M.: Synchronization in small-world systems. Phys. Rev. Lett. **89**, 054101 (2002).

14. Hong, H., Choi, M.Y., Kim, B.J.: Synchronization on small-world networks. Phys. Rev. E **69**, 026139 (2004).
15. Atay, F.M., Bıyıkoğlu, T., Jost, J.: Network synchronization: Spectral versus statistical properties. Physica D **224**, 35–41 (2006).
16. Chen, G., Duan, Z.: Network synchronizability analysis: A graph-theoretic approach. Chaos **18**, 037102 (2008).
17. Pecora, L.M., Carroll, T.L.: Master stability functions for synchronized coupled systems. Phys. Rev. Lett. **80**, 2109–2112 (1998).
18. Jost, J., Joy, M.P.: Spectral properties and synchronization in coupled map lattices. Phys. Rev. E **65** 016201 (2002).
19. Lü, J., Yu, X., Chen, G., Cheng, D.: Characterizing the synchronizability of small-world dynamical networks. IEEE Trans. Circuits Syst. I **51**, 787–796 (2004).
20. Nishikawa, T., Motter, A.E.: Maximum performance at minimum cost in network synchronization. Physica D **224**, 77–89 (2006).
21. Newman, M.E.J.: Finding community structure in networks using the eigenvectors of matrices. Phys. Rev. E **74**, 036104 (2006).
22. Oh, E., Rho, K., Hong, H., Kahng, B.: Modular synchronization in complex networks. Phys. Rev. E **72**, 047101 (2005).
23. Zhou, C., Kurths, J.: Hierarchical synchronization in complex networks with heterogenous degrees. Chaos **16**, 015104 (2006).
24. Arenas, A., Díaz-Guilera, A., Pérez-Vicente, C.J.: Synchronization reveals topological scales in complex network. Phys. Rev. Lett. **96**, 114102 (2006).
25. Arenas, A., Díaz-Guilera, A., Pérez-Vicente, C.J.: Synchronization processes in complex networks. Physica D **224** 27–34 (2006).
26. Gómez-Gardeñes, J., Moreno, Y., Arenas, A.: Paths to synchronization on complex networks. Phys. Rev. Lett. **98**, 034101 (2007).
27. Gómez-Gardeñes, J., Moreno, Y., Arenas, A.: Synchronizability determined by coupling strengths and topology on complex networks. Phys. Rev. E. **75**, 066106 (2007).
28. Hung, Y.C., Huang, Y.T., Ho, M.C., Hu, C.K.: Paths to globally generalized synchronization in scale-free networks. Phys. Rev. E **77**, 016202 (2008).
29. Wu, W., Chen, T.P.: Partial synchronization in linearly and symmetrically coupled ordinary differential systems. Physica D **238**, 355–364 (2009).
30. Guan, S.G., Wang, X.G., Gong, X.F., Li, K., Lai, C.H.: The development of generalized synchronization on complex networks. Chaos **19**, 013130 (2009).
31. Chen, J., Lu, J., Wu, X., Zheng, W.X.: Generalized synchronization of complex dynamical networks via impulsive control. Chaos **19**, 043119 (2009).
32. Liu, H., Chen, J., Lu, J., Cao, M.: Generalized synchronization in complex dynamical networks via adaptive couplings,. Physica A **389**, 1759–1770 (2010).
33. Mohar, B.: Graph Laplacians. In: Topics in Algebraic Graph Theory, pp. 113–136. Cambridge University Press, Cambridge (2004).
34. Atay, F.M., Bıyıkoğlu, T.: Graph operations and synchronization of complex networks. Phys. Rev. E **72**, 016217 (2005).
35. Biggs, N.: Algebraic Graph Theory. 2nd ed., Cambridge Mathematical Library, Cambridge (1993).
36. Fiedler, M.: Algebraic connectivity of graphs. Czechoslovak Mathematical Journal **23**, 298–305 (1973).
37. Anderson, W.N., Morley, T.D.: Eigenvalues of the Laplacian of a Graph. Linear and Multilinear Algebra **18**, 141–145 (1985) (Widely circulated in preprint form as University of Maryland technical report TR-71-45, October 1971).
38. Merris, R.: A note on Laplacian graph eigenvalues. Linear Algebra and its Applications **285**, 33–35 (1998).
39. Rojo, O., Sojo, R., Rojo, H.: An always nontrivial upper bound for Laplacian graph eigenvalues. Linear Algebra and its Applications **312**, 155–159 (2000).

40. Li, J., Pan, Y.: A note on the second largest eigenvalue of the Laplacian matrix of a graph. Linear and Multilinear Algebra **48**, 117–121 (2000).
41. Merris, R.: Laplacian matrices of graphs: a survey. Linear Algebra and its Applications **197/198** 143–167 (1994).
42. Duan, Z., Liu, C., Chen, G.: Network synchronizability analysis: The theory of subgraphs and complementary graphs. Physica D **237**, 1006–1012 (2008).
43. Wang, X.F., Chen, G.: Synchronization in scale-free dynamical networks: Robustness and fragility. IEEE Trans. on Circ. Syst.-I **49**, 54–62 (2002).
44. Wang, X.F., Chen, G.: Synchronization in small-world dynamical networks. Int. J. of Bifur. Chaos **12**, 187–192 (2002).
45. Erdös, P., Rényi, A.: On the evolution of random graphs. Pul. Math. inst. Hung. Acad. Sci. **5**, 17–60 (1960).
46. Newman, M.E.J., Watts, D.J.: Renormalization group analysis of the small-world network model. Phys. Lett. A **263**, 341–346 (1999).
47. Barabási, A.-L., Albert, R.: Emergence of scaling in random networks. Science **286**, 509–512 (1999).
48. Jalan, S., Bandyopadhyay, J.N.: Random matrix analysis of network Laplacians. Physica A **387**, 667–674 (2008).
49. Zhan, C., Chen, G., Yeung, L.F.: On the distributions of Laplacian eigenvalues versus node degrees in complex networks. Physica A **389** 1779–1788 (2010).
50. Acebrón, J.A., Bonilla, L.L., Pérez-Vicente, C.J., Ritort, F., Spigler, R.: The Kuramoto model: A simple paradigm for synchronization phenomena. Rev. Mod. Phys. **77**, 137–185 (2005).
51. Brede, M.: Local vs. global synchronization in networks of non-identical Kuramoto oscillators. European Phys. J. B **62** 87–94 (2008).
52. Lu, W.L., Liu, B., Chen, T.P.: Cluster synchronization in networks of coupled non-identical dynamical systems. Chaos **20** 013120 (2010).

Chapter 5
Growing Networks Driven by the Evolutionary Prisoner's Dilemma Game

J. Poncela, J. Gómez-Gardeñes, L.M. Floría, and Yamir Moreno

Abstract In this chapter, we present a model of growing networks in which the attachment of nodes is driven by the dynamical state of the evolving network. In particular, we study the interplay between form and function during network formation by considering that the capacity of a node to attract new links from newcomers depends on a dynamical variable: its evolutionary fitness. The fitness of nodes are governed in turn by the payoff obtained when playing a weak Prisoner's Dilemma game with their nearest neighbors. Thus, we couple the structural evolution of the system with its evolutionary dynamics. On the one hand, we study both the levels of cooperation observed during network evolution and the structural outcome of the model. Our results point out that scale-free networks arise naturally in this setting and that they present non-trivial topological attributes such as degree-degree correlations and hierarchical clustering. On the other hand, we also look at the long-term survival of the cooperation on top of these networks, once

J. Poncela
Institute for Biocomputation and Physics of Complex Systems, Zaragoza, Spain
e-mail: poncela@unizar.es

J. Gómez-Gardeñes
Institute for Biocomputation and Physics of Complex Systems, Zaragoza, Spain
e-mail: gardenes@gmail.com

L.M. Floría
Departamento de Física de la Materia Condensada, Universidad de Zaragoza, Zaragoza, Spain

Institute for Biocomputation and Physics of Complex Systems, Zaragoza, Spain
e-mail: mario.floria@gmail.com

Y. Moreno (✉)
Instituto de Biocomputación y Física de Sistemas Complejos (BIFI), Universidad de Zaragoza, Zaragoza, Spain
e-mail: yamir.moreno@gmail.com

M.T. Thai and P.M. Pardalos (eds.), *Handbook of Optimization in Complex Networks: Theory and Applications*, Springer Optimization and Its Applications 57,
DOI 10.1007/978-1-4614-0754-6_5,

the growth has finished. This mechanism points to an evolutionary origin of real complex networks and can be straightforwardly applied to other kinds of dynamical networks problems.

5.1 Introduction

It is well established that the pattern of interactions among the constituents of many complex systems can not be accurately described neither by lattices or other uniformly distributed spatial models, nor using mean-field formulations. Instead, they need to be characterized by what is generally known as a complex network [1, 2]. In many of these networks, the distribution of the number of interactions, that is the degree k, that an individual shares with the rest of the elements of the system, it is to say, $P(k)$, is found to follow a power-law, $P(k) \sim k^{-\gamma}$, with an exponent $2 < \gamma < 3$ in most cases. This implies a high heterogeneity in the degree distribution. The ubiquity in Nature of these so-called scale-free (SF) networks has led scientists to propose many models aimed at reproducing the SF degree distribution [1, 2]. Nonetheless, most of the existing approaches are based on growth rules that depend solely on the topological properties of the network and therefore neglect the connection between the structural evolution and the particular function of the network or the dynamics that takes place on it. This is the case of the well-known Barabási–Albert (BA) model [3], based on two fundamental ingredients: growth and preferential attachment. In this model, the new nodes are sequentially added to the network attaching preferentially to those who have the highest connectivity. However, it is important to recall that accumulated evidences suggest that form follows function [4] and that the formation of a network is also related to the dynamical states of its components through a feedback mechanism that shapes its structure. Taking these facts into consideration, one should not ignore the particular dynamics evolving on top of a network when trying to propose a model for its growth. On the contrary, the outcome of that dynamics should affect somehow the development of the structure.

A paradigmatic case study of the structure and dynamics of complex systems is that of social networks. In these systems, it is particularly relevant to understand how cooperative behavior emerges. The mathematical approach to model the cooperative versus defective interactions is usually addressed under the general framework of Evolutionary Game Theory [5–7] through diverse social dilemmas [8]. In the general case, it is the individual benefit rather than the overall welfare what drives the system evolution. The emergence of cooperation in natural and social systems has been the subject of intense research recently [9–19]. (see also the recent reviews [20, 21]). These works are based either on the assumption of an underlying, given static network (or two static, separate networks for interaction and imitation, respectively) or on a coevolution and rewiring process, starting from a fully developed network that already includes all the participating elements [22–27] (see also the recent review [28]). As we already know, it has been shown that if the well-mixed population hypothesis is abandoned, so that individuals only interact with their neighbors, cooperation is promoted by heterogeneity, specifically on SF networks. However, the main questions remain unanswered: Are cooperative behavior and structural properties of networks related or linked in any way? If so, how? Moreover, if SF networks are best suited to support cooperation, then, where did they come from? What are the mechanisms that shape the structure of the system?

To contribute answering those questions, in this chapter we analyze the growth and formation of complex networks by *coupling* the network formation rules to the dynamical states of the elements of the system. As we have already mentioned, many mechanisms have been proposed for constructing complex scale-free networks similar to those observed in natural, social, and technological systems from purely topological arguments (for instance, using a preferential attachment rule or any other rule available in the literature [1, 2]). As those works do not include information on the specific function or origin of the network, it is very difficult to discuss the origin of the observed networks on the basis of those models, hence motivating the question we are going to address. The fact that the existing approaches consider separately the two directions of the feedback loop between the function and form of a complex system demands a new mechanism where the network grows coupled to the dynamical features of its components. Our aim here is to discuss a recent attempt in this direction, by linking the growth of the network to the dynamics taking place among its nodes.

The model combines two ideas in a novel manner: preferential attachment and evolutionary game dynamics. Indeed, with the problem of the emergence of cooperation as a specific application in mind, we consider that the nodes of the network are individuals involved in a social dilemma and that newcomers are preferentially linked to nodes with high fitness, the latter being proportional to the payoffs obtained in the game. In this way, the fitness of an element is not imposed as an external constraint [29, 30], but rather it is the result of the dynamical evolution of the system. At the same time, the network is not exogenously imposed as a static and rigid structure on top of which the dynamics evolves, but instead it grows from a small seed and acquires its structure during its formation process.

Finally, we stress that this is not yet another preferential attachment model, since the quantity that favors linking of new nodes has no direct relation with the instantaneous topology of the network. In fact, as we will see, the main result of this interplay is the formation of homogeneous or heterogeneous networks (depending on the values of the parameters of our system) that share a number of topological features with real world networks such as a high clustering and degree-degree correlations. Thus, the model we propose not only explains why heterogeneous networks are appropriate to sustain cooperation, but also provides an evolutionary mechanism for their origin. On the other hand, we will find that there are some important and quite surprising differences between the networks we build using this model, and SF topologies. In particular, we will show that the microscopic organization of cooperation is quite different from that observed when studying the

Prisoner's Dilemma game in static networks. This new organization of cooperation appears during network growth and its fingerprint is still observed when analyzing again the microscopic patterns formed by cooperators when the network has ceased its growth.

5.2 The Model

Our model naturally incorporates an intrinsic feedback between dynamics and topology. In this way, the growth of the network starts at time $t = 0$ with a core of m_0 fully connected nodes. New elements are incorporated to the network and attached to m of the existing nodes with a probability that depends on the dynamical states of these nodes. In particular, the growth of the network proceeds by adding a new node at equally spaced time intervals (denoted by τ_{T}), and the probability that a node i in the network receives one of the m new links is

$$\Pi_i(t) = \frac{1 - \varepsilon + \varepsilon f_i(t)}{\sum_{j=1}^{N(t)}(1 - \varepsilon + \varepsilon f_j(t))}, \tag{5.1}$$

where $f_i(t)$ accounts for the dynamical state of a node i, namely its instant fitness [37] (see below), and $N(t)$ is the size of the network at time t. The parameter $\varepsilon \in [0,1)$ controls the weight that newcomers assign to the fitness $f_i(t)$ of the existing nodes in order to decide the attachment probabilities. Therefore, when $\varepsilon > 0$ those nodes with large fitness $f_i(t)$ are preferentially chosen.

How does the dynamical fitness of nodes change in time? The fitness of a node at time t, $f_i(t)$, is defined by the payoffs obtained when playing the Prisoner's Dilemma (PD) game [31] with its $k_i(t)$ neighbors (note that the connectivity k_i of a node i depends on time since it increases due to the attachment of the nodes as the network grows). In particular, each of the nodes present in the network play a round of the PD game at equally spaced time intervals (denoted by τ_{D}). Each of these rounds consist in playing once with each of its $k_i(t)$ neighbors. The sum of the payoffs obtained in the last round of the game constitutes its instant evolutionary fitness, $f_i(t)$, that appears in the attachment probability (5.1). Obviously, when a new round of the PD game is played the fitness changes, so that $f_i(t)$ is updated every τ_{D} time steps.

What is the payoff obtained by a node after playing with a neighbor? The payoff that a node receives when playing with one of its neighbors depends on their instant strategies. The strategy of a node can take two values: Cooperation (C) or Defection (D). According to the PD game there are three possible situations for each pair of nodes linked together in the network, as far as the payoffs obtained by them are concerned:

- If two cooperators meet, both receive R (*Reward*).
- If two defectors play, both receive P (*Punishment*).
- If a cooperator and a defector compete, the former receives S (*Sucker's payoff*) and the latter obtains T (*Temptation*).

In the PD game, the ordering of the four payoffs (T, R, P and S) is the following: $T > R > P > S$. In the following, we will fix the value of three of these parameters to $R = 1$ and $P = S = 0$, and we will leave the Temptation payoff as the unique free parameter $T = b > 1$ of the game [9, 32, 33]. With this choice we are considering the so-called weak PD game.

Are the strategies of nodes fixed or do they change in time? After playing a round of the PD game (at each τ_D time steps), every node i compares its evolutionary fitness with that corresponding to a randomly chosen neighbor j. Then, if $f_i(t) \geq f_j(t)$, node i will keep its strategy for the next round of the game, but if $f_j(t) > f_i(t)$ node i will adopt the strategy of player j with a probability proportional to the payoff difference, $f_j(t) - f_i(t)$ [6, 7, 9, 11, 34–36]. The specific form of this probability is given by

$$P_i = \frac{f_j(t) - f_i(t)}{b \cdot \max\left[k_i(t), k_j(t)\right]}. \tag{5.2}$$

Note that the denominator of the above expression provides a proper normalization so that $0 \geq P_i \leq 1$. To this aim, the denominator, $b \cdot \max\left[k_i(t), k_j(t)\right]$, is the maximum possible fitness difference between two nodes of degree k_i and k_j.

To complete the description of our model we have to set the initial strategy of the initial core of m_0 nodes and newcomers. Those nodes in the initial core are initially set as cooperators. Therefore, our model can be seen as a test for the survival of this small initial core of cooperation. On the other hand, newcomers adopt with the same probability one of the two available strategies, cooperation or defection, since these nodes have not played any round of the game before being added to the network .

Having introduced how the network grows (at each τ_T time steps) and how the fitness of nodes evolve (at each τ_D time steps) we have settled the basis of a model in which both processes evolve entangled. In other words, the growth of the network as defined above is coupled to the evolutionary dynamics of the PD game that simultaneously evolves in the system. The strength of this coupling is controlled by the parameter ε and by the two associated time scales (τ_T and τ_D). Therefore, (5.1) can be viewed as an "Evolutionary Preferential Attachment" (EPA) mechanism. Depending on the value of ε, we can have two extreme situations:

(1) When $\varepsilon \simeq 0$, referred to as the *weak selection limit* [16], the network growth is independent of the evolutionary dynamics as all nodes have roughly the same probability of attracting new links.
(2) Alternatively, in the *strong selection limit*, $\varepsilon \to 1$, the fittest players (highest payoffs) are much more likely to attract the links from newcomers.

Between the above two situations there is a continuum of intermediate values that will give rise to a wide range of in-between behaviors.

We have carried out numerical simulations of the model exploring the (ε, b)-space. It is worth mentioning that we have also explored different time relations $\tau_D - \tau_T$, but we will focus on the results obtained when $\tau_D/\tau_T > 1$, that is when the network growth is faster than the evolutionary dynamics. Since $\tau_D > \tau_T$, the linking procedure is done with the payoffs obtained after the last round of the game. All results reported have been averaged over at least 10^2 realizations, and the number of links of a newcomer is taken to be $m = 2$ whereas the size of the initial core is $m_0 = 3$. Note that, since a fixed number of m links are added with each new node, during network growth the average degree of the nodes remains constant to $\langle k \rangle = 2m$.

5.3 Degree Distribution and Average Level of Cooperation

The dependence of the degree distribution on ε and b is shown in Fig. 5.1. As it can be seen, the weak selection limit produces homogeneous networks characterized by a tail that decays exponentially fast with k. Alternatively, when ε is large, scale-free networks arise. Although this might a priori be expected from the definition of the growth rules, this needs not be the case: indeed, it must be taken into account that in a one-shot PD game (i.e. when the PD game is played only once), defection

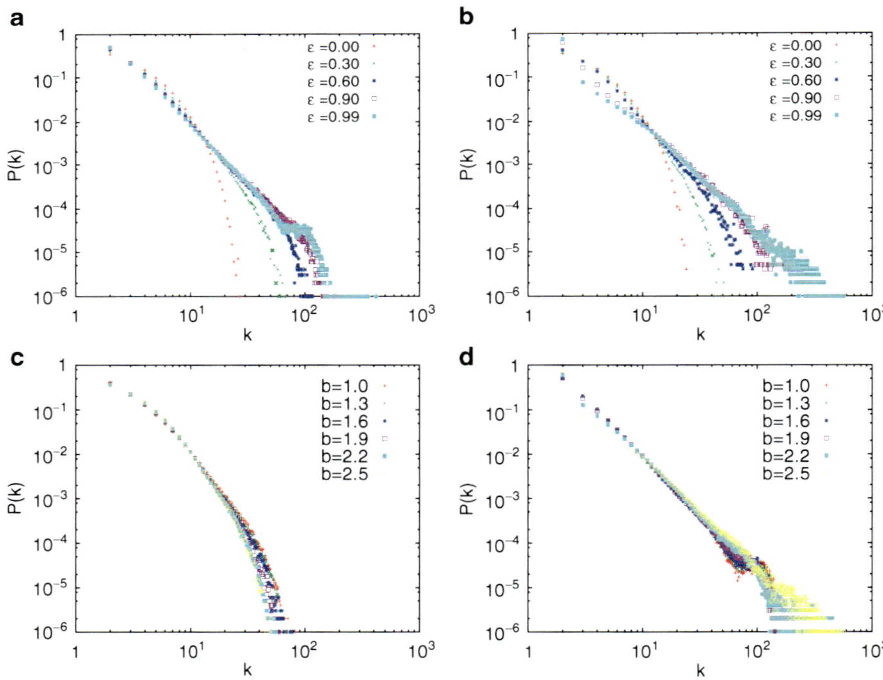

Fig. 5.1 Degree distribution of the topologies created for fixed values of $b = 1.5$ (*Top left*) and $b = 2.5$ (*Top right*), and fixed values of $\varepsilon = 0.3$ (*Bottom left*) and $\varepsilon = 0.99$ (*Bottom right*). The networks are made up of $N = 10^3$ nodes, with $\langle k \rangle = 4$, and $\tau_D = 10\tau_T$. Every point is the average of 300 independent realizations

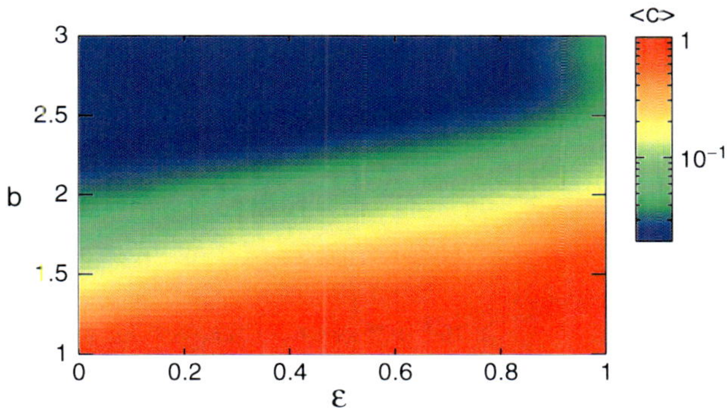

Fig. 5.2 Color-coded average level of cooperation in the system $\langle c \rangle$ right at the end of the EPA procedure, it is to say, when the final size is achieved as a function of the temptation to defect b and the selection pressure ε. The networks are made up of 10^3 nodes with $\langle k \rangle = 4$ and $\tau_D = 10\tau_T$. Reprinted from [38]

is the best strategy regardless of the opponent's choice. However, if the network dynamics evolves into a state in which all players (or a large part of the network) are defectors, they will often play against themselves and their payoffs will be reduced (we recall that $P = 0$). The system's dynamics will then end up in a state close to an all-D configuration (i.e. with all the nodes playing as defectors) rendering $f_i(t) = 0$ $\forall i \in [1, N(t)]$ in (5.1). From this point on, new nodes would attach randomly to other existing nodes (see (5.1)) and therefore no hubs can come out. This turns out not to be the case, which indicates that for having some degree of heterogeneity, a nonzero level of cooperation is needed. Conversely, the heterogeneous character of the system provides a feedback mechanism for the survival of cooperators that would not overcome defectors otherwise.

In Fig. 5.1, we also show the dependence of the degree of heterogeneity of the networks with the temptation to defect, and we found out that only in the strong selection limit, it depends slightly on b. On the other hand, for small values of ε, there is not any dependence of the degree distribution on b, because in this scenario the dynamics does not play a relevant role on the attachment, on the contrary, it is almost random.

Regarding the outcome of the dynamics, we have also studied the average level of cooperation $\langle c \rangle$. The average level of cooperation is defined as the fraction of nodes that play as cooperators in the stationary state of the evolutionary dynamics. We have checked that, although the network keeps growing, the fraction of nodes that play as cooperators reaches a stationary value after a transient time. The Fig. 5.2 shows the dependence of $\langle c \rangle$, as a function of the two model parameters ε and b. As shown in the figure, for a fixed value of $b \gtrsim 1$, the level of cooperation increases with ε. In particular, in the strong selection limit $\langle c \rangle$, the system attains its maximum value. This is a somewhat counterintuitive result as in the limit $\varepsilon \to 1$, new nodes are preferentially linked to those with the highest payoffs, which for the PD game, should correspond to defectors. However, the population achieves the highest value of $\langle c \rangle$. On the other hand, higher levels of cooperation are achieved in heterogeneous rather than in homogeneous topologies, which is consistent with previous findings [9–11].

5.4 Degree Distribution Among Cooperators

In this section we want to study the dependence between strategy and degree of connectivity, comparing the results with those obtained for the static SF scenario, where we recall that cooperators occupy always the highest and medium classes of connectivity, while defectors are not stable when seating on the hubs ([39]). As we will show, the interplay between the local structure of the network and the hierarchical organization of cooperation seems to be highly nontrivial, and quite different to what has been reported for static scale-free networks [9, 11]. In Fig. 5.3 one can see that, surprisingly enough, as the temptation to defect increases, the likelihood that cooperators occupy the hubs decreases. Indeed, during network growth, cooperators are not localized neither at the hubs nor at the lowly connected nodes, but in intermediate degree classes. It is important to realize that this is a new effect that arises from the competition between network growth and the evolutionary dynamics. In particular, it highlights the differences between the microscopic organization in the steady state for the PD game in static networks and that found when the network is evolving.

To address this interesting and previously unobserved phenomenon, we have developed a simple analytical argument that can help understand the reasons behind it. Let k_i^c be the number of cooperator neighbors of a given node i. Its fitness is $f_i^d = bk_i^c$, if node i is a defector, and $f_i^c = k_i^c$, if it is a cooperator. The value of k_i^c

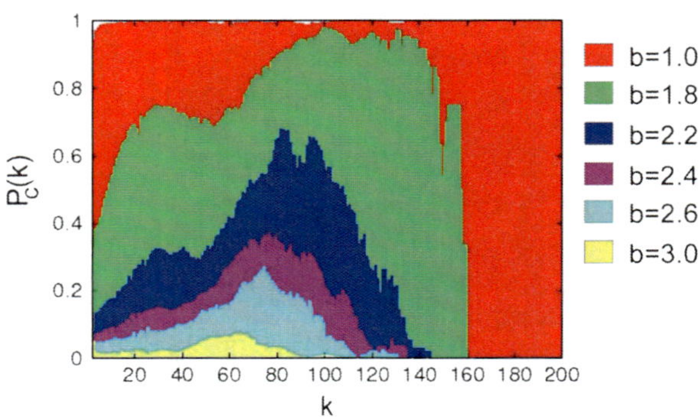

Fig. 5.3 Probability $P_c(k)$ that a node with connectivity k plays as a cooperator for different values of b in the strong selection limit ($\varepsilon = 0.99$) at the end of the growth of a network with $N = 10^3$ nodes and $\langle k \rangle = 4$. Reprinted from [38]

is expected to change because of two factors. On the one hand, due to the network growth (node accretion flow, at a rate of one new node each time unit τ_T) and on the other hand, due to imitation processes dictated by (5.2), that take place at a pace τ_D. As it has been mentioned before, we will focus on the case in which τ_D is much larger than τ_T, for now. Thus, the expected increase of fitness is

$$\Delta f_i = \Delta_{\text{flow}} f_i + \Delta_{\text{evol}} f_i, \tag{5.3}$$

where $\Delta_{\text{flow}} f_i$ means the variation of fitness in node i due to the newcomers flow, and $\Delta_{\text{evol}} f_i$ stands for the change in fitness due to changes of neighbors' strategies. Note that both $\Delta_{\text{flow}} f_i$ and $\Delta_{\text{evol}} f_i$ are related to the change of k_{i}^{c} (the former due to the attachment of new cooperator neighbors to node i and the latter due to the old neighbors that changed their strategy). Therefore, the above expression leads to an expected increase in k_{i}^{c} given by

$$\Delta k_{\text{i}}^{\text{c}} = k_{\text{i}}^{\text{c}}(t+\tau_D) - k_{\text{i}}^{\text{c}}(t) = \Delta_{\text{flow}} k_{\text{i}}^{\text{c}} + \Delta_{\text{evol}} k_{\text{i}}^{\text{c}}. \tag{5.4}$$

On the other hand, the expected increase of degree of node i in the interval of time $(t, t+\tau_D)$ only has the contribution from newcomer flow and recalling that new nodes are generated with the same probability to be cooperators or defectors, i.e, $\rho_0 = 0.5$, it will take the form

$$\Delta k_i = \Delta_{\text{flow}} k_i = 2\Delta_{\text{flow}} k_{\text{i}}^{\text{c}}. \tag{5.5}$$

If the fitness (hence connectivity) of node i is high enough to attract a significant part of the newcomer flow, the first term in (5.3) dominates at short time scales, and then the hub degree k_i increases exponentially. Connectivity patterns are then dominated by the growth by preferential attachment, ensuring, as in the BA model [3], that the network will have a SF degree distribution. Moreover, the rate of increase of the connectivity:

$$\Delta_{\text{flow}} k_{\text{i}}^{\text{c}} = \frac{1}{2} m \tau_D \frac{f_i}{\sum_j f_j} \tag{5.6}$$

is larger for a defector hub by a factor b, because of its larger fitness, and then one should expect hubs to be mostly defectors, as confirmed by the results shown in Fig. 5.3. This small set of most connected defector nodes attracts most of the newcomer flow.

On the contrary, for nodes of intermediate degree, say of connectivity $m \ll k_i \ll k_{\max}$, the term $\Delta_{\text{flow}} f_i$ in (5.3) can be neglected, in other words, the arrival of new nodes is a rare event, so for a large time scale, we have that $\dot{k}_i = 0$. Note that if $\dot{k}_i(t) = 0$ for all t in an interval $t_0 \leq t \leq t_0 + T$, the size of the neighborhood is constant during that whole interval T, and thus the evolutionary dynamics of strategies through imitation is exclusively responsible for the strategic field configuration in the neighborhood of node i. During these periods, the probability distribution of strategies in the neighborhood of node i approaches that of a static network. Thus, recalling that the probability for this node i of intermediate degree to be a cooperator is large in the static regime [11], we then arrive to the conclusion that for these nodes the density of cooperators must reach a maximum, in agreement with Fig. 5.3. Of course, it is clear that this scenario can be occasionally subject to sudden avalanche-type perturbations following "punctuated equilibrium" patterns in the rare occasions in which a new node arrive.

Furthermore, our simulations show that these features of the shape of the curve $P_c(k)$ are indeed preserved as time goes by, giving further support to the above argument based on time scale separation and confirming that our understanding of the mechanisms at work in the model is correct.

5.5 Clustering Coefficient and Degree–Degree Correlations

Apart from the degree distribution, we are also interested in exploring other topological features emerging from the interaction between network growth and the evolutionary dynamics in our EPA networks. Namely, we will focus on two important topological measures that describes the existence of nontrivial two-body and three-body correlations: the degree–degree correlations and the clustering coefficient respectively. We will show that the networks generated by the EPA model display both hierarchical clustering and disassortative degree–degree correlations.

5.5.1 Clustering Coefficient

The clustering coefficient of a given node i, cc_i, expresses the probability that two neighbors j and m of node i, are also connected. The value of cc_i is obtained by counting the actual number of edges, denoted by e_i, in $\mathcal{G}_i$, the subgraph induced by the k_i neighbors of i, and dividing this number by the maximum possible number of edges in $\mathcal{G}_i$:

$$cc_i = \frac{2e_i}{k_i(k_i - 1)}. \tag{5.7}$$

The clustering coefficient of a given network, CC is calculated by averaging the individual values $\{cc_i\}$ across the network nodes, $CC = \sum_i cc_i/N$. Therefore, the clustering coefficient CC measures the probability that two different neighbors of a same node, are also connected to each other. In the left panel of Fig. 5.4, we show the value of CC as a function of b and ε. In this figure, we observe that it is in the strong selection limit where the largest values of CC are obtained. Therefore, in this regime, not only highly heterogeneous networks are obtained but the nodes also display a large clusterization into neighborhoods of densely connected nodes. In the right panel of Fig. 5.4 we show the scaling of the clustering with the network

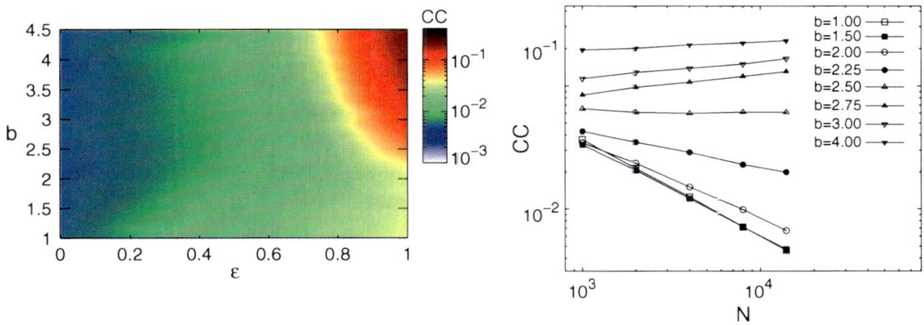

Fig. 5.4 (*Left*) Clustering coefficient CC as a function of b and ε. (*Right*) Scaling of CC with the network size for several values of b in the strong selection limit ($\varepsilon = 0.99$). The networks are made up of $N = 10^3$ nodes and have $\langle k \rangle = 4$

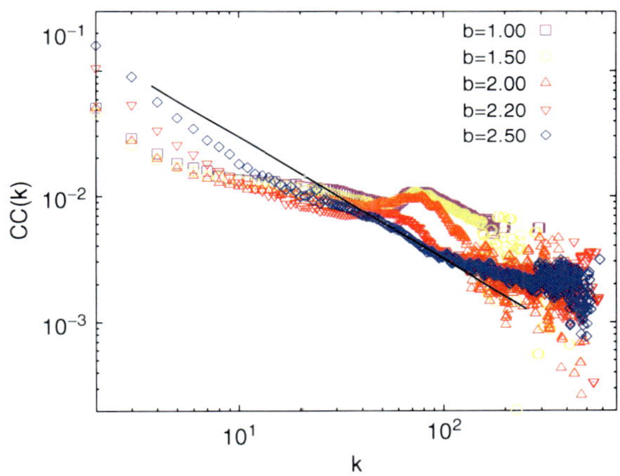

Fig. 5.5 Dependence of the clustering coefficient $CC(k) \sim k^{-\beta}$ with the nodes' degree for different values of b in the strong selection limit ($\varepsilon = 0.99$). The networks are made up of $N = 10^3$ nodes and have $\langle k \rangle = 4$. The straight line is an eye guide that corresponds to k^{-1}. Reprinted from [38]

size $CC(N)$ in the strong selection limit. In this case, we observe that for $b \geq 2.5$ the value of CC is stationary while when $b < 2.5$ the addition of new nodes in the network tends to decrease its clustering.

We now focus on the dependence of the clustering coefficient CC with the degree of the nodes, k, in the strong selection limit ($\varepsilon = 0.99$). Interestingly enough, we show in Fig. 5.5 that the dependence of $CC(k)$ is consistent with a hierarchical organization expressed by the power law $CC(k) \sim k^{-\beta}$, a statistical feature found to describe many real-world networks [2]. The behavior of $CC(k)$ in Fig. 5.5 can be understood by recalling that in scale-free networks, cooperators are not extinguished even for large values of b if they organize into clusters of cooperators that provide the group with a stable source of benefits [11]. To understand this feature in detail, let's assume that a new node j arrives to the network: since the attachment probability depends on the payoff of the receiver, node j may link either to a defector hub or to a node belonging to a cooperator cluster. In the first scenario, it will receive less payoff than the hub, so it will sooner or later imitate its strategy, and therefore will get trapped playing as a defector with a payoff equal to $f_j = 0$. Thus, node j will not be able to attract any links during the subsequent network growth. On the other hand if it attaches to a cooperator cluster and forms new connections with m elements of the cooperator cluster, two different outcomes are possible, depending on its initial strategy: if it plays as a defector, the triad may eventually be invaded by defectors and may end up in an state where the nodes have no capacity to receive new links. But if it plays as a cooperator, the group will be reinforced, both in its robustness against defector attacks and in its overall fitness to attract new links.

To sum up, playing as a cooperator while taking part in a successful (high fitness) cooperator cluster reinforces its future success, while playing as a defector undermines its future fitness and leads to dynamically and topologically frozen structures (it is to say, with $f_i = 0$), so defection cannot take long-term advantage from cooperator clusters. Therefore, cooperator clusters that emerge from cooperator triads to which new cooperators are attached can then continue to grow if more cooperators are attracted or even if defectors attach to the nodes whose connectivity verifies $k > mb$. Moreover, the stability of cooperator clusters and its global fitness grow with their size, specially for their members with higher degree, and naturally favors the formation of triads among its components. Thus, it follows from the above mechanism that a node of degree k is a vertex of $(k-1)$ triangles, and then

$$CC(k) = \frac{k-1}{k(k-1)/2} = 2/k, \tag{5.8}$$

which is exactly the sort of functional form for the clustering coefficient we have found (Fig. 5.5).

5.5.2 Degree–Degree Correlations

Now we turn the attention to the degree–degree correlations of EPA networks. Degree–degree correlations are defined by the conditional probability, $P(k^{'}|k)$, that a node of degree k is connected with a node of degree $k^{'}$. However, since the computation of this probability yields very noisy results, it is difficult to assess whether degree–degree correlations exist in a given network topology. A useful measure to overcome this technical difficulty is to compute the average degree of the neighbors of nodes with degree k, $K_{\rm nn}(k)$, that is related with the probability $P(k|k^{'})$ as

$$K_{\rm nn}(k) = \sum_{k^{'}} k^{'} P(k^{'}|k). \tag{5.9}$$

In networks without degree–degree correlations the function $K_{\rm nn}(k)$ is flat whereas for degree–degree correlated networks the function is approximated by $K_{\rm nn} \sim k^{\nu}$ and the sign of the exponent ν reveals the nature of the correlations. For assortative

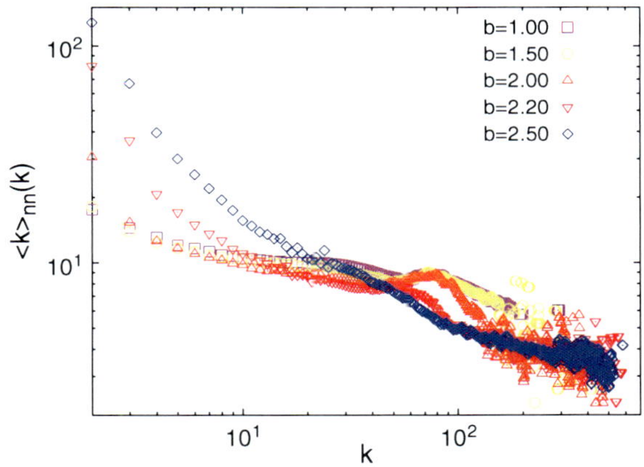

Fig. 5.6 Degree–degree correlations among the nodes of the EPA networks. We plot the average nearest-neighbors degree $K_{\rm nn}(k)$ of a node of degree k for several values of the parameter b used to generate the networks. The networks are generated with $\varepsilon = 0.99$, and have $N = 4.10^3$ nodes and $\langle k \rangle$. Note that negative correlations imply that hubs are not likely to be connected to each other

networks $\nu > 0$ and nodes are connected to neighbors with similar degrees. On the other hand, for disassortative networks $\nu < 0$, high degree nodes tend to be surrounded by low degree nodes.

In Fig. 5.6, we plot several functions $K_{\rm nn}(k)$ corresponding to different values of b in the strong selection limit. We observe that for all the cases there exist negative correlations that make highly connected nodes to be more likely connected to poorly connected nodes and viceversa. Therefore, the EPA topologies are disassortative while this behavior is enhanced as the temptation to defect, b, increases as observed from the slope of the curves in the log-log plot. This disassortative nature of EPA networks will be of relevance when analyzing the results presented in the following section.

5.6 Dynamics on Static Networks Constructed Using the EPA Model

Up to this section we have analyzed the topology and the dynamics of the EPA networks while the growing process takes place. Now we adopt a different perspective by considering the networks as static substrates while studying the evolutionary dynamics of the nodes. This approach will be done in different ways allowing us to have a deeper insight on the EPA networks and their robustness.

5.6.1 Stopping Growth and Letting Evolutionary Dynamics Evolve

To confirm the robustness of the networks generated by Evolutionary Preferential Attachment, let us consider the realistic situation in which after incorporating a

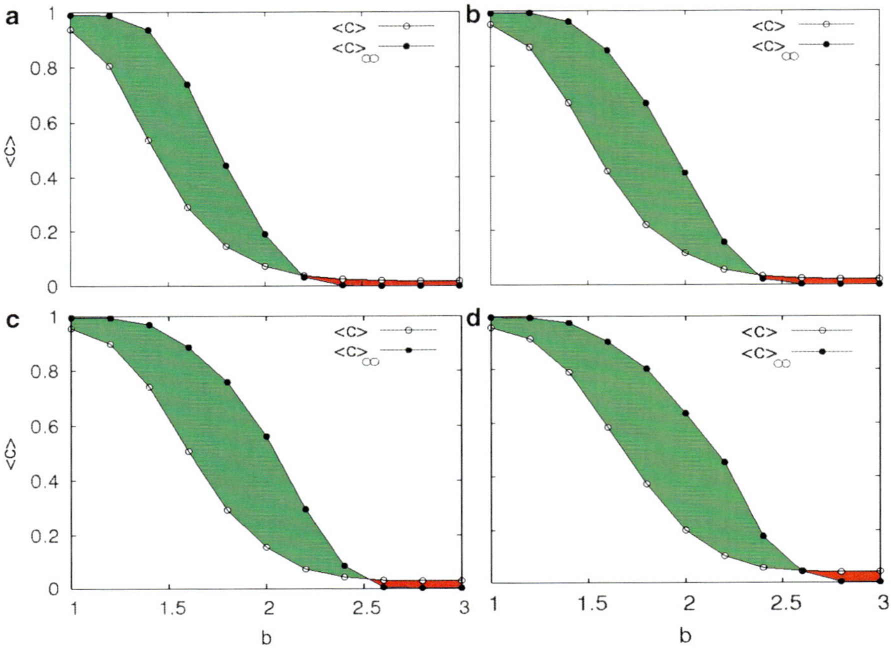

Fig. 5.7 Degree of cooperation when the last node of the network is incorporated, $\langle c \rangle$, and the average fraction of cooperators observed when the system is time-evolved $\langle c \rangle_\infty$ after the network growth ends. The four panels show these measures for several values of ε. From top to bottom and left to right we show $\varepsilon = 0.5$, 0.75, 0.9, and 0.99 (strong selection limit). The networks are made up of $N = 10^3$ nodes with $\langle k \rangle = 4$ and $\tau_{\mathrm{D}} = 10\tau_{\mathrm{T}}$. Every point is the average over 10^3 realizations

large number of participants, the network growth stops when a given size N is reached, and after that, only evolutionary dynamics takes place. The question we aim to address here is: will the cooperation observed during the coevolution process resist?

In Fig. 5.7, we compare the average level of cooperation $\langle c \rangle$ when the network just ceased growing with the same quantity computed after allowing the evolutionary dynamics to evolve many more time steps without attaching new nodes, $\langle c \rangle_\infty$. The green area indicates the region of the parameter b where the level of cooperation increases with respect to that at the moment the network stops growing. On the contrary, the red zone shows that beyond a certain value, b_c, of the temptation to defect the cooperative behavior does not survive and the system dynamics evolves to an all-D state. Surprisingly, the cooperation is enhanced by the growth stop for a wide range of b values pointing out that the cooperation levels observed during growth are very robust. Moreover, the value of b_c appears to increase with the intensity of selection ε in agreement with the increase of the degree heterogeneity

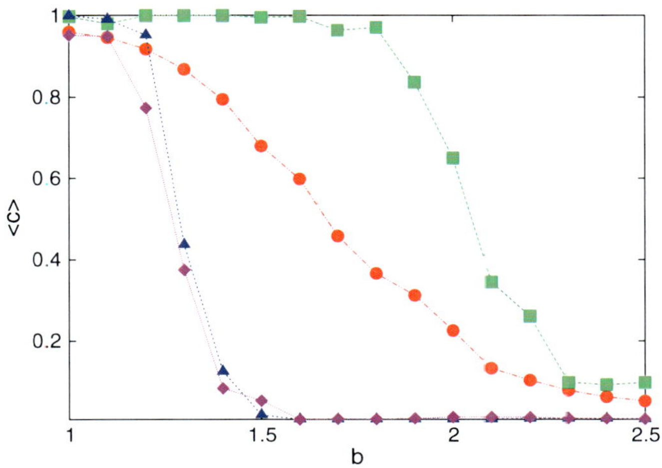

Fig. 5.8 Cooperation levels at the end of the growth process and after letting the network relax as a function of b. The original network was grown up to $N = 4.10^3$ nodes with $\varepsilon = 0.99$ and $\langle k \rangle = 4$, and the asymptotic cooperation levels are computed 10^7 time steps afterwards. Full circles show the cooperation level when the network stops growing. The other curves show the asymptotic cooperation when the structure of the network has been randomized (*triangles*), when the strategies of the nodes have been reassigned randomly (*squares*) and with both randomizations processes (*diamonds*). Reprinted from [40]

of the substrate network. These results highlight the phenomenological difference between playing the PD game simultaneously to the growth of the underlying network and playing on fixed static networks.

5.6.2 *Effects of Randomizations in the Evolutionary Dynamics*

Now, in order to gain more insight in the relation between network topology and the supported level of cooperation, we study the evolution of cooperation when network growth is stopped and we make different randomizations of both the local structure and the strategies of the nodes. In particular, in Fig. 5.8, we show the asymptotic level of cooperation when the following randomizations are made after the growth is stopped: (1) the structure of the EPA network is randomized by rewiring its links while preserving the degree of each node; (2) the structure of the network is kept intact but the strategies of the nodes are reassigned while preserving the global fraction of cooperation (strategy randomization); and (3) when the two former randomization procedures are combined. Note that the randomization of the network structure is made by interchanging pair of links. This randomization, although preserves the degree distribution of the network destroys the degree correlations of the original EPA network and decreases significantly its clustering coefficient.

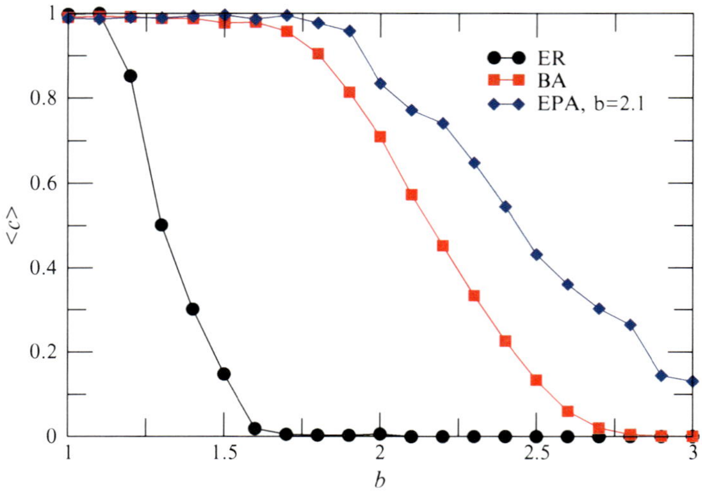

Fig. 5.9 Cooperation levels in ER, BA, and our Evolutionary Preferential Attachment network models, as a function of the temptation parameter b. The EPA network is built up using the model described in the main text for $b = 2.1$ and $\varepsilon = 0.99$. All networks are made up of $N = 10^3$ nodes, with $\langle k \rangle = 4$, and every point shown is the average over 10^3 independent realizations. Reprinted from [40]

As it can be seen from Fig. 5.8, the crucial factor for the cooperation increment during the size-fixed period of the dynamics is the structure of these EPA networks, since its randomization leads to an important decrease of cooperation at levels far away from those of the original one. This drop of cooperation when randomizing the structure is in good agreement with previous findings in complex topologies, specifically, for static BA networks [35, 41]. On the other hand, the strategy randomization procedure does not prevent high levels of cooperation, thus confirming that the governing factor of the network behavior is the structure arising from the co-evolutionary process. Moreover, the asymptotic level of cooperation in this case (squares in Fig. 5.8) is larger that those observed when the network is simply let to evolve without any randomization (C_∞ in Fig. 5.7). This result points out that using a random initial condition for the strategies differs strongly from starting from a configuration where degrees and strategies are correlated as a result of the EPA model (Fig. 5.3). We will come back to this point in Sect. 5.8.

5.6.3 EPA Networks as Substrates for Evolutionary Dynamics

The high levels of cooperation observed when applying a random initial configuration for the strategies to EPA networks motivate the question on whether EPA networks are best suited to support cooperative behavior than other well-known models. In order to answer this question, we consider our EPA networks when used as static substrates for the evolutionary dynamics and compare with the cases of both Barabási–Albert [3] and Erdős–Reńyi (ER) [42] graphs. To this aim, we take a particular example of our model networks, grown with $b = 2.1$ and $\varepsilon = 0.99$, and run the evolutionary dynamics starting from an initial configuration with 50% cooperators and defectors placed at random. The average level of cooperation as a function of the temptation to defect is represented in Fig. 5.9 together with the diagrams for BA and ER networks. Surprisingly, the plot shows that the EPA network remarkably enhances the survival of cooperation for all the values of b studied. Therefore, the attachment process followed by EPA networks is seen to be more efficient than the BA preferential attachment model studied in [9, 11, 14]. Obviously, the roots of this behavior cannot be found in the degree distribution, $P(k)$, but in the high levels of clustering [43] and the disassortative mixing [44] shown above.

5.7 Time Evolution of the $P_c(k)$ After Network Growth

As we have already mentioned, it is widely known that SF topologies are able to sustain higher levels of cooperation than random structures, due to the microscopical organization of the strategies [9, 11]. In particular, it has been shown that in those heterogeneous settings the hubs always play as cooperators being surrounded by a unique cluster of cooperators, while defectors cannot take advantage of high connectivity, and thus occupy medium and low degree classes. Nonetheless, in our EPA structures, we have observed (Sect. 5.4) that during network grows, some hubs play as defectors, thus implying a very different scenario than that of static heterogeneous networks.

In this section, we turn again to the situation in which the network growth is stopped (and no randomization is made) to study the roots of the increment of the asymptotic level of cooperation observed in Fig. 5.7. To this aim, we look at the temporal evolution of the probability that a node of degree k is a cooperator, $P_c(k)$, once the network growth has ceased. As we have observed in Sect. 5.4, the growth process leads to a concentration of cooperators at intermediate degree nodes, explained from the fact that while the network is growing, newcomers join in with the same probability of being cooperators or defectors. In this situation, defectors have an evolutionary advantage as they get higher payoffs from cooperator newcomers. Although these cooperators will subsequently change into defectors and stop providing payoff to the original defector, the stable source of fresh cooperator nodes entering the network compensates for this effect. However, when the growth stops while the dynamics continues, we observe from Fig. 5.10 that low degree nodes are rapidly taken over by cooperators, and after 10^4 time steps they are mainly cooperators. On the contrary, hubs are much more resistant to change, and even after 10^7 time steps not all of them have changed into cooperators (revealed by those values $P_c(k) = 0$ in Fig. 5.10).

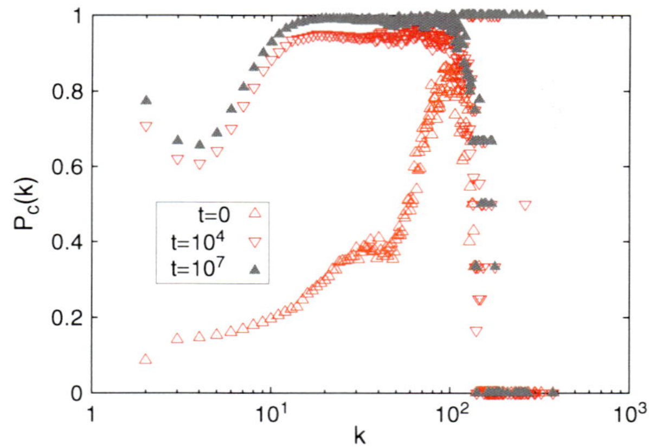

Fig. 5.10 Probability of being a cooperator as a function of the degree at the end of the Evolutionary Preferential Attachment process, 10^4 time steps later, and 10^7 time steps later, for $b = 2.2$ and $\varepsilon = 0.99$. Reprinted from [40]

The persistence of hub defectors is a very striking observation, in contrast with previous findings in static SF networks [9, 11, 41], for which hubs are always cooperators or, in other words, a defector hub is unstable. This occurs because a defector seating on a hub will rapidly convert its neighbors to defectors, which in turn leaves it with zero payoff; subsequently, if one of its neighbors turns back to cooperation, the hub will eventually follow it. It seems, however, that the coupling of evolutionary game dynamics with the network growth leads to a structural configuration that stabilizes defection on hubs. The unexpected result that Fig. 5.10 shows is that defector hubs can also be asymptotically stable once the network growth has ceased. Indeed, we have observed in our simulations that hubs are defectors for as long as the dynamics continues (at least, $t = 10^7$ generations after finishing growing the network). However, it is important to stress that not all realizations of the process end up with defector hubs. For low values of b, this is practically never the case and almost no realizations produce defectors at the hubs, but, as b increases, the percentage of realizations where this phenomenon is observed increases rapidly.

In Sect. 5.4, we have discussed why a hub can be a defector while the network is growing, because it takes advantage of the newcomer flow, getting high benefits from them. Nevertheless, the surprising fact that defector hubs may have very long lives on the static regime, may be the relevant feature for the cooperative behavior of the network resulting from the growth process, and thus it is important to fully understand the reason for such a slow dynamics. We next analyze this in detail.

5.8 Microscopic Roots of Cooperation After Network Growth

Having identified the coexistence of cooperator and defector hubs, we next study why this configuration seems to be asymptotically stable and why the hubs are not invaded by opposite strategies. In Fig. 5.11, we present the payoffs of

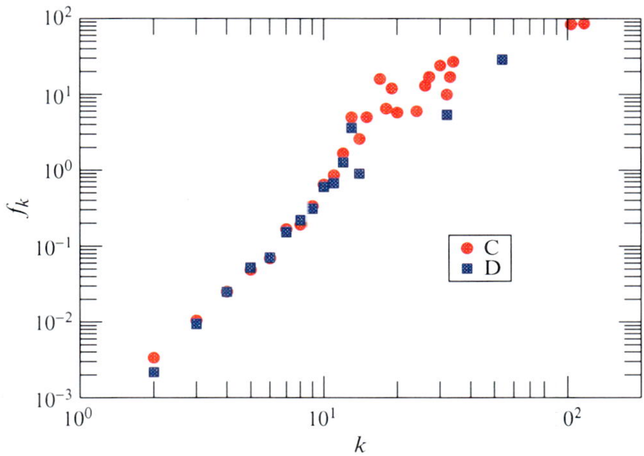

Fig. 5.11 Average payoffs of cooperators and defector nodes at the end of network growth ($t = 0$) as a function of their degrees, k, for a realization of the Evolutionary Preferential Attachment model with $b = 1.8$. Note that the similarity between cooperators' and defectors' payoffs implies that imitation events take place on a long time scale. Reprinted from [40]

cooperators and defectors as a function of their degree. This plot is taken from a single realization of the dynamics in which defector and cooperator hubs coexist asymptotically. As can be seen, the payoff grows approximately as a power law, $f_k \sim k^{\alpha}$; however, the key point here is not this law but the fact that the payoffs for defectors and cooperators of the same degree are very similar. In view of the strategy update rule (5.2), it becomes clear that the evolution must be very slow. Moreover, if we take into account the role of the degree in that expression, we see that hubs have a very low probability to change their strategies, whatever they may be.

Considering now the disassortative nature of the degree–degree correlations (Fig. 5.6) we can explain how these dynamical configurations can be promoted by the structure of the network. The large dissasortativity of EPA networks suggests that hubs are mostly surrounded by low degree nodes and not directly connected to other hubs. Instead, the connection with hubs is made in two steps (i.e. via a low degree node). This local configuration resembles that of the so-called Dipole Model [45], a configuration in which two hubs (not directly connected) are in contact with a large amount of common neighbors which in turn are low degree nodes. In this configuration, it can be shown analytically that the two hubs can coexist asymptotically with opposite strategies, provided that the hub playing as cooperator is in contact with an additional set of nodes playing as cooperators, for this will provide the hubs with a stable source of benefits. On the contrary, defector hubs are only connected to the set of nodes that are also in contact with the cooperator hubs. In this setting, the low degree individuals attached to both hubs experience cycles of cooperation and defection (we call them *fluctuating individuals*, because their strategies can never get fixed) due to the high payoffs obtained by the hubs. If such

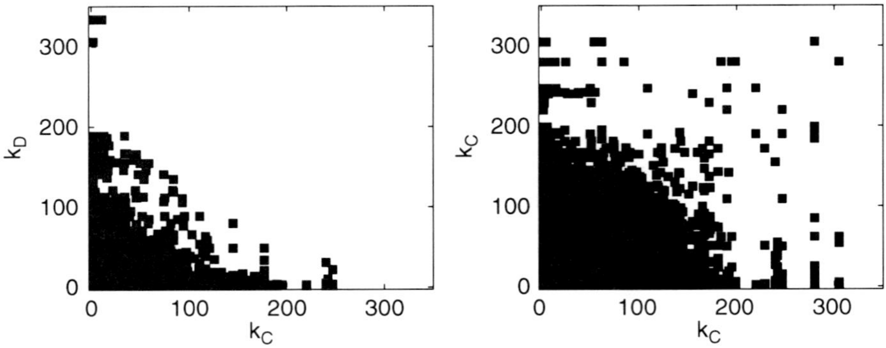

Fig. 5.12 Connectivity matrix of cooperators with defectors (*Left*) and of cooperators with themselves (*Right*). The element (i, j) is set to 1 (black square in the figure) when a link between a defector (cooperator) of degree i and a cooperator (cooperator) of degree j exists, respectively. Reprinted from [40]

a local configuration for the strategies of hubs and their leaves arises, neither of the two hubs will take over the set of fluctuating individuals, nor the latter will invade the hubs as they are mainly lowly connected nodes with small payoffs.

In order to test if the grown networks exhibit local dipole-like structures, we have measured the connectivity of the neighbors of defector and cooperator hubs, which we represent in Fig. 5.12. The figure undoubtedly shows that highly connected nodes playing as defectors are mainly connected to poorly connected cooperators (acting as the set of fluctuating strategists), whereas cooperator hubs are connected to each other and also to a significant fraction of lowly connected nodes. This fully confirms that, in contrast to all previous results, there is a structure allowing the resilience of defector hubs, and moreover, it gives rise to a situation quite similar to that described by the Dipole Model.

5.9 Conclusions

In this chapter we have presented a model in which the rules governing the formation of the network are linked to the dynamics of its components. This model provides an evolutionary explanation for the origin of the two most common types of networks found in natural systems: when the selection pressure is weak, homogeneous networks arise, whereas strong selection pressure gives rise to scale-free networks. A remarkable fact is that the proposed evolution rule gives rise to complex networks that share some topological features with those measured in real systems, such as the power law dependence of the clustering coefficient with the degree of the nodes. Interestingly, our results shows that the microscopic dynamical organization of strategists in EPA networks is very different from the case in which the population evolves on static networks. Namely, there can be hubs playing as defectors during network growth, while cooperators occupy mainly the middle classes. It is worth stressing that the level of cooperation during network growth reach the largest values for the strong selection limit in which the newcomers launch their links to those fittest elements of the system.

Furthermore, the generated networks are robust in the sense that after the growth process stops, the cooperative behavior remains. Moreover, we have shown that for most cases the cooperative behavior increases when network growth is stopped. We have also shown that the non-trivial topological patterns of EPA networks are the roots for such enhancement of the cooperation. In particular, we have shown that rewiring the links while keeping the degree distribution (thus destroying any kind of correlations between nodes) yields a dramatic decrease of the levels of cooperation. On the other hand, a randomization of the strategies does not affect the asymptotic levels of cooperation. Therefore, the ability of EPA networks to promote the resilience of cooperation is rooted in the correlations created during network formation via the coevolution with evolutionary dynamics.

Maybe the most important difference we have found between the evolutionary dynamics on top of EPA networks and that on top of well-known model networks is the dynamic stabilization of defectors on hubs, long after the growth has finished. We have shown that these defector hubs can be extremely long-lived due to the similarity of payoffs between cooperators and defectors arising from the co-evolutionary process. Moreover, we have been able to link the payoff distribution to the network structure. In particular, we show that the disassortative nature of EPA networks together with the formation of local dipole-like structures during network growth is responsible for the fixation of defection in hubs.

Finally, the coevolutionary perspective presented in this chapter has focused on the formation of a complex system rather than being applied to the rewiring of links in already formed systems. Given the simplicity of the formulation presented here we thus expect that the model will contribute to explain other realistic scenarios in which the dynamical states of the constituents of a complex system coevolve with its formation.

References

1. M. Newman, SIAM Review **45**, 167 (2003)
2. S. Boccaletti, V. Latora, Y. Moreno, M. Chavez, D. Hwang, Phys. Rep. **424**, 175 (2006)
3. A. Barabási, R. Albert, Science **286**, 509 (1999)
4. R. Guimerá, M. Sales-Pardo, Mol. Sys. Biol. **2**, 42 (2006)
5. J. Maynard Smith, G. Price, Nature **246**, 15 (1973)
6. H. Gintis, *Game theory evolving.* (Princeton University Press, Princeton, NJ, 2000)
7. J. Hofbauer, K. Sigmund, *Evolutionary games and population dynamics.* (Cambridge University Press, Cambridge, UK, 1998)
8. M. Nowak, *Evolutionary dynamics: exploring the equations of life.* (Harvard University Press., Cambridge, MA, 2006)
9. F.C. Santos, J.M. Pacheco, Phys. Rev. Lett. **95**, 098104 (2005)

10. E. Lieberman, C. Hauert, M.A. Nowak, Nature **433**, 312 (2005)
11. J. Gómez-Gardeñes, M. Campillo, L.M. Floría, Y. Moreno, Phys. Rev. Lett. **98**, 108103 (2007)
12. H. Ohtsuki, E.L. C. Hauert, M.A. Nowak, Nature **441**, 502 (2006)
13. V.M. Eguíluz, M.G. Zimmermann, C.J. Cela-Conde, M. San Miguel, American Journal of Sociology **110**, 977 (2005)
14. F.C. Santos, J.M. Pacheco, T. Lenaerts, Proc. Natl. Acad. Sci. USA **103**, 3490 (2006)
15. F.C. Santos, J.M. Pacheco, T. Lenaerts, PLos Comput. Biol. **2(10)**, e140 (2006)
16. M. Nowak, Science **314**, 1560 (2006)
17. R. Jiménez, H. Lugo, J. Cuesta, A. Sánchez, J. Theor. Biol. **250**, 475 (2008)
18. S. Lozano, A. Arenas, PLos ONE **3(4)**, e1892 (2008)
19. H. Ohtsuki, M.A. Nowak, J.M. Pacheco, Phys. Rev. Lett. **98**, 108106 (2007)
20. G. Szabó, G. Fáth, Phys. Rep. **446**, 97 (2007)
21. C.P. Roca, J. Cuesta, A. Sánchez, Phys. Life Rev. **6**, 208 (2009)
22. M.G. Zimmermann, V.M. Eguiluz, M.S. Miguel, Phys. Rev. E **69**, 065102(R) (2004)
23. H. Ebel, L.I. Mielsch, S. Bornholdt, Phys. Rev. E **66**, 056118 (2002)
24. A. Szolnoki, M. Perc, New J. Phys. **10**, 043063 (2008)
25. J.M. Pacheco, A. Traulsen, M.A. Nowak, Phys. Rev. Lett. **97**, 258103 (2006)
26. A. Szolnoki, M. Perc, **67**, 337 (2009)
27. A. Szolnoki, M. Perc, **86**, 3007 (2009)
28. M. Perc, A. Szolnoki, Biosystems **99**, 109 (2010)
29. G. Bianconi, A.L. Barabási, Europhys. Lett. **54**, 436 (2001)
30. G. Caldarelli, A. Capocci, P.D.L. Rios, M.A.M. noz, Phys. Rev. Lett. **89**, 258702 (2002)
31. A. Rapoport, A.M. Chammah, *Prisoner's Dilemma.* (Univ. of Michigan Press, Ann Arbor, 1965)
32. K. Lindgren, M. Nordahl, Physica D **75**, 292 (1994)
33. M.A. Nowak, R.M. May, Nature **359**, 826 (1992)
34. C. Hauert, M. Doebeli, Nature **428**, 643 (2004)
35. F.C. Santos, F.J. Rodrigues, J.M. Pacheco, Proc. Biol. Sci. **273**, 51 (2006)
36. J. Hofbauer, K. Sigmund, Bull. Am. Math. Soc. **40**, 479 (2003)
37. M. Nowak, A. Sasaki, C. Taylor, D. Fudenberg, Nature **428**, 646 (2004)
38. J. Poncela, J. Gómez-Gardeñes, L. Floría, A. Sánchez, Y. Moreno, PLos ONE **3**, e2449 (2008)
39. J. Poncela, J. Gómez-Gardeñes, L. Floría, Y. Moreno, J. Theor. Biol. **253**, 296 (2008)
40. J. Poncela, J. Gómez-Gardeñes, L. Floría, A. Sánchez, Y. Moreno, Europhys. Lett. **88**, 38003 (2009)
41. F.C. Santos, J.M. Pacheco, J. Evol. Biol. **19**, 726 (2006)
42. P. Erdős, A. Reńyi, Publicationes Mathematicae Debrecen **6**, 290 (1959)
43. S. Assenza, J. Gómez-Gardeñes, V. Latora, Phys. Rev. E **78**, 017101 (2008)
44. A. Pusch, S. Weber, M. Porto, Phys. Rev. E **77**, 036120 (2008)
45. L.M. Floría, C. Gracia-Lázaro, J. Gómez-Gardeñes, Y. Moreno, Phys. Rev. E **79**, 026106 (2009)

Part II
Structure and Dynamics of Complex Networks

Chapter 6
Defining and Discovering Communities in Social Networks

Stephen Kelley, Mark Goldberg, Malik Magdon-Ismail, Konstantin Mertsalov, and Al Wallace

6.1 Introduction

The categorization of vertices in a network is a common task across a multitude of domains. Specifically, identifying structural divisions into internally well connected sets have been shown to be useful in computer science, social science, and biology. In each of these areas, grouping vertices using structural boundaries helps one to understand the underlying processes of a network. Identifying such groupings is a non-trivial task and has been a subject of intense research in recent years.

In general, identifying groups of vertices in a network based on structural properties alone is known as *community detection*. Methods to identify such groups take a wide variety of approaches, mirroring the diversity in domains where an accurate view of structural communities is useful. Depending on the definition of a community used, one could discover groups that maximize a global quality function, contain a specific set of substructures, or satisfy a set of local criteria. Each of these definitions has resulted in a number of methods which aim to produce the "best" set of communities relative to the definition chosen.

Rather than focusing on a number of features which differentiate these definitions and methods from each other, this text will focus on perhaps the most fundamental question in the field of community detection; should groups be disjoint or should they be allowed to overlap?

In the past, the field of community detection has primarily focused on identifying a set of groups such that each vertex in the network is assigned to a single group. This requirement results in a set of disjoint groups covering the entire

S. Kelley (✉)
Oak Ridge National Laboratory, 1 Bethel Valley Road, Oak Ridge TN 37830-8050, USA
e-mail: kelleyjs@ornl.gov

M. Goldberg • M. Magdon-Ismail • K. Mertsalov • Al Wallace
Rensselear Polytechnic Institute, 110 Eighth Street, Troy, NY, USA

M.T. Thai and P.M. Pardalos (eds.), *Handbook of Optimization in Complex Networks: Theory and Applications*, Springer Optimization and Its Applications 57,
DOI 10.1007/978-1-4614-0754-6_6,

network. However, with the explosion of social network and on-line communication data available, research has expanded towards methods which consider overlapping groups.

In the remainder of this text, we will first include a brief discussion on the intuition behind disjoint and overlapping communities and provide the reader with a basic understanding of a small sample of commonly used methods for community detection. Further into the text, we will present the difficulties involved when detecting overlapping communities and introduce a method for discovering overlapping communities which avoids these common pitfalls. This algorithm will be presented with results on real and synthetic benchmark networks. Finally, we will show that in real data, communities that do overlap are natural and necessary to capture many of the associations between vertices in a network.

6.2 Methods for Detecting Community Structure

The most fundamental division between community definitions is whether or not vertices can belong to a single community or any number of communities. Justifications exist for each approach, and ultimately, the selection of which definition to use is likely domain and application dependent. For instance, when analyzing biological protein interaction networks, if an analyst wishes to generate a taxonomy of proteins, a hierarchical disjoint method is desired. When analyzing social networks, due to the variety of affiliations and interests that an individual may have, an overlapping method may be more appropriate.

We begin with a brief examination of some of the previous work in the area of community detection to give the reader a sense of current methods. This examination is far from complete; it is intended to serve only as a brief introduction. For a more comprehensive survey covering a variety of methods in depth, please see [8].

6.2.1 Disjoint Community Detection

The majority of current methods treat the problem of locating communities as a hierarchical partitioning problem. According to this approach, the community structure of a network is assumed to be hierarchical; individuals form disjoint groups which become subgroups of larger groups until one group, comprising the whole society, is formed. Such methods form a tree of subgroup relations called a dendrogram. A dendrogram allows the community structure of a network to be analyzed at various resolutions. An example of this structure, which is commonly used as a visual tool for hierarchical clustering methods, is given in Fig. 6.1.

Originally, the method for detecting a hierarchical grouping in networks was to repetitively identify edges which do not belong to the same dense subgraph [9, 21].

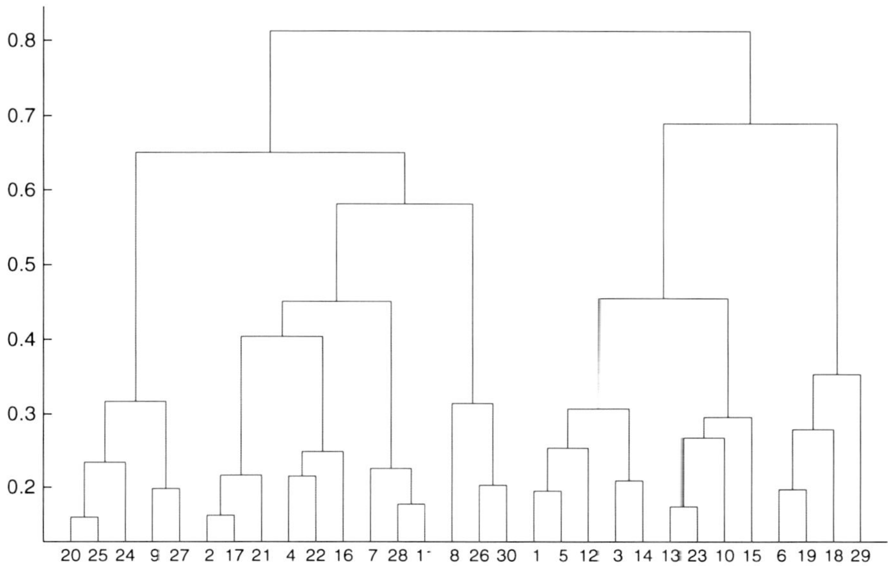

Fig. 6.1 Dendrogram visualization detailing the merging or splitting communities until the entire society is contained in a single group or until each community consists of a single individual

If we consider a group containing all individuals and for each edge compute the centrality according to one of a number of definitions (information, shortest path, circuit, betweenness, etc), edges with higher centrality scores will be ones which link, rather than compose, dense areas of the network. Such edges are repetitively removed. Those edges removed first will be edges that form a significant connection between two dense areas of a network. This process of calculation and removal is performed until the graph becomes disconnected. Upon disconnection, a single group splits into two groups containing each component. This process is continued until each vertex is contained in a group by itself. As a result, a hierarchy of splits is produced, showing the relationship between small groups and larger ones.

This analysis can be quite useful for networks where visual inspection of the dendrogram provides an accurate picture of the underlying community structure. However, this method lacks the ability to point out precisely at which level of the hierarchy the "best" groups have been discovered. For large networks where visual inspection is impossible or for networks in which there exists no intuition to suggest the best set of groups, this fact is problematic. In order to determine the best split in an automated manner, the notion of *modularity* [17] has been proposed. This measure can be expressed as

$$Q = \frac{1}{2m} \sum_{i,j \in V} \left[A_{i,j} - \frac{k_i k_j}{2m} \right] \delta(c_i, c_j), \tag{6.1}$$

where m is the number of edges in a network, $A_{i,j}$ is the edge weight connecting vertex i and j, k_i is the degree of vertex i, and $\delta(c_i, c_j)$ is a function returning 1 if the community assignments of vertex i and vertex j are the same and 0 otherwise. Intuitively, the measure expresses the difference between the number of edges inside communities and the number which are expected to be within a community, given a community's degree. With this measure, one can compare the modularity of all levels of the hierarchy and identify the most well defined set of groups compared to the null model.

The introduction of modularity as an evaluation measure of group quality has resulted in a number of methods which attempt to optimize this value. The most well known of these methods is a greedy agglomerative method originally proposed by Clauset, Newman, and Moore [5]. This algorithm begins by placing vertices in unique communities and merging those that produce the largest increase in modularity. Additional methods have been proposed based on simulated annealing [12], extremal optimization [7], methods from statistical mechanics [23], and other heuristic optimizations [3]. Recent work has also identified a variety of non-hierarchical methods utilizing label propagation [22] and minimizing the amount of information needed to describe random walks in a network [24].

6.2.2 Overlapping Communities

While hierarchical grouping is valid for some types of networks, such as organizational networks or taxonomies, intuition and experience suggest that social networks contain pairs of communities that overlap. Consider an individual in a social network representing "friendship." He or she may have friendship relations across many different social circles, such as those formed in the workplace, by a family unit, by a religious group, or by social clubs. In this case, assuming the community structure of the network to be hierarchical might lead to missing important information about members' attachment to the numerous social circles with which they concurrently interact.

However, the shift from disjoint community assignments to non-disjoint assignments is not a simple one. Various interpretations exist for how vertices can be assigned to groups. Specifically, there is some debate as to whether the goal is to identify a weighted assignment from an individual to all groups or a set of binary assignments indicating an individual's membership. The former has been used in identifying fuzzy groups via probabilistic assignment [6, 27] and maximizing an overlapping version of modularity [18]. Additional work has been done on finding the best set of communities such that each individual can only associate with k sets. An interesting algorithm based on label propagation can be found in [10]. This text however, will examine only the problem of deriving a set of binary individual to group mappings without such constraints. Such a mapping allows communities to be discovered at a local level, where a vertex's association with a group is determined independently of any association with other groups.

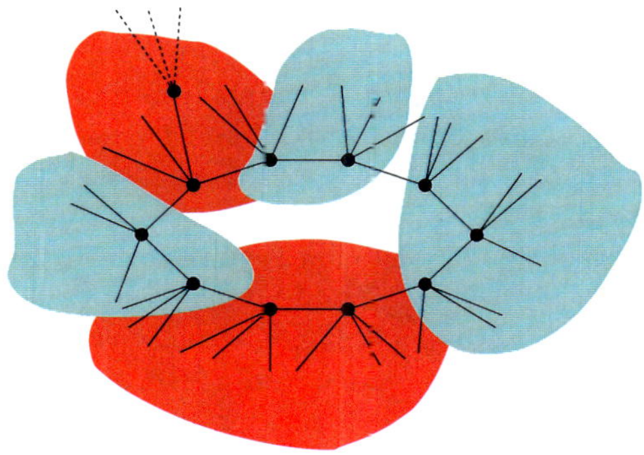

Fig. 6.2 A demonstration of local optimality

Methods which identify these non-fuzzy overlapping communities tend to be one of two types; either the algorithm attempts to identify instances of a specific structure in the network or a sets which maximize a localized quality function. It is important to notice that, unlike the global measure of modularity, each of these tasks is local in nature.

6.2.2.1 Clique Percolation

An algorithm that attempts to identify a defined, local substructure indicative of a community is the Clique Percolation Method (CPM), which was proposed in [20]. In a nutshell, the algorithm first finds all cliques of size k, called k-cliques and defines a graph such that each node represents one of the identified k-cliques. Two nodes are adjacent in the new graph if the corresponding cliques share $k-1$ nodes. The nodes in the union of the k-cliques corresponding to each connected component are declared to be a community. For $k = 2$, clique percolation defines the communities as the connected components in the network.

CPM attempts to discover communities by identifying complete subgraphs of size k. One can claim that, for reasonably sized values of k, such substructure is clearly an instance of community structure. However, this definition sets a very rigid definition for a community. If one edge of a otherwise complete subgraph is missing or if two k-cliques overlap by only $k-2$ nodes, it is not considered a community. Clique percolation would not, for example, be able to find the group illustrated in the toy community in Fig. 6.2. The main problem with such a definition is that it is too rigid and is uniform over the whole network, requiring all communities to share the same structural composition. Additionally, identifying k-cliques of arbitrary sizes can be expensive computationally.

6.2.2.2 Local Optimization

In an effort to identify communities of various composition, new methods have been proposed based on the notion of local optimality. Generally, these methods begin with some set of seed groups which are then optimized relative to a local density function. The seed groups are considered communities when a single vertex addition or removal does not increase the group's quality.

Despite a large number of proposed methods for detecting communities via local optimization [2, 4, 16], there has been a general agreement in the form of the density function used to optimize seed groups. Intuitively, the search for community structure can be viewed as a search for sets of individuals that are intensely connected relative to their isolation with the rest of the network. Specifically, this can be expressed in a manner representative of the functions in previous literature as the ratio of edges internal to the set over all edges connected to the set. This can be given as

$$d(S) = \frac{w_{\text{in}}}{w_{\text{in}} + w_{\text{out}}}, \tag{6.2}$$

where w_{in} is the number of edges internal to the set S and w_{out} is the number of edges connecting the set S to the rest of the network. This and similar density functions are essentially local modularity measures which attempt to maximize internal while minimizing external edges.

Methods based on local optimization add and remove a vertex relative to a set's density when the vertex is evaluated. The implications of this will be discussed at length later in the chapter. However, for now it is only important to realize that locally optimal sets are constructed relative to only their neighborhood. This allows a wide range of communities with both high and low densities to be discovered.

To motivate why this is important, consider the stylized example in Fig. 6.2. This figure depicts some form of organized/coordinated ring-group which would intuitively pass as a community (e.g., a committee of NSF-reviewers). Since we allow overlapping groups, a node could belong to multiple communities, as illustrated by the shaded areas. A node belongs simultaneously to this ring-community as well as to other communities. By virtue of belonging to those other communities, the node communicates extensively outside the ring-group (especially if the node belongs to many other communities). This means that the node displays *more extra-group* similarity than intra-group similarity with respect to the ring-group. There is no flaw with the intuition that a community should display intra-group similarity; the reason the extra-group similarity can be larger is because the communities can overlap. Note that the ring itself in our example, though it is *connected* and appears structured, is not particularly dense; in fact, if each member connects to δ external nodes, then $d(S) = 1/(\delta + 1)$, which can be sufficiently small. Other communities may not have as low a density as this.

We can go further in claiming that this subset should be considered a community independent of the nature of the other communities in the network. Accepting the *locality* property of the communities suggests that the methods that define a global objective function (e.g., modularity [17]) and optimize it to identify all the communities might fail to discover the ring-community. Such methods have found success in partitioning a network, but when overlap is allowed and essential, it is not even clear how to properly define global objective functions.

In the toy ring group shown in Fig. 6.2, the density of our ring-community is $d(S) = 1/(\delta+1)$. One can easily verify that if we remove a node u from the group, its density drops to

$$d(S-u) = \frac{1}{\delta+1+\delta/(|S|-2)}. \tag{6.3}$$

Alternatively, suppose we try to add one of the neighboring nodes z to S. To illustrate, assume that this node has one connection into S and β connections to other nodes. In this case, adding z changes the density to

$$d(S+z) = \frac{1+1/|S|}{\delta+1+\beta/|S|}, \tag{6.4}$$

which is smaller than $d(S)$, when z has more connections to the outside world than the average for nodes already in S. This means that S is *locally optimal* with respect to single node moves. Thus, the requirement of local optimality can capture S as a community.

The main benefits of defining communities as locally optimal sets are that sets with vastly different structural properties can be locally optimal, with varying densities and that locally optimal communities can overlap. Not being able to improve a community (as measured by the density d) is intuitive; this does *not* require a high density or a specific structure of the community. The unified idea of the discussion is that a community is a *locally* defined object. A community in one part of the network should not rely on what is going on in another part of the network. Further, community structure can vary over the network – communication in some communities can be more intense than in others; their structures can be different.

6.3 Local Optimality Examined

The benefits of local optimality as a mechanism to discover overlapping communities have not been lost on researchers. However, despite general agreement that locally optimal sets of vertices form reasonable communities, there is a lack of consensus as to the specifics of the notion of local optimality. Further, additional issues that present themselves when identifying local communities have largely ignored. In this section, we begin by examining the notions of local optimality and density functions. Consolidating this discussion, the section is concluded with a set of axioms which we suggest to be the simplest, smallest set of criteria that any local, overlapping groups should satisfy.

6.3.1 Vertex Removals and Connectivity

As previously stated, various methods have been proposed which attempt to optimize local density functions to identify potentially overlapping communities. However, methods define optimality with respect to different processes. In the process of optimization, some methods allow vertices to be added and removed while others allow only additions. This results in two different notions of local optimality.

An additional problem, which exists with any algorithm allowing vertices to be removed during the optimization, involves the connectivity of communities. Whether a vertex is added to a group or not is determined by the distribution of the vertex's degree as well as the community's density at the time of consideration. This may cause a cut vertex, which was previously inserted into the set based on an earlier, lower density to be removed, thereby disconnecting the set. Producing a disconnected set of vertices in a grouping algorithm is clearly a problem and can affect any local optimization algorithms allowing vertex removal. Clauset's algorithm in [4] successfully avoids this problem by only adding to the group during the optimization, and [25] only merges candidate groups, ensuring the connectivity of the resulting set.

Examining Fig. 6.3, a graph is shown that demonstrates this problem. Consider a candidate group being optimized containing only vertex 1. Initially, the set's density is 0, as there are no internal edges. Upon iterating through all vertices in order of increasing degree, vertex 2 is added to the cluster. This results in an increase in density due to the addition of an internal edge. Proceeding to Fig. 6.3c, the group expands to contain the chains and triangles connected to vertex 2. At this point, however, the density has increased such that the community would have a higher density without vertex 2 being a member. This will result in the removal of vertex 2 and the disconnection of the set. Vertex 1 will also be removed producing a locally optimal, disconnected set.

6.3.2 Tuning Parameters

Examining the previously defined density function in (6.2), we wish to determine the conditions by which a vertex is added or removed from the set. Consider the situation detailed in Fig. 6.4. Here, some vertex i is being considered for addition into the set C. The vertex's degree k_i is split into α and β such that $\alpha = \sum_{j \in C} w_{i,j}$, $\beta = \sum_{j \notin C} w_{i,j}$, and $k_i = \alpha + \beta$. For the vertex i to be added to the set, the density of $C \cup \{i\}$ must be greater than the density of C alone. Therefore, we have

$$\frac{w_{\text{in}}}{w_{\text{in}} + w_{\text{out}}} < \frac{w_{\text{in}} + \alpha}{w_{\text{in}} + w_{\text{out}} + \beta}, \tag{6.5}$$

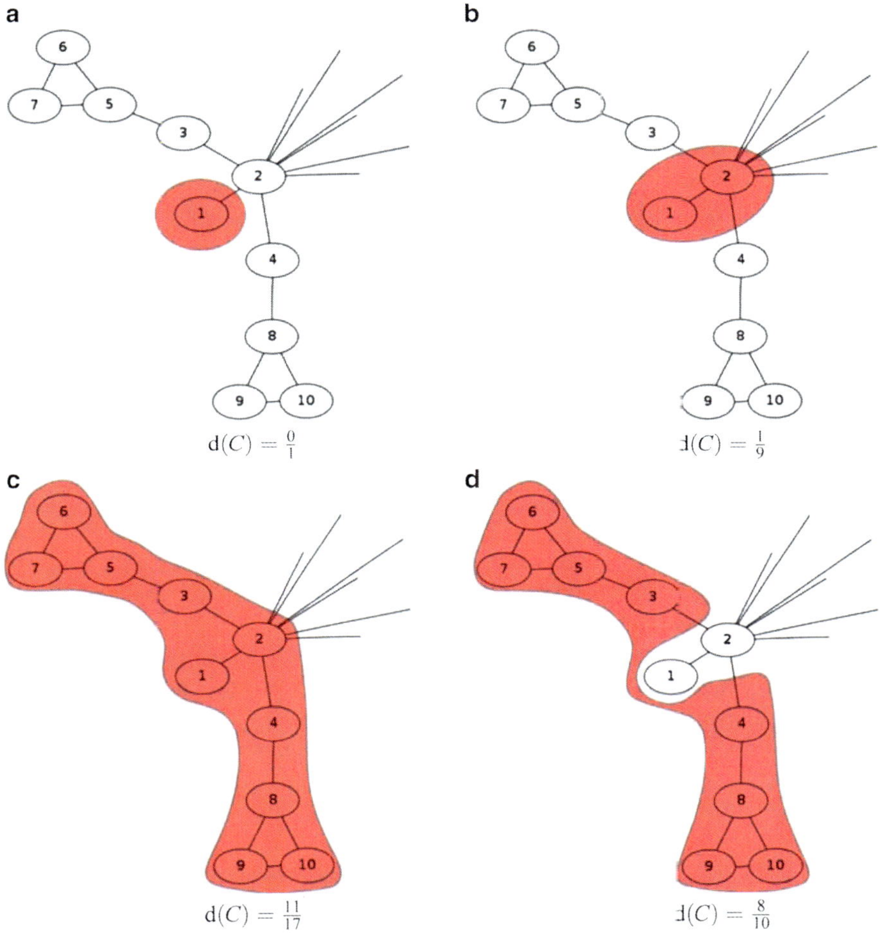

Fig. 6.3 A sample graph demonstrating the generation of a locally optimal, disconnected group. The density function being used for this examination is (1)

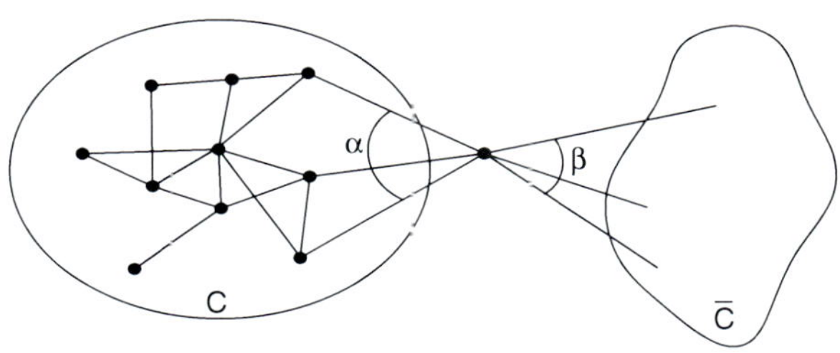

Fig. 6.4 The breakdown of α and β for the addition of a vertex to community C

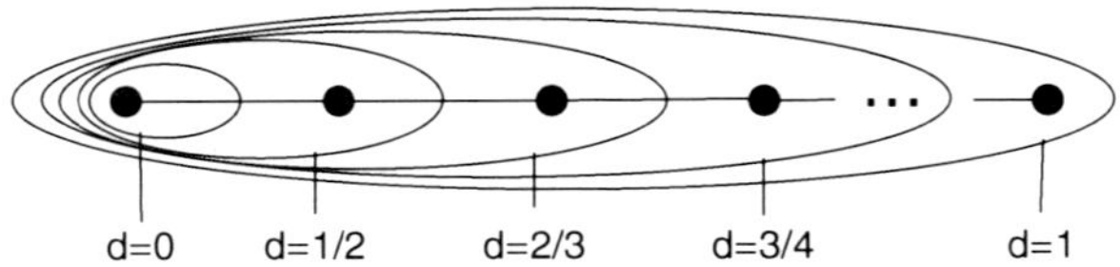

Fig. 6.5 A sample graph demonstrating the performance of local optimization on a chain of vertices

This can be simplified to

$$\frac{\alpha}{\beta} > \frac{w_{\text{in}}}{w_{\text{in}} + w_{\text{out}}}. \tag{6.6}$$

Performing a similar procedure for removals, we arrive at

$$\frac{\alpha}{\beta} < \frac{w_{\text{in}}}{w_{\text{in}} + w_{\text{out}}}. \tag{6.7}$$

It is clear to see from these two relations, that additions and removals occur relative to the density of the set at the time of consideration. It is worth examining how this metric behaves when sparse areas of the graph are encountered. Consider a vertex with degree 2, adjacent to the set being optimized, where $\alpha = \beta = 1$. Since there is at least one edge cut by the community's boundary (implying a density < 1), vertices matching this description will always be added to the group. In practice, this results in groups with a large amount of edges forming a "core" and expanses of sparse vertices. This is a problem primarily when dealing with low degree graphs or social networks whose degree distribution is scale free. This effect is shown in Fig. 6.5. The d values show how density increases until the entire chain is contained within the set. For many applications, such a grouping would be inaccurate, since vertices on the left and right of the chain are very distant and can be presumed to be dissimilar.

It is unintuitive how a community detection algorithm should handle sparse chains of vertices. At one end of the spectrum, one could imagine each pair of vertices in the chain the most salient communities. However, there could also be an argument made that the entire chain should compose a group. This can be controlled by adding a parameter to the density function, introducing a penalty for additions which significantly reduce the edge probability of the community. The following density function is proposed

$$d(C) = \frac{w_{\text{in}}}{w_{\text{in}} + w_{\text{out}}} + \lambda e_p. \tag{6.8}$$

where e_p is the edge probability within the group C

$$e_p = \frac{\sum_{i,j \in C} e_{i,j}}{|C| \times (|C| - 1)} \tag{6.9}$$

and λ is a parameter allowing the results to be fine-tuned. Setting $\lambda = 0$ will produce the same results as (6.2), while larger values will increase the amount of significance the internal edge probability of the set has. This also has the advantage of producing smaller groups for larger values of λ which allows groups to be produced across a wide variety of resolutions. As suggested by Lancichinetti et al. in [16], this and other, similar parameters could also be used to determine the significance of groups. Groups which are structurally significant could maintain their local optimality across numerous values of λ.

6.3.3 Local, Overlapping Axioms

Based on the above observations, as well as previous literature, a set of axioms can be described that any local, community detection method should aim to satisfy. We now state the minimum requirements of a community.

Connectedness. A community should induce a connected subgraph in the network. If the only way to get from one node to another in the community is via some external node, it suggests that the community is incomplete or trivially divisible.

Local Optimality. According to an appropriate density metric $d(C)$, predefined on all subsets of nodes, the density of a community cannot be improved with the removal or addition of a single node.

Note, that the local optimality requirement, but not the connectivity requirement, was first introduced in [1, 2]. Examples can be easily developed of locally optimal sets that induce disconnected subgraphs. Our community axioms posit, in particular, that communities are identified "locally," within one-hop distance from the set. Specifically, we require local optimality with respect to the addition or removal of a single vertex. Previously proposed methods have suggested identifying locally optimal sets with respect to addition only. However, it can be argued that if a community can be improved relative to some density function via removal, it is less meaningful than one constructed via addition and removal. Additionally, one could suggest further notions of local optimality which are relative to a larger number of removals or additions. These other notions of optimality are left for future work. As we will see, these two axioms alone are sufficient for discovering communities which overlap and satisfy the intuitive properties we expect of a community.

It is important to note that this definition is quite different from many previous notions such as those of a "strong" or "weak" community suggested by Raddichi in [21] as well as the definition of modularity which was previously discussed. Rather, this definition focuses on a localized approach that eschews globally formulated null models and strict edge-based requirements.

Algorithmically, it is not easy to identify all communities satisfying these properties. Thus, we resort to a simple heuristic which we discuss next. Our goal is to show that the communities discovered using this heuristic identify salient communities in both common benchmark data as well as real, observed on-line associations.

6.4 Connected Iterative Scan

In [2], the authors describe a community detection algorithm, termed Iterative Scan. Here we describe a modification of IS to discover communities satisfying the previously identified axioms of optimality and connectedness.

Iterative Scan consists of repeated "scans" each starting with an initial set developed by the previous scan (a "seed-set" for the first iteration). It examines each node of the network once, adding or removing it if such an action increases the current density of the set. The scans are repeated until the set is locally optimal with respect to a defined density metric. The choice of seed-sets is not predetermined; seeds can consist of any combination of nodes in the network. A heuristic for seeding, called LinkAggregate, is presented in [1]. LinkAggregate efficiently produces seed-sets that form a cover (with some overlap) of the entire vertex set. The nodes are evaluated by IS from low to high degree. Iterative Scan in this form had been used for a variety of interesting applications such as modeling dynamic networks [11]. A similar method, implementing the idea of the greedy local optimization (as a replacement of a scan in IS) was later given in [16]. For every iteration, the algorithm examines all vertices in order to find the one which causes the maximum increase of the density. That vertex is used to update the current set and any density improving removals are then performed.

The density metric itself can be defined in a number of ways; our analysis uses a modification of the standard density function in Equation 6.2. Rather than using w_{in}, recent literature [16] has proposed using using the internal and external degree of all vertices in the group rather than the number of edges. This is a slight modification, resulting in the the use of $2 * w_{\mathrm{in}}$ in place of w_{in}. For the sake of comparison to previous work, we will optimize using this density function. Our experiments show that in many social networks, there is a very large set of potential communities. Thus, filtering of candidate sets is often necessary and should be done as dictated by the specifics of the application in which community structure is useful. One possibility is to order the candidates by $d(S)$, and consider as most "interesting" those communities which had more internal than external communication ($d(S) > \frac{1}{3}$). This filter is consistent with the notion of a "weak" community as defined by Raddicchi *et al* in [21] and is done in this work to restrict the scope of the analysis for computational reasons. Note that when overlap is allowed, this additional requirement might not be satisfied by all communities. The other possibility of filtering is to look at the communities for which $d(S) < \frac{1}{3}$, as these communities are still connected and locally optimal, even though their members communicate outside of the community a significant fraction of time, which results in sparse internal communication.

To ensure the connectivity of the identified communities, we modify IS and term the resulting algorithm Connected Iterative Scan, CIS. Psuedocode for this algorithm is presented in Algorithm 1. As is the case with IS, CIS consists of a number of scans that are repeated for each current set until no change of the set occurs. The set is then declared to be a community. Every scan proceeds through

Algorithm 1: Connected Iterative Scan

Require: $G = (V, E), S \neq \emptyset$
Ensure: density(S) $\geq$density($S \cup \{v\}$) & density(S) $\geq$density($S \backslash \{v\}$), $\forall v \in V$

```
improved ← true
while improved == true do
   improved ← false
   for all v ∈ V do
      if v ∈ S then
         if density(S\{v}) >density(S) then
            S ← S\{v}
            improved ← true
         end if
      else
         if density(S ∪ {v}) >density(S) then
            S ← S ∪ {v}
            improved ← true
         end if
      end if
   end for
   S ←maxComponent(S)
end while
```

the nodes in the order of increasing node degree. Once a scan is finished, the set's connectivity is examined. If the set consists of multiple connected components, it is replaced by the connected component with the highest density, after which the next scan starts. Note that selecting only the highest density component effectively sidesteps the issue of repeatedly optimizing to the same, disconnected cluster. The specific selection of this rule for identifying connected, locally optimal sets is motivated by the desire to generate as many groups as possible. The running time of the algorithm however, suffers from repetitive connectivity evaluations. For applications where running time is important, one can simply discard those sets that are not connected as a additional post-processing step. Finally, the seeding is this text is done by placing each vertex in its own initial seed community.

The disadvantage of CIS is the same as that of IS; both methods may produce a large number of highly overlapping communities. However, this problem can be managed by effective post-processing of results and merging of highly similar communities. Sample results of CIS for a community analysis of Zachary's Karate Club data set [26] are given in Fig. 6.6. This network represents a set of friendships with in a collegiate martial arts club. Performing analysis on the data, which was collected while the group was undergoing a fissure, provides interesting insight into the set of individuals for whom selecting which splinter group to join was not a trivial choice. Using CIS, these individuals exist in the overlap between the two larger groups in the network. These groups are clearly salient and similar results are found across a variety of literature in community detection.

The complexity of CIS is difficult to analyze due to its dependence on the number and quality of the seeds being optimized as well as the underlying graph structure.

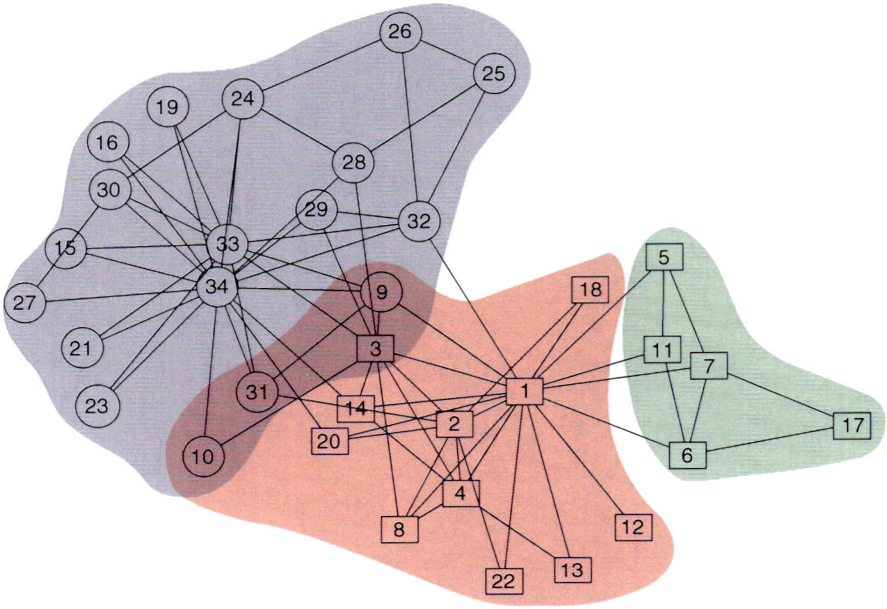

Fig. 6.6 Overlapping groups found in Zachary's Karate Club dataset. Different shapes identify the eventual group division. Groups were ordered to correspond to the number of distinct seeds which produced them. Groups were then selected until the graph was covered. Additional examination of groups which are produced by fewer seeds offers insight into potentially overlapping subgroups of the primary groups presented here

However, similar optimization techniques have previously [1, 16] been empirically shown to have a running time on the order of $O(n^2)$. For many graphs, running time can likely be reduced by introducing higher quality seeds, utilizing a simpler density function, or simply throwing out locally optimal, disconnected sets rather than checking for connectivity at each iteration. Additionally, since the optimization process is independent for each seed, the algorithm is highly parallelizable.

6.4.1 Benchmark Performance

Quantifying the performance of the algorithm is difficult due to the approach. Namely, few other methods aim to produce a large set of locally optimal groups. Rather, they tend to focus on finding partitionings or covers which best express the data. In addition, methods that allow for overlap tend to be insufficient due to the unsatisfied community axioms. In this section, numerous benchmarks will be examined. First, a small, toy graph with uniform degree proposed by Girvan and Newman will be considered. Then, random scale free networks with embedded community structure will be explored for both the overlapping and the non-overlapping case. Each of these experiments will be evaluated via the Normalized Mutual Information measure proposed in [16].

6.4.1.1 GN Benchmark

One of the first benchmarks proposed for community detection algorithms was proposed by Girvan and Newman in [9]. This benchmark dataset, consists of 128 vertices divided into four groups of 32. Each vertex has a degree of 16. The strength of the community associations are given by a mixing parameter which indicates the probability that an edge is placed between two communities rather than internal to a single community. Specifically, this mixing parameter is given by

$$\mu_k = \frac{k_o}{k_i + k_o} \tag{6.10}$$

where k_o is the number of edges connecting a vertex to a vertex in another community and k_i are the number of edges connecting a vertex to other vertices within a community. It should be explicitly noted that this benchmark assigns each vertex to exactly one community during network generation. Despite this, it is important that methods which identify non-disjoint communities be capable of producing accurate communities even when the underlying structures are disjoint.

For Connected Iterative Scan, the results are given in Fig. 6.7. Each point represents the average normalized mutual information over 25 graphs with a given mixing parameter. Seeds are generated by placing each vertex in a candidate cluster. The results shown are a reflection of what is considered to be the "base" settings of the algorithm. This configuration is the density function previously described in the text, vertices ordered by increasing degree, and seeding done by placing each vertex into a seed group by itself. Unless otherwise noted, there is no additional weighting placed on the internal edge probability of the community being optimized.

The two curves in Fig. 6.7 show the result of taking all locally optimal sets discovered by the algorithm as well as using some domain knowledge to filter out the four most frequently discovered sets. It should be noted in the results that the curve is similar to those produced via other methods, though with slightly less accuracy for networks with well defined group structure.

6.4.1.2 LFR Disjoint Benchmark

A more realistic set of benchmark graphs can be tested using the LFR benchmark. Here, a scale free graph is generated with communities of varying sizes. This benchmark was first used in [15] to compare methods of community detection on a more complex network than the GN benchmark. For the experiments contained within this text, graphs are generated matching a power-law degree distribution with

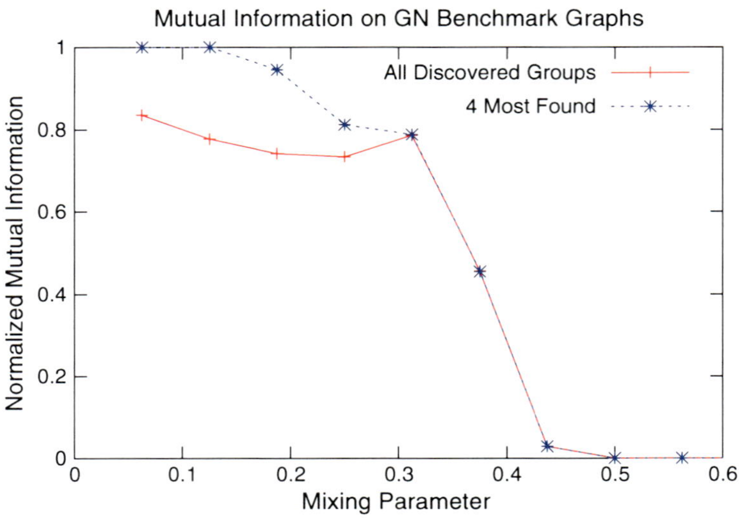

Fig. 6.7 Normalized mutual information for Connected Iterative Scan on GN benchmark graphs

$\alpha_d = 2$ and a power-law community size distribution with $\alpha_c = 1$. For all networks, the average degree of each vertex is 20 and the max degree 50. Community sizes are limited to 10–50 for runs marked "S" and 20–100 for runs marked "B." The output of CIS is processed for evaluation by removing duplicate communities and removing those communities which contain the entire graph. Each data-point represents the average of 25 trials.

The results of this analysis using CIS and CPM are given in Fig. 6.8. Figure 6.8a clearly shows the limitations of identifying a specific structure when compared to Fig. 6.8b–d. Identifying overlapping cliques is much less accurate as group size increases. While CPM produces better results for networks with well defined, small communities, Connected Iterative Scan produces better results in networks with larger community sizes as well as those networks with less well defined communities. The quality of the communities produced via CIS are comparatively stable in the face of changing community and graph properties.

6.4.1.3 LFR Overlapping Benchmark

The LFR benchmark software also allows groups to be embedded such that a given portion of individuals exist in a specified number of groups. This allows algorithms to be compared on networks with known community overlap. Taking the same degree and community size distributions as the previous set of experiments, Connected Iterative Scan and CPM can be compared at varying levels of overlap. Figures 6.9 and 6.10 detail the results of this comparison for 10% and 30% of the

a
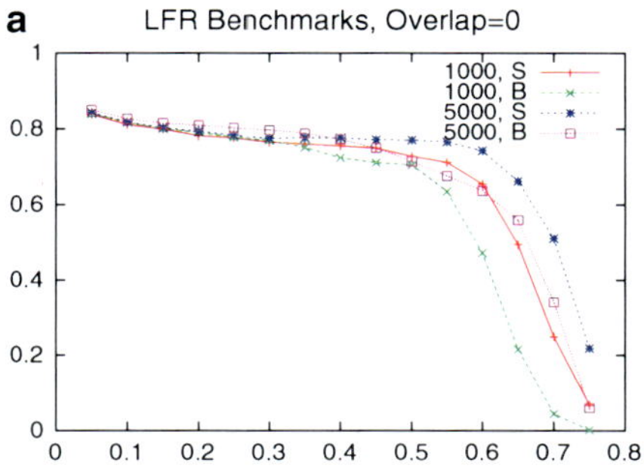

Normalized mutual information for Connected Iterative Scan on LFR benchmark graphs

b
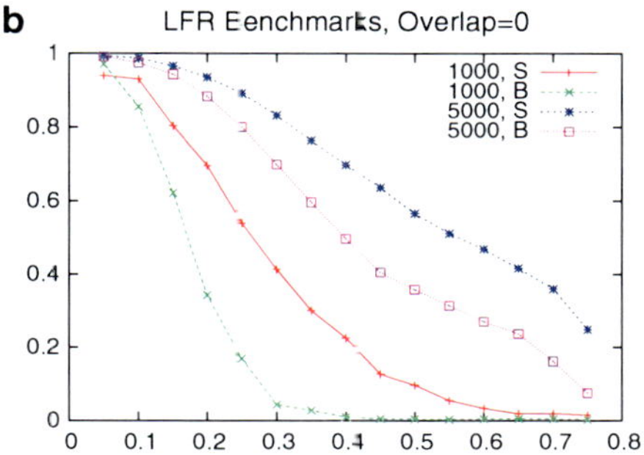

Normalized mutual information for CFinder on LFR benchmark graphs with k=3

c
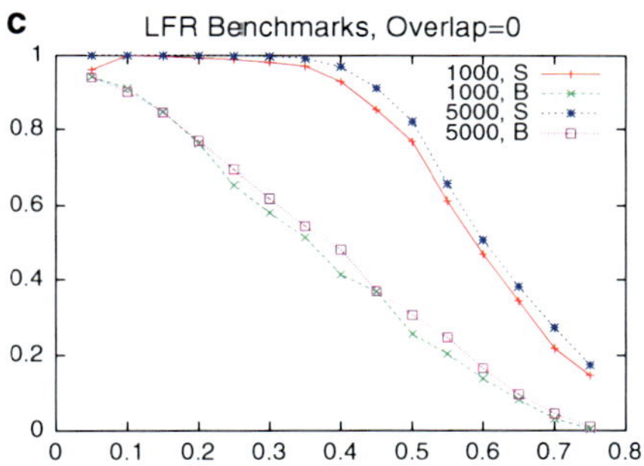

Normalized mutual information for CFinder on LFR benchmark graphs with k=4

d
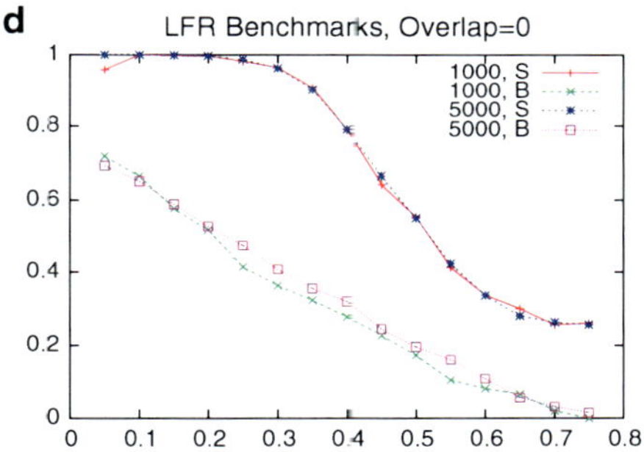

Normalized mutual information for CFinder on LFR benchmark graphs with k=5

Fig. 6.8 Connected Iterative Scan vs CFinder for LFR benchmark graphs with disjoint embedded comunities

vertices existing in 2 communities. Again, the same general trend exists; identifying communities by looking for a set of rigid structural traits fails to identify larger embedded communities, while those produced by CIS are discovered with the same accuracy regardless of community composition.

6.4.2 λ Value

Intuitively, inclusion of the internal edge probability in the density function for Connected Iterative Scan allows the algorithm to be tuned to discover different types of communities. It introduces a criteria for addition different from what was initially proposed during the development of Iterative Scan. When $\lambda > 0$, the vertex being considered for addition must strike a balance between the change in the original density value and the change in edge probability.

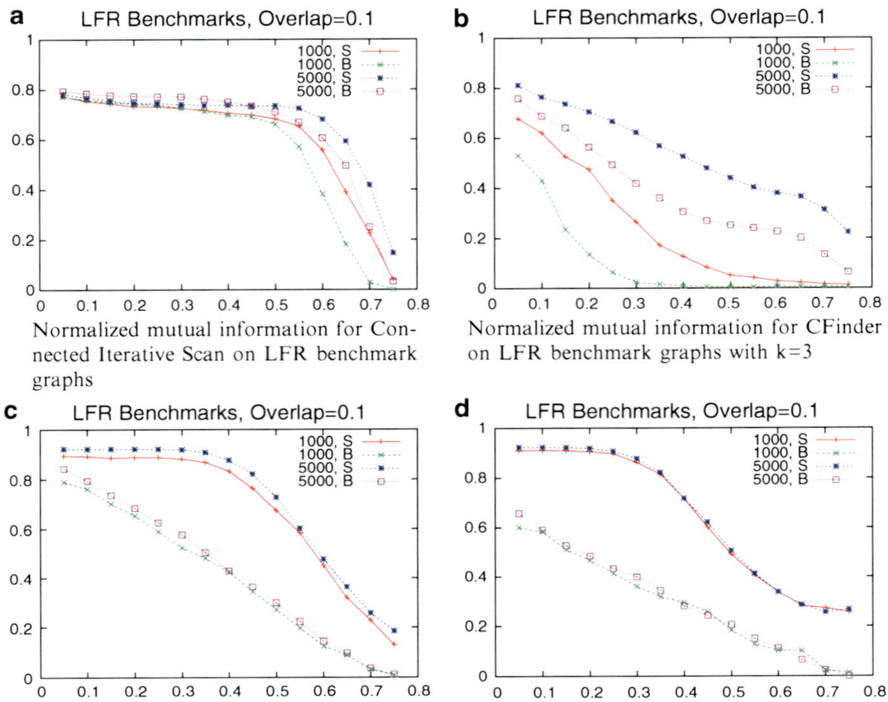

Normalized mutual information for Connected Iterative Scan on LFR benchmark graphs

Normalized mutual information for CFinder on LFR benchmark graphs with k=3

Normalized mutual information for CFinder on LFR benchmark graphs with k=4

Normalized mutual information for CFinder on LFR benchmark graphs with k=5

Fig. 6.9 Connected Iterative Scan vs CFinder for LFR benchmark graphs with overlapping embeded communities where 10% of the vertices associate with 2 communities

This effect can be seen in real networks as well. In this analysis we consider a network in which vertices represent football teams affiliated with universities within the United States. Typically, teams are members of conferences, within which they play a significant portion of their games. Edges in the network indicate that two teams played each other. Groupings produced by Connected Iterative Scan can be compared to the natural divisions created by conferences.

Groupings were performed using a number of different values of λ and filtering the communities by taking only the most discovered groups. The normalized mutual information between the true grouping and the discovered grouping are plotted in Fig. 6.11. The peak at $\lambda = 0.125$ indicates the grouping which most closely matches the underlying conference structure of the network. Qualitatively, the difference between $\lambda = 0$ and $\lambda = 0.125$ is an increased focus on small, tight-knit cores.

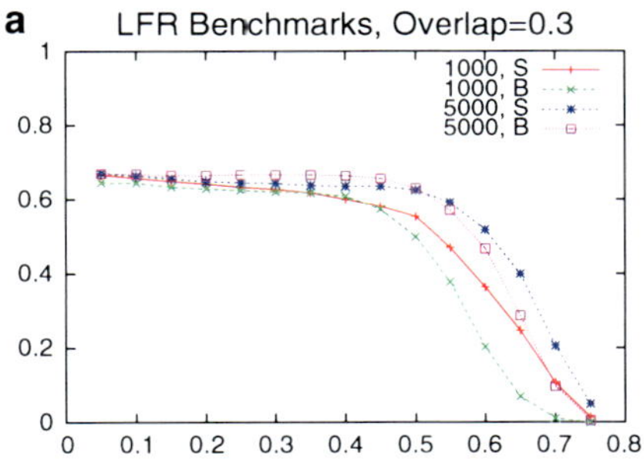

Normalized mutual information for Connected Iterative Scan on LFR benchmark graphs

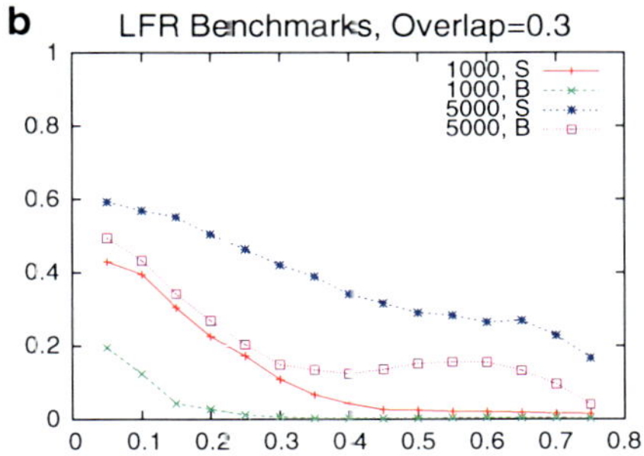

Normalized mutual information for CFinder on LFR benchmark graphs with k=3

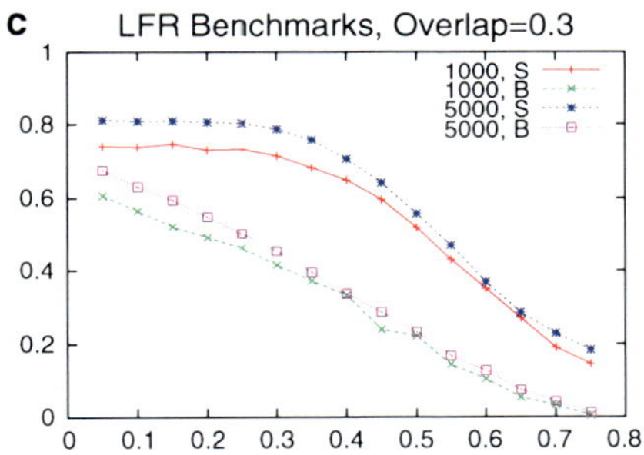

Normalized mutual information for CFinder on LFR benchmark graphs with k=4

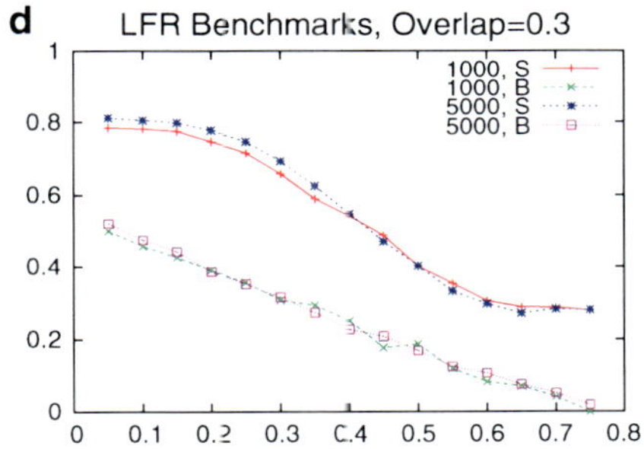

Normalized mutual information for CFinder on LFR benchmark graphs with k=5

Fig. 6.10 Connected Iterative Scan vs CFinder for LFR benchmark graphs where 30% of the vertices associate with 2 communities

6.5 Significance of Overlap

In order to demonstrate that group overlap is a significant feature of some social networks, it is important first to consider the features which pairs of groups should have to indicate that the overlap between them is significant. Consider the overlapping groups presented in Fig. 6.12. Here, group A consists of white and grey vertices, and group B consists of the the black and grey vertices. By this definition, individuals represented by vertices colored grey are members of both group A and B.

For a pair of overlapping groups to have *significant* overlap, and thus be considered a *non-separable pair*, the groups and their overlap must fit certain criteria. In general, each criterion serves to identify a quality of overlapping groups that cannot be expressed via a single group (the union), or two, or three partitions. These criteria can be described conceptually as follows.

a

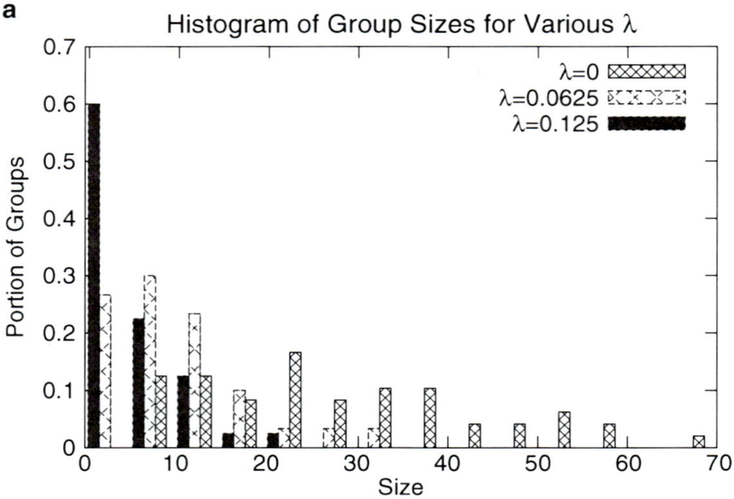

Size distribution for groups produced with various values of λ on the college football dataset.

b

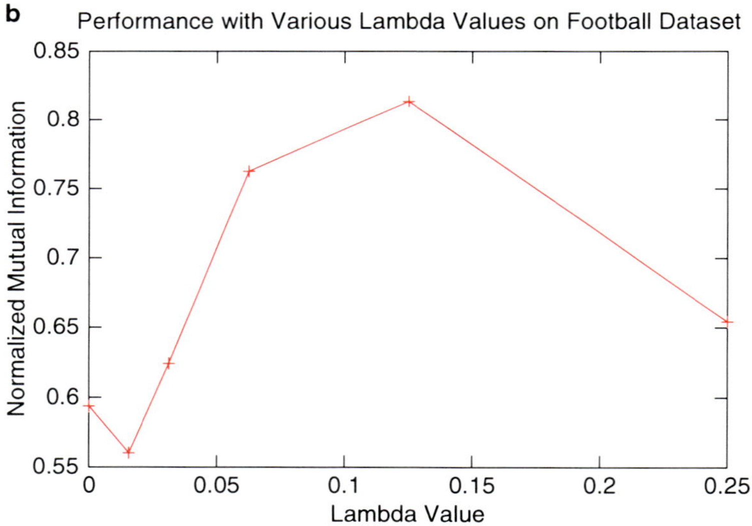

Plot showing the peak in NMI between the discovered groups and the ground truth and λ 0.125

Fig. 6.11 Performance of various λ values on the college football dataset

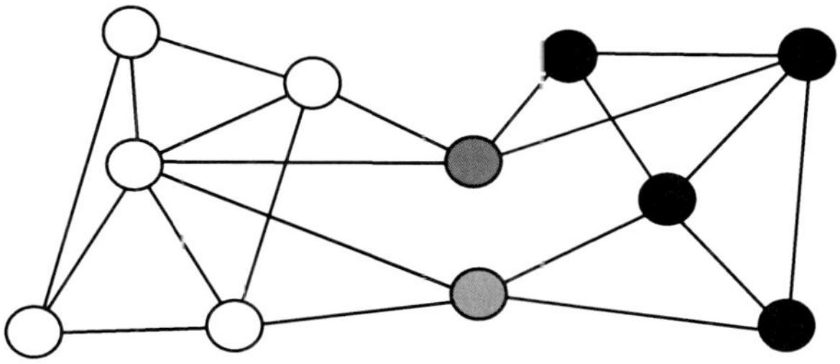

Fig. 6.12 An example of a pair of groups that overlap. The overlap is identified by the grey vertices while individuals in only one group are colored black or white depending on the group of which they are a member

6.5.1 *Structural Significance*

The existence of overlap between a pair of groups should enhance the "quality" of each of the groups individually. For example, if the quality of each group is measured by the ratio of edges internal to the group to those which are cut by the boundary of the group, removing $A \cap B$ from A and B in the groups expressed in Fig. 6.12 would result in a decrease in the quality of each group. The two vertices in the intersection $A \cap B$ have the same degree within each group as they have external to each group. Thus, relative to the previous quality metric, the vertices should be a part of each group since they increase the numerator while holding the denominator constant. Therefore, the overlap is the key to the structural significance of both groups in Fig. 6.12.

6.5.2 *Group Validity*

It is also important that each group be somehow verifiable using a reasonable method relative to the input data. Ideally, using some underlying traits of the individuals in the network being analyzed, groups should have higher trait similarity between members than one would expect if membership in groups were determined at random. Examples of this type of validation have been used in various previous literature, using age and location as traits of the individuals [19]. Group validity is essential in filtering out groups that are products of random structures in the underlying communication graph and serves to ensure that the group detection is accurate.

6.5.3 *Overlap Validity*

Using the same notion of trait similarity, the individuals within the overlap must have some similarity with the remainder of each group of which they are a member.

In Fig. 6.12, the graph is divided into three groups, $A - B$, $B - A$, and $A \cap B$ (white, black, and grey respectively). For overlap to be important, $A - B$ and $A \cap B$ must be similar, $B - A$ and $A \cap B$ must be similar, and $A - B$ and $B - A$ must be dissimilar relative to certain significant traits in the data. Individuals in the overlap need to be similar to the remainder of either group. However, it is necessary that the remaining individuals in each group be dissimilar to those in the other group. If this dissimilarity does not exist, the overlapping pair can be captured in a single partition and overlap is not necessary to explain the relationships in the data.

Pairs of groups that satisfy each of these criteria are fundamentally sound communities due to their structural significance and their group validity. Conceptually, the existence of overlap validity restricts how the individuals can be placed in a partitioning. If all users of the three groups are placed in a single partition, dissimilar vertices in $A - B$ and $B - A$ are associated. If the vertices are placed in three partitions according to color, a strong association between $A \cap B$ and both $A - B$ and $B - A$ is missed. The vertices may be placed in a pair of disjoint groups only if the similarity between $A \cap B$ and both $A - B$ and $B - A$ is highly unbalanced. If the two similarities are comparable, however, one does not have justification to place the users in one group or the other. A detailed description of each of these cases is given further in the text. Significant numbers of non-separable pairs indicate that overlap is an essential component of communities within the network.

6.5.4 Measures

It becomes necessary to formulate a set of measures to indicate whether the notions of group validity and overlap validity are satisfied for a given community or pair of communities. We begin by identifying the set of data used in the analysis.

Due to the implementation of the Friend Feed provided by LiveJournal, friendship declarations can serve as an indicator of interest. By declaring a friendship, the declaring user is notified whenever his or her friend makes a post. It can be assumed that individuals which attract a large number of these friend declarations are highly important to the discourse on some set of topics. Thus, friendship declarations serve as a proxy for some set of declared interests from each user. In this analysis, an individual is defined as influential if he or she has a friendship in-degree of 300 or more. This criteria marks approximately 4,800 bloggers as influential.

The selection of a subset of the friendship relations was done for purely computational reasons, cutting the set of possible friend relations from 500,000 to 5,000. Additionally, interest declarations could be used as validation data. However, within LiveJournal, this data is entered via comma separated values, resulting in a much larger set of possible declarations. Additionally, the popular declared interests, such as "books", "movies", or "music", are much more universal than the most popular friendships. Further, words typed with spelling errors, abbreviations, slang, and the use of synonyms can all be indicative of the same set of topics. The friendship relationship is used in this situation because of its concreteness.

Now, given that each vertex i has a set of declared friendships F_i, we can describe our validation measures. The group validity requirement claims that there should be more similarity within the group than one would find at random. To measure this, we define the notion of *internal pairwise similarity* (denoted *IPS*). For a given community C, the internal pairwise similarity can be computed as

$$\text{IPS}(C) = \frac{\sum_{i \in C} \sum_{j \in C, j \neq i} J(F_i, F_j)}{|C|^2 - |C|}, \tag{6.11}$$

where $J(F_i, F_j)$ is the Jaccard index [13] between the two sets. This value can be expressed as

$$J(F_i, F_j) = \frac{|F_i \cap F_j|}{|F_i \cup F_j|}. \tag{6.12}$$

The value $J(F_i, F_j)$ will be maximized ($J(F_i, F_j) = 1$) if the sets F_i and F_j are identical and will be minimized ($J(F_i, F_j) = 0$) if the two sets are disjoint. Intermediate values of $J(F_i, F_j)$ indicate shared friendships and is normalized by the number of possible shared friendships between the two individuals. Thus, the *IPS* value measures the average similarity between the friendship declarations of pairs within the community. This value is utilized in place of Normalized Mutual Information discussed earlier due to the fact that the "ground truth" in this situation is unknown.

Revisiting the notion of overlap validity, it becomes apparent that a method comparing sets of friendship declarations are needed. Given a pair of overlapping communities A and B, three friendship declaration vectors can be computed. These vectors, denoted L_{A-B}, L_{B-A}, and $L_{A \cap B}$, give the probability that a vertex within each set indicated by the subscript will declare a given individual in the popular friend set as a friend. Formally, $L^i_{A \cap B}$ can be defined for each of the elements of $L_{A \cap B}$ as

$$L^i_{A \cap B} = \frac{|\{x | x \in A \cap B, i \in F_x\}|}{|A \cap B|}, \tag{6.13}$$

where F_x is the set of friends declared for vertex x. Similar vectors can be defined for L_{A-B} and L_{B-A}.

Once these vectors are constructed, the similarity between each of them can be calculated via the *cosine* similarity. Formally, this can be given, relative to two equal dimension vectors X and Y, as

$$\cos(\theta_{X,Y}) = \frac{X \cdot Y}{||X|| \, ||Y||}. \tag{6.14}$$

A low value of $\cos(\theta_{X,Y})$ indicates that the vectors X and Y are close to orthogonal. High values indicate that the vectors have similar values across many dimensions.

Table 6.1 Statistics of groups from CNM and CIS. Q shows the modularity value of the grouping generated by CNM and "Cov" indicates the portion of vertices which are in at least one group

Statistics of groups found via CNM and CIS

	Groups	AvSize	AvDens	Q	Cov
CNM	264	1190	0.745	0.485	100%
CIS	14903	168.8	0.455	–	47.5%

Given the three friendship declaration vectors described previously, the *cosine* similarity between them can give an indication as to whether or not the overlapping group satisfies the overlap validity requirement. Namely, that the inter-group similarity $\cos(\theta_{L_{A-B},L_{B-A}})$ be less than the intra-group similarities $\cos(\theta_{L_{A-B},L_{A\cap B}})$ and $\cos(\theta_{L_{B-A},L_{A\cap B}})$.

In order to simplify this notion, the intra-group and inter-group similarities can be combined into a single statistic representing the relative similarity between the three sets. For the sake of notation, let the inter-group similarity $\cos(\theta_{L_{A-B},L_{B-A}})$ be given by the variable $inter$ and let each of the intra-group similarities $\cos(\theta_{L_{A-B},L_{A\cap B}})$ and $\cos(\theta_{L_{B-A},L_{A\cap B}})$ be given by $intra_A$ and $intra_B$, respectively. These values can be combined into a measure of overlap validity as

$$\mathrm{OV}(A,B) = \frac{intra_A + intra_B}{2} - inter, \tag{6.15}$$

for values of $\mathrm{OV}(A,B) > 0$, the intersection is more similar to each group than the remainder of each group is with each other, indicating that the overlap is split in its association with each set.

6.5.5 Results on LiveJournal

We applied the Connected Iterative Scan algorithm, CIS, to the LiveJournal dataset to produce a set of communities that satisfy the axioms. We also partitioned this graph using the algorithm CNM designed by Clauset, Newman, and Moore ([5]) to give the reader a point of reference and to demonstrate the difference in community sets produced by the two methods. Statistics demonstrating the number of groups, average size, average density, modularity (Q, only applicable for the partitioning), and the number of vertices which are placed in at least one community are given in Table 6.1.

The partitioning produces a small number of sets across a wide variety of sizes while the overlapping group detection produces a much larger number of smaller groups which do not cover the entire graph. Coverage is not a requirement; it is not necessary for every node to belong to a cluster. Rather, we are interested in finding those groups which naturally overlap and studying the significance of this overlap.

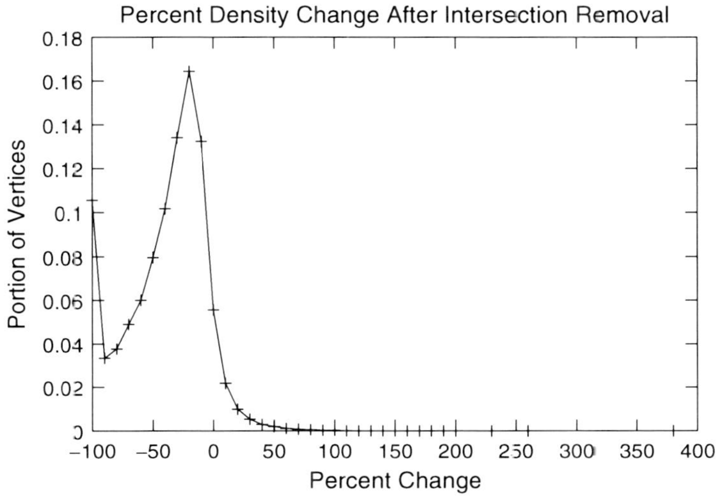

Fig. 6.13 Portion of clusters that experience a given percentage change in density when the intersection of an overlapping pair is removed. Portions are collected in bins of size 10%. This plot contains 50 data points.

If the overlapping groups detected fit the requirement of having structural significance, removal of a pair's overlap will produce a decrease in group quality, as measured by the density d. Overlapping groups are more compelling when the overlap is structurally necessary for each group. After filtering out subset inclusion (a trivial form of overlap), the remaining groups display a high degree of structural significance for the overlap. Specifically, for 80.8% of the overlapping pairs, both groups in the pair experience a decrease in density if the intersection is removed. Figure 6.13 shows more details of the exact distribution of changes in density when the overlap is removed. Even though we observed that some groups are improved by the removal of intersection, the overwhelming majority of groups experience a significant decrease in density. We conclude that in general, within this grouping, that the overlap is structurally significant.

We now investigate the validity of the groups found with respect to user traits. Figure 6.14a shows the average internal pairwise similarity between users within a community as well as the average similarity between users in connected random groups as a function of size. The figure shows that groups produced by CIS have much larger amounts of similarity between users than the random case for sizes greater than 10. This value appears lower than random for sizes less than 10 due to the number of groups which have undefined friendship declarations. The portion of these groups discovered by CIS and at random are given in Fig. 6.14b. Figure 6.14c shows the same information as Fig. 6.14a but with these undefined friendships removed.

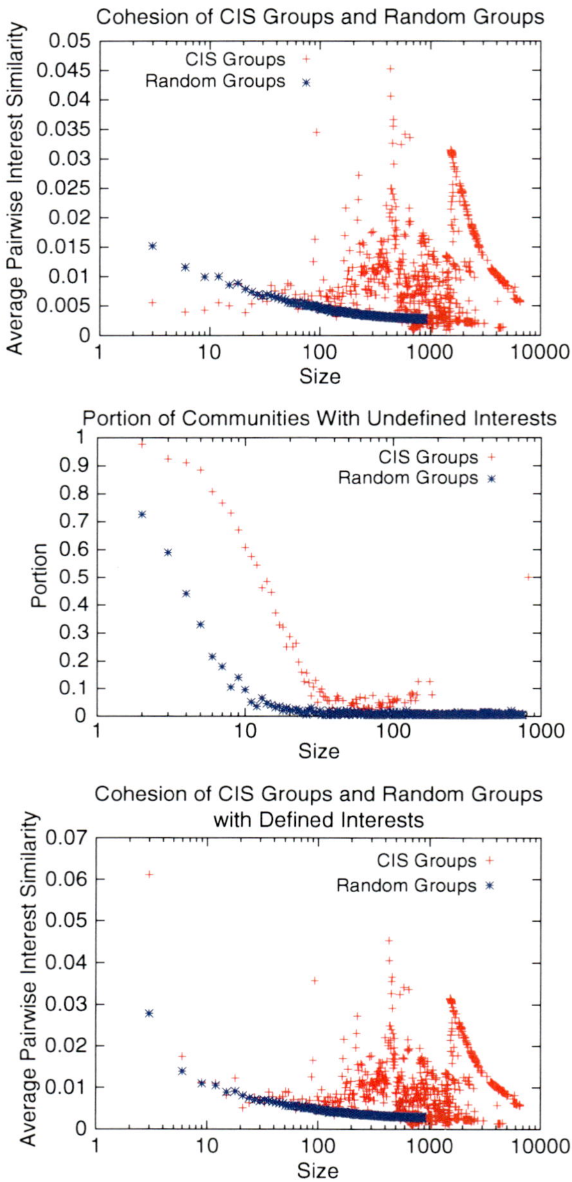

Fig. 6.14 Plot showing the average pairwise Jaccard Index of vertex friendships for all pairs within discovered communities of the same size and values found in randomly generated connected groups of the same size. The plot indicates that there is more similarity in a majority of the discovered groups than one would expect at random

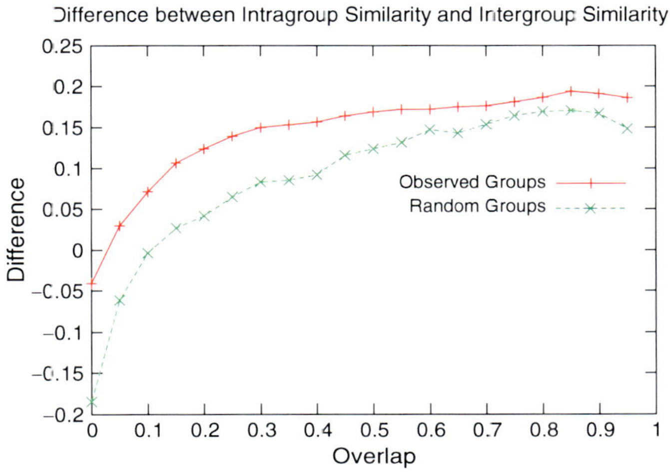

Fig. 6.15 Curves showing the average overlap validity measure $\mathrm{OV}(A,B)$ for identified, non-subset overlapping pairs and random groups of the same size and overlap

Figures 6.14 and 6.15 show the overlap validity measure over pairs of groups with a given overlap. This value is compared with the overlap validity measure for randomly selected groups with the same size and overlap. The x-axis denotes the overlap of the pair, where overlap is defined as the Jaccard index of the two sets. Clearly, there is a larger difference in similarity between the groups identified via CIS and those generated at random.

For the 14,903 unique groups that were discovered, 6,373(∼42%) of them overlap with at least one other group such that the pair can be considered justified by the three conditions previously described. These pairs are composed of 125740 unique users, a very significant portion of the graph.

Further, a significant portion of the non-separable groups have comparable intra-group similarity between the intersection $A \cap B$ and both of the sets $B - A$ and $A - B$. If the similarities are considered comparable when they are within 5% of each other, 3,544 of the non-separable pairs have an overlap that is associated equally with the remainder of each group. These groups consist of 100,000 unique users. The existence of these groups is particularly significant in justifying overlap between communities. They clearly show that many sets of users are equally associated with distinct groups. Using a partition-based method for the detection of communities would either merge the entire pair into one group, failing to recognize the relative dissimilarity between the vertices in sets $A - B$ and $B - A$, or place the intersection with $A - B$ or $B - A$, missing the connection between the intersection and the other set.

6.6 Summary

Detecting communities in networks is a highly useful and non-trivial task. In certain domains, it is reasonable to expect that community structure overlaps. This necessitates defining the fundamental notions of what overlapping communities should look like. The axioms laid out in this chapter attempt to fulfill that need, while at the same time being as minimal as possible to allow for methodological and application specific variations.

Additionally, this chapter has shown that having a loosely defined definition of community structure is often a better choice compared to more restrictive methods that attempt to discover very specific structural formations in networks. The ability of a method and definition to produce quality communities across a wide array of network types is quite important. The axioms laid out in this text provide a framework for such methods to be proposed within. We have also shown that in some networks, the best set of communities will only be found via some additional parameter tuning, particularly those parameters that relate to the size of the groups discovered.

Previous attempts at developing algorithms for the detection of overlapping communities have been primarily intuitive and were developed without first examining to what degree overlap occurs in naturally occurring networks. A large amount of justified overlap indicates that the added complexity of new methods is essential to capturing all relationships expressed in the data. As a test network, we examined a social network composed of communications in a popular blogging service.

The overlap between groups must satisfy certain criteria to be considered significant. First, the inclusion of the common region in either group should enhance the quality of the groups by some metric. In addition, the groups themselves should be verifiable as significant through the use of a set of relevant user traits. Finally, the similarity between components of both groups involved in the overlap must be such that the intersection is more similar with the remainder of each group than the remainder of the groups are with each other. If each of these criteria is satisfied, placing the members of the group in some partitioning will not capture the subtle associations present in the data.

6.7 Future Directions

The use of overlapping community structure has significant potential to aid in the comprehension of underlying processes in an increasingly interconnected world. Intuition and the empirical observations contained in this chapter suggest that the associations contained within such communities capture essential and meaningful relationships which are implicit in the data. The field is far from mature, and various questions have arisen throughout research which remain open problems.

Community detection algorithms have tended to focus on static networks. However, real world data has the potential to be quite dynamic. As a result, new methods will need to be proposed to handle network ties with a temporal component. One simple extension to the work described in this text would be a sociologically grounded edge weight function. Such a function would take the age of a network association into account and decrease edge weight accordingly. The introduction of edge decay creates a potentially interesting area of study involving repetative reoptimization of sets over time.

An additional open area is the identification of additional methods of validating and quantifying the correctness of community detection methods. Recent work has introduced new methods to compare sets of overlapping sets [16], however, more fundamental analysis techniques should be used for comparison. Additional validation techniques such as computing feature similarity of identified groups require data sets with additional, frequently self-reported, information. The problems which exist with self-reported information can clearly be seen in the lack of networks with a well defined, overlapping "ground truth." Often, overlapping communities tend to be more subtle than their disjoint counterparts. As such, it is difficult for individuals to list each of the groups with which they associate, as such groups may be ill defined in the minds of their members.

Another open problem is identifying a method or measure to determine the significance of a community among the set of those which have been discovered. As previously stated, using the minimal axioms described above, there are a vast number of sets which can be considered groups. In order for this type of analysis to be useful as a feature to some other mechanism, it is likely that the "best" groups with regard to application specific metrics will prove to be more useful than others. Significance measures have previously been explored somewhat with regards to disjoint community detection [14], but with the exception of a brief comment in [16], this discussion has largely been absent when examining the detection of overlapping communities.

Acknowledgements Research was sponsored by the Army Research Laboratory and was accomplished under Cooperative Agreement Number W911NF-09-2-0053 and by UT-Battelle, a contractor for the U.S. Government under Department of Energy Contract DE-AC05-00OR22725. The views and conclusions contained in this document are those of the authors and should not be interpreted as representing the official policies, either expressed or implied, of the Army Research Laboratory or the U.S. Government. The U.S. Government is authorized to reproduce and distribute reprints for Government purposes notwithstanding any copyright notation here on.

References

1. J. Baumes, M. Goldberg, and M. Magdon-ismail. Efficient identification of overlapping communities. In *In IEEE International Conference on Intelligence and Security Informatics (ISI)*, pages 27–36, 2005.
2. J. Baumes, M. K. Goldberg, M. S. Krishnamoorthy, M. Magdon-Ismail, and N. Preston. Finding communities by clustering a graph into overlapping subgraphs, 2005.
3. V. D. Blondel, J.-L. Guillaume, R. Lambiotte, and E. Lefebvre. Fast unfolding of communities in large networks. *Journal of Statistical Mechanics: Theory and Experiment*, 2008(10):P10008, 2008.

4. A. Clauset. Finding local community structure in networks. *Physical Review E (Statistical, Nonlinear, and Soft Matter Physics)*, 72(2):026132, 2005.
5. A. Clauset, M. E. J. Newman, and C. Moore. Finding community structure in very large networks. *Physical Review E*, 70(6):066111, 2004.
6. G. B. Davis and K. M. Carley. Clearing the fog: Fuzzy, overlapping groups for social networks. *Social Networks*, 30(3):201–212, 2008.
7. J. Duch and A. Arenas. Community detection in complex networks using extremal optimization. *Physical Review E*, 72:027104, 2005.
8. S. Fortunato. Community detection in graphs. *Physics Reports*, 486(3-5):75–174, 2010.
9. M. Girvan and M. E. J. Newman. Community structure in social and biological networks. *Proc Natl Acad Sci USA*, 99(12):7821–6, 2002.
10. S. Gregory. Finding overlapping communities in networks by label propagation. *New J. Phys.*, 12, 103018, 2010.
11. M. Goldberg, S. Kelley, M. Magdon-Ismail, K. Mertsalov, and W. A. Wallace. Communication dynamics of blog networks. In *The 2nd SNA-KDD Workshop '08 (SNA-KDD'08)*, August 2008.
12. R. Guimerà, M. Sales-Pardo, and L. A. N. Amaral. Modularity from fluctuations in random graphs and complex networks. *Phys. Rev. E*, 70(2):025101, 2004.
13. P. Jaccard. Étude comparative de la distribution florale dans une portion des alpes et des jura. *Bulletin del la Société Vaudoise des Sciences Naturelles*, 37:547–579, 1901.
14. B. Karrer, E. Levina, and M. E. J. Newman. Robustness of community structure in networks. *Physical Review E*, 77(4):046119+, Sep 2007.
15. A. Lancichinetti and S. Fortunato. Benchmarks for testing community detection algorithms on directed and weighted graphs with overlapping communities. *Phys. Rev. E*, 80(1):016118, Jul 2009.
16. A. Lancichinetti, S. Fortunato, and J. Kertesz. Detecting the overlapping and hierarchical community structure of complex networks. *New Journal of Physics*, 11, 2009.
17. M. E. Newman. Modularity and community structure in networks. *Proc. Natl. Acad. Sci. USA*, 103(23):8577–8582, 2006.
18. V. Nicosia, G. Mangioni, V. Carchiolo, and M. Malgeri. Extending the definition of modularity to directed graphs with overlapping communities. *J.STAT.MECH.*, page P03024, 2009.
19. G. Palla, A.-L. Barabasi, and T. Vicsek. Quantifying social group evolution. *Nature*, 446(7136):664–667, 2007.
20. G. Palla, I. Derenyi, I. Farkas, and T. Vicsek. Uncovering the overlapping community structure of complex networks in nature and society. *Nature*, 435:814, 2005.
21. F. Radicchi, C. Castellano, F. Cecconi, V. Loreto, and D. Parisi. Defining and identifying communities in networks. *Proceedings of the National Academy of Sciences of the United States of America*, 101(9):2658–2663, 2004.
22. U. N. Raghavan, R. Albert, and S. Kumara. Near linear time algorithm to detect community structures in large-scale networks. *Phys. Rev. E*, 76(3):036106, Sep 2007.
23. J. Reichardt and S. Bornholdt. Statistical mechanics of community detection. *Phys. Rev. E*, 74(1):016110, Jul 2006.
24. M. Rosvall and C. T. Bergstrom. Maps of random walks on complex networks reveal community structure. *Proceedings of the National Academy of Sciences*, 105(4):1118–1123, 2008.
25. H. Shen, X. Cheng, K. Cai, and M.-B. Hu. Detect overlapping and hierarchical community structure in networks. *Physica A: Statistical Mechanics and its Applications*, 388(8):1706–1712, 2009.
26. W. Zachary. An information flow model for conflict and fission in small groups. *Journal of Anthropological Research*, 33:452–473, 1977.
27. S. Zhang, R.-S. Wang, and X.-S. Zhang. Identification of overlapping community structure in complex networks using fuzzy c-means clustering. *Physica A: Statistical Mechanics and its Applications*, 374(1):483–490, 2007.

Chapter 7
Modularity Maximization and Tree Clustering: Novel Ways to Determine Effective Geographic Borders

D. Grady, R. Brune, C. Thiemann, F. Theis, and D. Brockmann

Abstract Territorial subdivisions and geographic borders are essential for understanding phenomena in sociology, political science, history, and economics. They influence the interregional flow of information and cross-border trade and affect the diffusion of innovation and technology. However, most existing administrative borders were determined by a variety of historic and political circumstances along with some degree of arbitrariness. Societies have changed drastically, and it is doubtful that currently existing borders reflect the most logical divisions. Fortunately, at this point in history we are in a position to actually measure some aspects of the geographic structure of society through human mobility. Large-scale transportation systems such as trains and airlines provide data about the number of people traveling between geographic locations, and many promising human mobility proxies are being discovered, such as cell phones, bank notes, and various online social networks. In this chapter we apply two optimization techniques to a human mobility proxy (bank note circulation) to investigate the effective geographic borders that emerge from a direct analysis of human mobility.

D. Grady
Northwestern University, Evanston, IL, USA
e-mail: d-grady@northwestern.edu

R. Brune • C. Thiemann
Max Planck Institute for Dynamics and Self-Organization, Göttingen, Germany
e-mail: r-brune@northwestern.edu; thiemann@northwestern.edu

F. Theis
Group leader CMB Professor for Mathematics in Systems Biology,
Institute of Bioinformatics and Systems Biology, Helmholtz Zentrum München,
Ingolstädter Landstraße 1, 85764 Neuherberg, Germany
e-mail: fabian.theis@helmholtz-muenchen.de

D. Brockmann (✉)
Northwestern University, Evanston, IL, USA
e-mail: brockmann@northwestern.edu

M.T. Thai and P.M. Pardalos (eds.), *Handbook of Optimization in Complex Networks: Theory and Applications*, Springer Optimization and Its Applications 57,
DOI 10.1007/978-1-4614-0754-6_7, © Springer Science+Business Media, LLC 2012

7.1 Introduction

The geographic compartmentalization of maps into coherent territorial units is not only essential for the management and distribution of administrative responsibilities and the allocation of public resources. Territorial subdivisions also serve as an important frame of reference for understanding a variety of phenomena related to human activity. Existing borders frequently correlate with cultural and linguistic boundaries or topographical features [17, 34], they represent essential factors in trade and technology transfer [19, 29], and they indirectly shape the evolution of human-mediated dynamic processes such as the spread of emergent infectious diseases [11, 12, 21, 32].

The majority of existing administrative and political borders, for example in the United States and Europe, evolved over centuries and typically stabilized many decades ago, during a time when human interactions and mobility were predominantly local and the conceptual separation of spatially extended human populations into a hierarchy of geographically coherent subdivisions was meaningful and plausible.

However, modern human communication and mobility has undergone massive structural changes in the past few decades [30, 34]. Efficient communication networks, large-scale and widespread social networks, and more affordable long-distance travel generated highly complex connectivity patterns among individuals in large-scale human populations [3, 9]. Although geographic proximity still dominates human activities, increasing interactions over long distances [7, 25, 38] and across cultural and political borders amplify the small-world effect [31, 41] and decrease the relative importance of local interactions.

Human mobility networks epitomize the complexity of multi-scale connectivity in human populations (see Fig. 7.1). More than 17 million passengers travel each week across long distances on the United States air transportation network alone. However, including all means of transportation, 80% of all traffic occurs across distances less than 50 km [7, 8]. The coexistence of dominant short-range and significant long-range interactions handicaps efforts to define and assess the location and structure of effective borders that are implicitly encoded in human activities across distance. The paradigm of spatially coherent communities may no longer be plausible, and it is unclear what structures emerge from the interplay of interactions and activities across spatial scales [7, 8, 25, 40]. This difficulty is schematically illustrated in Fig. 7.2. Depending on the ratio of local versus long-range traffic, one of two structurally different divisions of subpopulations is plausible. If short-range traffic outweighs long-range traffic, local, spatially coherent subdivisions are meaningful. Conversely, if long-range traffic dominates, subdividing into a single, spatially de-coherent urban community and disconnected suburban modules is appropriate and effective geographic borders are difficult to define in this case.

Although previous studies identified community structures in long-range mobility networks based on topological connectivity [28, 36], this example illustrates that the traffic intensity resulting from the interplay of mobility on all spatial scales must

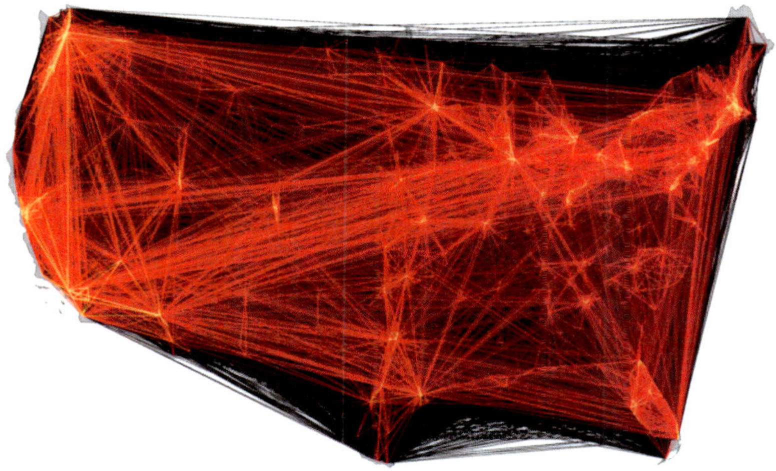

Fig. 7.1 The Where's George? network. Multi-scale human mobility is characterized by dominant short-range and significant long-range connectivity patterns. The illustrated network represents a proxy for human mobility, the flux of bank notes between 3,109 counties in the lower 48 United States. Each link is represented by a line, the color scale encodes the strength of a connection from small (*dark red*) to large (*bright yellow*) values of w_{ij} spanning four orders of magnitude

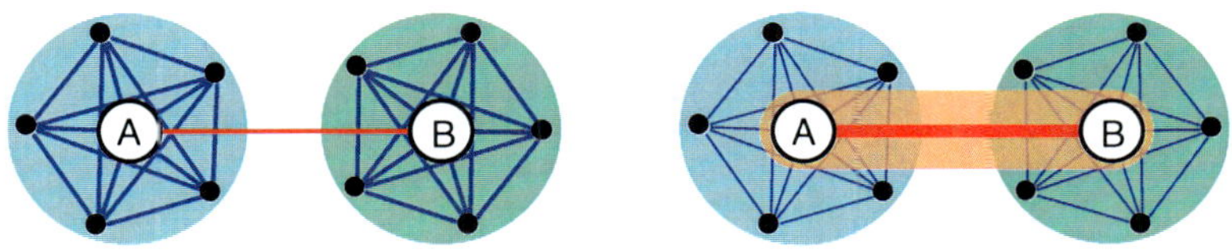

Fig. 7.2 A simplified illustration of generic traffic patterns between and within metropolitan mobility hubs (A and B), with two types of connections w_L and w_D, local traffic connecting individual hubs to smaller nodes in their local environment (*blue*) and long-distance links connecting the hubs (*red*). Depending on the ratio of local and long-range flux magnitude, two qualitatively different modularizations are plausible. If $w_L \gg w_D$, two spatially compact communities are meaningful (*left*), whereas if $w_L \ll w_D$, the metropolitan centers belong to one geographically delocalized module (*orange*), effectively detached from their local environment, yielding three communities altogether (*right*)

be taken into account. Obtaining comprehensive, complete, and precise datasets on human mobility covering many spatial scales is a difficult task, and recent studies have followed a promising alternative strategy based on the analysis of proxies that permit indirect measurement of human mobility patterns [7, 15, 16, 25, 30, 38].

We focus on one human mobility proxy, a dataset collected at the website www.wheresgeorge.com. This website hosts a bill-tracking game called Where's George? in which participants can tag an individual US banknote of any denomination by logging in to the website and entering the bill's serial number along with their location. Subsequent participants who receive the bill may do the same, thereby recording a part of the spatial trajectory the bill follows during its lifetime. We use this information to construct a network whose nodes represent counties in the continental United States and whose edges encode the number of bills exchanged between pairs of counties; details of this and a discussion of some statistics of the data are given in Sect. 7.3.

Both of our analyses rest on the idea of finding community partitions of the network, that is, dividing all of the nodes into a set of mutually disjoint groups or communities. A community of nodes can be defined in many different ways, but all definitions try to capture some aspect of the intuitive idea of a community: a set of nodes that belong together, or are more similar to one another than they are to the rest of the population.

Our first analysis in Sect. 7.2 uses a modularity maximization technique to identify community partitions. Modularity is a method of scoring any given community partition in a network. A partition with a high modularity score has many more intra-group links, and fewer inter-group links, than expected by random chance. Our optimization algorithm searches for high-modularity partitions through a stochastic, simulated annealing process.

We go on to determine community partitions in Sect. 7.6 by searching for nodes with similar topological features, namely their shortest-path tree. Each node is the root of a shortest-path tree that comprises a minimal set of the strongest links connecting that node to the rest of the network. By looking for topological similarities between shortest-path trees, we identify groups of nodes that have similar patterns of connectivity.

With both methods, once a community partition is identified, a corresponding geographic border structure is produced simply by drawing borders between counties that do not belong to the same community, and in Sect. 7.5 we discuss how a superposition of border structures alleviates some of the long-standing weaknesses of modularity maximization. The fact that communities tend to be spatially compact is one of the most surprising findings of this research, and we conclude in Sect. 7.7 by developing a method for comparing border structures and examining the degree to which effective mobility borders line up with various existing borders, such as state boundary lines, census areas, and economic areas.

7.2 Network Modularity

This section introduces the modularity measure and describes the simulated annealing algorithm we use for finding maximal-modularity community partitions.

We assume here that W is a square, symmetric matrix that represents a symmetric, weighted network; the elements w_{ij} are nonnegative and measure the strength of the connection between nodes i and j. Based on the idea that two nodes i and j are effectively proximal if w_{ij} is large, we search for a community partition of the nodes that has a high value of modularity [13, 24, 35]. This standard network-theoretic measure of community structure prefers partitions such that the intra-connectivity of the modules in the partition is high and inter-connectivity between them is low as compared to a random null model. Given a partition P of the nodes into k modules M_n, the modularity $Q(P)$ is defined as

$$Q = \sum_n \Delta F_n \tag{7.1}$$

in which $\Delta F_n = F_n - F_n^0$ is the difference between F_n, the fraction of total mobility within the module M_n, and the expected fraction F_n^0 of a random network with an identical weight distribution $p(w)$. Q cannot exceed unity; high values indicate that a partition successfully groups nodes into modules, whereas random partitions yield $Q \approx 0$. Maximizing Q in large networks is an NP-hard problem [6], but a variety of algorithms have been developed to systematically explore and sample the space of possible divisions in order to identify high-modularity partitions [13, 22].

7.2.1 *Finding Optimal Partitions*

As discussed in more detail in Sect. 7.4, our method relies on finding several different high-modularity partitions, which restricts the range of applicable algorithms. For example, the deterministic divisive algorithms described by Newman and Girvan [35] cannot find several different local maxima of the modularity function. In contrast, Monte Carlo algorithms return different partitions with probabilities that monotonically increase with the corresponding modularity values, one of which is the simulated annealing algorithm described by Guimerà and Amaral [27]. Additionally, this algorithm has been found to perform the best in terms of correctly identifying modules in networks with artificial community structure in a survey by Danon et al. [13], which lead us to choose this algorithm for our work.

The partition vector P is initialized such that each of the N nodes is in its own module, $P_i = i$. Alternatively, one could randomly assign each node to one of a few modules to form the initial partition. We found, however, that in this case the algorithm will split these few large modules into a large number of very small modules before slowly merging them into the final result. Since splits of large modules, involving a recursive simulated annealing run, are computationally very expensive, we avoid them by starting with a partition of single-node modules.

A small modification of the partition is then made (see below) to obtain a new partition P' and its effect on the modularity value, $\Delta Q = Q(P') - Q(P)$. If $\Delta Q > 0$, the new partition is better than the old one and we replace $P = P'$. If $\Delta Q < 0$, the partition is only accepted with probability $p_T(\Delta Q) = \exp(\Delta Q/T)$, where T is a "temperature" that controls the typical penalty on Q we are willing to accept with the new partition P'.

This procedure is repeated a number of times, initially with a high $T = T_0$ accepting modifications with large negative impact on modularity and therefore allowing to sample multiple local maxima. After $O(N^2)$ modifications, the temperature is lowered by a *cooling factor* c. When T is small enough, worse partitions are not accepted anymore and the partition P has "annealed" into a local maxima of the modularity landscape $Q(\cdot)$.

During each temperature step, we intersperse fN^2 local with fN global modifications of the partitions, where f is a tuning parameter. A local modification is a switch of one node to another, randomly selected, module, while a global modification can be a merge of two or a split of one randomly selected module. Finding a suitable split of a module that is not immediately rejected is done by recursively running a simplified version of the simulated annealing algorithm on it: the module in question is extracted and treated as an independent network, initially randomly partitioned into two modules. Only local modifications are allowed while annealing this bipartition into a local modularity maximum. Afterwards, the split module is replaced into the full network and evaluated against the modularity value of the full partition.

We observed that the global structure of the partition is found quickly by the algorithm and mostly only local modifications are accepted at low temperatures. Since the split operations are computationally intensive, we therefore track the number of rejected split modifications in each temperature step and reduce the probability of future trials if that number is high.

To generate the large ensemble of partitions discussed in Sect. 7.5, we used $T_0 = 2.5 \cdot 10^{-4}$ as initial temperature, $c = 0.75$ as the cooling factor, and $f = 0.05$. We abort the procedure and accept the partition as "optimal" if no better partition is found in three consecutive temperature steps.

The run time of this stochastic algorithm depends in a complex way on both the size *and* the structure of the the input, and therefore the time complexity does not scale with a simple function of the input size. However, we found the algorithm to perform very well and in acceptable runtime (60–90 minutes on a 2.8 GHz processor for most runs) with the configuration given above, although these parameters are less conservative than those proposed by Guimerà et al. [27]. The large ensemble of resulting partitions (Fig. 7.15) has a tight distribution of modularity values, indicating the algorithm tends to converge onto a stable maximum.

7.3 A Proxy for Multiscale Human Mobility Networks

Here, we construct a proxy network for human mobility from the geographic circulation of banknotes in the United States. Movement data was collected using the online bill-tracking game at www.wheresgeorge.com. Individuals participating in this game can mark individual bills and return them to circulation; other individuals who randomly receive bills can report this find online along with their current location (zip code). Our analysis is based on the intuitive notion that the coupling strength between two locations i and j increases with w_{ij}^H, the number of individuals that travel between a pair of locations per unit time, and furthermore that the flux of individuals in turn is proportional to the flux of bank notes,

Table 7.1 Denominations in the WG dataset

Denomination	$1	$2	$5	$10	$20	$50	$100
Number of bills	9,931,261	36,639	1,069,427	401,101	461,076	24,209	26,526
Fraction [%]	83.11	0.31	8.95	3.36	3.86	0.20	0.22

Table 7.2 Absolute number and relative fraction of bills based on Federal Reserve Bank

FRB Code	Location	Count	Fraction [%]
A	Boston	799,537	6.69
B	New York City	1,325,942	11.10
C	Philadelphia	822,340	6.88
D	Cleveland	661,278	5.53
E	Richmond	948,516	7.94
F	Atlanta	1,565,732	13.10
G	Chicago	1,207,448	10.10
H	St. Louis	472,930	3.96
I	Minneapolis	360,194	3.01
J	Kansas City	713,393	5.97
K	Dallas	869,866	7.28
L	San Francisco	2,203,063	18.44

denoted by w_{ij}. Evidence for the validity of this assumption has been obtained previously [7, 8, 25] and we provide further evidence below.

As of January 15th, 2010 a total of 187,925,059 individual bills are being tracked at the website www.wheresgeorge.com. Approximately 11.24% of those have had "hits", that is they were reported a second time at the site after initial entry. The current analysis is based on a set of $N_0 = 11,950,239$ bills that were reported at least a second time. For each bill n we have a sequence of pairs of data

$$B_n = \{Z_{n,i}, T_{n,i}\} \quad i = 0, \ldots, L_n \quad n = 1, \ldots, N_0$$

of zip codes $Z_{n,i}$ and times $T_{n,i}$ at which the bill was reported. Each B_n reflects a geographic trajectory of a bill with L_n individual legs. In total, we have 14,612,391 single legs in our database. Note that the majority (81.78%) of trajectories are single-legged reflecting a reporting probability of $\approx$20% during the lifetime of a bill.

The set of B_n represents the core dataset of our analysis. For each bill we have additional information:

1. Denomination: $1, $2, $5, $10, $20, $50, or $100. The fraction of each denomination is depicted in Table 7.1.
2. The Federal Reserve Bank code, A through L, corresponding to one of 12 of the United States Federal Reserve Banks that issued the bill. The fraction of bills as a function of FRB origin is provided in Table 7.2.

We restrict the analysis to the lower 48 states and the District of Columbia (thus excluding Hawaii and Alaska) and consider only legs with origin and destination locations in these states, reducing the original dataset to 11,759,420 bills (98.40% of the original data) and 14,376,232 trajectory legs (98.38%).

The spatial resolution of the dataset is given by 41,106 zip codes, with mean linear extent of 14 km. The mean linear extent of the lower 48 states is 2,842 km defining the bounds of the system. For each zip code Z_i we use centroid information to associate with each report a longitude/latitude location $\mathbf{x} = (\Theta, \phi)$, such that each trajectory n corresponds to a sequence of geographic locations $\mathbf{X}_i$ with $i = 1, \ldots, L_n$:

$$t_n : \quad \{\mathbf{X}_{n,0}, \Delta T_{n,1}, \mathbf{X}_{n,1}, \ldots, \Delta T_{n,L_n}, \mathbf{X}_{n,L_n}\} \quad \text{with} \quad n = 1, \ldots, N_0, \tag{7.2}$$

where $\mathbf{X}_{n,0}$ is the initial entry location, and $\Delta T_{n,i} = T_{n,i} - T_{n,i-1}$ are inter-report times.

7.3.1 Geographical Distributions

Based on these trajectories we define the density of initial entries as

$$p_{\mathrm{IE}}(\mathbf{x}) = \frac{1}{N} \sum_{n=1}^{N} \delta(\mathbf{x} - \mathbf{X}_{n,0}), \tag{7.3}$$

and the density of reports as

$$p_R(\mathbf{x}) = \frac{1}{N} \sum_{n=1}^{N} \frac{1}{L_n} \sum_{i=1}^{L_n} \delta(\mathbf{x} - \mathbf{X}_{n,i}), \tag{7.4}$$

where δ is the Dirac delta function, equal to 1 when its argument is 0 and equal to 0 otherwise.

In order to assess the spatial distribution of reports and initial entries and to quantify the correlation with the population density we compute the number of reports and initial entries for each of the $M = 3{,}109$ counties in the lower 48 states. Defining for each county k a characteristic function

$$\chi_k(\mathbf{x}) = \begin{cases} 1 & \text{if} \quad \mathbf{x} \in P_k \\ 0 & \text{otherwise} \end{cases} \tag{7.5}$$

where P_k is the polygon defining the county's interior, the number of reports and initial entries in county k are given by

$$m_{\mathrm{R}}(k) = \langle \chi_k \rangle_R = \int \chi_k(\mathbf{x})\, p_R(\mathbf{x})\, \mathrm{d}\mathbf{x} \quad \text{and} \quad m_{\mathrm{IE}}(k) = \langle \chi_k \rangle_{\mathrm{IE}} = \int \chi_k(\mathbf{x})\, p_{\mathrm{IE}}(\mathbf{x})\, \mathrm{d}\mathbf{x},$$

respectively. Figure 7.3 compares the distribution of reports $m_{\mathrm{R}}(k)$, initial entries $m_{\mathrm{IE}}(k)$ and the population $P(k)$ of the 3,109 counties. As all three quantities are

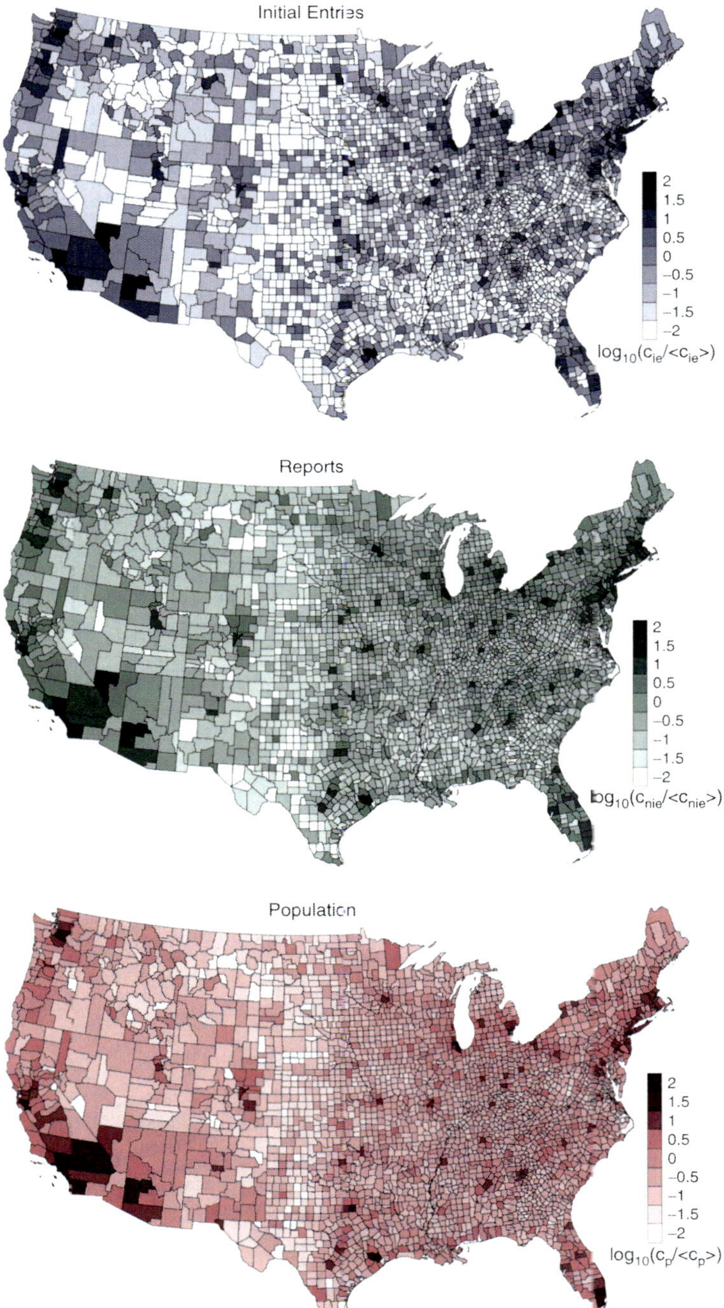

Fig. 7.3 The frequencies of reports (*top*) and initial entries (*middle*) correlate with the county population (*bottom*) in the lower 48 states

positive and vary over many orders of magnitude, the maps depict $\log_{10}(m_{\mathrm{R}})$, $\log_{10}(m_{\mathrm{IE}})$ and $\log_{10}(P)$. Qualitatively, reports and initial entries correlate strongly with the population density. Computing the correlation coefficient of the logarithmic quantities yields $c(R,P) = 0.933$ and $c(IE,P) = 0.819$. Despite the expected increase of $m_{\mathrm{R}}(k)$ and $m_{\mathrm{IE}}(k)$ with $P(k)$, only the report count increases approximately linearly with population size, whereas initial entries show a deviation for small populations. We believe that this deviation is a consequence of the social difference between the subpopulation of "Georgers" that are responsible for initiating bills and entering them into the system, "actively" playing the game, and the larger group of people that randomly receive a bill and report it, "passively" participating. This hypothesis could explain that areas with higher population densities contain a larger proportion of internet-savvy communities that are inclined to become Georgers and initiate bills. In order to exclude a potential bias caused by this effect we exclude all the legs in (7.6) that contain an initial entry as the origin, i.e. we only investigate the reduced set

$$t_{2,n}: \quad \{\mathbf{X}_{n,1}, \Delta T_{n,2}, \mathbf{X}_{n,2}, \ldots, \Delta T_{n,L_n}, \mathbf{X}_{n,L_n}\} \quad \text{with} \quad n = 1, \ldots, N_0, \tag{7.6}$$

that excludes the first legs of all t_n. Excluding the first leg reduces the number of bills to 4,743,330. However, the key results, for example the border structures discussed in Sect. 7.5, are robust against the inclusion of initial entries. Computing mobility networks based on either set, t_n or $t_{2,n}$ does not change the observed pattern significantly.

7.3.2 *Distance and Time: Spatially Averaged Quantities*

From $t_{2,n}$ we extract pairs of spatio-temporal leg distances $\{d_{\mathrm{s}}(\mathbf{X}_{n,i}, \mathbf{X}_{n,i-1}), \Delta T_{n,i}\}$, where $d_{\mathrm{s}}(\cdot,\cdot)$ denotes the distance on a sphere (shorter segment of the great circle that passes through both points). This type of dataset was first investigated in 2006 based on a much smaller core dataset of bill trajectories [7]. In particular, the combined probability density (pdf)

$$p(r,t) = \langle \delta\,(r - d_{\mathrm{s}}(\mathbf{X}_{n,i}, \mathbf{X}_{n,i-1}))\,\delta\,(t - \Delta T_{n,i}) \rangle\,, \tag{7.7}$$

was estimated as well as marginal pdfs $p(r)$ and $p(t)$. The central finding of the 2006 study was that $p(r) \sim r^{-(1+\beta)}$ and that the time evolution of the density (7.7) can be described by a bi-fractional diffusion equation. Here, we reproduce some of the properties before we construct the mobility network used in the main text. Figure 7.4 shows the short time pdf of a bill traversing a distance r in a time $t < \tau$ where we chose $\tau = 4$ days. Using maximum likelihood we find this function can be described by a power-law

$$p(r) \sim \frac{1}{r^{1+\beta}} \quad \text{with} \quad \beta = 0.7056 \pm 0.0659.$$

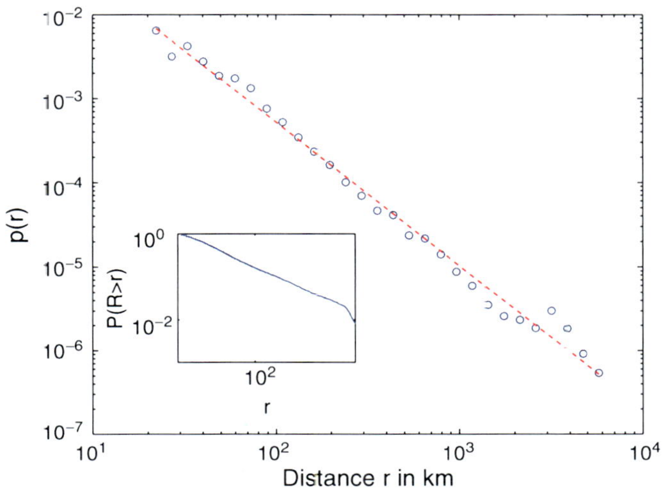

Fig. 7.4 The estimated probability $p(r|t < \tau)$ of a bill traversing a distance r in time $t < \tau$ where $\tau = 4$ days. In red a maximum likelihood fit of the the function $p(r) \sim r^{-(\ +\beta)}$ with $\beta = 0.7056$

This power law describes the dispersal characteristics on a population-averaged level. The short-time distance pdf represents a dispersal kernel and for small times t approximates the instantaneous rate of traversing a distance r.

Complementary to this, temporal aspects of the process can be revealed by computing the pdf for the time t between reporting events given that these occur within a small radius $r > r_0$. Figure 7.5 depicts $p(t)$ for all legs with $r < 10\,\text{km}$ and a minimal inter-report time of $t_{\text{min}} = 1$ day. The inter-report times are described well by a power law moderated by an exponential factor

$$p(t) \sim t^{-\alpha} \mathrm{e}^{-t/T_0} \quad \text{with} \quad T_0 = 248 \pm 27, \quad \alpha = 0.99 \pm 0.05. \tag{7.8}$$

The observed power-law decay $\sim t^{-1}$ for times $t \ll T_0$ is intriguing. These type of decays have been observed in a multitude of contexts involving human activity, for instance the time between consecutive phone calls [25], emails, [5] and the number of words between two identical words in texts [1]. A consequence of this law is bursting behavior, i.e. given an event occurred at time t_0 the probability rate that an event occurs immediately after the first is higher than expected from ordinary Poisson statistics. This behavior is best illustrated by the so-called hazard function $h(t)$ that quantifies the instantaneous probability rate of an event happening at t given that the last event occurred at $t = 0$. If we let

$$P(\tau > t) = \int_t^\infty p(s)\,\mathrm{d}s,$$

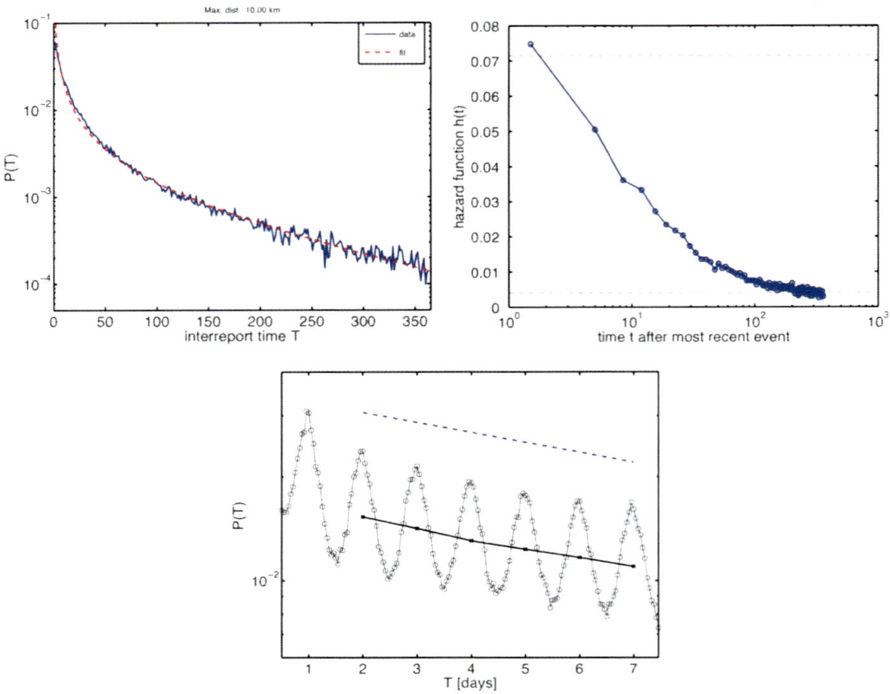

Fig. 7.5 Inter-report time statistics. (*Left*) The function $p(t|r < r_0)$ for $r_0 = 10\,\text{km}$. The observed function can be accounted for by an initial algebraic decay t^{-1} moderated by an exponential function for large arguments. The red dashed curved is a fit obtained from maximum-likelihood estimation. (*Right*) The hazard function $h(t)$ that represents the instantaneous rate of an event at time t provided that an event occurred at $t = 0$. The dashed lines represent reporting rates of once per 2 weeks (*top*), once per month (*middle*) and once per $T_0 = 248$ days (*bottom*). (*Bottom*) $p(t)$ for very short times. A zoom-in resolves daily oscillations modulated by the decay observed on the left. These oscillations indicate that users tend to report to the website at the same time of the day with the highest probability

be the cumulative probability that the second event occurs at a time τ later than t, the hazard function is defined by

$$P(\tau > t) = \mathrm{e}^{-\int_0^t h(s)\,\mathrm{d}s}.$$

For a Poisson process with rate γ we have

$$h(t) = \gamma \quad \Rightarrow \quad P(\tau > t) = \mathrm{e}^{-\gamma t}.$$

The hazard function can be computed according to

$$h(t) = -\frac{\mathrm{d}}{\mathrm{d}t}\log\left[P(\tau > t)\right] = \frac{p(t)}{P(\tau > t)}.$$

Figure 7.5 depicts the function $h(t)$ for inter-report times in the WG data. For small times ($t < 1$ week) the probability rate for a report is of the order of one report per two weeks, which is also the expected time between two reports in this time window. For larger times ($t > 100$ days) the constant value of $1/T_0$ is approached, equivalent to one report in 3/4 of a year. Possible explanations of the bursting behavior and the initial algebraic decay in $p(t)$ are a strong behavioral heterogeneity of players that participate in the game or an effective queueing in the system, i.e. bills may enter shops and initially have a comparatively high likelihood of leaving, being "on top of the stack." As time passes these bills may "get stuck" and equilibrate to the long time scale present in the system.

7.3.3 *Definition of the Mobility Network*

From the trajectories defined by (7.6) and the characteristic functions of the counties (7.5) we construct a matrix $\tilde{w}_{ij}$ that counts the number of legs which originate at county i and terminate at j,

$$\tilde{w}_{ij} = \sum_{n=1}^{N} \sum_{k=2}^{L_n} \chi_i \left(\mathbf{X}_{n,k-1}\right) \chi_j \left(\mathbf{X}_{n,k}\right) \Theta(T - \Delta T_{n,k}),$$

where $\Theta(\cdot)$ is the Heaviside step-function. In order to exclude potential biases induced by initial entries we ignore the first leg of all trajectories ($k = 2$ in the above sum). This choice is motivated by the fact that the community of individuals that initiate bills might be less representative than those that find bills and report them. Indications that this might have an effect are supported by the different scaling behavior of initial entry frequencies with population as compared to report frequencies with population. The factor $\Theta(T - \Delta T_{n,k})$ excludes legs that have an inter-event time larger than time T. The matrix $\tilde{w}_{ij}$ need not to be symmetric, as the flux of bills from $i \to j$ need not equal those that travel $j \to i$. However, as Fig. 7.6 indicates the flux matrix is statistically symmetric. Plotting $\tilde{w}_{ij}$ against $\tilde{w}_{ji}$ indicates a clear mean linear relationship. Since we base our analysis on the flux of money between two given counties we symmetrize the network and use w_{ij} in our analysis defined by

$$w_{ij} = \frac{1}{2} \left(\tilde{w}_{ij} + \tilde{w}_{ji}\right),$$

which of course also depends on the time threshold parameter T. Choosing the optimal value for T is a trade-off between trying to estimate instantaneous flux, i.e. choosing T as small as possible, and using as many legs as possible to decrease fluctuations, i.e. choosing large values for T. Choosing a value for $T < 30$ days for instance rules out bills that visit a Federal Reserve Bank in between reports in counties i and j, as bills that enter FRBs do not return to circulation until approximately 3–4 weeks after entering the FRB. To make sure that our results do

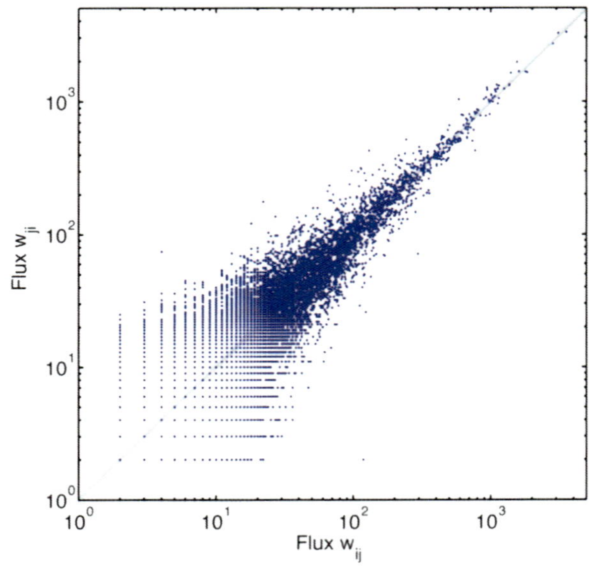

Fig. 7.6 Symmetry of flux network $\tilde{w}_{ij}$

not significantly change as the parameter T is varied we performed the analysis for various values of T ranging from a few days to $T = 1$ year. The computed border structure does not significantly depend on the value of T. Decreasing T thins out the network and reduces the overall connectivity, yet the effects are similar to bootstrapping the network randomly, a process that also does not change our results and is discussed in Sect. 7.7.1.

7.3.4 Gravity as a Null Model

In addition to the empirical data described above, we also construct a synthetic mobility network based on the well-known gravity law hypothesis [2,10,42] to serve as a null model. In gravity models, the interaction strength between a collection of sub-populations with geographic positions x_i, sizes N_i (obtained from census data,[1]) and distances $d_{ij} = |x_i - x_j|$ is given by

$$p_{ij} \propto \frac{N_i^{\alpha} N_j^{\beta}}{{d_{ij}}^{1+\mu}} \tag{7.9}$$

in which α, β, and μ are non-negative parameters.

[1] http://www.census.gov.

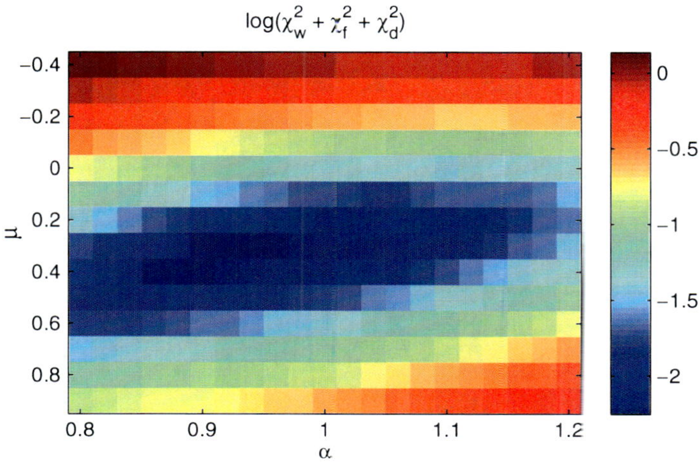

Fig. 7.7 χ^2 goodness-of-fit for different parameters of the gravity law. The minimum is at $(\alpha,\mu) = (0.96, 0.3)$

To create a model network comparable to our data, we first compute p_{ij} for all counties i and j in the continental U.S. and normalize them such that $\sum_{i,j} p_{ij} = 1$. We then interpret these values as probabilities for a travel event to happen between the two counties (or, speaking in terms of the original data source, a dollar bill report). Thus, starting with all-zero link weights w_{ij}, we repeatedly draw a pair of nodes according to p_{ij} and increase the corresponding w_{ij} by one, until approximately the same connectivity (number of non-zero w_{ij}) as in the real-data network is reached.

We generated gravity networks for different parameter values and gauged them against our real data by comparing the distributions of first-order network statistics to find the best fit to our data. Distributions have been compared by log-binning the values and computing the χ^2 statistic

$$\chi^2 = \sum_{i}^{n} \frac{(N_i^G - N_i^R)^2}{N_i^R}$$

where n is the number of bins and N_i^G (N_i^R) is the number of values from the gravity (real-data) network in bin i.

Our real data is symmetric and node fluxes are proportional to population sizes, therefore we assume $\alpha = \beta \approx 1$ to narrow down the search volume in parameter space. We computed χ^2 for the distribution of link weights, node fluxes and geographical distances and used the sum of them, $\chi_w^2 + \chi_f^2 + \chi_d^2$, as the goodness-of-it measure. Figure 7.7 shows this quantity for $(\alpha,\mu) \in [0.8, 1.2] \times [-0.4, 0.9]$, from which we concluded that $\alpha = \beta = 0.96$ and $\mu = 0.3$ are the best parameter choices. The resulting network and first-order statistics are shown in Fig. 7.8.

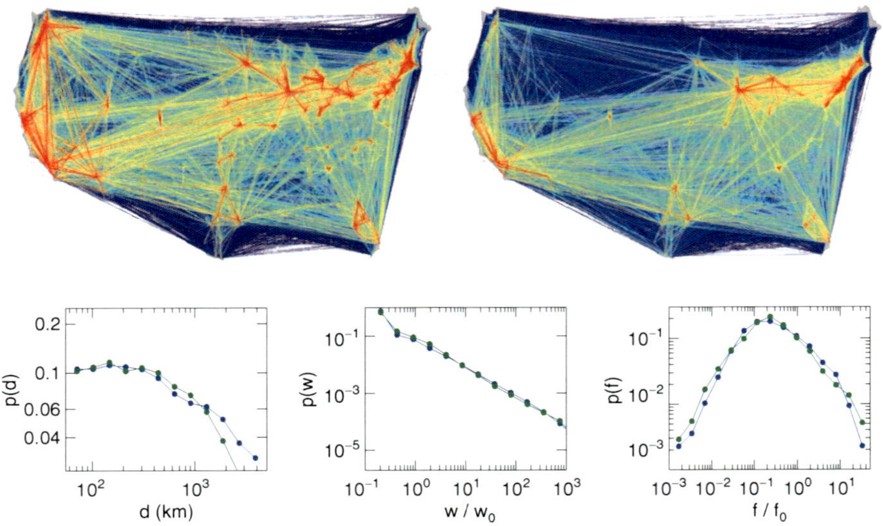

Fig. 7.8 Comparison of the real-data network (*top left*) and the gravity model network with $\alpha = \beta = 0.96$ and $\mu = 0.3$ (*top right*). The bottom plot shows the distributions of geographical distances d, link weights w, and node fluxes f in the real-data network (*blue lines*) and the gravity model (*green lines*)

Similar to the bootstrapping procedure described in Sect. 7.7.1, we tested the robustness of the community structure of the model network by generating snapshots of the network at different connectivities and computing an ensemble of 80 high-modularity partitions for each snapshot. We found that the modularity statistics are stable around the target connectivity of 0.0765 (Fig. 7.9).

7.4 Degeneracy and Superposition

Given a mobility network constructed from the Where's George data, we then apply the optimization algorithm described in Sect. 7.2 to generate community partitions. Since the optimization process is stochastic, the resulting partition varies between realizations of the process. Two representative examples of high-modularity partitions are displayed in Fig. 7.10. Note that, although modularity only takes into account the structure of the weight matrix W and is explicitly blind to the geographic locations of nodes, the effective large-scale modules are spatially compact in every map. Consequently, although long-distance mobility plays an important role, the massive traffic along short distances generates spatial coherence of community patches of mean linear extension $l = 633 \pm 250\,\text{km}$. Note however that although each maps exhibits qualitative similarities between detected large-scale subdivisions and although each of the maps possess a high modularity score,

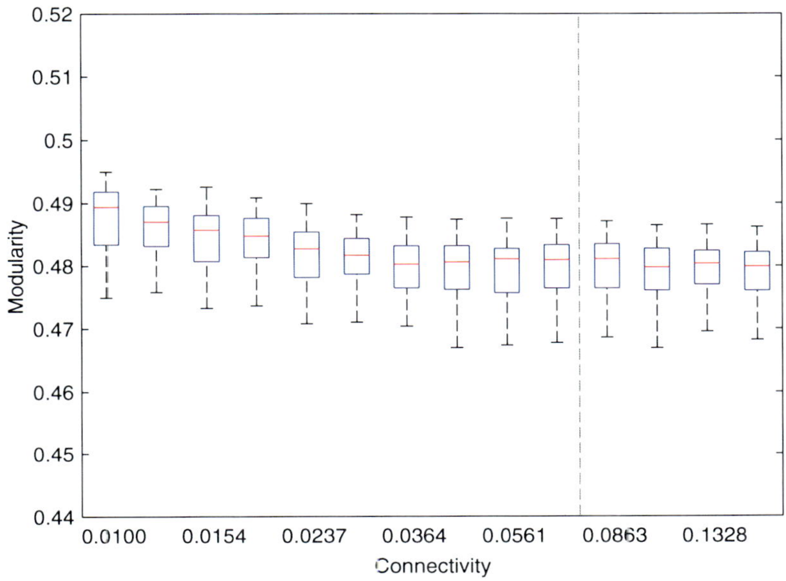

Fig. 7.9 Distributions of modularity values for an ensemble of 80 partitions each computed for snapshots of the model network at different connectivities. The dashed line corresponds to 0.0765, the connectivity of the real-data mobility network

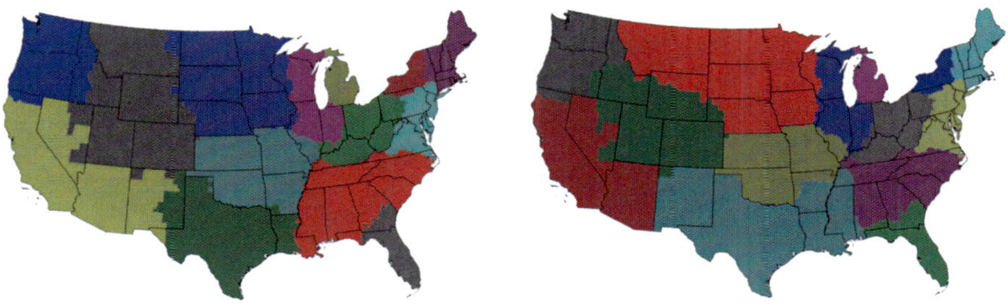

Fig. 7.10 High-modularity community partitions of the WG mobility network. The stochastic algorithm produces different partitions when run many times; these are two representative examples. Modularity values are 0.6808 (*left*) and 0.6807 (*right*)

obvious structural differences exist; in fact, even if it were possible to determine a partition with maximal modularity, any such partition is not in principle unique. It is thus questionable whether any single effective map can be considered the most plausible partition.

Theoretical concerns aside, recent work [23] has identified practical issues with modularity maximization, in particular the so-called *resolution limit*. We demonstrate that a superposition of community partitions can alleviate these issues

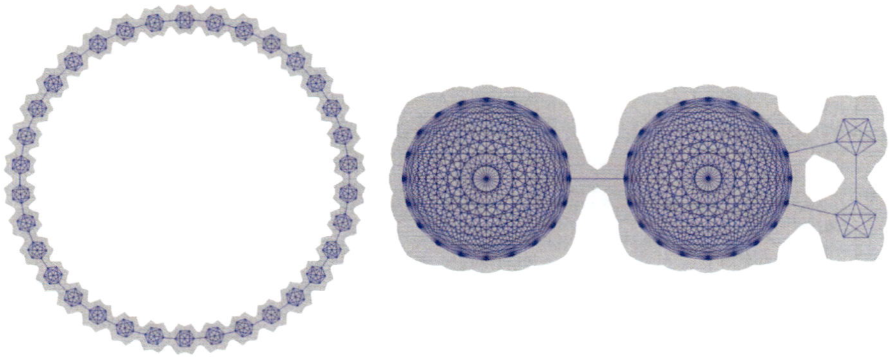

Fig. 7.11 Two networks that expose the resolution limit problem with modularity. The shaded areas indicate an artificial geography for nicer visualization of the boundaries in the next figures. (*Left*) A ring of 34 cliques, each of 6 nodes and connected to their neighbors by single links. (*Right*) A network of two 20-node cliques and two five-node cliques

with the modularity score, and to this end discuss its known shortcomings in more detail.

In fact, it is straightforward to construct networks of which several distinct partitions with equal and maximum modularity value exist. This *degeneracy of modularity* was independently found by Good et al. [26] and marked as a drawback of the modularity measure.

Fortunato and Barthélemy [23] also report on the resolution limit of modularity. The authors present two artificial, unweighted networks that exhibit an intuitively very clear community structure, yet partitions exist that do not reflect this structure but have a higher modularity value than the partition that does. In particular, these networks are constructed by connecting multiple fully connected graphs ("cliques") with single links (Fig. 7.11). It is clear that every clique should be grouped into one module, but the best partition according to modularity will group multiple cliques together. This only occurs if the cliques are small (in terms of number of links) compared to the full network, thus the modularity measure cannot detect communities below a certain resolution limit.

Our proposed method combines an ensemble of partitions by focusing on the boundaries of a partition ("Which adjacent nodes are separated into different modules?") rather than its volumes ("Which nodes are grouped together?)," and then computing for each boundary the fraction of partitions in which it exists. Because we are interested in geographically embedded networks and modules are virtually always spatially compact in our case, we can restrict ourselves to boundaries that are also real geographical borders between nodes. However, the idea can be easily generalized to non-geographical networks, at the expense of convenient straightforward visualization. Since all partitions in the ensemble have a high modularity value, this method highlights similarities and differences in degenerated partitions,

Fig. 7.12 (*Left*) The optimal partition in the clique ring groups pairs of cliques together (the same color is used for multiple modules). (*Center*) Example of a partition found by the modularity optimization algorithm. (*Right*) Superposition reveals boundaries in the clique ring between every clique. Color codes the fraction of partitions in which the boundary was found. We use $T = 2.5 \cdot 10^{-4}$, $c = 0.75$, and $f = 0.5$ for this example and the next

yielding a unique "partition" (or to be more precise, a map) of the network and thus overcoming the degeneracy problem.

In our method, any single partition obviously suffers from this limitation as well. However, the resolution limit can be alleviated by looking at an ensemble, if enough small modules exist to create degeneracies. To illustrate this, we applied our method to the two example networks from Fortunato and Barthélemy [23]. Figure 7.11 (left) shows a ring of 34 6-cliques, all connected to their neighbors by a single link. The intuitive partition in which each clique is in its own module has modularity $Q_{\text{real}} = 0.9081$ while a partition that groups pairs of cliques together has $Q_{\text{opt}} = 0.9099$. However, two distinct partitions exist that group pairs of cliques. Thus, an ensemble of optimal partitions will be composed out of those two partitions, yielding a boundary map in which every boundary between two cliques appears in 50% of the ensemble partitions. For nicer visualization, we created an artificial geography for this network and computed partitions and boundaries shown in Fig. 7.12. Due to the nature of our algorithm, the resulting partitions contain a few n-tuples of cliques and single-clique modules that have not been split or merged into perfect clique-pairs before the termination criterion, and thus the observed boundaries are stronger than expected.

The second network proposed in Fortunato and Barthélemy is constructed from two 20-cliques and two 5-cliques (Fig. 7.11 (right)). Here, the two smaller cliques are merged into one module by the optimal partition ($Q_{\text{opt}} = 0.5426$), although one would again expect each of them to be in its own module ($Q_{\text{real}} = 0.5416$). Our method is not able to capture the intuitive community structure in this case (Fig. 7.13), because no degeneracy exists (the partitions in which only one of the small cliques are grouped with the large one, but not the other, are too far from the optimum to be produced by the algorithm, $Q_{\text{deg}} = 0.4959$).

But if we extend the network such that four small cliques exist, the partition which groups all cliques into their own modules is still suboptimal to any partition that groups together more than one of the small cliques, but degeneracies are created

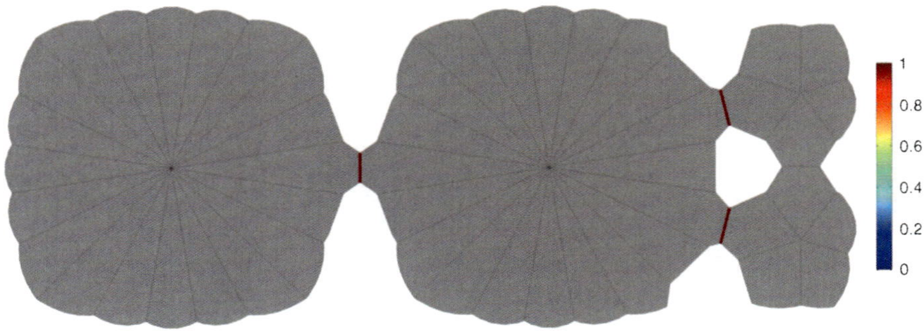

Fig. 7.13 Boundaries found in the clique network shown in Fig. 7.11 (*right*). Our algorithm is not able to find a boundary between the two small cliques

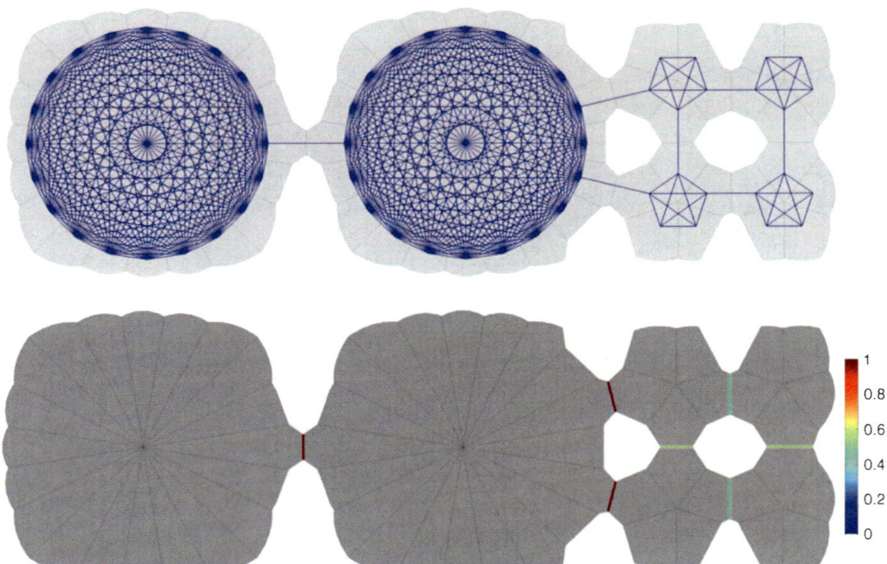

Fig. 7.14 Modification of the clique network in Fig. 7.11 (*right*). Because there are multiple high-modularity partitions that group the smaller cliques into pairs, our method can detect the correct community structure in this case

and the ensemble of partitions reveals the true community structure in this network (Fig. 7.14).

In conclusion, our method is able to dissolve both the degeneracy and resolution limit problems if enough small modules exist to create degeneracies. In fact, we will observe small "building blocks" in the WG data that are not seen in single partitions but emerge from the superposition of a partition ensemble.

7.5 Assessment of Border Structures

Using the algorithm described in Sect. 7.2, we compute an ensemble of 1,000 partitions of the WG mobility network, all exhibiting a high modularity ($Q = 0.6744 \pm 0.0026$, see also Fig. 7.15 for the distribution of modularity values) and spatially compact modules, and perform a linear superposition of the set of maps. This method extracts features that are structural properties of the entire ensemble. The most prominent emergent feature is a complex network of spatially continuous geographic borders (Fig. 7.16). These borders are statistically significant topological features of the underlying multi-scale mobility network. An important aspect of this method is the ability to not only identify the location of these borders but also to quantify the frequency with which individual borders appear in the set of partitions, a measure for the strength of a border.

Investigating this system of effective mobility borders more closely, we see that although they correlate significantly with territorial state borders ($p < 0.001$, see Sect. 7.7) they frequently occur in unexpected locations. For example, they effectively split some states into independent patches, as with Pennsylvania, where the strongest border of the map separates the state into regions centered around Pittsburgh and Philadelphia. Other examples are Missouri, which is split into two halves, the eastern part dominated by St. Louis (also taking a piece of Illinois) and the western by Kansas City, and the southern part of Georgia, which is effectively allocated to Florida. Also of note are the Appalachian mountains. Representing a real topographical barrier to most means of transportation, this mountain range only partially coincides with state borders, but the effective mobility border is clearly

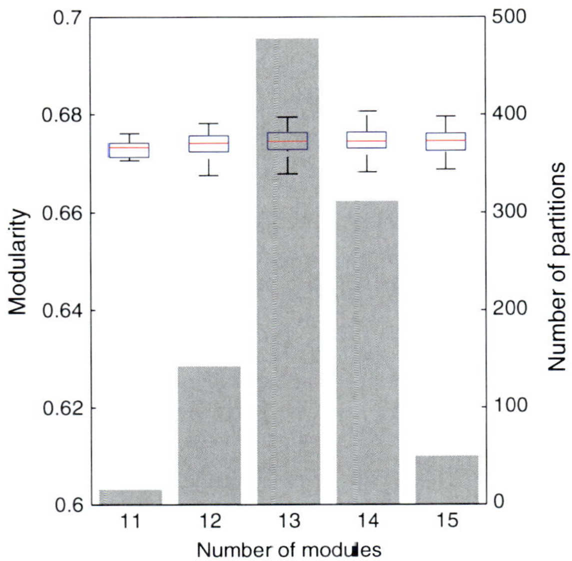

Fig. 7.15 Ensemble statistics of geographic subdivisions for a set of $N = 1{,}000$ partitions. The number of modules k in each subdivision is narrowly distributed around 13 (*grey bars*), and so are the conditional distributions of modularity (*superimposed whisker plots*). The ensemble mean is $\overline{Q} = 0.674 \pm 0.0026$

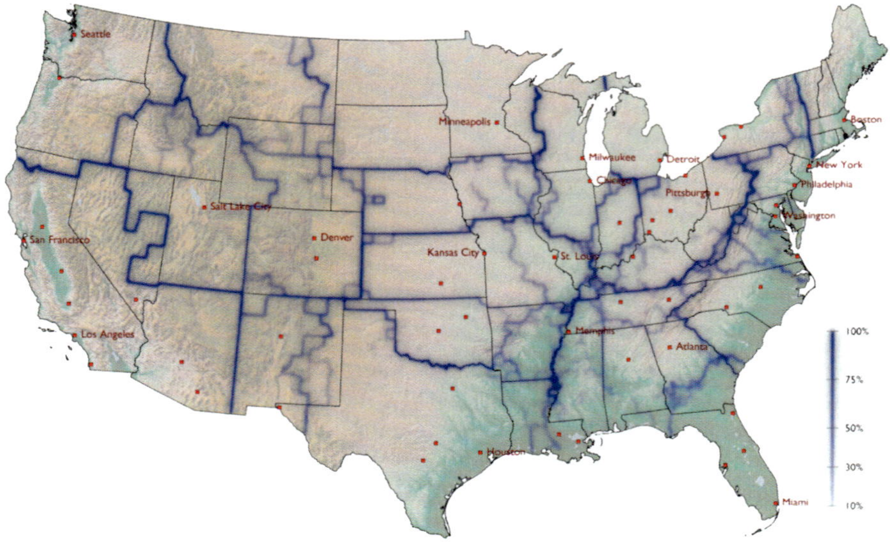

Fig. 7.16 Effective borders emerge from linear superposition of all maps in the ensemble (*blue lines*). Intensity encodes border significance (i.e. the fraction of maps that exhibit the border). Black lines indicate state borders. Although 44% of state borders coincide with effective borders (*left pie chart*), approximately 64% of effective borders do not coincide with state borders. These borders are statistically significant features of the ensemble of high modularity maps, they partially correlate with administrative borders, topographical features, and frequently split states

correlated with it. Finally, note that effective patches are often centered around large metropolitan areas that represent hubs in the transportation network, for instance Atlanta, Minneapolis and Salt Lake City. We find that 44% of the administrative state borders are also effective boundaries, while 64% of all effective boundaries do not coincide with state borders.

7.5.1 Comparison to Gravity Models

We also investigate whether the observed pattern of borders can be accounted for by the prominent class of gravity models [2,10,42], frequently encountered in modeling spatial disease dynamics [42]. In these phenomenological models, it is assumed that the interaction strength w_{ij} between a collection of sub-populations is given by (7.9), and we construct such a model according to the procedure described in Sect. 7.3.4. Although their validity is still a matter of debate, gravity models are commonly used if no direct data on mobility is available. The key feature of a gravity model is that w_{ij} is entirely determined by the spatial distribution of sub-populations. We, therefore, test whether the observed patterns of borders (Fig. 7.16) are indeed determined by the existing multi-scale mobility network or rather indirectly by the

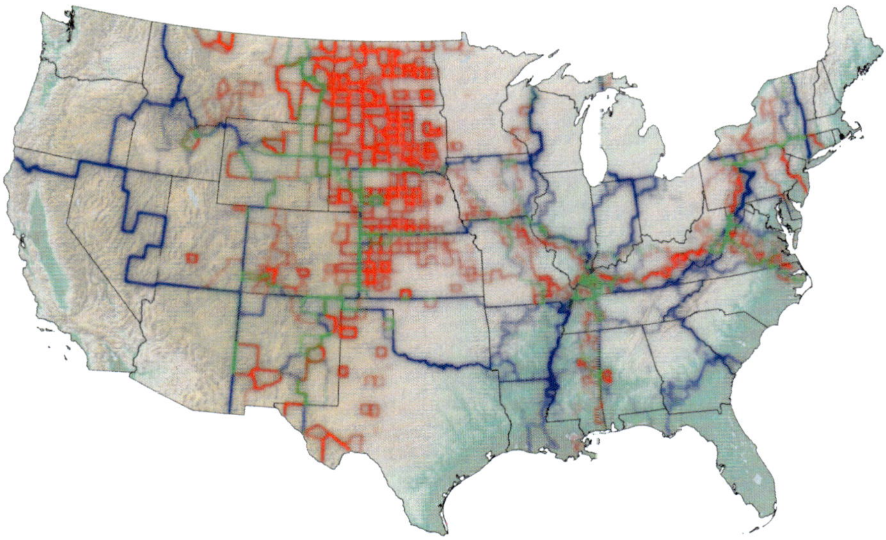

Fig. 7.17 The border structure of the gravity network (*red*) partially coincides with the borders in the original data (*blue*), but not significantly. The overlap is shown in green, for significance tests see Sect. 7.7

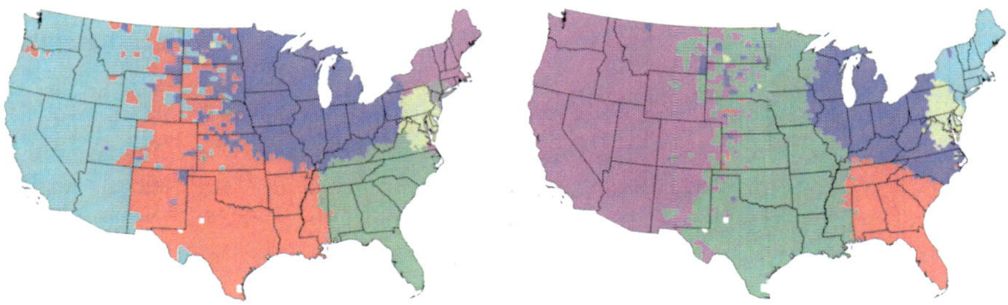

Fig. 7.18 Sample partitions of the gravity network. Although they share qualitative features with those from the original network (Fig. 7.10), generic partitions of the gravity model network are structurally different, typically exhibiting fewer modules per partition, in different locations and with less spatial compactness

underlying spatial distribution of the population in combination with gravity law coupling. Figure 7.17 illustrates the borders we find in a network that obeys (7.9).

Comparing this model network to the original multi-scale network we see that their qualitative properties are similar, with strong short-range connections as well as prominent long-range links. However, maximal modularity maps typically contain only five subdivisions with a mean modularity of only $\bar{Q} = 0.4791$. Because borders determined for the model system are strongly fluctuating (Fig. 7.18), they yield much less coherent large-scale patches. Some specific borders, e.g. the Appalachian rim, are correctly reproduced in the model. The difference between the borders of the model system and the empirical data is statistically significant (see Sect. 7.7), and we conclude that the sharp definition of borders in the original multi-scale mobility network and the pronounced spatial coherence of the building blocks are an intrinsic feature of the real multi-scale mobility network and cannot be generated by a gravity model that has a maximum first-order statistical overlap with the original mobility network.

7.6 Shortest-Path Tree Clustering

The methods already discussed successfully extract the structure of geographic borders inherent in multi-scale mobility networks. Bootstrapping the network indicates that these structures are surprisingly stable in response to perturbations of the network, but neither the modularity measure nor the stochastic algorithm we use to discover partitions provide specific information about the substructures in the network that make these borders so robust. What feature of the network, more specifically which subset of links if any, generates the observed borders? In order to address this question and further investigate the structural stability of the observed patterns, we developed a new and efficient computational technique based on the concept of shortest-path trees (SPT). Like stochastic modularity maximization, this technique identifies a structure of borders that encompass spatially coherent regions (Fig. 7.19), but unlike modularity this structure is unique. More importantly, it identifies a unique set of connections in the network, a network backbone, that correlates strongly with the observed borders.

This second method for identifying community partitions, based on topological features of the analyzed network, has three parts. Given a network with N nodes containing a single connected component, we first compute a shortest-path tree for each node in the network. At least three widely-known algorithms are applicable (Dijkstra, Floyd–Warshall, and Bellman–Ford) and various optimizations are possible; in addition, if the input is sparse some of these algorithms improve in time complexity. In the worst case, however, this can be computed in $O(N^3)$ time.

Second, we compute a dissimilarity score for each pair of shortest-path trees, and using the dissimilarity functions described below, this can also be accomplished in $O(N^3)$ time.

Third and last, we apply hierarchical clustering to the table of dissimilarity scores, which also takes $O(N^3)$ time for a naive implementation (because we compute the smallest element of an at-largest-N-by-N table N times). Therefore the entire suggested procedure takes $O(N^3)$ time.

As mentioned, various optimizations are possible for computing shortest-path trees and hierarchical clustering, and these algorithms are so widely used that high-quality, efficient implementations are easily available. In fact, we find that the second step, computing dissimilarity scores, actually dominates the running time although it is by far the simplest computation; this is due to the fact that we use interfaces to pre-compiled, canned routines for steps 1 and 3, while step 2 is a naive MATLAB script. In practice the entire analysis can be run start to finish in under a half hour for our network of $N = 3{,}109$ nodes on a circa-2008 laptop.

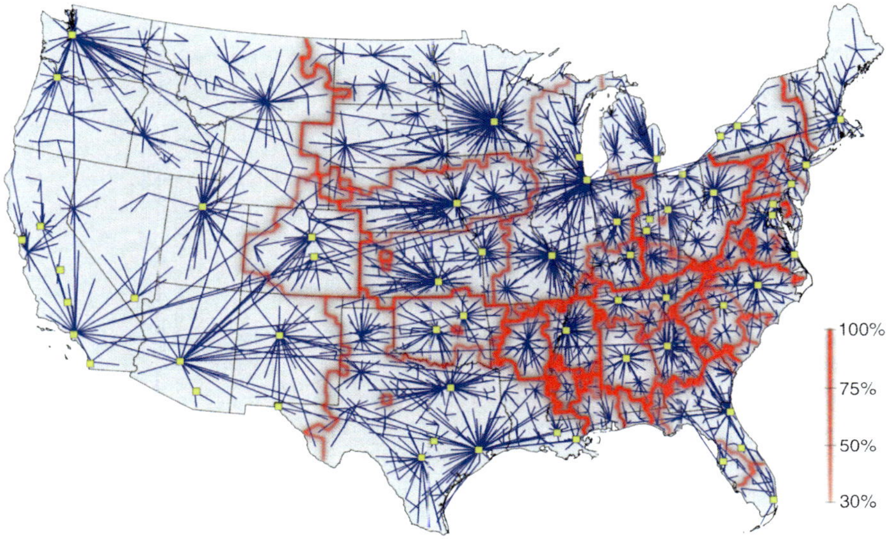

Fig. 7.19 By comparing the border structure from SPT clustering with the ensemble of significant links (those that appear in at least half of the shortest-path trees) we identify topological structures which reveal the core of the network that explains the majority of border locations. This core is represented by the network in blue consisting of star-shaped modules centered around large cities (*yellow squares*)

7.6.1 Computing Shortest-Path Trees

The shortest path from vertex i to vertex j is the series of edges that minimizes the total effective distance $d = \sum 1/w_{ij}$ along the legs of the path [14]. The distance along an edge for us is the inverse of the edge weight, as a highly-weighted edge indicates that two vertices are effectively proximal. (There are no edges with an infinite distance, because we do not define an edge between vertices if there is zero weight.)

The shortest-path tree T_i rooted at node i is the union of all shortest paths originating at i and ending at other nodes. We use the MATLAB interface[2] to the Boost Graph Library[3] to compute shortest-path trees. To prevent random fluctuations in our data from overwhelming the signal, we add a weak link between neighboring counties.

[2]http://www.stanford.edu/~dgleich/programs/matlab_bgl/.

[3]http://www.boost.org/doc/libs/1_41_0/libs/graph/doc/index.html.

7.6.2 Measuring Tree Distance

A shortest-path tree can be easily represented as a vector of vertex labels $T = [t_k], k = 1 \dots N$, such that t_k is the label of the parent of vertex k, with a special symbol (perhaps 0) used to indicate the root. There are no disconnected nodes in the mobility network, thus each tree vector represents a single tree and not a forest. This representation lends itself to straightforward and meaningful comparisons between two trees.

We define two related measures of the dissimilarity between two trees. The first, called *parent dissimilarity*, asks the question, how many of the vertices in T_A do not have the same parent in T_B? We denote this by $z_p(T_A, T_B)$, and it is exactly the general Hamming distance of two symbol sequences, that is, the number of places where corresponding labels in T_A and T_B do not match. The second, called *overlap dissimilarity*, asks the question, how many edges do the two trees *not* share? It is defined as $z_o(T_A, T_B) = s_{\max} - s(T_A, T_B)$. Here, $s_{\max}$ is the largest number of edges two trees could share, which is the number of vertices less one (since the root does not contribute an edge). $s(T_A, T_B)$ is the number of edges that T_A and T_B *do* share, and where z_p asks essentially the same question considering edges to be directed, z_o considers edges to be undirected. Also note that although we consider only the topology of trees when measuring their dissimilarity, the topology is determined by the weight of edges in the original graph and thus the mobility dynamics. For both measures, possible z values range from 0 (completely identical trees) to N, the number of nodes in the network.

We compute both measures for each distinct pair of trees in our network and find that they are highly correlated (the Pearson correlation coefficient of the two sets is 0.9980). For this reason, and because of the more straightforward interpretation, we focus exclusively on z_p. The parent dissimilarity values in our data range from 2 to 240.

To test the stability of this measure we also added various amounts of noise to the original weight matrix; for example, adding 1% noise means that we adjusted each entry by a random number such that its perturbed value is within 1% of its original value. We then compute the set of shortest-path trees for the perturbed weight matrix, calculate the tree dissimilarities, and then compute the Pearson correlation of the original dissimilarities and the perturbed. The results (0.9995 for 0.1% noise, 0.9984 for 1% noise, 0.9937 for 5% noise) indicate the method is robust against small perturbations, and in addition we do not observe significant changes in the structure of borders determined by the perturbed matrices.

7.6.3 Hierarchical Clustering and Borders

The measures described above produce a dissimilarity matrix well-suited for use with hierarchical clustering [20]. This technique iteratively groups data points together into clusters that are less and less similar; it begins by identifying the

Table 7.3 Cophenetic correlation coefficients [37] for various linkage functions using parent dissimilarity of trees and inverse weight of links

Linkage	$z_P(T_i, T_j)$	$1/w_{ij}$
Single	0.6584	0.1883
Average	0.8048	0.3757
Complete	0.7197	0.1400

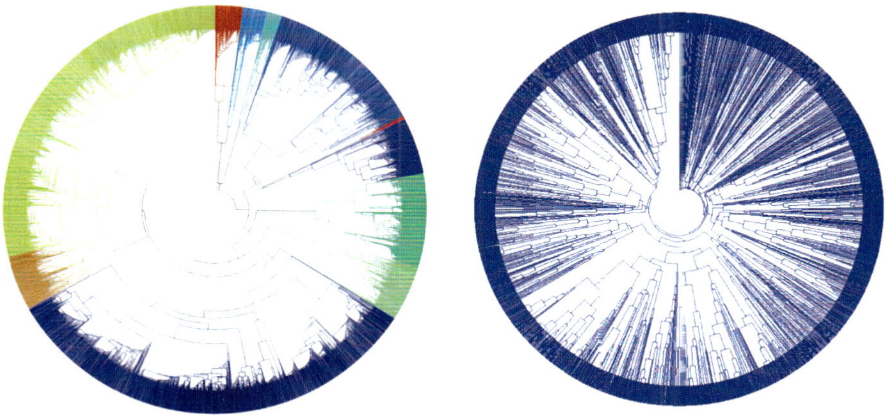

Fig. 7.20 Dendrograms from hierarchical clustering. (*Left*) Using the parent dissimilarity matrix and average linkage. Colors correspond to a particular community partition depicted in Fig. 7.21. (*Right*) Using the inverse weight matrix with noise and average linkages. Even inspection by eye reveals immediately that clustering of the inverse weight matrix produces a poor fit to the data, pointing to the need for some type of "pre-conditioning," here provided by SPT dissimilarity

two points with the lowest dissimilarity and grouping them together, then finding the next-most-similar data point or group, and so on When it is necessary to compare the dissimilarity of one point (or group of points) with another group of points, a linkage function is used. There are several commonly-used linkage functions; we compute single linkages (comparing the shortest distance between two groups), average linkages (the average distance between two groups), and complete linkages (the greatest distance between two groups) and find that the average linkage produces the best fit to our data (Table 7.3).

The result of the hierarchical clustering algorithm is a linkage structure that can be represented graphically with a dendrogram (Fig. 7.20). The radial lines in the dendrogram represent vertices in our network or groups of vertices, and the arcs represent a link that joins groups together in the hierarchy. The nearer an arc is to the center of the circle, the greater the dissimilarity between the groups joined by the arc.

Each arc corresponds to a geographic border between a set of counties, and the closer the arc is to the center of the circle, the more significant the border. At the outermost level, the dendrogram necessarily puts a border around each individual

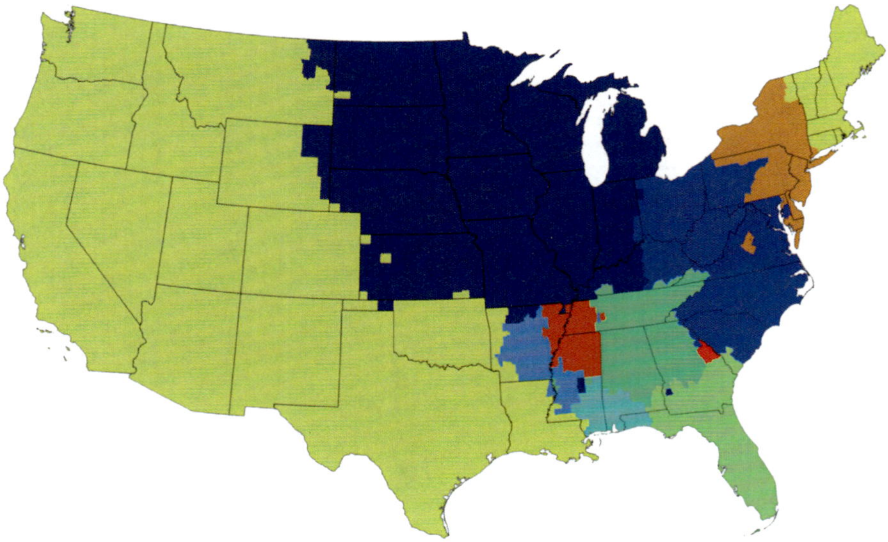

Fig. 7.21 The geographic partition determined by cutting the dendrogram of Fig. 7.20 at a height of 95

county, and we threshold at 30% of the height of the tree (corresponding to a dissimilarity $z_p = 41.6019$) for the analysis of the WG network.

As you can see in Fig. 7.21, the high-level groups identified by this procedure are spatially coherent, but may be divided into spatially disjoint regions at certain heights in the dendrogram.

Hierarchical clustering is also sometimes applied directly to the inverse of the weights, $1/w_{ij}$. We have investigated this method as well and find that it has several shortcomings. First, to apply a hierarchical clustering algorithm requires computing a dissimilarity for every pair of data points; since many pairs of counties are not directly connected by a link in our network (w_{ij} is zero), the inverse does not exist and it is consequently necessary to add some noise to the weight matrix at the very first step, representing pairs of vertices that are "extremely distant" but not disconnected. Second, the linkage structures produced from this approach fit the data poorly (Table 7.3). Last, one can see by visual inspection of the dendrograms in Fig. 7.20 that this approach does not yield significant information. Comparing the dendrograms for the z_p and $1/w$ matrix, we see that in the shortest-path tree approach most of the links that appear higher in the tree (closer to the center) are linking together two groups that are strongly dissimilar from one another (seen by comparing the height of the parent link to the heights of the children links). In the inverse weight method, this is not true: links high in the tree are linking groups that are quite similar; that is, inverse weight clustering does not identify groups of strongly dissimilar vertices.

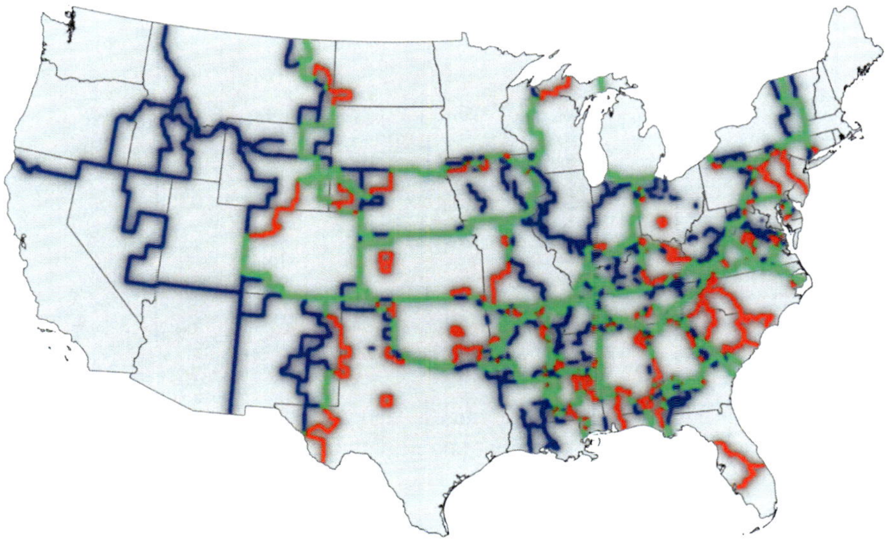

Fig. 7.22 Comparing borders from modularity maximization (*blue*) with SPT clustering (*red*) reveals a significant overlap (*green*). The cumulative topological overlap (see Sect. 7.7) is 0.5282 indicating that the SPTD method represents an alternative computational approach to border extraction

Although the method yields a unique sequence of topological segmentations, the observed geographic borders exhibit a strong correlation with those determined by modularity maximization (Fig. 7.22).

7.6.4 Link Significance

The key advantage of this method is that it can systematically extract properties of the network that match the observed borders. A way to demonstrate this is to measure the frequency σ at which individual links appear in the ensemble of all SPTs, which is conceptually related to their link betweenness [35]. Computing this *link significance* σ for each connection, we find that the distribution $P(\sigma)$ of the network is bimodally peaked (Fig. 7.23). This is a promising feature of $P(\sigma)$ as it allows labeling links as either significant or redundant without introducing an arbitrary threshold which is necessary for more continuously distributed link centrality measures. Extracting the group of significant links and constructing a subnetwork from these links only we observe that this subnetwork matches the computed border structure. By virtue of the fact that the most frequently shared links between SPTs are local, short-range connections we see that the SPT boundaries enclose local neighborhoods and that the boundaries fall along lines where SPTs

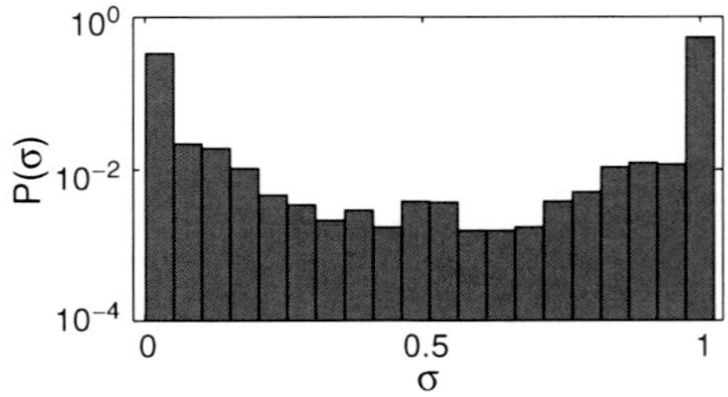

Fig. 7.23 The distribution of link significance σ, defined for each link as the number of shortest-path trees the link appears in, exhibits a strong bimodal distribution. This implies that SPTD can sort links into important or not, and that σ is approximately a binary variable

do not share common features. Note that effective metropolitan areas around cities can be detected with greater precision than modularity, although the western US is detected as effectively a single community.

Finally, we performed statistical analyses that quantify the overlap of the effective, mobility-induced borders with those provided by census-related systems. We choose the set of borders separating the states, the borders defined by the districts of the 12 Federal Reserve Banks, and the borders of Economic Areas [39]. We discuss this analysis in more detail in Sect. 7.7, but briefly, we find a significant correlation with economic boundaries ($p < 0.001$, z-score 8.024 for the modularity borders and $p < 0.001$, z-score 13.29 for the SPT borders).

7.7 Significance and Comparison of Border Structures

7.7.1 Bootstrapping the Where's George Data

In order to test the robustness of our method against random data removal, we performed the following bootstrapping analysis. Starting with the full dollar bill dataset, and the resulting network weight matrix W with elements w_{ij}, we randomly remove single dollar bill reports until the total flux $f = \sum_{i,j} w_{ij}$ is reduced by a factor γ. Using this method we constructed several networks for $0 \leq \gamma \leq 0.95$ and computed an ensemble of 100 partitions for every value of γ, using the simulated annealing algorithm described in Sect. 7.2. We find that the modularity value is unaffected by bootstrapping even if 95% of the total flux is removed, although the number of modules in each partition rises as the network is thinned out more than 85% (Fig. 7.24). Also, the boundary structure emerging from superposition of all partitions is very robust under this procedure (Fig. 7.25). At 20% of the original flux ($\gamma = 0.8$) virtually all of the boundaries found in the complete network are still identified, although the sparsity of the data evokes some singular counties. Even with only 5% of the flux, when boundaries become more fuzzy, some of the original structures are still detected.

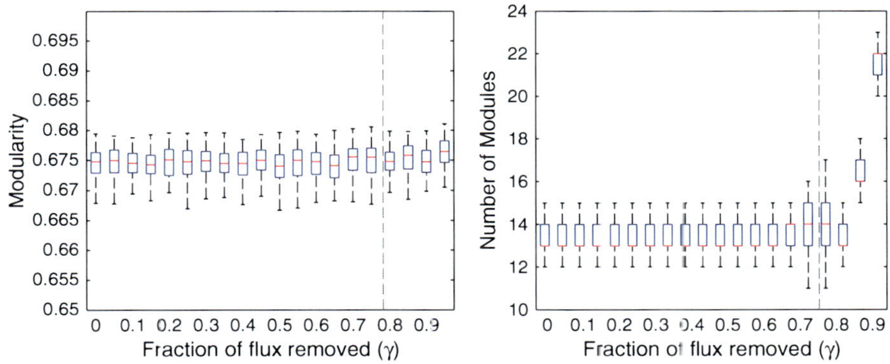

Fig. 7.24 Distributions of modularity values and number of modules for an ensemble of 100 partitions computed for each value of the bootstrapping parameter γ. The dashed line corresponds to 78.2%, the amount of flux ignored if all links shorter than 400 km would be removed

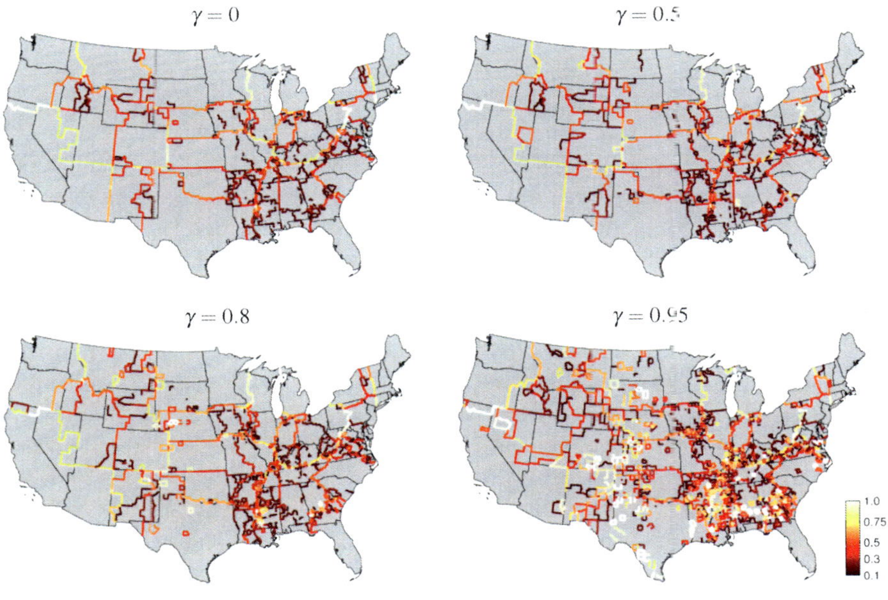

Fig. 7.25 Linear superposition of 100 partitions for four different values of the bootstrapping parameter γ, color-coded according to the fraction of partitions they appear in

7.7.2 Measuring Overlap of Two Boundary Networks

In this section, we describe how to compare boundary networks defined on a planar graph, in our case the county network of the continental US excluding Alaska.

A *boundary network* b is simply given by assigning a nonnegative number w to each edge between adjacent counties: If the two counties are not divided by b, then

$w = 0$. Otherwise, $w > 0$ implies that the border shared between the two counties has the strength w. In Sect. 7.4, we described how to generate such a boundary network by superposition of many partitions of the Where's George money travel network. We denote this boundary network by the modularity boundaries b_{M}.

We want to quantify how much information the modularity boundaries b_{M} shares with e.g. a state network, a random network, or a boundary network generated with another method.

For this we essentially need to determine the cross-correlation between two boundary networks b and b'. However, cross-correlation itself is not well-suited for dealing with the non-negativity of the edge weightings, so we calculate a non-centered version of it. The *absolute cross-correlation* of the two boundary networks b and b' is then given by the normalized scalar product of their edge weightings, i.e. by

$$a(b,b') := \frac{(1/|E|)\sum_{e\in E} b(e)b'(e)}{\sqrt{(1/|E|)\sum_{e\in E} b(e)^2}\sqrt{(1/|E|)\sum_{e\in E} b'(e)^2}},$$

where E denotes the set of edges connecting adjacent counties. This quantity lies between 0 and 1 and equals 1 if and only if the two boundaries are identical up to scaling.

Apart from the upper bound, this quantity however is difficult to judge. In particular, we cannot compare right away two cross-correlations between different networks since $a(\cdot,\cdot)$ might depend on the number of clusters and inhomogeneity of weights etc. We avoid finding a direct interpretation of the absolute cross-correlation by instead considering deviation of observed values against cross-correlations with a null model.

Such null models are used to tell random occurrences of structures from true information. One typically wants to keep some statistics of the network fixed while at the same time randomly sampling from its representational class. This results in the notion of random graphs with certain additional properties such as Erdös–Rényi [18] or Barabási–Albert [4]. The key idea is to generate a random network preserving planarity and possible additional information by using the original structure and iteratively changing it by a random local modification. For instance for unweighted networks, a random graph can be generated by "rewiring": two distinct edges and two different vertices contained in either of the two are randomly selected and then swapped. Clearly this operation keeps both degree distributions fixed. After a certain number of iterations, the thus-generated Markov chain produces independent samples of the underlying random graph with given degree distributions [33]. This concept has been generalized to weighted graphs [43]; in this case, it is debatable whether to swap the whole weighted edge or to split up the weight.

In our case, we search for a randomization of a boundary network i.e. of a planar, weighted graph. Rewiring as above is not possible since it would destroy planarity. Instead, we propose to locally modify the graph at a random county: select a subpath

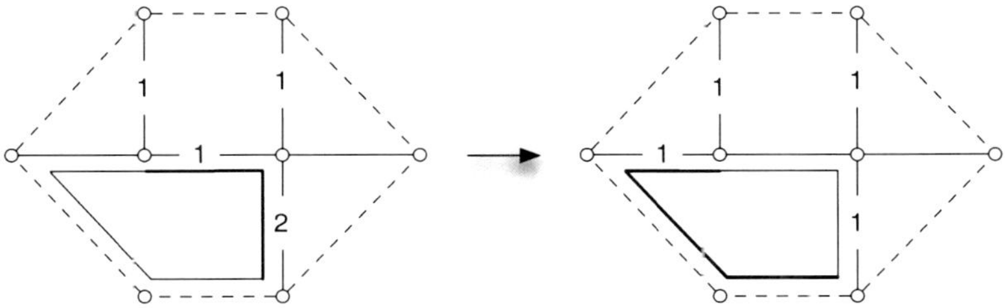

Fig. 7.26 Local modification of a planar graph. We select the bottom left county to modify. The selected path to modify is shown in bold in the left figure. Its minimal weight is 1. This is subtracted in the right hand figure, where the complementary path is shown

of its boundary and flip it to its complement. In the case of non-trivial weights, we reassign a random number between 0 and the minimal edge weight on the subpath. We have illustrated this procedure on an example in Fig. 7.26.

This procedure is now repeated multiple times until sufficiently de-correlated samples from the original network are produced. In practice, it is common to choose iterations in the range of the number of edges in the network or more.

7.7.3 *Randomization of the Mean Partition Boundary of the Where's George Network*

In order to test for significances of calculated similarities, we build a random model of the mean partition boundary by generating 1,000 random networks using the above algorithm with >15,000 successful iterations for each random network. The corresponding maps for the first 900 iterations are shown in Fig. 7.27. Clearly, the original structure in the boundary network is increasingly diluted, and after >10,000 iterations becomes stably random.

This can be seen by calculating the absolute cross-correlation $a(b_{\mathrm{M}}, b_{\mathrm{R}})$ of the modularity boundary network b_{M} with the random networks b_{R}, when increasing the number of iterations, see Fig. 7.28. We observe convergence to roughly 0.5 after about 10,000 steps. This lies well in the range of random correlation with a mean of 0.49 and a standard deviation of 0.028, see histogram in Fig. 7.29(a). This implies that the randomization procedure converges to a set of random boundary networks, which can now be used to put calculated autocorrelations into perspective against this null model.

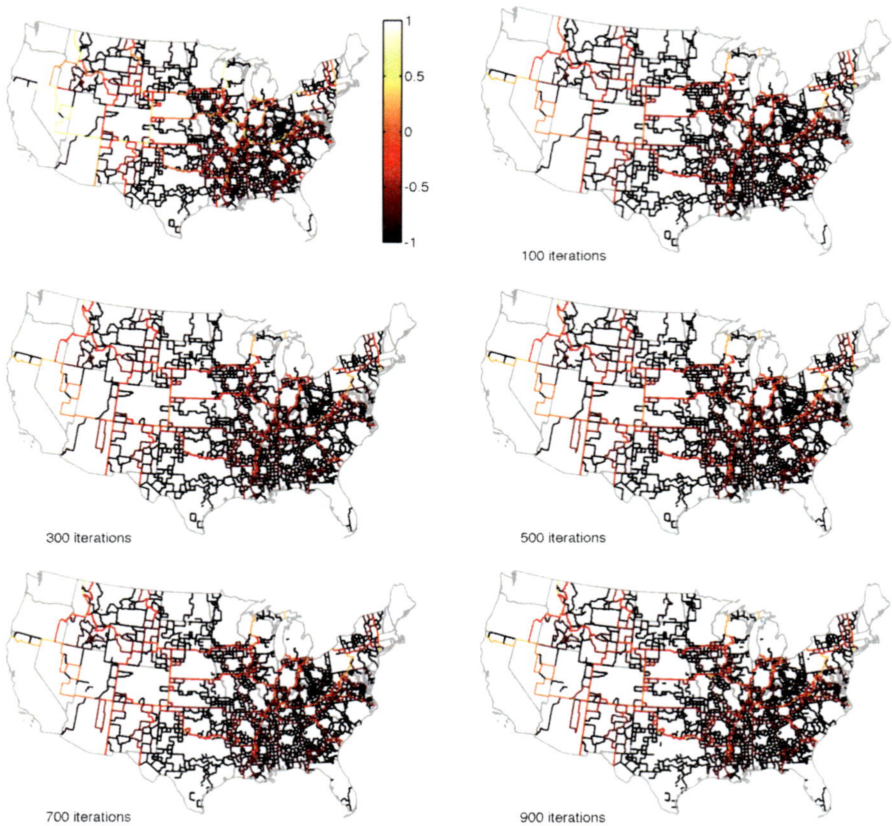

Fig. 7.27 Randomization of the modularity boundary network b_M. The original network and the first 900 iterations are shown

7.7.4 *Significances when Comparing Boundary Networks with the Null Model*

We describe and quantify overlap of the estimated modularity boundaries b_M with other political or social boundaries. As described before, we can quantify overlap by determining the absolute cross-correlation $a(b, b_M)$. In order to determine interpretable numbers, we compare this value to correlations with random boundaries b_R from a null model.

We now determine significance of coincidence of the modularity boundary network b_M and the SPT boundary network b_S with:

- Modularity boundaries b_M
- State boundaries

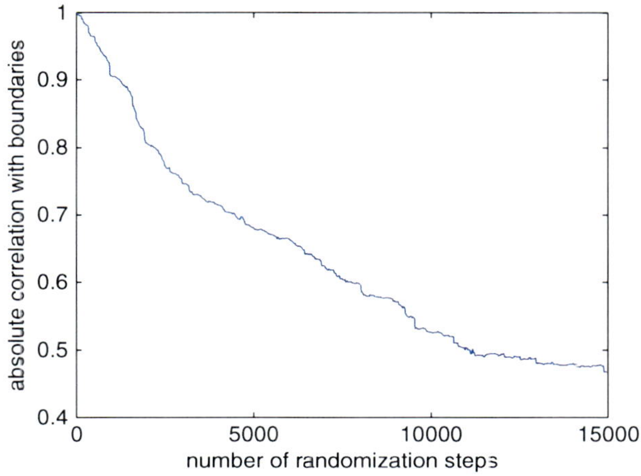

Fig. 7.28 Absolute cross-correlation of the randomized network after the given number of iterations with the modularity boundary network b_M

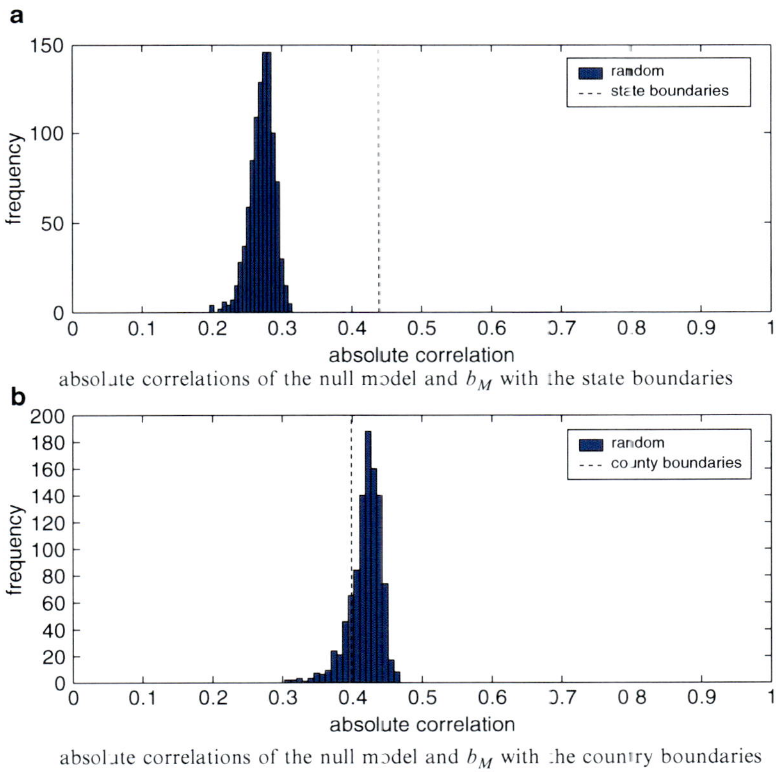

Fig. 7.29 Absolute cross-correlation of state and county boundaries when compared with a null model based on the modularity boundary network b_M

Table 7.4 Comparing boundary overlaps for various boundary networks with the modularity boundaries b_{M} and the corresponding null model b_{R} using absolute cross-correlation a

Boundary network	$a(\cdot, b_{\mathrm{M}})$	$a(\cdot, b_{\mathrm{R}})$	p-value	z-score
Modularity boundaries	1.000	0.495 ± 0.028	$<10^{-3}$	18.15
SPT communities	0.552	0.385 ± 0.024	$<10^{-3}$	7.03
State boundaries	0.439	0.272 ± 0.018	$<10^{-3}$	9.46
County boundaries	0.398	0.419 ± 0.023	0.84	0.90
Gravity boundaries	0.260	0.253 ± 0.019	0.35	0.40
Large-range network boundaries	0.198	0.181 ± 0.017	0.14	1.02
Federal reserve district boundaries	0.377	0.227 ± 0.019	$<10^{-3}$	7.91
Economic area boundaries	0.452	0.307 ± 0.018	$<10^{-3}$	8.024

- County boundaries (to test for sensitivity of the method against number of communities)
- Boundaries resulting from the SPT algorithm b_{S}
- Boundaries determined on the gravity model
- Boundaries determined on long-range distances only
- Federal reserve district boundaries (FRB)
- Economic area boundaries (http://www.bea.gov)

The significance is calculated by replacing b_{M} and b_{S}, respectively, by elements from the corresponding null model.

For illustration we show two histograms and actual values for state and county boundaries in Fig. 7.29. Clearly, the random cross-correlations are quite different, which means that we have to interpret the actual values of 0.439 and 0.398 differently as well. Indeed it turns out that the state value is far from the mean random cross-correlation 0.272 ± 0.018, whereas the county one is not (0.419 ± 0.023). Indeed, the empirical p-values, determined as the fraction of random correlations above the observed true one, is 0 in the former and 0.84 in the latter case.

In order to compare cases with large deviation from the distribution, we determine the z-score that is the distance of the absolute cross-correlation from the mean of the null model normalized by the standard deviation:

$$z(b) := \frac{a(b, b_{\mathrm{M}}) - E(a(b, b_{\mathrm{R}}))}{\operatorname{std}(a(b, b_{\mathrm{R}}))},$$

where E denotes mean and std standard deviation. In the state case, this z-score is very high, 9.46, which means that the observed correlation is more than 9 standard deviations away from the random mean. In contrast the county z-score is 0.90, which means that the observation is within one standard deviation and hence not significant.

We summarize the calculated cross-correlations in Tables 7.4 and 7.5 for b_{M} and b_{S}.

Table 7.5 Comparing boundary overlaps for various boundary networks with the SPT-based boundary b_S and the corresponding null model b_R using absolute cross-correlation a

Boundary network	$a(\cdot, b_S)$	$a(\cdot, b_R)$	p-value	z-score
Modularity boundaries	0.552	0.251 ± 0.013	$<10^{-3}$	22.55
SPT communities	1.000	0.367 ± 0.0164	$<10^{-3}$	40.63
State boundaries	0.358	0.220 ± 0.0138	$<10^{-3}$	10.99
County boundaries	0.569	0.562 ± 0.016	0.36	0.44
Gravity boundaries	0.305	0.260 ± 0.016	0.002	2.73
Large-range network boundaries	0.257	0.199 ± 0.015	$<10^{-3}$	3.94
Federal reserve district boundaries	0.307	0.159 ± 0.013	$<10^{-3}$	11.79
Economic area boundaries	0.492	0.318 ± 0.013	$<10^{-3}$	13.29

7.7.5 Discussion

For the state and the SPT boundaries we observe a strong deviation from the null model when comparing against the modularity boundaries. So we can conclude that both state boundaries and SPT boundaries are more similar to b_M than expected by chance with a p-value $<10^{-3}$.

This is not the case for the gravity model, the county boundaries and the long-range model. In these cases, the cross-correlation with b_M is not larger than with a random model (p-value ≈ 0.44, ≈ 0.84 and ≈ 0.14). This means that they do not significantly coincide with b_M.

The absolute cross-correlation of the FRB boundaries with b_M is $a(b_F, b_M) = 0.38$, which is significantly high when compared with the null model, which exhibits cross-correlations of only $a(b_F, b_R) = 0.23 \pm 0.019$. We observe a strong deviation from the null model and can therefore conclude that the FRB boundaries are more similar to b_M than expected by chance with a p-value $<10^{-3}$.

The corresponding z-score equals 7.91, which is lower than the one for states (9.46). This implies that the modularity boundaries' overlap with the states is larger than the one with the FRB boundaries.

We interpret the results on the FRB boundaries when compared with b_M as follows:

- The structure of b_M may be (partially) due to political structure i.e. result from b_S or due to additional money transport within FRB districts i.e. correlate with b_F. Since both b_S and b_F share strong similarities, in each of the two situations, we would see overlap with both boundaries, so we can only judge strength of overlap with respect to the other boundary.
- We quantified strength of overlap by deviation from the null model, and the corresponding z-score was more than 1.5 standard deviations higher for the state model. This stronger overlap of states with b_F therefore favors the first hypothesis i.e. the situation that political boundaries are a stronger factor for the pattern observed in b_M. In the case of dominance of the second hypothesis, we would instead expect to still see overlap with state boundaries, but less overlap than with the FRB ones.

References

1. Altmann, E.G., Pierrehumbert, J.B., Motter, A.E.: Beyond word frequency: Bursts, lulls, and scaling in the temporal distributions of words. PLoS ONE **4**(11), e7678 (2009). DOI 10.1371/journal.pone.0007678. URL http://dx.doi.org/10.1371%2Fjournal.pone.0007678
2. Anderson, J.: A theoretical foundation for the gravity equation. The American Economic Review **69**(1), 106–116 (1979). URL http://www.jstor.org/stable/1802501
3. Balcan, D., Colizza, V., Goncalves, B., Hu, H., Ramasco, J.J., Vespignani, A.: Multiscale mobility networks and the spatial spreading of infectious diseases. P Natl Acad Sci USA **106**(51), 21,484–21,489 (2009). DOI 10.1073/pnas.0906910106
4. Barabási, A., Albert, R.: Emergence of scaling in random networks. Science **286**, 509–512 (1999)
5. Barabási, A.L.: The origin of bursts and heavy tails in human dynamics. Nature **435**, 207 (2005)
6. Brandes, U., Delling, D., Gaertler, M., Goerke, R., Hoefer, M., Nikoloski, Z., Wagner, D.: On modularity clustering. IEEE T Knowl Data En **20**(2), 172–188 (2008). DOI 10.1109/TKDE.2007.190689. URL http://ieeexplore.ieee.org/search/wrapper.jsp?arnumber=4358966&tag=1
7. Brockmann, D., Hufnagel, L., Geisel, T.: The scaling laws of human travel. Nature **439**(7075), 462–465 (2006). DOI 10.1038/nature04292
8. Brockmann, D., Theis, F.: Money circulation, trackable items, and the emergence of universal human mobility patterns. IEEE Pervas Comput **7**(4), 28–35 (2008)
9. Buzna, L., Peters, K., Ammoser, H., Kühnert, C., Helbing, D.: Efficient response to cascading disaster spreading. Phys. Rev. E **75**(5), 056,107–056,114 (2007). URL http://pre.aps.org/abstract/PRE/v75/i5/e056107
10. Cochrane, R.: A possible economic basis for the gravity model. Journal of Transport Economics and Policy **9**(1), 34–49 (1975). URL http://www.jstor.org/stable/20052391
11. Colizza, V., Barrat, A., Barthélemy, M., Valleron, A.J., Vespignani, A.: Modeling the world-wide spread of pandemic influenza: Baseline case and containment interventions. PLoS Med **4**(1), 95–110 (2007). DOI 10.1371/journal.pmed.0040013
12. Colizza, V., Barrat, A., Barthélemy, M., Vespignani, A.: The modeling of global epidemics: Stochastic dynamics and predictability. B Math Biol **68**(8), 1893–1921 (2006). DOI 10.1007/s11538-006-9077-9
13. Danon, L., Díaz-Guilera, A., Duch, J., Arenas, A.: Comparing community structure identification. J Stat Mech-Theory E p. P09008 (2005). DOI 10.1088/1742-5468/2005/09/P09008. URL http://www.iop.org/EJ/abstract/1742-5468/2005/09/P09008/
14. Dijkstra, E.: A note on two problems in connexion with graphs. Numerische Mathematik **1**(1), 269–271 (1959). URL http://www.springerlink.com/index/UU8608U0U27K7256.pdf
15. Eagle, N., Macy, M., Claxton, R.: Network diversity and economic development. Science **328**(5981), 1029–1031 (2010). DOI 10.1126/science.1186605
16. Eagle, N., Pentland, A.S., Lazer, D.: Inferring friendship network structure by using mobile phone data. P Natl Acad Sci USA **106**(36), 15,274–15,278 (2009). DOI 10.1073/pnas.0900282106. URL http://www.pnas.org/content/106/36/15274
17. Eaton, J., Kortum, S.: Technology, geography, and trade. Econometrica **70**(5), 1741–1779 (2002)
18. Erdos, P., Rényi, A.: On random graphs. i. Publicationes Mathematicae **6**, 290–297 (1959)
19. Ernst, D., Kim, L.: Global production networks, knowledge diffusion, and local capability formation. Research Policy **31**(8-9), 1417–1429 (2002). URL http://linkinghub.elsevier.com/retrieve/pii/S0048733302000720
20. Everitt, B., Landau, S., Leese, M.: Cluster analysis p. 237 (2001). URL http://books.google.com/books?id=htZzDGlCnQYC&printsec=frontcover
21. Ferguson, N., Cummings, D., Cauchemez, S., Fraser, C., Riley, S., Meeyai, A., Iamsirithaworn, S., Burke, D.: Strategies for containing an emerging influenza pandemic in southeast asia. Nature **437**(7056), 209–214 (2005). DOI 10.1038/nature04017

22. Fortunato, S.: Community detection in graphs. Physics Reports **486**(3-5), 75–174 (2010). DOI doi:10.1016/j.physrep.2009.11.002. URL http://dx.doi.org/10.1016/j.physrep.2009.11.002
23. Fortunato, S., Barthélemy, M.: Resolution limit in community detection. P Natl Acad Sci USA **104**(1), 36–41 (2007). DOI 10.1073/pnas.0605965104. URL http://www.pnas.org/content/URL 104/1/36
24. Girvan, M., Newman, M.: Community structure in social and biological networks. P Natl Acad Sci USA **99**(12), 7821–7826 (2002). DOI 10.1073/pnas.122653799. URL http://www.google.com/search?client=safari&rls=en-us&q=Community+structure+in+social+and+biological+networks&ie=UTF-8&oe=UTF-8
25. González, M.C., Hidalgo, C.A., Barabási, A.L.: Understanding individual human mobility patterns. Nature **453**(7196), 779–782 (2008). DOI 10.1038/nature06958
26. Good, B.H., de Montjoye, Y.A., Clauset, A.: The performance of modularity maximization in practical contexts. arXiv Preprint Server p. 0910.0165v1 (2009)
27. Guimerà, R., Amaral, L.A.N.: Cartography of complex networks: modules and universal roles. Journal of Statistical Mechanics **2**, P02,001 (2005). DOI 10.1088/1742-5468/2005/02/P02001
28. Guimerà, R., Mossa, S., Turtschi, A., Amaral, L.A.N.: The worldwide air transportation network: Anomalous centrality, community structure, and cities' global roles. P Natl Acad Sci USA **102**(22), 7794–7799 (2005). DOI 10.1073/pnas.0407994102
29. Keller, W.: International technology diffusion. Journal of Economic Literature **42**(3), 752–782 (2004). URL http://www.atypon-link.com/AEAP/doi/abs/10.1257/0022051042177685
30. Lazer, D., Pentland, A., Adamic, L., Aral, S., Barabási, A.L., Brewer, D., Christakis, N., Contractor, N., Fowler, J., Gutmann, M., Jebara, T., King, G., Macy, M., Roy, D., Alstyne, M.V.: Computational social science (2009). DOI 10.1126/science.1167742. URL http://www.sciencemag.org/cgi/content/summary/323/5915/721
31. Liben-Nowell, D., Novak, J., Kumar, R., Raghavan, P., Tomkins, A.: Geographic routing in social networks. P Natl Acad Sci USA **102**(33), 11,623–11.628 (2005). DOI 10.1073/pnas.0503018102
32. Lloyd, A.L., Jansen, V.A.A.: Spatiotemporal dynamics of epidemics: synchrony in metapopulation models. Mathematical Biosciences **188**(1-2), 1–15 (2004). DOI DOI:10.1016/j.mbs.2003.09.003. URL http://www.sciencedirect.com/science/article/B6VHX-4B9K8TY-4/2/f79d91db6e0f3711169a9079e3cb9ca2
33. Maslov, S., Sneppen, K.: Specificity and stability in topology of protein networks. Science **296**(5569), 910–3 (2002). DOI 10.1126/science.1065103
34. Newman, D.: The lines that continue to separate us: borders in our 'borderless' world. Progress in Human Geography **30**(2), 143–161 (2006). DOI 10.1191/0309132506ph599xx. URL http://phg.sagepub.com/cgi/content/abstract/30/2/143
35. Newman, M., Girvan, M.: Finding and evaluating community structure in networks. Phys. Rev. E **69**(2), 026,113 (2004). DOI 10.1103/PhysRevE.69.026113. URL http://prola.aps.org/abstract/PRE/v69/i2/e026113
36. Sales-Pardo, M., Guimerà, R., Moreira, A.A., Amaral, L.A.N.: Extracting the hierarchical organization of complex systems. P Natl Acad Sci USA **104**(39), 15,224–15,229 (2007). DOI 10.1073/pnas.0703740104
37. Sokal, R., Rohlf, F.: The comparison of dendrograms by objective methods. Taxon **11**(2), 33–40 (1962). URL http://www.jstor.org/stable/1217208
38. Song, C., Qu, Z., Blumm, N., Barabási, A.L.: Limits of predictability in human mobility. Science **327**(5968), 1018–1021 (2010). DOI 10.1126/science.1177170. URL http://www.sciencemag.org/cgi/content/abstract/327/5968/1018
39. United States Bureau of Economic Analysis: Economic areas. http://www.bea.gov/regional/docs/econlist.cfm (2010). URL http://www.bea.gov/regional/docs/econlist.cfm
40. Vespignani, A.: Predicting the behavior of techno-social systems. Science **325**(5939), 425–428 (2009). DOI 10.1126/science.1171990

41. Watts, D., Strogatz, S.: Collective dynamics of 'small-world' networks. Nature **393**(6684), 440–442 (1998)
42. Xia, Y., Bjornstad, O., Grenfell, B.: Measles metapopulation dynamics: A gravity model for epidemiological coupling and dynamics. Am Nat **164**(2), 267–281 (2004)
43. Zlatic, V., Bianconi, G., Díaz-Guilera, A., Garlaschelli, D., Rao, F., Caldarelli, G.: On the rich-club effect in dense and weighted networks. European Physical Journal B **67**, 271–275 (2009)

Chapter 8
Emergence and Structure of Cybercommunities

Marija Mitrović and Bosiljka Tadić

Abstract We study topology of bipartite networks representing high-resolution data of the online communications of users on Blogs and similar Web portals. User communities occurring in connection with certain popular posts, movies etc., are detected by spectral analysis of these networks. Due to indirect nature of the online interactions between users, further information about the structure of the communities is inferred by text analysis of the related comments. We employ the emotion classifier based on machine-learning methods and trained for this type of data, to determine the emotional contents of text of each post and comment within a given community. Combined with the network theory, in this way we are able to unravel the role of emotion expressed in the text for the patterns of user behavior, which leads to the emergence of collective states with the appearance of communities, and their internal structure and evolution. *All data are fully anonymized. No information about user IDs are given.*

8.1 Introduction

Mechanisms underlying online interactions of users at Web portals may lead to new social phenomena in Cyberspace, with certain features which are not known in the conventional social grouping [7, 18]. The emergent social phenomena among Web users are the primary concern of the multidisciplinary area known as science of the Web [2], apart from developing the technology and algorithms for safe and efficient information processing and understanding the Web structure [10] and its evolution mechanisms [36].

M. Mitrović • B. Tadić (✉)
Department of Theoretical Physics, Jožef Stefan Institute, Ljubljana, Slovenia
e-mail: marija.mitrovic@ijs.si; bosiljka.tadic@ijs.si

M.T. Thai and P.M. Pardalos (eds.), *Handbook of Optimization in Complex Networks: Theory and Applications*, Springer Optimization and Its Applications 57,
DOI 10.1007/978-1-4614-0754-6_8,

The online interactions have several features which make them different from conventional face-to-face interactions [5]. In particular, user interactions are fast and virtually no distance is involved. Moreover, the user interactions are *indirect*, i.e., mediated by the posted material (text of comments, movie, music, etc.). In contrast to conventional social interactions, users interacting via posted material are not associated in real life, which has several implications in their conduct and thus influences the emergent collective phenomena [5, 34]. The dynamics of the online interactions is a large degree self-driven, i.e., user's action (posting a comment) on previously posted material induces more actions that eventually lead to a burst of activities [1, 2, 6, 18, 28]. Recent studies of the data from Blogs [5, 15, 22, 23, 30, 35, 38], Diggs [28], Forums [19], movie or music sharing databases [16, 20, 21, 24, 40] suggest very rich structures of interactions, leading to temporary occurrence of user communities in Cyberspace.

Mapping the interaction data onto networks and using the theory of complex networks [3] provides a way for the quantitative analysis of various complex systems and detecting the topological community structure [13] of the networks. In the case of online interactions, for example Blogdata, a suitable representation is by *bipartite networks* [30], where the *users* are represented as one partition having no direct interactions with each other, while the *posts and comments* are the other natural partition thorough which the users are connected. The communities occurring on these networks thus consist of *users linked via certain posts and comments*, and can be identified by formal graph theory methods [13, 21, 31].

Further insight into a particular community structure and its dynamics is achieved by analysis of contents in the text of the related posts and comments. It has been recognized recently that beside conveyed information, the *emotional content* of the text has profound effects on user emotional conduct, which can be measured at different scales [9, 14, 33, 37]. Text analysis methods are being developed in the computer science [25, 26], independently from the theory of complex systems and phenomena of the collective user behavior. The theoretical foundations for the analysis of the emotional contents contained in the text, seen as a classification problem, is described in [27].

In this chapter, we review a combined methodology for quantitative analysis of user communities appearing on Blogs, Diggs, and movie datasets: Full data mapping onto bipartite graphs and the use of the network theory is combined with the machine-learning methods of text analysis for the emotional contents. We present some results of the emergent networks, their topology and community structure as well as the patterns of user emotional behavior, which underly the occurrence and the evolution of the observed Cybercommunities.

8.2 Networks of Users' Online Interactions

High-resolution data of user online interactions via different Web portals can be mapped onto different types of techno-social networks. In particular, bipartite networks are suitable representations of the indirect interactions between users

Table 8.1 Size and structure of the analysed data from Blogs and Diggs

Name	Emotional content	Number of posts	Number of comments	Number of users	Comment-on-comment
BBC Blogs	Yes	$N_P = 3972$	$N_C = 80873$	$N_U = 21462$	No
B92 Blogs	No	$N_P = 4784$	$N_C = 406527$	$N_U = 4598$	Yes
digg.com	Yes	$N_P = 1195808$	$N_C = 1646153$	$N_U = 484986$	Yes

which are mediated by the posts (and comments) on Blogs and similar Web portals. By definition, in bipartite graphs no link exists between two nodes of the same partition [4]. Below we describe several types of networks, by which we represent the interactions contained in the datasets from Blogs and Diggs [28, 30] and the movie database [17].

The data structures from Blogs and Diggs that we consider have high-resolution (1 sec) of the temporal occurrence of the events (posting a comment) and full information about both *Users* and *Posts-and-Comments*, as well as the full *Text* of the Posts and Comments. It should be stressed that, although at all considered blogsites, user registration with a unique ID is requested, different policy is practiced in storing the information, especially regarding the information about *comment-on-comment*. In view of our precise mapping, as described below, such information affects the structure of the emergent network. For this reason, we also collected and analysed data from different blogsite, B92 Blogs, where such information is available. On the other hand, the emotion classifier of text data is currently available [27] for English language only. Summary of the data is given in Table 8.1.

In the movie data, somewhat similar structure with the texts of comments is available. However, we do not analyse the text of these comments. For the present analysis, we only consider topology of the emergent networks for small sets of movies, selected according to the number of votes [17].

8.2.1 Data Mapping and Types of Networks

The datasets from Blogs and Diggs, which we analyse have high resolution in time of the data on both users and posts and comments, which we map onto a *bipartite network*. On these networks *users* are recognized as one natural partition, while the other partition are *posts and comments* together. As discussed above, this representation is particularly suitable for the users in the Cyberspace of Blogs (and similar Web portals, see also [17, 20]). Depending on the type of user activity, i.e., reading the existing text or posting a new text, we distinguish the directions of the links as follows: *post (comment)*$\rightarrow$*user* represents *the user reading the post (comment)*, while *user*$\rightarrow$*post (comment)* indicates that *the user is writing a new post (comment)*.

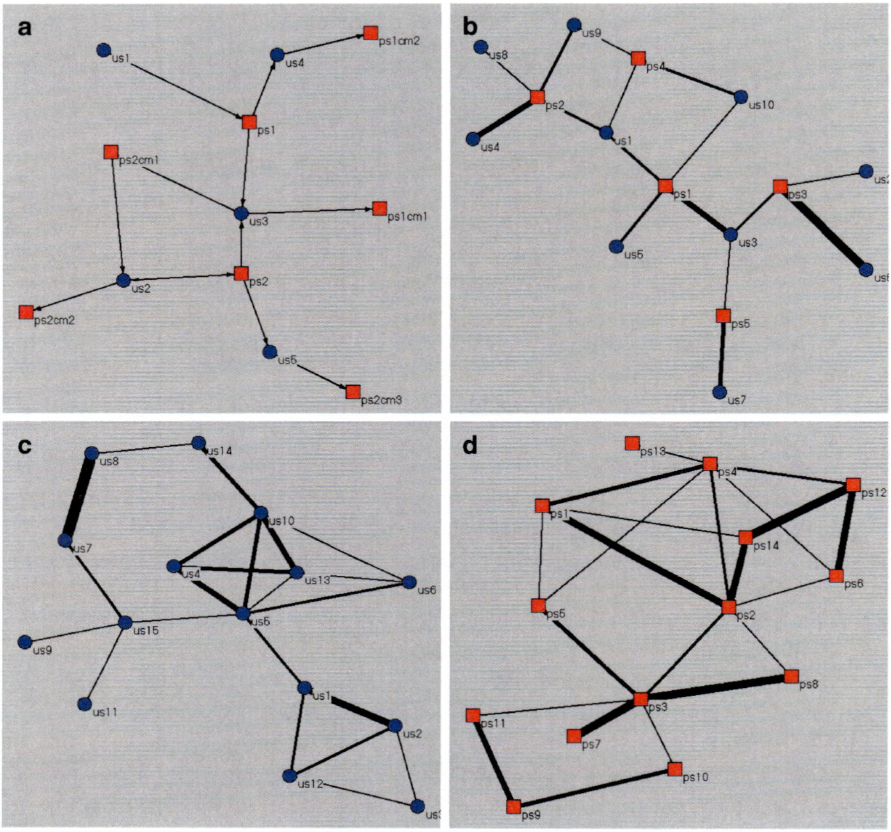

Fig. 8.1 Illustration of data mapping onto a directed bipartite network with users represented by *bullets*, and posts and comments, represented by *squares* (**a**), and a bipartite network of users and posts, while the weighted links represent the number of related comments (**b**). Weighted monopartite networks of users connected via number of common posts (**c**) and network of posts connected via number of common users (**d**)

Starting with a given dataset, several types of networks can be derived, depending on the degree of information that we want to keep. In particular, we consider following types of networks:

- *Full bipartite network*, as described above, with users (U) and posts & comments (P + C), as two partitions. Together with the direction of the links and times of their appearance, this network contains full information stored in the data. It is most suitable for visually presenting a *network of an individual post with all its users and their comments* as separate nodes. Notice that in the considered datasets such network can be very large (cf. Table 8.1 for the size of data). The mapping is illustrated in Fig. 8.1a.

- *Weighted bipartite network* is a compressed bipartite network with users and posts as two types of nodes (U + P), while the weights of links W_{ij} represent the number of comments of the user i on the post j, as illustrated in Fig. 8.1b. These networks are undirected and more suitable for the spectral analysis, compared to full bipartite networks.
- *Weighted networks in user-projection or post-projection*, are undirected monopartite versions obtained by suitable projections from the full bipartite network of the data. In the projection onto user networks, two users are connected directly with a weighted link, representing the number of common posts per pair of users, C_{ij}^{B}, or similarly, the number of common users per pair of posts, C_{ij}^{U}, in the post-projection networks, as illustrated in Fig. 8.1c,d.

Four examples of the networks which are generated from our datasets of Diggs, Blogs, and movies are shown in Fig. 8.2. They represent different levels of the data mapping described above as directed and weighted bipartite networks and the weighted networks of two monopartite projections. Most of these networks also exhibit the community structure, which will be further discussed in the Sect. 8.2.3.

8.2.2 *Topological Properties of the Emergent Networks*

We have explored in detail the topological properties of the networks derived from Blogs and Diggs [27, 30] and movie datasets [17]. As discussed below, these networks are topologically inhomogeneous at the microscopic scale, at the level of node connectivity, and appear to have specific mixing properties, at the level of neighboring nodes. The projected monopartite networks appear to have high clustering coefficient. Moreover, a striking topological property of these networks is their *mesoscopic* inhomogeneity, or the occurrence of *communities* of nodes with stronger connections among each other [12, 13, 31]. These communities play an important role in the collective dynamics that we are interested in. The methodology to systematically detect the communities in these types of networks based on the *eigenvalue spectral analysis of the weighted bipartite graphs* is outlined below in Sect. 8.2.3. Details of the spectral analysis of modular monopartite networks can be found in recent work [31] and references cited therein.

Degree of a node q, is given by the number of its first neighbors, the number of links between that node and the rest of the network [3]. In directed networks, one can distinguish between in- and out-degree of the nodes. The degree distribution $P(q)$ is a probability that a randomly chosen node in the network has degree q, while *cumulative degree distribution* represents probability that a randomly chosen node has degree $> q$ [3]. In our bipartite networks, the node degree has a particular meaning for each partition. Specifically, for the user partition, the in-degree $q_{\rm in}^{\rm U}$ of a node represents the number of posts and comments read (and commented), while the

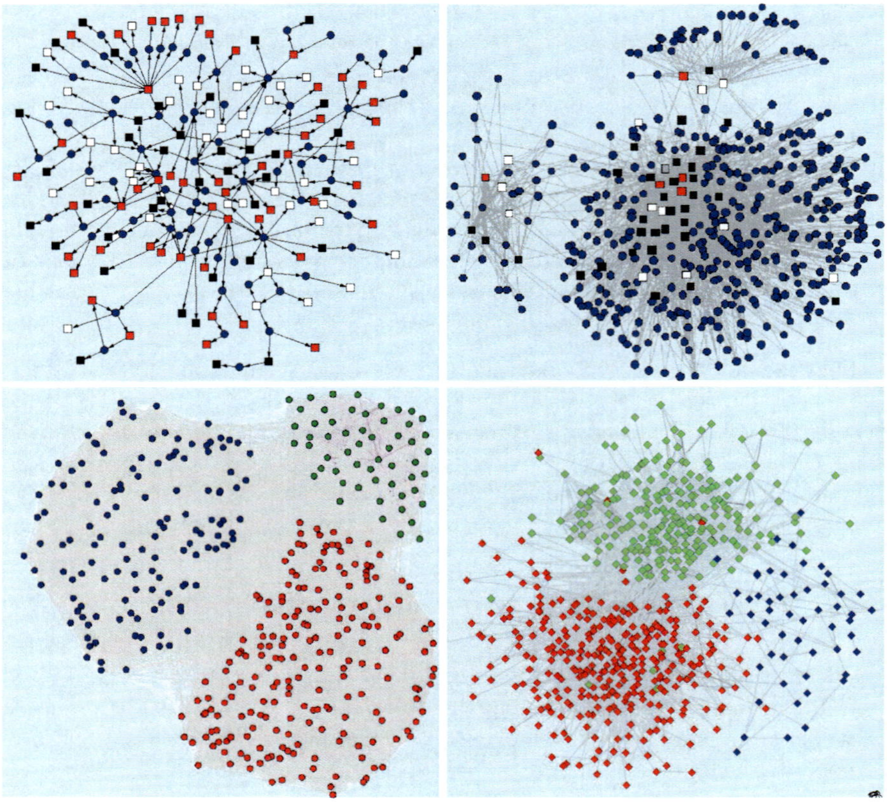

Fig. 8.2 Examples of networks studied in this work, which are obtained by mapping from Blogs, Diggs, and Movie datasets. (*top left*) Directed bipartite network related to one Digg story with *comments*, shown as *colored boxes*, and *users*, shown as *bullets*. *Color* of the comments denotes their emotional content: *white*-neutral, *red*-positive and *black*-negative. (*top right*) Weighted bipartite networks of popular BBC Blogs with *posts* as c*colored boxes*, and *users* as *bullets*. Weights of the links indicate multiple comments by the user to the post, while the color of posts indicates their overall emotional contents, computed from all comments made at that post. (*bottom left*) Monopartite weighted network of *users* from popular Digg stories. *Colors* of the nodes where indicate that these nodes belong to the same user community. (*bottom right*) Weighted network of movies with number of votes between 1,000 and 2,000. Only part of the network of size 596 nodes is shown, corresponding to commons above $C_{ij}^{\mathrm{M}} > 5$. *Colors* indicate different modules identified by the weighted maximum-likelihood algorithm [29]. The nodes in *red* group appear to be predominantly marked as "horror," *green* nodes as "drama," and *blue* nodes as "romance" movies [17] in the dataset. [Data replotted from [17]]

out-degree $q_{\mathrm{out}}^{\mathrm{U}}$ represents the number of posts and comments written by that user. Note that in the data the same user often comments a particular post (or comment) more than once, which is denoted by multiple link. Within the posts-and-comments partition, the out-degree q_{out}^{B} of a node represents the number of users who left a

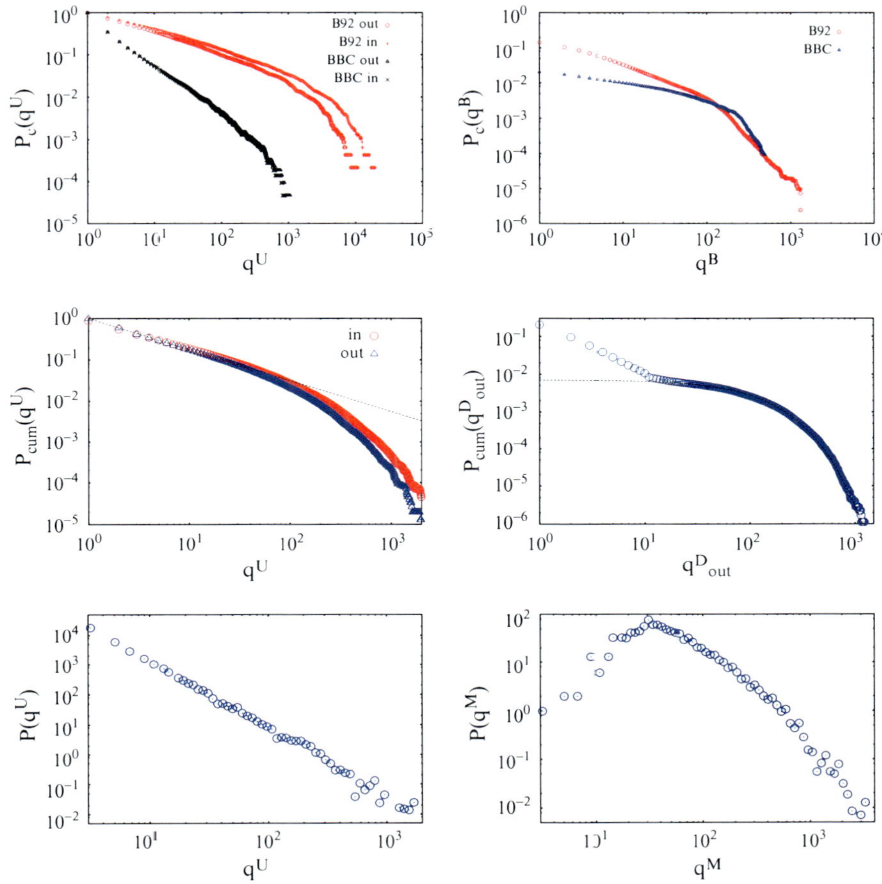

Fig. 8.3 Cumulative degree distributions for *users* and *posts* from the bipartite networks representing data from (*top*) Blogs [Reprinted from [30], ©Springer 2009.]; (*middle*) Diggs data, and (*bottom*) data for popular movies with more than 1,000 votes

comment on that particular post (comment), whereas the in-degree $q_{\rm in}^B$ is the number of authors of the post (comment), which is always equal to one in the Blogs and Diggs data.

Cumulative distributions for in- and out-degree for both partitions of the networks corresponding to the Blog and Digg data listed in Table 8.1 are shown in Fig. 8.3. Broad degree distributions indicate large inhomogeneity both in the user- and post-partition. In particular, for the user partition, we have power-law dependences for both in- and out-degree over two orders of magnitude, before the cut-off. Note that when the information about comment-on-comment is not available, that is BBC Blogs data, in- and out-degree distributions coincide, otherwise they are different. The observed exponent is compatible with the differential distributions of user degree as

$$P(q) \sim (q)^{-\tau} \exp\left[-\left(\frac{q}{q_0}\right)^{\sigma}\right], \tag{8.1}$$

with the exponent is found in the range $\tau \in [1.5, 2]$ and the cut-off q_0 and stretching σ are depending on the dataset.

The degree distributions of the posts-and-comments partition, on the other hand, show a characteristic bending around $q^B \sim 100$. The posts with the degree above this point represent a different category of *popular posts*, which follow another dynamical pattern, as discussed in [27]. Because of the different data structure in the case of movie networks, the user degree q^{U} and movie degree q^{M}, which we define here, are affected differently by the applied filter with the number of votes above 1,000. The results are shown in Fig. 8.3 (bottom panels). Again, we observe high heterogeneity in both partitions (see also [17]). The user degree distribution exhibits prominent power-law with the slope close to 1.5, while the applied filter affects the movie distribution for small values of q^{M}.

Mixing pattern (or dis-assortativity) is another local topology measure, defined at the level of node's neighborhood, which in our bipartite techno-social networks differs from the conventional social networks. The (dis)-assortativity measures the node's preferences to attach to (dis-)similar neighbors [32] and can be quantified through the average degree of its neighbors $< q >_{nn}$. In the case of our bipartite networks, we have

$$< q_{\kappa} >_{nn} = \sum_{i}^{q_{\kappa}} \frac{q^i_{\kappa'}}{q_{\kappa}}, \tag{8.2}$$

where (κ, κ') indicate neighboring nodes, which belong to the different partitions. For instance, in the user–movie network when q_{κ} indicates degree of a given user-node, then $q^i_{\kappa'}$ stands for the degree of a movie-node i, which is directly linked to that user. Plotting the average degree of nearest-neighbor nodes $< q_{\kappa} >_{nn}$ against node's degree q_{κ} marked in (8.2), leads to a specific pattern, which can be recognized either as disassortative mixing (descending curve, indicating that rich nodes link to poor nodes), or the opposite, assortative mixing, where rich nodes link to similarly rich nodes. The respective power-law dependences with small negative exponents are found in a number of technological networks, while positive exponents are characteristic for conventional social networks [32]. Whereas a number of networks exhibit neither the assortative nor disassortative behavior (zero exponent).

The corresponding plot in the case of our bipartite movie network is shown in Fig. 8.4 (right). Descending curve indicates a *disassortative* mixing for both user and movie nodes. This indicates that in the techno-social interactions the underlying technological network dominates the social mixing of users. However, the functional dependence is logarithmic rather than power-law. Similar formula applies for the weighted bipartite networks of posts and users, where instead of the degree we rather

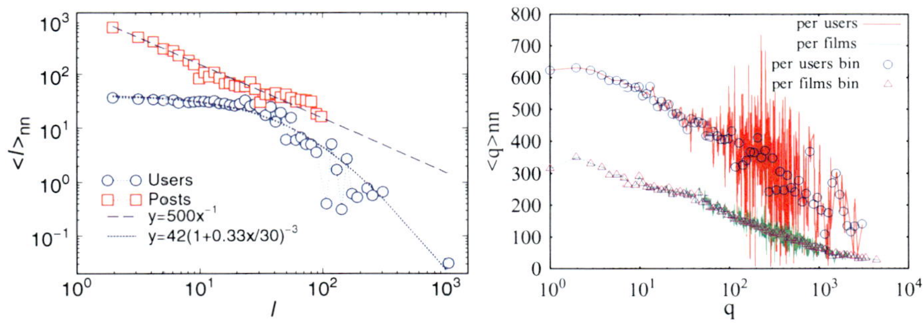

Fig. 8.4 Mixing properties of user – post nodes from the weighted bipartite networks of Blogs of normal popularity (*left*). Mixing properties in bipartite user – movie networks (*right*) [Figure right reprinted with permission from [17], ©IEEE 2009.]

consider the strength of the corresponding node, i.e., $q^i_{\kappa'} \to \ell^i_{\kappa'}$, where $\ell^i_{\kappa'} \equiv \sum_j C^\mu_{ij}$ is a sum of weights of all links at that node. The results are also shown in Fig. 8.4 (left). In this case a disassortative mixing is also found in both user and post partitions. However, a power-law decay is observed for the posts partition, while in the case of user partition we have a power-law tail for node degree above $q_0 \sim 30$ and no assortativity at smaller degree.

In the directed bipartite networks representing our data from Blogs and Diggs, one can distinguish between several combinations of the node's degrees in (8.2), which have particular meaning. Besides typical *in–out* degree correlations in (8.2) one can also consider *in–in* and *out–out* degree correlations or next-neighborhood of a node. For example, consider a user node *i*, whose out-degree is given by the number of posts and comments written by that user, while out-degree of these posts and comments is given by the number of users who read them, possibly 'fans' or a community. Similarly, in-degree is given by all posts and comments that the user *i* read, while their out-degree is given by the number of other users who read them, and thus might share similar interests with the user *i*. The mixing at the next-neighborhood level in our bipartite networks thus indicates the linking patterns that may lead to the formation of communities, which we discuss below.

Distributions of commons, that is the number of common users per pair of posts and vice versa, the common number of posts per pair of users, is a particular property of our bipartite networks, which further characterizes their local topology at the level of pairs of nodes. It is thus related to both the degree and the mixing properties of nodes. Moreover, a systematic study of the commons on a bipartite network provides a natural way to project the networks to either one of the monopartite versions with the weights of links given exactly by the *commons*. Note that on such monopartite projections, the sum of commons of a node with all its neighbors determines the topological *strength* of that node. The distributions of common number of posts C^B_{ij} and C^P_{ij} per pair of users from our data on Blogs and Diggs, respectively, are shown in top row of Fig. 8.5. We find power-law dependences of these distributions, with

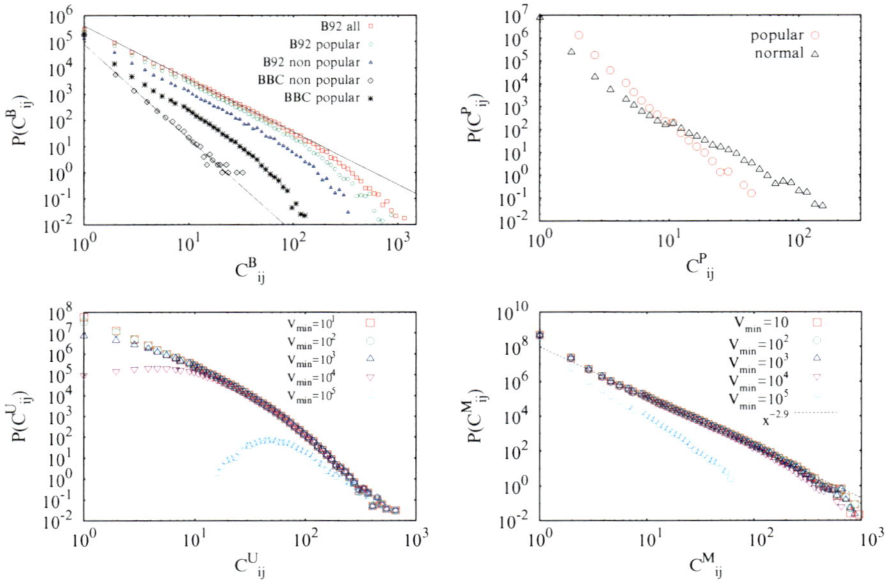

Fig. 8.5 Distribution of common posts per pair of users for BBC and B92 Blog (*top left*) and Digg (*top right*). Distribution $P(Cij)$ of the common number C_{ij}^{U} of users per pair of movies (*bottom left*) and of the common number C_{ij}^{M} of movies per pair of users (*bottom right*) found in the bipartite networks of different sizes, selected according to number of votes. [*Top left* Figure reprinted from [30], ©Springer 2009. *Bottom* Figures reprinted from [17], ©IEEE 2009.]

the exponents and the cut-offs depending on the dataset and popularity of the posts. In the case of movie dataset, we have a fat-tail distribution both in the movie and user partition. The distribution depends on the popularity or the number of votes of the movies (see Fig. 8.5, lower panels).

8.2.3 *Occurrence of Communities in Blogs, Diggs, and MDb Users*

Having the networks constructed from the data and put into a suitable forms, as discussed above, we look for the *community structure* of these networks. In the topological sense, a community is a mesoscopic inhomogeneity with a group of nodes, which are better connected among themselves than to the rest of the network, and thus can be detected by suitable methods (for a recent review of different methods, see [13]). Our networks representing the user interactions over Blogs, Diggs and movies, have several specific features, that need to be taken into account when the communities are searched on them. In particular,

- High clustering coefficient of the monopartite projections: note that k-links of a user node project to a k-clique in the post-projection network. Having the power-law degree distributions in both partitions, Fig. 8.3, thus leads to a high clustering and an overlap of cliques of different sizes in the projected networks;
- Weights on links, both in the projected and compressed bipartite versions, are given by power-law distributions, e.g., commons in Fig. 8.5.

For these reasons the standard methods of community structure analysis, based on the max-flow–min-cut principle, are often less efficient. We use two methods, based on the spectral analysis of graphs [31], and the weighted maximum-likelihood method [29], which are suitable for strongly clustered and weighted networks. Here we describe the basic features of the spectral method and show some results of the community structure in our networks derived from the Blogs, Diggs and movie datasets, as the networks shown in Fig. 8.2.

Spectral analysis method for finding communities is based on specific properties of eigenvalues and eigenvectors of normalized Laplacian for mesoscopically inhomogeneous graphs [11, 31], which is related to the adjacency matrix of the underlying (weighted) network. For all cases considered in this work, the Laplacian operator can be written in the form

$$L^U_{ij} = \delta_{ij} - \frac{W_{ij}}{\sqrt{\ell_i \ell_j}} \; ; \qquad \ell_i \equiv \sum_j W_{ij} \, . \tag{8.3}$$

where the matrix representing the network to be considered is given by W_{ij} and ℓ_i is so called *strength* of a node (that can be defined both for user or post nodes). As mentioned above, in the cases that we consider here the matrix W_{ij} is weighted, i.e., its nonzero elements are real numbers or integers larger than 1, for instance representing the commons, as further discussed below. In the limit of binary graphs, the elements $W_{ij} = \{0,1\}$, whereby the strength ℓ_i reduces to the number of links (degree) of the node i.

In each particular network, the matrix W_{ij} is specified. For instance, in the case of user-projected networks, which is more suitable for the community analysis in normally popular Blogs discussed in [30], the matrix W_{ij} is replaced by the commons C^B_{ij}. Similarly, in the monopartite networks projected onto posts-and-comments partition, the weighted network connections W_{ij} is suitably given by the number of common users linking them, i.e., C^U_{ij}. Whereas in the weighted bipartite networks of posts and users, suitable for the analysis of popular posts [27,28], one can use the number of comments of a user to a post as the elements of the matrix W_{ij}. Below we discuss several of these cases in detail.

The standard algorithms for solving the eigenvalue problem are then used to determine the complete spectrum of the eigenvalues $\{\lambda_0, \lambda_1, \ldots, \lambda_N\}$ and the corresponding eigenvectors $\{V_0, V_1, \ldots, V_N\}$. The size of the network N can be sometimes a limiting factor for the numerical algorithm. In such cases, we reduce the network by using its topological properties, for example node's connectivity, centrality or strengths, or by removing the links below certain weight. The spectrum

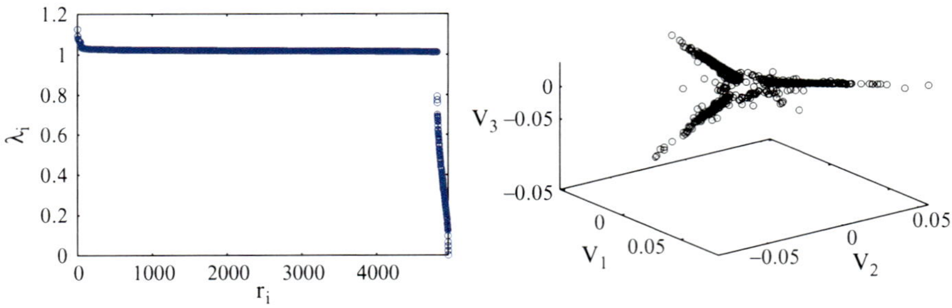

Fig. 8.6 Spectrum (*left*) and 3D scatter plot of eigenvectors corresponding to smallest non-zero eigenvalues (*right*) for weighted projection of bipartite network of BBC normal posts on user partition. [Figure on right reprinted from [30], ©Springer 2009.]

of normalized Laplacian is bounded and all eigenvalues are in the interval $[0,2]$ regardless of network size and its structure. The number of zero eigenvalues is equal to the number of disconnected components. For connected network there is only one zero eigenvalue with the eigenvector $V_0 = (\sqrt{l_1}, \ldots, \sqrt{l_N})$. Since the operator given in (8.3) is symmetric and has real elements, the eigenvalues are real while the corresponding eigenvectors are orthogonal. In addition, the spectrum of the Laplacian of a modular network has specific structure, with a separate set of the eigenvalues occurring between zero and the main part of the spectrum (corresponding to an extra peak in the spectral density of such graphs, see [31]). Furthermore, the eigenvectors corresponding to these eigenvalues appear to be *localized* on the network modules, in view of the orthogonality to the eigenvector V_0. This means that the nonzero (positive or negative) components of these eigenvectors carry indexes of the nodes belonging to a particular network subgraph. Consequently, scatter plots in the space of the eigenvectors of the lowest nonzero eigenvalues, for example in the (V_1, V_2)-plane, exhibit a characteristic *branched* structure, with different branches indicating a different substructure on the graph. A typical example of such branched structure in 3-dimensional space of the eigenvectors and the corresponding spectrum of the eigenvalues obtained from the user-projected network of normally popular Blogs is shown in Fig. 8.6.

Having the IDs of nodes which belong to each community, the structure of the communities can be further examined using additional information contained in the original data. For example, in the communities shown in Fig. 8.6, we find the lists of posts that the users in each community commented. They appear to have strong *subject similarity*, for example sports, economy and business, etc. However, the user communities identified with the same approach on Diggs and the popular Blogs do not entirely follow the subject-related structure. For example, among the three communities found in the user-projected network on Diggs, shown in Fig. 8.2 (bottom left), the smallest one contain users who commented a particular subject (politics). While the other two groups are related with a mixture of different subjects. The difference between the two groups is that the users in one of the groups are in the average three times more active than in the other.

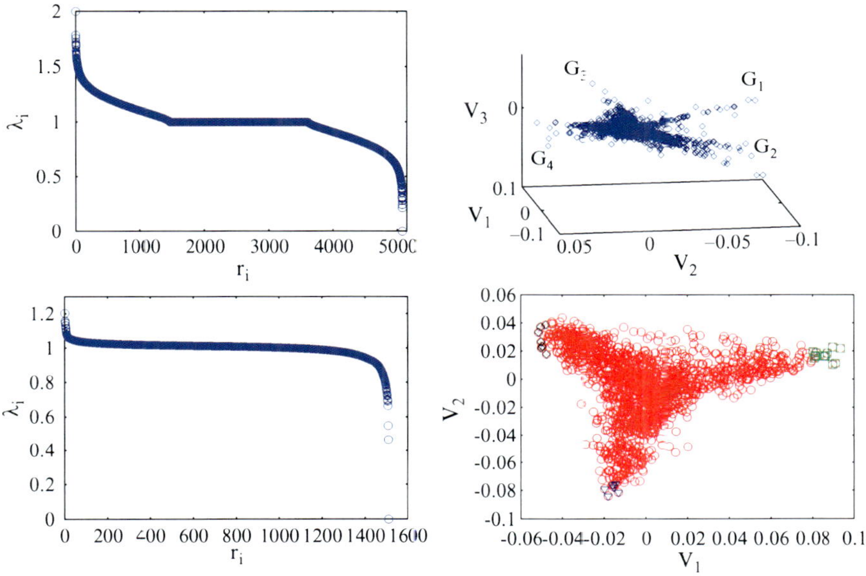

Fig. 8.7 Spectrum (*top left*) and scatter plot (*top right*) for weighted bipartite network of popular B92 Posts and Users related to them. Spectra (*bottom left*) and scatter plot of two eigenvectors for smallest non-zero eigenvalues (*bottom right*) for weighted network of movies with number of votes in the interval $[1000, 2000]$. Points with different color represent movies of different genre: ($\triangle$) drama, ($\diamond$) romance and ($\square$) horror movies. [*Top right* Figure reprinted from [30], ©Springer 2009. *Bottom panels* Figures reprinted from [17], ©IEEE 2009.]

In the case of *popular* Blogs, we find the community structure on the weighted bipartite networks with users and posts, while the weights W_{ij} are given by the number of comments, as explained above. An example of the spectrum and the corresponding scatter plot with a rich structure of the communities is shown in Fig. 8.7, top row. Note that in this case the points in each branch indicate the nodes of both partitions, i.e., users and posts. An example of the network with three such communities found on popular Blogs is shown in Fig. 8.2 (top right), where user nodes within a community are linked via several posts. As mentioned above, the subjects of the posts within the same community vary, which indicates that another feature of these posts makes them popular.

In Fig. 8.7, lower panels show the spectrum and the corresponding two-dimensional scatter plot of the eigenvectors obtained for the movie-projection network from the bipartite network of the movie data. Note that in this case the weights of the links W_{ij} are given by the number of common users per pair of movies, C_{ij}^U, which also exhibits a power-law distribution (cf. in Fig. 8.5). As the scatter plot indicates, three groups of movies can be distinguished, based on the community structure analysis, in which the numbers of common users per pair of movies play a role. Following the IDs of the movies in each group (away from the center of the plot), further information in the database suggests their strong similarity according to the genre.

8.3 Robust Features of User's Emotional Behavior on Blogs

Beyond the topological features with the community structure, the origin and the evolution of the user communities is a question for which both the *patterns of user behavior* and the *emotional contents* of the posts and comments play an important role. In this section, we address this question in connection with the community detection, that we have studied above (see also [27, 28]). With the help of the Emotion Classifier, described in [27], we can take into account the emotional contents of the texts of posts and comments made by the users within identified communities, and thus unravel the role of the expressed emotions for the evolution and the structure of that community. As mentioned above, the communities related with the posts of normal popularity that we studied in Blogs data, are predominantly subject-related, or author-related [30]. Applying the emotion classifier to the texts of comments at these posts, we find that they are often classified as "positive" or "neutral," with few texts that are classified as "negative." In Fig. 8.8, we show how a smaller community emerges on such posts, the snapshots of the weighted bipartite networks are shown at several time intervals. As the Fig. 8.8 shows, among many posts written by the two very productive authors, only few of them become commented and even smaller number of them attract more users. The weights of the links indicate that few users comment the particular post multiple times. The emotional contents of the posts is not shown.

In the evolution of popular posts, with the number of comments above 100, however, the role of the emotions is much more pronounced. Closer inspection in the fluctuations of the number of users (size of the community) reveals strong correlations with the emotional contents of their comments made to the posts through which they are linked. The results are shown in Fig. 8.9 for two communities which are visible on the network in Fig. 8.2 (top right). As the figure shows, the excess of negative comments (critique) is strongly correlated with the increase of the size of community, and vice versa, the community size decreases when the overall *charge* of the comments is balanced around zero. Similar correlations are found for other communities on popular Blogs and in the case of popular Digg stories [27, 28].

Patterns of user behavior at posts is another feature that can be studied from the available Blog data [30]. The temporal patterns of user activity have a fractal structure, as shown for instance for the blogsdata which we consider here in [30]. This means that the distance between two successive points in time Δt, that is user inactivity time, obeys a power-law distribution

$$P(\Delta t) \sim (\Delta t)^{-\tau_\Delta} \exp\left[-\frac{\Delta t}{\Delta_0}\right], \tag{8.4}$$

with the exponent $\tau_\Delta \gtrsim 1$ and the cut-off length depending on the dataset. Combined with our emotion classifier, we now consider the times of user actions which resulted in an emotional (positive/negative) comment on the popular posts in our Blogdata. The distributions $P(\Delta t)$ averaged over all 3,000 most active users in our dataset are

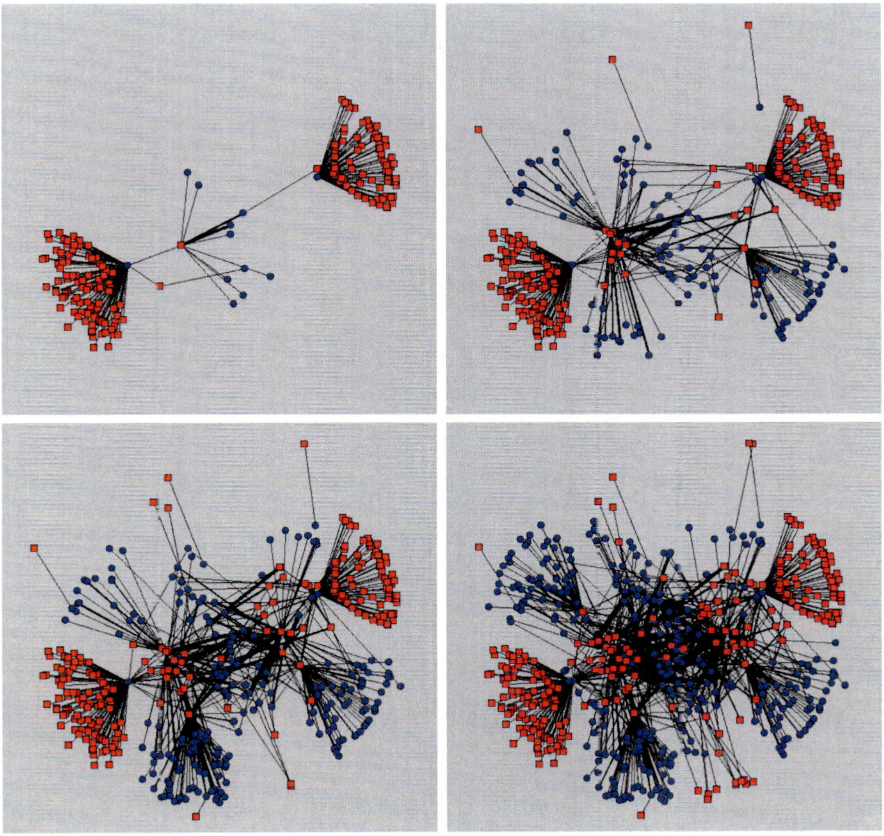

Fig. 8.8 Snapshots in the evolution of the community of users on normally popular Blogs. Weighted bipartite network featuring users (*circles*) and posts (*squares*), while links are weighted by the number of comments of a user to the post

given in Fig. 8.10a. The power-law distributions up to a cut-off are found for both positive and negative emotional comments, with the exponent $\tau_\Delta \sim 1$, compared with $\tau_\Delta = 1.15$ found for the distribution of all types of comments in the entire dataset [30], indicating the robustness in the user's activity.

Another situation is found on posts, i.e., when one considers the reaction time to the posted material. The distribution $P(t - t_{P0})$, where t stands for the action time of anyone of the users, and t_{P0} is the time of posting of a given post, is shown in Fig. 8.10b for our dataset of popular posts. We also consider here only actions which resulted in emotional comments. The delay of the user's emotional action to the posted text obeys a power-law distribution, shown in Fig. 8.10b for both positive and negative comments. Such distributions are fitted with the q-exponential expression

$$P(t - t_{P0}) = B\left(1 - (1-q)\frac{t - t_{P0}}{T_0}\right)^{1/1-q}, \tag{8.5}$$

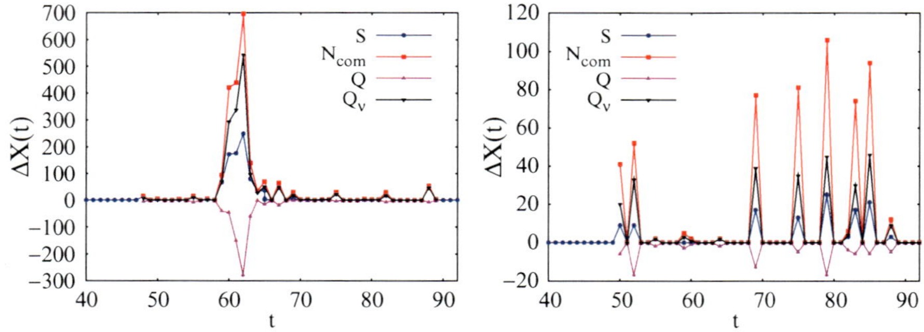

Fig. 8.9 Evolution of size of a community and charge of comments on popular Blog posts for two of the communities shown in Fig. 8.2 (*top right*). Time is measured in weeks. Also shown are the fluctuations in the number of all comments and the number of comments marked as subjective within the same community

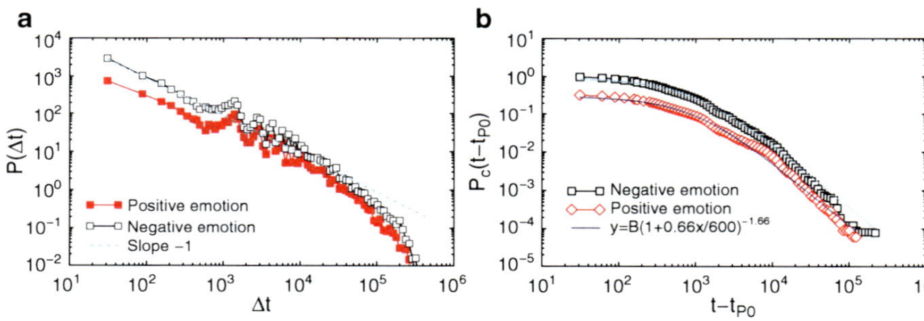

Fig. 8.10 (**a**) Distribution of the delay between successive emotional comments (positive/negative) of a user on popular Blogs, averaged over 3,000 most active users in the dataset. (**b**) Cumulative distribution of the delay of the emotional comments at a given post with respect to its posting time, averaged over all popular posts in the dataset. [Fig. (a) reprinted from [27], ©Springer 2010]

which is often found in nonergodic dynamical systems [39]. The fits are also shown in Fig. 8.10b, where $q = 1.60$. The power-law tail has the decay exponent $\tau = 1.66$ of the cumulative distribution (or 2.66 for the differential distribution), compared with 2.3 found in all comment types in the entire dataset from Blogs in [30]. The reaction time with the emotional comments thus represents another robust feature of human behavior, which belongs to class of events leading to $\tau > 2$, predicted in [8].

8.4 Conclusions

Datasets from Blogs and similar Web portals contain valuable information from which the social interaction in Cyberspace can be studied. We have demonstrated a systematic methodology for detecting Cybercommunities of users linked over different movies or textual posts, and study of their internal structure and the evolution. In our approach we emphasise the combined methods of the network theory and the machine-learning algorithms of the text analysis, specifically the emotion-classifier, which reveals the role of emotions expressed in particular posts (and related comments) in the occurrence and the evolution of user communities.

The bipartite networks, which are suitable representations of the datasets from Blogs, Diggs, movie databases, and similar Web portals, have several specific features, which we studied here. In particular:

- *Networks topology* is rich, characterized with power-law dependences of various quantities, including the weights of links. High clustering and community structure occur in the appropriate monopartite projections. Furthermore, the mixing patterns with a dis-assortative nature in both user and post partitions, induced by technology-mediated interactions, are found in contrast to conventional social networks.
- *Cybercommunities*, identified as topological mesoscopic inhomogeneities on these emergent networks, consist of users grouped around certain popular posts. The contents of these posts and to them related comments thus play an important role in user grouping.
- *Emotional content*, which we classify with the machine-learning methods, of the posts and comments related with the communities, reveal the role of emotions in the appearance and evolution of these communities.
- *Patterns of user behaviors* underlies the community evolution. We have studied the response time to the posts, especially in respect to the posts with emotional contents. The results reveal robust behavior characteristic to a larger class of human response patterns, which suggests that the emotional comments are not contemplated but provoked by the posted material within a self-organized dynamics.

The idea of the methodology that we demonstrated here on Blogdata, which combines the network theory with the appropriate machine-learning methods, can be useful for study online communications in different other media.

Acknowledgments The research leading to these results has received funding from the European Community's Seventh Framework Programme FP7-ICT-2008-3 under grant agreement n^o231323 (CYBEREMOTIONS). B.T. also thanks support from the national program P1-0044 (Slovenia). Special thanks to G. Paltoglou for providing user-friendly emotion classifier for Blogs.

References

1. Adrianson, L.: Gender and computer-mediated communication: group processes in problem solving. Computers in Human Behavior **17**(1), 71–94 (2001). DOI DOI:10.1016/S0747-5632(00)00033-9. URL http://www.sciencedirect.com/science/article/B6VDC-423HJ18-5/2/3cc96b3953897a77fb44412ebddf7f53d

2. Berners-Lee, T., Hall, W., Hendler, J., Shadbolt, N., Weitzner, J.: Creating a Science of the Web. Science **313**, 769–771 (2006)
3. Boccaletti, S., Latora, V., Moreno, Y., Chavez, M., Hwang, D.U.: Complex networks: Structure and dynamics. Physics Reports **424**, 175–308 (2006). DOI 10.1016/j.physrep.2005.10.009
4. Bollobas, B.: Modern Graph Theory. Springer-Verlafg, Berlin Heidelberg (1998)
5. Brumfiel, G.: Breaking the convention? Nature **459**, 1050–1051 (2009)
6. Cattuto, C., Barrat, A., Baldassarri, A., Schehr, G., Loreto, V.: Collective dynamics of social annotation. PNAS **106**, 10511–10515 (2009)
7. Cho, A.: Ourselves and Our Interactions: The Ultimate Physics Problem? Science **325** (2009)
8. Crane, R., Schweitzer, F., Sornette, D.: Power Law Signature of Media Exposure in Human Response Waiting Time Distributions. Phys. Rev. E **81**(5), 056101 (2010)
9. Dodds, P., Danforth, C.: Measuring the Happiness of Large-Scale Written Expression: Songs, Blogs, and Presidents. Journal of Happiness Studies **11**(4), 441–456 (2009)
10. Donato, D., Leonardi, S., Millozzi, S., Tsaparas, P.: Mining the inner structure of the Web graph. Journal of Physics A: Mathematical and Theoretical. **41**, 224,017 (2008)
11. Donetti, L., Muñoz, M.A.: Detecting network communities: a new systematic and efficient algorithm. Journal of Statistical Mechanics: Theory and Experiment **10** (2004)
12. Evans, T.S., Lambiotte, R.: Line Graphs, Link Partitions and Overlapping Communities. Phys. Rev. E **80**(1), 016105 (2009). DOI 10.1103/PhysRevE.80.016105
13. Fortunato, S.: Community detection in graphs. Physics Reports **486**(3-5), 75–174 (2010). DOI DOI:10.1016/j.physrep.2009.11.002
14. Fowler, J.H., Christakis, N.A.: Dynamic spread of happiness in a large social network: longitudinal analysis over 20 years in the framingham heart study. British Medicine Journal **337**, a2338 (2008). DOI 10.1136/bmj.a2338. URL http://www.bmj.com/cgi/content/abstract/337/dec04_2/a2338
15. Fu, F., Liu, L., Yang, K., Wang, L.: The structure of self-organized blogosphere. arXiv:0607361 (2006)
16. Grujić, J.: Movies recommendation networks as bipartite graphs. Lecture Notes in Computer Science **5102** (2008)
17. Grujić, J., Mitrović, M., Tadić, B.: Mixing patterns and communities on bipartite graphs on web-based social interactions. In: Digital Signal Processing, 2009 16th International Conference on, pp. 1–8 (2009). DOI 10.1109/ICDSP.2009.5201238
18. Kleinberg, J.: The Convergence of Social and technological Networks. Communications of the ACM **51**, 66–72 (2008)
19. Kujawski, B., Holyst, J., Rodgers, G.: Growing trees in internet news groups and forums. Phys. Rev. E **76**, 036103–+ (2007)
20. Lambiotte, R., Ausloos, M.: Uncovering collective listening habits and music genres in bipartite networks. Physical Review E **72**, 066107 (2005). URL doi:10.1103/PhysRevE.72.066107
21. Lambiotte, R., Ausloos, M.: On the genre-fication of music: a percolation approach (long version). The European Physical Journal B **50**, 183 (2006)
22. Leskovec, J., McGlohon, M., Faloutsos, C., Glance, N., Hurst, M.: Cascading behavior in large blog graphs (2007). URL http://arxiv.org/abs/0704.2803
23. Liu, L., Fu, F., Wang, L.: Information propagation and self-organized consensus in the blogosphere: a game theoretical approach (2007) http://arxiv.org/pdf/physics/0701316
24. Lorenz, J.: Universality of movie rating distributions. Eur. Phys. Journal B **71**(2), 251–258 (2008). DOI 10.1140/epjb/e2009-00283-3
25. Manning, C.D., Raghavan, P., Schütze, H.: Introduction to Information Retrieval, 1 edn. Cambridge University Press (2008). URL http://www.amazon.com/exec/obidos/redirect?tag=citeulike07-20&path=ASIN/0521865719
26. Manning, C.D., Schüetze, H.: Foundations of Statistical Natural Language Processing, 1 edn. The MIT Press (1999). URL http://www.amazon.com/exec/obidos/redirect?tag=citeulike07-20&path=ASIN/0262133601

27. Mitrović, M., Paltoglou, G., Tadić, B.: Networks and emotion-driven user communities on popular blogs. Eur. Phys. Journal B **77**, 597–609 (2010)
28. Mitrović, M., Paltoglou, G., Tadić, B.: Quantitative analysis of bloggers colelctive behavior powered by emotions. Journal of Statistical Mechanics: Theory and Experiment **2011**(02), P02005 (2011)
29. Mitrović, M., Tadić, B.: Search of Weighted Subgraphs on Complex Networks with Maximum Likelihood Methods. Lecture Notes in Computer Science **5102**, 551–558 (2008)
30. Mitrović, M., Tadić, B.: Bloggers behavior and emergent communities in blog space. Eur. Phys. Journal B **73**(2), 293–301 (2009). DOI 10.1140/epjb/e2009-00431-9
31. Mitrović, M., Tadić, B.: Spectral and dynamical properties in classes of sparse networks with mesoscopic inhomogeneities. Phys. Rev. E **80**(2), 026123–+ (2009). DOI 10.1103/PhysRevE.80.026123
32. Newman, M.E.J.: Mixing patterns in networks. Phys. Rev. E **67**, 026126 (2003). URL http://www.citebase.org/abstract?id=oai:arXiv.org:cond-mat/0209450
33. Pang, B., Lee, L.: Opinion Mining and Sentiment Analysis. Now Publishers Inc (2008). URL http://www.amazon.com/exec/obidos/redirect?tag=citeulike07-20&path=ASIN/1601981503
34. Panzarasa, P., Opsahl, T., Carley, K.: Patterns and dynamics of users' behavior and interactions: network analysis of and online community. Journal of the American Society for Information Science and Technology **60**, 911–932 (2009)
35. Sano, Y., Takayasu, M.: Macroscopic and microscopic statistical properties observed in blog entries. Journal of Economic Interaction and coordination **5**(2), 10 (2009)
36. Tadić, B.: Dynamics of directed graphs: the world-wide Web. Physica A **293**, 273–284 (2001)
37. Thelwall, M., Buckley, K., Paltoglou, G., Cai, D., Kappas, A.: Sentiment Strenght Detection in Short Informal Text. Journal of the American Society for Information Science and Technology **61**(12), 2544–2558 (2010)
38. Thelwall, M., Byrne, A., Goody, M.: Which types of news story attract bloggers? Informetion-research **12**, 327+–22 (2007)
39. Tsallis, C., Bukman, D.J.: Anomalous diffusion in the presence of external forces: Exact time-dependent solutions and their thermostatistical basis. Phys. Rev. E **54**(3), R2197–R2200 (1996). DOI 10.1103/PhysRevE.54.R2197
40. Zhou, T., Jiang, L., Su, R., Zhang, Y.: Effect of initial configuration on network-based recommendation. Eyrophys. Lett. **81**, 58004 (2008)

Chapter 9
k-Core Organization in Complex Networks

G.J. Baxter, S.N. Dorogovtsev, A.V. Goltsev, and J.F.F. Mendes

Abstract We analyse the complex network architectures based on the k-core notion, where the k-core is the maximal subgraph in a network, whose vertices all have internal degree at least k. We explain the nature of the unusual "hybrid" phase transition of the emergence of a giant k-core in a network, which combines a jump in the order parameter and a critical singularity, and relate this transition to critical phenomena in other systems. Finally, we indicate generic features and differences between the k-core problem and bootstrap percolation.

9.1 Introduction

The first natural description of the global organization of a random network is based on connected components [1–4]. The emergence of a giant connected component in a network is the most important structural change in the network architecture. In a connected component, there exists a path between every pair of nodes of this component. By definition the giant connected component consists of a finite fraction of vertices (nodes) in an infinite network. Although this notion formally assumes that the network is infinite, this transition is still observable in large yet finite networks. The emergence of a giant component is a continuous phase transition which usually takes place when a network is sparse, that is, when the average degree of a node (the average number of connections) is finite. As the average degree of a node $\langle q \rangle$ increases, at the critical point for this phase transition, q_c,

G.J. Baxter (✉) • J.F.F. Mendes
Departamento de Física, I3N, Universidade de Aveiro, Campus Universitário de Santiago, 3810-193 Aveiro, Portugal
e-mail: gjbaxter@ua.pt

S.N. Dorogovtsev • A.V. Goltsev
A. F. Ioffe Physico-Technical Institute, 194021 St. Petersburg, Russia

M.T. Thai and P.M. Pardalos (eds.), *Handbook of Optimization in Complex Networks: Theory and Applications*, Springer Optimization and Its Applications 57,
DOI 10.1007/978-1-4614-0754-6_9,

the giant component emerges from zero continuously, without a jump. Usually, the relative size S of the giant connected component (the fraction of nodes within this component) grows continuously above the critical point q_c, $S \propto (\langle q \rangle - q_c)^{\beta}$ at $\langle q \rangle - q_c \ll 1$, where the critical exponent β equals, for example, 1 for Erdős–Rényi networks.

The k-core is the maximal subgraph in a network, whose vertices all have internal degree at least k. Using k-cores, that is, the straightforward generalization of connected components, allows one to understand the network architecture in more detail [5,6]. A network can be considered as a hierarchical set of its k-cores. We will demonstrate that, in contrast to the giant connected component, a giant k-core with $k \geq 3$ in a network emerges discontinuously at a critical value $q_c(k)$ of the average degree. Its relative size M_k emerges from zero with a jump $M_{kc} > 0$ but also has a critical singularity, $M_k - M_{kc} \propto (\langle q \rangle - q_c(k))^{\beta_k}$ where the critical exponent $\beta_k = 1/2$ is universal for this transition. This so called "hybrid" (or "mixed") transition differs sharply from both first-order and continuous phase transitions. This kind of phase transition is far less well known than first-order and continuous transitions, although it is present in various cooperative systems, see, e.g., [7] for the Kuramoto model. We will discuss the nature and features of this transition in detail.

The notion of the k-core has proven to be a useful tool giving insight into the deep structure of real complex networks [4, 8–13]. Alvarez-Hamelin et al in [9] used this k-core architecture to produce a set of beautiful visualizations of various networks. The notion of the k-core has found applications in diverse areas, from magnetic systems [5], rigidity [14] and jamming [15] transitions to neural networks [16, 17] and evolution [18] . The k-core has been extensively studied on tree-like networks, starting with Bethe lattices [5] and random graphs [19,20], before finally being extended to arbitrary degree distributions [11,12,21,22]. Results on non tree-like graphs have been largely numerical [23–26], although some analytic results incorporating clustering have recently been obtained [27,28].

The k-cores can be more generally defined for heterogeneous thresholds k, where k may vary from node to node in a network. This generalization will allow us to establish a clear relation between the heterogeneous-k-core problem and so called bootstrap percolation. Bootstrap percolation is an activation process, in which a node becomes active if at least k of its neighbors are active. We will show that the heterogeneous-k-core problem and bootstrap percolation admit a uniform description.

9.2 Tree Ansatz

Here, we discuss only large uncorrelated networks, that is random networks, in which correlations between degrees of neighboring nodes are absent. Conveniently, in the infinite size limit, they have very few finite loops if the networks are sparse [2–4], which is the most interesting case for us. Since trees are graphs without loops, the absence of finite loops means that the uncorrelated random networks have a locally tree-like structure, and so only the infinite loops should be taken into account. One of the consequences of this tree-like structure is the absence of loops in finite connected components, which are thus perfect trees. The second consequence is the absence of finite k-cores with $k \geq 3$. The locally tree-like structure of the networks dramatically simplifies calculations for a wide class of problems for complex networks.

9.3 Giant Components and Percolation

Let us outline the calculation of the giant connected component in an uncorrelated network with an arbitrary degree distribution $P(q)$ [29]. This is the so-called configuration model, that is a uniformly random network with a given degree sequence, in which degrees of nodes are given and otherwise the network is random, see the review [4] and references therein. The calculation is based on the tree ansatz. Introducing the probability x that, following a randomly chosen edge to one of its end vertices, one arrives at a finite connected component, we obtain the equation for x and the expression for the relative size of the giant connected component:

$$x = \sum_q \frac{qP(q)}{\langle q \rangle} x^{q-1}, \tag{9.1}$$

$$1 - S = \sum_q P(q) x^q. \tag{9.2}$$

Figure 9.1 presents these equations in graphic form. In this problem, $1 - x$ plays the role of the order parameter.

The giant connected component exists if (9.1) has a nontrivial solution $x < 1$ in addition to the solution $x = 1$, see Fig. 9.2. Equations (9.1) and (9.2) change only slightly if we, in addition, remove some of vertices (the percolation problem).

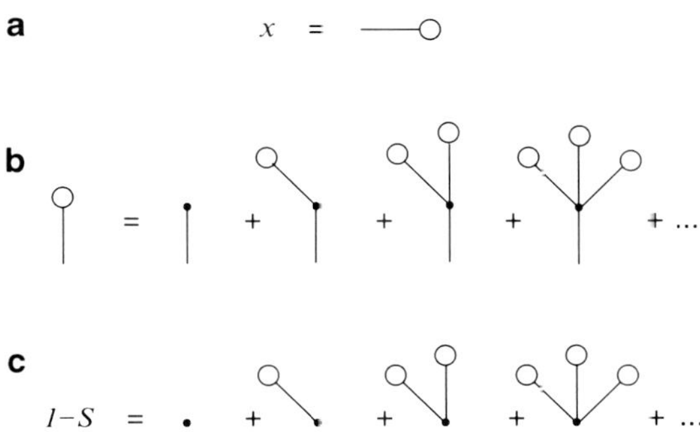

Fig. 9.1 (**a**) The graphical notation for the probability x that, following a randomly chosen edge to one of its end vertices, we arrive at a finite connected component. (**b**) Equation (9.1) in graphical form. (**c**) The graphical representation of formula (9.2) for the relative size S of the giant connected component

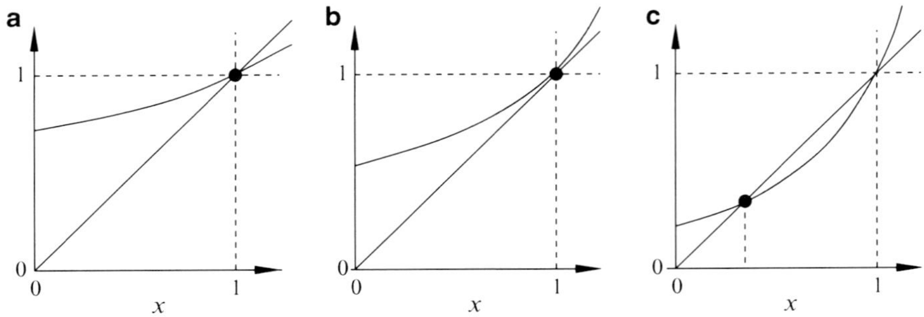

Fig. 9.2 Graphical solution of (9.1). The *lines* show the left- and the right-hand sides of (9.1) as functions of x. (**a**) The normal phase, the giant component is absent, the only intersection is at $x = 1$, so the solution is $x = 1$. Here, the derivative of the right-hand side of the equation is less than 1 at $x = 1$. (**b**) The critical point above which the giant component emerges. Here the derivative of the right-hand side of the equation equals 1 at $x = 1$. (**c**) The giant component is present. The derivative of the right-hand side of the equation exceeds 1 at $x = 1$, and so the non-trivial solution is present at $x < 1$

If a fraction p of vertices retained, then for the giant connected components the following results are obtained. The percolation threshold is

$$p_c = \frac{\langle q \rangle}{\langle q(q-1) \rangle}, \tag{9.3}$$

where $\langle q \rangle$ and $\langle q^2 \rangle$ are the first and the second moments of the degree distribution. This expression shows that the threshold p_c approaches zero if the second moment of the degree distribution diverges. Let us assume that the network is scale-free with a degree distribution $P(q) \propto q^{-\gamma}$. (1) If $\gamma > 4$, then $S \propto p - p_c$. (2) If $3 < \gamma < 4$, then $S \propto (p - p_c)^{1/(\gamma-3)}$. (3) If $\gamma < 3$, then $p_c = 0$, and the giant component is present at any finite p (hyper-resilience against random damage). For a more detailed discussion of these results and literature, see, for example, [2–4].

9.4 Statistics of Finite Components in a Network

Any continuous phase transition must demonstrate a power-law divergence of the corresponding susceptibility. This power law must be present both below and above the phase transition. For the emergence of the giant connected component in these problems, the average size of a finite component to which a randomly chosen vertex belongs plays the role of susceptibility. One can show that this quantity diverges as $1/|p - p_c|$, as it should for infinite-dimensional systems [4]. At the critical point, $p = p_c$, the size distribution of finite components is power-law, $\mathscr{P}(s) \propto s^{-\tau}$. Here, (1) if the exponent $\gamma > 4$, then $\tau = 5/2$, as in classical random graphs, and (2) if the exponent $3 < \gamma < 4$, then $\tau = 2 + 1/(\gamma - 2)$, see, for example, [4].

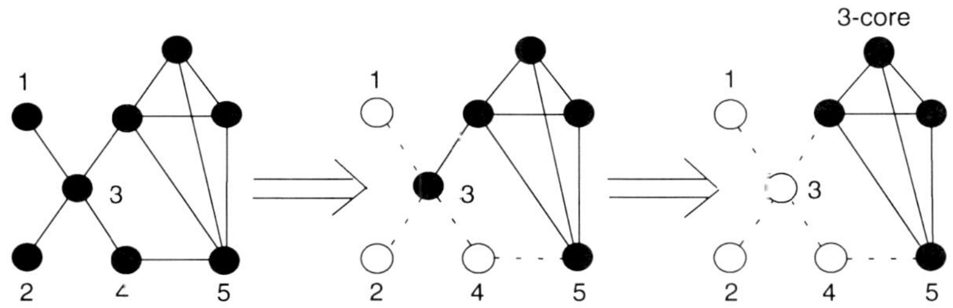

Fig. 9.3 Construction of the 3-core of a graph. First we remove vertices 1, 2 and 4 together with their links because they have degrees smaller than 3. In the obtained graph, vertex 3 has degree 1. Removing it, we get the 3-core of the graph

9.5 *k*-Cores: Pruning Algorithm

Recall that the k-core is the maximal subgraph whose vertices all have internal degree at least k [5, 6]. The k-core of a graph may be obtained by the "pruning algorithm" which is as follows (see Fig. 9.3). Remove from the graph all vertices of degrees less than k. Some of remaining vertices may now have less than k edges. Prune these vertices, and so on until no further pruning is possible. The result, if it exists, is the k-core. The "pruning algorithm" is a kind of optimization procedure which permits us to remove "weak nodes." It is obvious that the $k+1$-core is a subgraph of the k-core. Thus, one can represent a network as a hierarchically organized set of successfully enclosed k-cores, similar to a Russian nesting doll – "matrioshka." The first nontrivial k-core is the $k = 2$-core which can obtained by pruning dangling branches attached to the giant connected component of a network (see below). Below we will study giant connected k-cores which consist of a finite fraction of nodes in an infinite sparse uncorrelated random network. For this purpose we will use the configuration model of sparse uncorrelated random networks. In graph theory, these networks are also called random labeled graphs with a given degree sequence, see, for example, [4]. In a finite network, finite k-cores with a given k may also emerge due to a non-zero clustering coefficient and numerous finite loops, see Sect. 9.2. However, in the infinite size limit, sparse uncorrelated random networks have a tree-like structure and finite k-cores are absent. Note also that in these networks, there may be only one giant k-core at any k.

9.6 Order Parameter and Sizes of *k*-cores

The k-core percolation implies the breakdown of the giant k-core at a threshold concentration of vertices or edges removed at random from an infinite network. Recently, [19, 20], it was mathematically proved that the k-core organization of the configuration model is asymptotically exactly described in the framework of

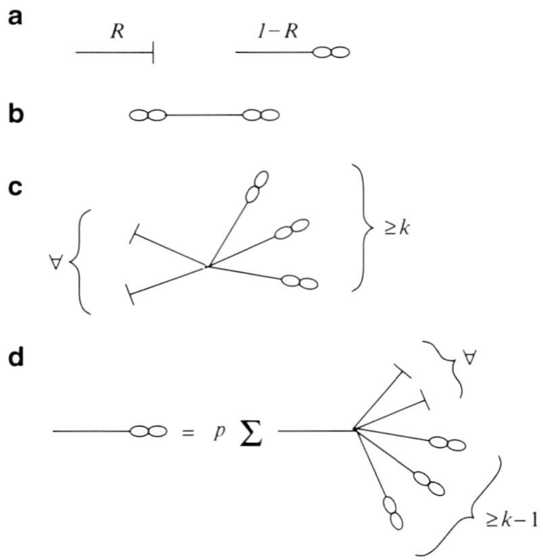

Fig. 9.4 Diagrammatic representation of (9.4)–(9.6). (**a**) Graphical notations for the order parameter R and for $1 - R$. (**b**) The probability that both ends of an edge are in the k-core, (9.4). (**c**) Configurations contributing to M_k, which is the probability that a vertex is in the k-core, (9.5). The symbol $\forall$ here indicates that there may be any number of the nearest neighbors which are not trees of infinite $(k-1)$-ary subtrees. (**d**) A graphical representation of (9.6) for the order parameter

a simple tree ansatz. The validity of the tree ansatz here is non-trivial since in this theory it is applied to a giant k-core which has loops. Note that in tree-like networks, $(k \geq 3)$-cores (if they exist) are giant – finite $(k \geq 3)$-cores are impossible. In contrast to the giant connected component problem, the tree ansatz in application to higher k-cores fails far from the k-core birth points. It is interesting to note that the tree ansatz gives very accurate results for networks even with the presence of significant clustering, for example, the AS Internet network and Facebook social networks [8–11, 30].

Let us discuss the k-core percolation in the configuration model with degree distribution $P(q)$ by using intuitive arguments based on the tree ansatz [11–13]. We assume that a vertex in the network is present with probability p. In this locally tree-like network, the giant k-core coincides with the infinite $(k-1)$-ary subtree. By definition, the m-ary tree is a tree where all vertices have branching at least m. We introduce the order parameter R in the problem as follows. R is the probability that a given end of an edge of a network is not the root of an infinite $(k-1)$-ary subtree [11]. (Of course, R depends on k.) Note that R defined in this way is 1 in the phase without the k-core, and $R < 1$ if the k-core is present. An edge is in the k-core if both ends of this edge are roots of infinite $(k-1)$-ary subtrees, which happens with probability $(1 - R)^2$. In other words,

$$(1 - R)^2 = \frac{L_e(k)}{L} \tag{9.4}$$

which expresses the order parameter R in terms of observables: the number of edges in the k-core, $L_e(k)$, and the total number of edges, L, in the network. $L_e(k)$ and L can be measured in real networks. Figure 9.4 graphically explains this and the following two relations. A vertex is in the k-core if at least k of its neighbors are roots of

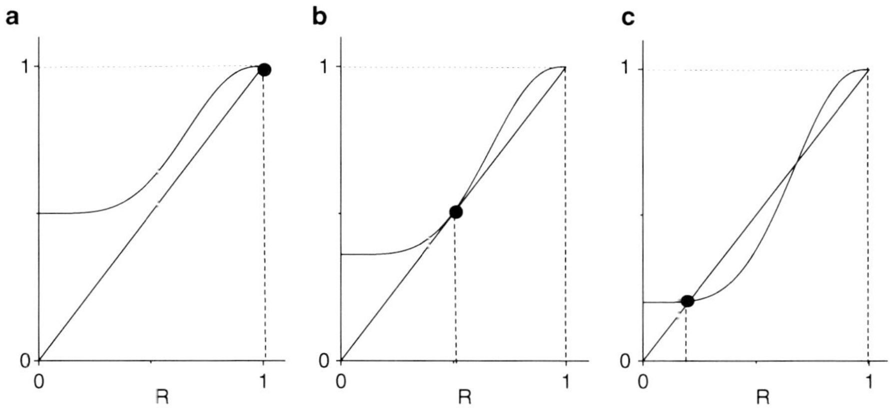

Fig. 9.5 Graphic solution of (9.6), compare with Fig. 9.2. For the sake of convenience, we rewrite (9.6) in the form $R = f(R)$ and omit the index k in R_k. (**a**) The k-core is absent. The only solution is trivial $R = 1$. (**b**) The point of emergence of the order parameter and the k-core, that is the point of singularity. (**c**) Solution above the critical point

infinite $(k-1)$-ary trees. So, the probability M_k that a random vertex belongs to the k-core (the relative size of the k-core) is given by the equation:

$$M_k = p \sum_{n \geq k} \sum_{q \geq n} P(q) C_n^q R^{q-n} (1-R)^n, \tag{9.5}$$

where $C_n^q = q!/[(q-n)!n!]$. To obtain the relative size M_k of the k-core, one must substitute the physical solution of (9.6) (see below) for the order parameter R into (9.5). We can write the equation for the order parameter, noticing that a given end of an edge is a root of an infinite $(k-1)$-ary subtree if it has at least $k-1$ children which are roots of infinite $(k-1)$-ary subtrees. Therefore, we obtain

$$1-R = p \sum_{n=k-1}^{\infty} \sum_{i=n}^{\infty} \frac{(i+1)P(i+1)}{z_1} C_n^i R^{i-n} (1-R)^n \equiv 1 - f(R). \tag{9.6}$$

This equation strongly differs from that for the order parameter in the ordinary percolation, compare with (9.2). (R should be compared with x.) The solution of (9.6) for $k \geq 3$ indicates a quite unusual critical phenomenon. See Fig. 9.5 for the graphical solution of this equation. This figure shows that the order parameter (and also the size of the k-core) has a jump at the critical point like a first order phase transition, compare with Fig. 9.2 for the connected component problem. On the other hand, the order parameter has a square root critical singularity:

$$R_c - R \propto [p - p_c(k)]^{1/2} \propto M_k - M_{kc}, \tag{9.7}$$

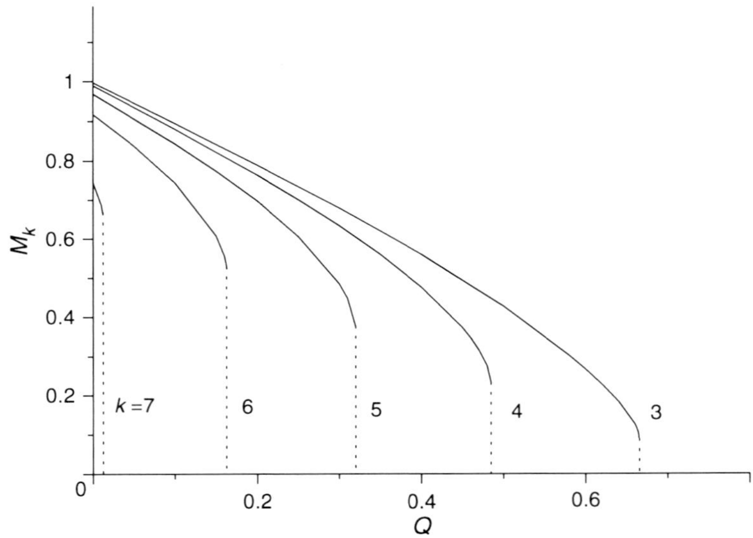

Fig. 9.6 Relative sizes of the k-cores, M_k, in classical random graphs with the mean degree $z_1 = 10$ vs. the fraction $Q = 1 - p$ of randomly removed vertices. The highest k-core, $k_{\rm h}$, in classical random graphs is determined by the mean degree, for example, $k_{\rm h}$=7 at $z_1 = 10$ [11]

see Fig. 9.6. This intriguing critical phenomenon is often called *a hybrid phase transition* [15, 26]. Relations (9.7) are valid if the second moment of the degree distribution is finite.

Figure 9.6 shows the dependence of relative sizes of k-cores on the fraction $Q = 1 - p$ of the removed vertices. As we have mentioned above, a complex network can be represented as a hierarchically organized set of successively enclosed k-cores. With increasing Q, first, the highest k_h-core disappears discontinuously with a critical singularity (9.7) at a critical concentration $Q_c(k_h) = 1 - p(k_h)$. Then the $k_h - 1$-core disappears and so on. Lastly, a giant connected $k = 2$-core disappears. At large Q there are only finite clusters. If the second moment of the degree distribution diverges, the picture is very similar to what we observed for ordinary percolation. In this case, the k-cores, even of high order, practically cannot be destroyed by the random removal of vertices from an infinite network. In other words, the k-core architecture is robust against a random damage of this class of complex networks.

The 2-core of a graph can be obtained from the giant connected component of this graph by pruning dangling branches. At $k = 2$, (9.6) for the order parameter is identical to one for the ordinary percolation. Therefore, the birth point of the 2-core coincides with that of the giant connected component, and the phase transition is continuous. According to (9.5) the size M_2 of the 2-core is proportional to $(1 - R)^2$ near the critical point, and so it is proportional to the square of the size of the giant connected component. This gives $M_2 \propto (p - p_c)^2$ if the degree distribution decays rapidly.

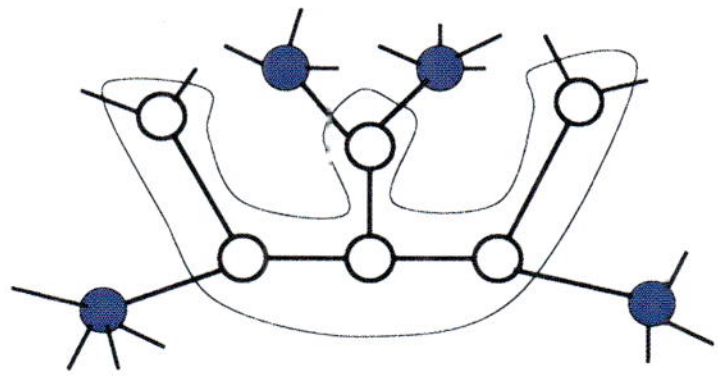

Fig. 9.7 A finite corona cluster of the $(k=3)$-core. This cluster consists of vertices represented by *open circles* with exactly three edges. Removal of any vertex on this figure results in pruning all vertices which belong to the corona cluster

In stark contrast to ordinary percolation, the emergence of $(k\geq 3)$-cores, that occurs, for example, at critical values of the mean degree, is not related to the divergence of corresponding finite components which are absent in tree-like networks. Then, is there any divergence associated with this hybrid transition? The answer is yes. To unravel the nature of this divergence, let us introduce a new notion. The k-core's *corona* is a subset of vertices in the k-core (with their edges) which have exactly k neighbors in the k-core, that is the minimum possible number of connections. One may see that the corona itself is a set of disconnected clusters (see Fig. 9.7). Let N_{crn} be the mean total size of corona clusters attached to a vertex in the k-core. It turns out that it is $N_{\mathrm{crn}}(p)$ which diverges at the birth point of the k-core,

$$N_{\mathrm{crn}}(p) \propto [p - p_c(k)]^{-1/2} \tag{9.8}$$

[12, 13, 15]. Moreover, the mean intervertex distance in the corona clusters diverges by the same law as $N_{\mathrm{crn}}(p)$ [12]. It looks like the corona clusters "merge together" exactly at the k-core percolation threshold $p_c(k)$ and simultaneously disappear together with the k-core, which, of course, does not exist at $p < p_c(k)$.

By using (9.5) and (9.6), we can easily find the k-core sizes, M_k in the important range $2 < \gamma < 3$:

$$M_k = p^{1/(3-\gamma)}(q_0/k)^{(\gamma-1)/(3-\gamma)}, \tag{9.9}$$

where q_0 is the minimal degree in the scale-free degree distribution [11]. The exponent of this power law agrees with the observed one in a real-world network – the Internet at the Autonomous System level and the map of routers [8, 9]. In the infinite scale-free networks of this kind, there is an infinite sequence of k-cores (9.9). All these cores have a practically identical architecture – their degree distributions asymptotically coincide with the degree distribution of the network in the range of high degrees.

Above we considered networks of an infinite size. Let us discuss networks of a finite but sufficiently large size. If a network is finite, then in (9.6) the summation over degree has a cutoff, $i+1 \leq q_{\mathrm{cut}}$. If $\gamma > 3$, then the finiteness of networks does not qualitatively change the results obtained above. The finiteness of networks with fat tailed degree distributions, i.e., for $2 < \gamma \leq 3$, restricts the k-core sequence to some maximum number k_{h} for the highest k-core. In [11–13], k_{h} was estimated

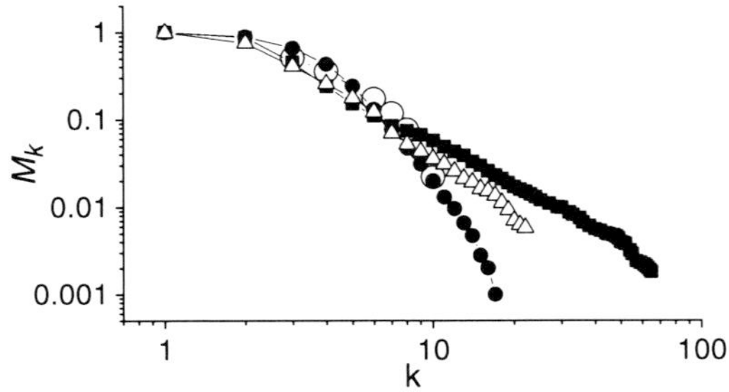

Fig. 9.8 Relative size, M_k, of the k-cores vs. k in several networks: (1) ○, M_k calculated neglecting correlations, by using the degree distribution of Internet router network, $N \approx 190{,}000$, adapted from [11]; (2) △, measurements for the Autonomous System network (CAIDA map), $N = 8{,}542$, adapted from [9]; (3) •, results for a maximally random scale-free ($\gamma = 2.5$) network of 10^6 vertices; (4) ■, for a clustered network with the same $\gamma = 2.5$, but with very large clustering coefficient $C = 0.71$, adapted from [24]

by substituting empirical degree distributions into the equations for uncorrelated networks. Unfortunately, the resulting $k_{\rm h}$ turned out to be several (3) times smaller than the observed values [8, 9]. In [24, 25], a much more realistic $k_{\rm h}$ was obtained by taking into account high clustering (see Fig. 9.8). (A maximally random network with a given degree distribution and a given clustering was simulated in [24, 25].) Sizes of k-cores and the maximum number $k_{\rm h}$ for some real networks (the Internet at the AS level, Facebook social networks, the PGP network, and the power grid) were found in [30]. There is also another way to diminish $k_{\rm h}$: random damaging first destroys the highest k-core, then the second highest, and so on.

9.7 Hybrid Transition as a Limiting Metastable State

Let us briefly discuss the nature of hybrid phase transitions. These transitions form a specific new kind of phase transition which combines a discontinuity like a first order phase transition with a critical singularity like a continuous phase transition, see (9.7) for the k-core problem and (9.27) for bootstrap percolation. Strong analogies between bootstrap percolation and metastable behavior in systems with first order phase transitions were already remarked upon in the early mathematical work [31]. Let us discuss this question from a more physical point of view.

In thermodynamics, where the changes of a control parameter (e.g., decreasing/increasing temperature) are assumed to be infinitely slow, a first order transition has no hysteresis. In reality, the changes always occur with a small but finite rate, and the "heating" and "cooling" branches of a first order transition do not coincide, which indicates the presence of a metastable state and hysteresis. The width of the hysteresis increases with this rate until some limit, which corresponds to the limiting metastable state. One can show that the resulting limiting curve for the

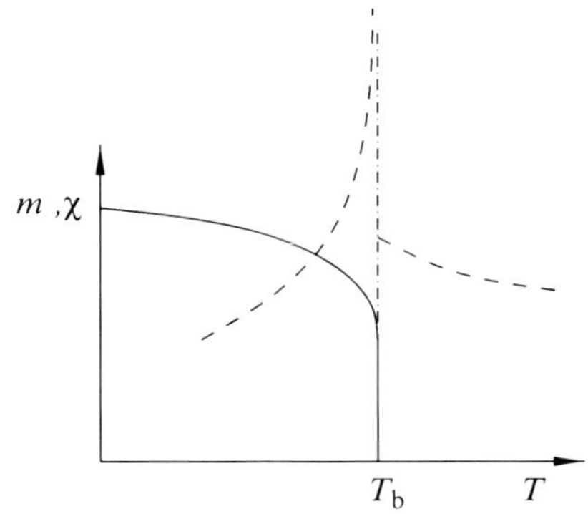

Fig. 9.9 The schematic view of the dependence of magnetization m (*solid line*) and zero-field susceptibility χ (*dashed line*) on temperature near the limiting metastable state for a first-order phase transitions in magnetic systems. Here, $m(T) - m(T_b - 0) \propto \sqrt{T_b - T}$ and $\chi(T < T_b) \propto 1/\sqrt{T_b - T}$

order parameter has a square-root singularity at the breakdown point, see Fig. 9.9. One can easily obtain this behavior by using, for example, the Landau theory with the free energy

$$F(m,H) = -mH + a(T - T_c)m^2 + bm^4 + cm^6, \qquad (9.10)$$

where the coefficients a, b, and c are positive and H is a magnetic field. A simple analysis shows that metastable states with $m \neq 0$ emerge below the critical temperature $T_b = T_0 + b^2/(3ac)$,

$$m(T) - m(T_b - 0) \propto \sqrt{T_b - T}, \qquad (9.11)$$

with the jump $m(T_b - 0) = \sqrt{b/3c}$, see Fig. 9.9. The thermodynamic first order phase transition takes place at a lower temperature. The susceptibility $\chi(T, H = 0) = dm/dH|_{H=0}$, also shows a singularity $\chi(T < T_b) \propto 1/\sqrt{T_b - T}$ but it has no singularity above T_b.

Compare these singularities with those of (9.7) and (9.8). In the k-core problem, it is N_{crn} that plays the role of susceptibility similarly to the mean size of a cluster to which a vertex belongs in ordinary percolation, see [12, 13] for more details. The exponent of the singularity in (9.8), $1/2$, dramatically differs from the standard mean-field value of exponent $\tilde{\gamma} = 1$. The only essential difference is that, in contrast to the k-core percolation, in ordinary thermodynamics this region is not approachable. In this sense, the hybrid (mixed) transition is a limiting metastable state for a first order phase transition. As we have mentioned in Sect. 9.1, a similar asymmetrical hybrid transition has been observed in the Kuramoto model of oscillator synchronization [7].

Importantly, the hybrid transition is asymmetrical. Namely, there are critical fluctuations and a divergent correlation length on only one side of the critical point. In contrast to that, continuous phase transitions demonstrate critical fluctuations and a divergent correlation length on the both sides of the transition, while in first order phase transitions, there are no critical fluctuations and the correlation length is finite everywhere.

9.8 Heterogeneous *k*-core

The concept of the k-core of a network may be generalized to allow each vertex in a network to have a different threshold. Each vertex of a network is assigned a variable $k_i \in \{0, 1, 2, ...\}$. The k_i values are assumed to be uncorrelated, selected from a distribution $Q_k(r)$. The heterogeneous k-core is then the largest subgraph of the network for which each vertex i has at least k_i neighbors. To find the heterogeneous k-core of a given network, we start with the full network, and prune any vertices whose degree is less than its value of k_i. We repeat the pruning until a stationary state is reached. The heterogeneous k-core may include finite clusters as well as any giant component. If a giant component is present we call it the giant heterogeneous k-core (giant-HKC). As we did for the ordinary k-core, we examine sparse random uncorrelated networks. These networks are completely determined by the degree distribution, and we make use of their locally tree-like property in the infinite size limit.

If all $k_i = 1$, then the HKC is simply the connected component of the network, and the giant-HKC is the giant connected component, exactly as in ordinary percolation. It appears with a continuous transition at the critical point p_c given by (9.3). If $k = 2$, we again have a continuous transition, similar to ordinary percolation. If all k_i are equal to $k \geq 3$, then we have the ordinary k-core. In this case the giant k-core appears with a discontinuous hybrid transition [11–13, 15].

In general, we might expect a combination of continuous and hybrid transitions. Let us define a simple representative example of the heterogeneous k-core (HKC) in which vertices have a threshold of either 1 or $k \geq 3$, distributed randomly through the network with probabilities f and $(1 - f)$, respectively:

$$Q_k(r) = \begin{cases} f & \text{if } r = 1, \\ 1 - f & \text{if } r = k, \\ 0 & \text{otherwise,} \end{cases} \tag{9.12}$$

for some integer $k \geq 3$. This parametrized HKC has as its two limits ordinary percolation ($f = 1$) and the original k-core ($f = 0$).

There are two transitions in the size of the giant heterogeneous k-core (giant-HKC): a continuous transition similar to that found in ordinary percolation and a discontinuous, hybrid, transition, similar to that found for the ordinary k-core. We find a complex phase diagram for this giant-HKC with respect to the proportion of each threshold and the amount of damage to the network, in which, depending on the parameter region, either transition may occur first, see Fig. 9.10. The giant-HKC is absent in the region labeled I. For p below the percolation threshold p_c, the giant-HKC never appears for any f. Above p_c, the giant-HKC appears with a continuous transition, growing linearly with f (or p for that matter) close to the critical point. The threshold is indicated by the thin black line in the Figure, which divides regions I and II. From p_{s-k} a second, discontinuous, hybrid transition appears. This is marked by the heavy black curve in Fig. 9.10.

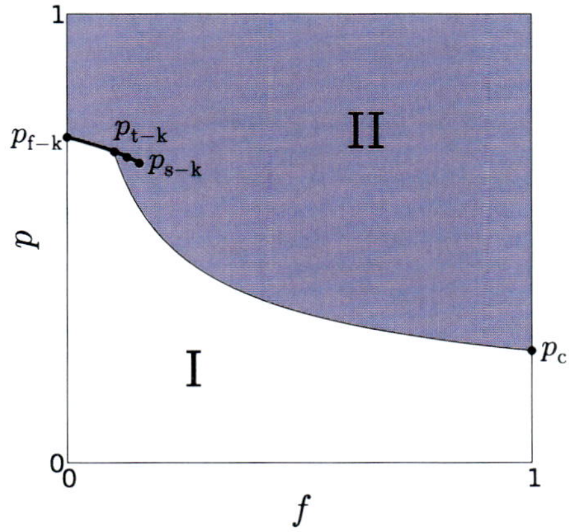

Fig. 9.10 Phase diagram for heterogeneous κ-core (*left*) in the f-p plane. The giant-HKC is present in region II, appearing continuously at the threshold marked by the *thin black curve*. The hybrid, discontinuous transition occurs at the points marked by the *heavy black line*, beginning from p_{s-k}. Above p_{t-k} the first appearance of the giant-HKC is with a discontinuous transition. Above p_{f-k} the giant-HKC appears discontinuously for any $f > 0$

Let $\mathscr{S}_k$ be the fraction of vertices that are in the heterogeneous k-core. That is, the total of all components, whether finite or infinite, that meet the threshold conditions. This is equal to the probability that an arbitrarily chosen vertex of the original network is in the heterogeneous k-core. The hybrid transition is discontinuous, and the size of the heterogeneous k-core grows as the square root of the distance above the critical point $f = f_{c2-k}$:

$$\mathscr{S}_k(f) = \mathscr{S}_k(f_{c2-k}) + a(f - f_{c2-k})^{1/2}. \tag{9.13}$$

See line 2 in Fig. 9.13. This is the same as the hybrid transition found in the ordinary k-core. For a given p the critical point f_{c2-k} can be found from self-consistent solution of (9.16) and (9.19). Solving (9.16) near the critical point f_{c2-k} and substituting it into (9.15), we arrive at (9.13). Note that f_{c2-k} may be greater than the critical point f_{c1-k} above which the giant heterogeneous-k-core appears continuously – so that the first appearance of the giant-HKC is similar to that found in ordinary percolation – or less than f_{c1-k}, with the giant-HKC appearing discontinuously from zero, as in the ordinary k-core [12, 13, 15] (see Fig. 9.13). The critical point f_{c1-k} for a given p can be found from (9.18), see below.

At the special point p_{s-k} (Fig. 9.10) the size of the discontinuity reduces to zero and the scaling near the critical point is cube root:

$$\mathscr{S}_k(f) = \mathscr{S}_k(f_{c2-k}) + a(f - f_{c2-k})^{1/3}. \tag{9.14}$$

Table 9.1 Symbols used in graphical representations of self-consistency equations for the heterogeneous k-core

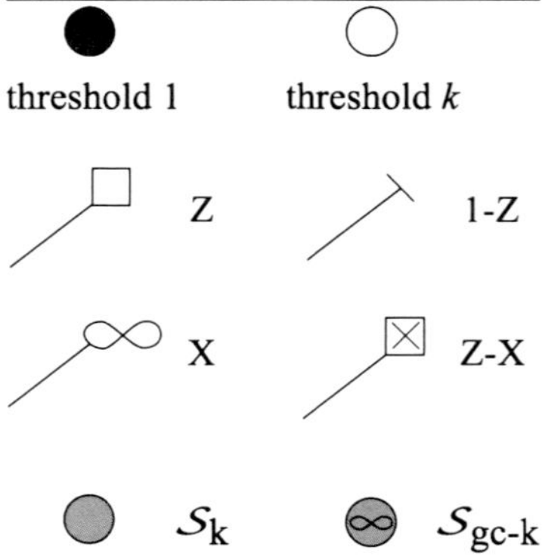

The probability $\mathscr{S}_k$ is the sum of the probabilities that it has $k_i = 1$ and at least one neighbor in the core, or $k_i = k$ and has at least k neighbors in the core. We can represent this diagrammatically as (see Table 9.1):

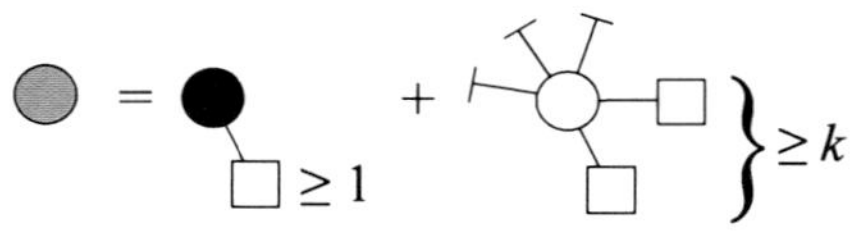

These conditions can be written as binomial terms, and summing over all possible values of the degree of i, this diagram can be written in mathematical form as:

$$\mathscr{S}_k = pf \sum_{q=1}^{\infty} P(q) \sum_{l=1}^{q} \binom{q}{l} Z^l (1-Z)^{q-l} + p(1-f) \sum_{q=k}^{\infty} P(q) \sum_{l=k}^{q} \binom{q}{l} Z^l (1-Z)^{q-l}, \tag{9.15}$$

where the factor p accounts for the probability that the vertex has not been damaged. A square represents the probability Z, which we define in terms of a "$(k_i - 1)$-ary tree," a generalization of the $(k-1)$-ary tree. A $(k_i - 1)$-ary tree is a sub-tree in which, as we traverse the tree, each vertex encountered has at least $k_i - 1$ child edges (edges leading from the vertex, not including the one we entered by). In our example, k_i is either 1, in which case the vertex does not need to have any children (though it may have them), and is only required to be connected to the tree, or $k_i = k$, in which case it must have at least $k - 1$ children. The probability Z can then be very simply stated as the probability that, on following an arbitrarily chosen edge in the network, we reach a vertex that is a root of a $(k_i - 1)$-ary tree. For the specific case considered in this chapter, the vertex encountered either has $k_i = 1$, or it has $k - 1$ children leading to the roots of $(k_i - 1)$-ary trees. The probability Z is represented by a square in the diagram. A bar represents the probability $(1-Z)$, a black circle represents a vertex with $k_i=1$, and a white circle a vertex with $k_i=k$ – see Table 9.1. To calculate Z, we construct a recursive (self consistency) expression in a similar way, based on the definition given above. This is represented by the diagram:

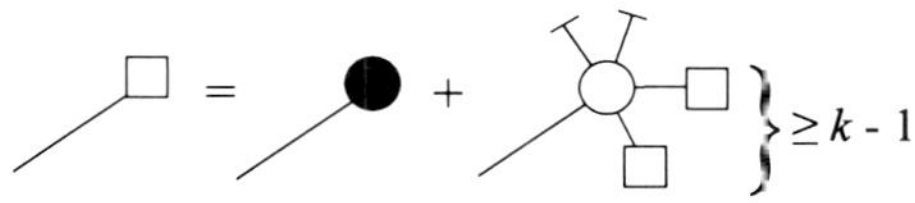

which, in equation form is:

$$Z = pf + p(1-f)\sum_{q\geq k}\frac{qP(q)}{\langle q\rangle}\sum_{l=k-1}^{q-1}\binom{q-1}{l}Z^l(1-Z)^{q-1-l}$$
$$\equiv \Psi(Z,p,f)\,. \qquad (9.16)$$

We have used that $qP(q)/\langle q\rangle$ is the probability that the vertex reached along an arbitrary edge has degree q. Solving (9.16) (usually numerically) for Z and then substituting into (9.15) allows the calculation of $\mathscr{S}_k$.

We follow a similar procedure to calculate $\mathscr{S}_{gc-k}$, the relative size of the giant heterogeneous k-core (that is, the subset of the heterogeneous k-core which forms a giant component) – also the probability that an arbitrarily chosen vertex is in the giant heterogeneous k-core. The fraction of vertices forming finite clusters is therefore $\mathscr{S}_k-\mathscr{S}_{gc-k}$. Note that in the standard k-core this is negligibly small. We denote by X the probability that an arbitrarily chosen edge leads to a vertex which is the root of an infinite (k_i-1)-ary tree. That is, the definition is similar to Z, but with the extra condition that the sub-tree reached must extend indefinitely. We represented X by an infinity symbol (Table 9.1). The diagram for $\mathscr{S}_{gc-k}$ is:

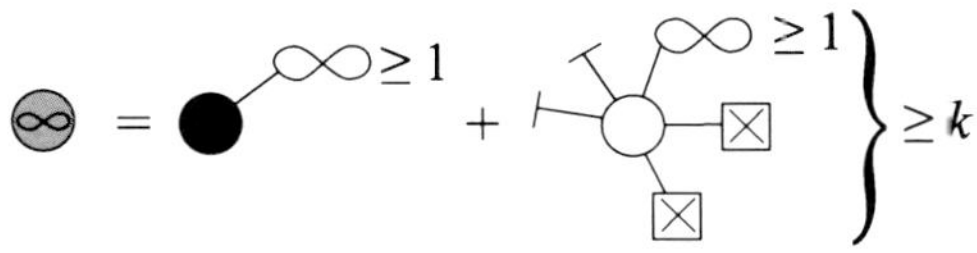

which is equivalent to the equation:

$$\mathscr{S}_{gc-k} = pf\sum_{q=0}^{\infty}P(q)\sum_{m=1}^{q}\binom{q}{m}X^m(1-X)^{q-m}$$
$$+\,p(1-f)\sum_{q=k}^{\infty}P(q)\sum_{l=k}^{q}\binom{q}{l}(1-Z)^{q-l}\sum_{m=1}^{l}\binom{l}{m}X^m(Z-X)^{l-m}\,. \qquad (9.17)$$

To find X, we construct a self-consistency equation from the diagram

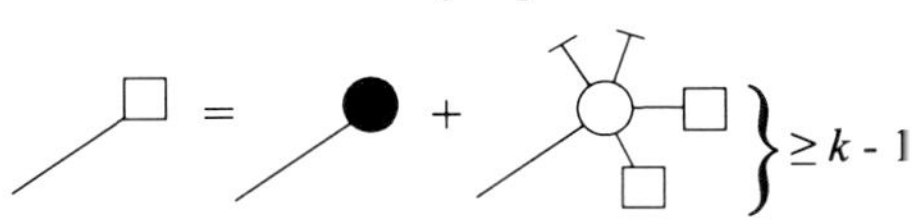

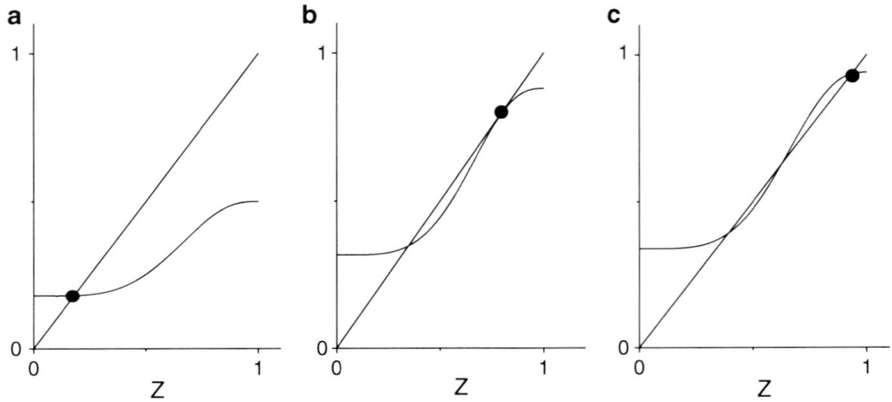

Fig. 9.11 Graphic solution of (9.16), compare with Fig. 9.5. (**a**) At small f, $f < f_{c2-k}$, there is only one solution of (9.16) which determines the parameter Z for the heterogeneous k-core. (**b**) At the critical point $f = f_{c2-k}$ a second solution with a larger Z (*black dot*) emerges. This is a stable solution. (**c**) At $f > f_{c2-k}$ only the solution with the largest Z (*black dot*) is stable

leading to

$$X = pf \sum_{q=0}^{\infty} \frac{qP(q)}{\langle q \rangle} \sum_{m=1}^{q-1} \binom{q-1}{m} X^m (1-X)^{q-1-m}$$
$$+p(1-f) \sum_{q=k}^{\infty} \frac{qP(q)}{\langle q \rangle} \sum_{l=k-1}^{q-1} \binom{q-1}{l} (1-Z)^{q-1-l} \sum_{m=1}^{l} \binom{l}{m} X^m (Z-X)^{l-m} . \tag{9.18}$$

Solution of (9.16) and (9.18) then allows the calculation of $\mathscr{S}_{gc-k}$ through (9.17). Note that when there are multiple solutions of Z, we choose the largest solution as the "physical" one. To find the appearance of the giant component for a given p, we find leading term $O(X)$ of the right hand side of (9.18) for $X \ll 1$, and solve (9.18) for f. This gives the critical point f_{c1-k}.

To calculate the location of the hybrid transition, i.e., the critical point f_{c2-k}, we note that at this critical point a second solution to (9.16) appears. This occurs when the function $\Psi(Z)$ just touches the line Z (see Fig. 9.11b), which must be at a local extremum of Ψ/Z:

$$\frac{\mathrm{d}}{\mathrm{d}Z}\left(\frac{\Psi}{Z}\right) = 0. \tag{9.19}$$

Expanding (9.16) about Z_c, the value of Z at the critical point (at the top of the jump), and using (9.19), we see that Z grows as the square-root of the distance from the critical point. Using (9.15) we find (9.13). In a general case, (9.16) may have three solutions, see Fig. 9.11c. Only the solution with the largest Z is stable.

Furthermore, at the special point p_{s-k} where the second transition disappears, by a similar argument, a further condition must also be satisfied:

$$\frac{\mathrm{d}^2}{\mathrm{d}Z^2}\left(\frac{\Psi}{Z}\right) = 0. \tag{9.20}$$

Thus, the critical point p_{s-k} is determined by simultaneous solution of (9.16), (9.19), and (9.20). This in turn leads to cube root scaling above the threshold; hence, (9.14). These conditions are very similar to those used for the ordinary k-core.

In this chapter, we have examined only a special case of the heterogeneous k-core, in which vertices have threshold either 1 or $k \geq 3$. For completeness, we now give the self-consistency equations for arbitrary threshold distribution $Q(r)$ where $Q(r)$ is the fraction of vertices with a threshold $r \geq 1$. The size $\mathscr{S}_k$ of the heterogeneous k-core is

$$\mathscr{S}_k = p \sum_{r\geq 1} Q(r) \sum_{q=r}^{\infty} P(q) \left[\sum_{l=r}^{q} \binom{q}{l} Z^l (1-Z)^{q-l} \right], \tag{9.21}$$

where, as above, Z is the probability of encountering a vertex i which is the root of a $(r_i - 1)$-ary tree:

$$Z = p \sum_{r\geq 1} Q(r) \sum_{q=r}^{\infty} \frac{(q)P(q)}{\langle q \rangle} \sum_{l=r-1}^{q-1} \binom{q-1}{l} Z^l (1-Z)^{q-1-l}. \tag{9.22}$$

We do not derive any results for this general case, but we can speculate that a more complicated phase diagram would appear. If any vertices have threshold less than 3, i.e. $Q(1) + Q(2) > 0$, we would find a continuous appearance of the giant-HKC. Thresholds of 3 or more, on the other hand, contribute discontinuous transitions, and it may be that there are multiple such transitions.

9.8.1 *Effect of Network Structure*

Network heterogeneity plays an important role. The results above and in Figs. 9.13 and 9.14, are qualitatively the same for networks with any degree distribution which has finite second and third moments. When only the second moment is finite, the phase diagram remains qualitatively the same, but the critical behavior is changed. Instead of a second order continuous transition, we have a transition of higher order. When the second moment diverges, we have quite different behavior.

To examine the behavior when the second and third moments diverge, we consider scale-free networks, with degree distributions tending to the form

$$P(q) \approx q^{-\gamma} \tag{9.23}$$

for large q. At present we consider only values of $\gamma > 2$.

To find the behavior near the critical points, we expand the right hand side of (9.18) near the appearance of the giant-HKC (that is, near $X = 0$, $X \ll 1$). When $\gamma < 4$, the third and possibly second moment of the degree distribution diverge. This means that coefficients of integral powers of X diverge, and we must instead find leading non-integral powers of X. When $\gamma > 4$, the second and third moments of the distribution are finite (9.18) leads to

$$X = c_1 X + c_2 X^2 + \text{higher order terms}, \tag{9.24}$$

which gives the critical behavior $X \propto (f - f_{c2-k})^{\beta}$ with $\beta = 1$. When $3 < \gamma \leq 4$, the linear term in the expansion of (9.18) survives, but the second leading power is $\gamma - 2$:

$$X = c_1 X + c_2 X^{\gamma-2} + \text{higher terms}, \tag{9.25}$$

where the coefficients c_1 and c_2 depend on the degree distribution, the parameters p and f, and the (non-zero) value of Z. The presence of the linear term means the giant-HKC appears at a finite threshold, but because the second leading power is not 2, the giant-HKC grows not linearly but with exponent $\beta = 1/(\gamma - 3)$. This means that the phase diagram remains qualitatively the same as Fig. 9.14, however, the size of the giant-HKC grows as $(f - f_{c1})^{\beta}$ with $\beta = 1/(\gamma - 3)$. This is the same scaling as was found for ordinary percolation [33].

For values of γ below 3, the change in behavior is more dramatic. When $2 < \gamma \leq 3$, the second moment of $P(q)$ also diverges, meaning the leading order in the equation for X is no longer linear but $\gamma - 2$:

$$X = d_1 X^{\gamma-2} + \text{higher terms}. \tag{9.26}$$

From this equation it follows that there is no threshold for the appearance of the giant-HKC (or giant-BPC). The giant-HKC appears immediately and discontinuously for any $f > 0$ (or $p > 0$), and there is also no upper limit to the threshold k. This behavior is the same for bootstrap percolation, so the (featureless) phase diagram is the same for both processes, even though the sizes of the giant-HKC and giant-BPC are different.

We can understand the hybrid transition in the heterogeneous k-core by considering the corona clusters. Corona clusters are clusters of vertices with threshold k that have exactly k neighbors in the HKC. These clusters are part of the heterogeneous k-core, but if any member of a cluster loses a neighbor, a domino-like effect leads to an avalanche as the entire cluster is removed from the HKC, see Fig. 9.12. The corona clusters are finite everywhere except at the discontinuous transition, where the mean size of corona clusters diverges as we approach from above [12, 13, 18]. Thus, an infinitesimal change in f (or p) leads to a finite fraction of the network being removed from the heterogeneous k-core; hence, a discontinuity in $\mathscr{S}_k$ (and also in $\mathscr{S}_{gc-k}$). The size distribution of corona clusters, and hence avalanches at this transition is determined by the size distribution $G(s)$, which at the critical point can be shown to follow $G(s) \sim s^{-3/2}$. This can be shown using a generating function approach, as demonstrated in [12, 13].

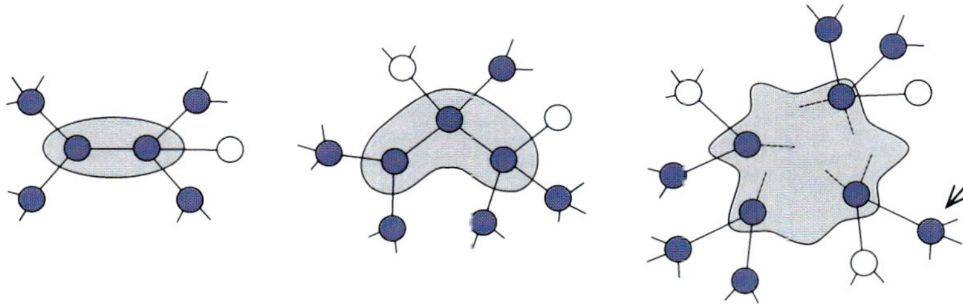

Fig. 9.12 Corona clusters of different sizes in heterogeneous k-core. *Left*: Because they are included unless pruned, two connected vertices (whose threshold is κ) form part of the heterogeneous k-core if each has $k-1$ other neighbors in the core, as each is "assisted" by the other. In general (*right*) a corona cluster consists of vertices (with threshold k) that each have exactly k active neighbors, either inside or outside the cluster. If one neighbor of any of the cluster vertices is removed from the core, an avalanche is caused as the entire cluster is pruned

9.9 k-Core vs. Bootstrap Percolation

The Heterogeneous k-core can be directly contrasted with another well known problem related to percolation, bootstrap percolation (BPC). In bootstrap percolation, vertices can be of two types: with probability f, a vertex is a "seed" and is initially active, and remains active. The remaining vertices (a fraction $1-f$) become active if their number of active neighbors reaches or exceeds a threshold value k. Once activated, a vertex remains active. The activation of vertices may mean that new vertices now meet the threshold criterion, and hence become active. This activation process continues iteratively until a stationary state is reached. The seed and activated vertices in bootstrap percolation are analogous to the threshold 1 and threshold k groups in the heterogeneous k-core. We will first smmarise the behavior of the bootstrap percolation process, before comparing with the heterogeneous k-core.

Again, two transitions are observed in the phase diagram of the giant component of active vertices in bootstrap percolation (giant-BPC). Above a certain value of p, p_c, the giant active component (giant-BPC) may appear continuously from zero at a finite value of f, f_{c1-b}, and grow smoothly with f, see lines 1 and 2 in the right panel of Fig. 9.13. For larger p, after the giant active component appears, there may also be a second discontinuous hybrid phase transition, at f_{c2-b}, as seen in line 3 of Fig. 9.13. There is a jump in the size of the giant active component $\mathscr{S}_{gc-b}$ from the value at the critical point (marked by a circle on dashed line 3). When approaching from below, the difference of $\mathscr{S}_{gc-b}$, from the critical value goes as the square root of the distance from the critical point:

$$\mathscr{S}_b(f) = \mathscr{S}_b(f_{c2-b}) - a(f_{c2-b} - f)^{1/2}. \tag{9.27}$$

At the special critical point p_{s-b}, the scaling becomes cube root:

$$\mathscr{S}_b(f) = \mathscr{S}_b(f_{c2-b}) - a(f_{c2-b} - f)^{1/3}. \tag{9.28}$$

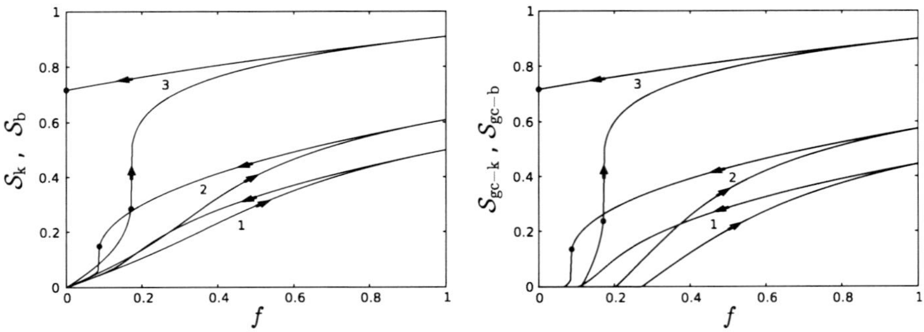

Fig. 9.13 *Left*: relative size $\mathcal{S}_k$ of the heterogeneous k-core (curves with *arrows* directed to the left, upper of each numbered pair), which is the subgraph including all vertices which meet the threshold requirements of (9.12), and fraction $\mathcal{S}_b$ of active vertices in bootstrap percolation (lower curves, with *arrows* directed to the right) as a function of f for the same network – an Erdős-Rényi graph of mean degree 5 – with the same $k = 3$, at three different values of p, corresponding to different regions of the phase diagrams. (1) $p = 0.5$, which is between p_{C} and p_{S} for both models. (2) $p = 0.61$, which is above $p_{\text{s-k}}$ but still below $p_{\text{s-b}}$. (3) $p = 0.91$, which is above $p_{\text{f-k}}$ and $p_{\text{s-b}}$. Each numbered pair forms the two branches of a hysteresis process. *Right*: Size $\mathcal{S}_{gc-k}$ of the giant heterogeneous k-core (*arrows* to the left) and size $\mathcal{S}_{gc-b}$ of the giant-BPC (*arrows* to the right) as a function of f for the same network and the same values of p

The overall behavior with respect to the parameters p and f of each model is summarized by the phase diagram Fig. 9.14. This diagram is qualitatively the same for any degree distribution with finite second moment.

Note the difference between (9.27) and (9.13) and between (9.28) and (9.14). In bootstrap percolation, the hybrid transition always occurs above the continuous one, and neither reaches $f = 0$. Note also that the special critical point p_{s-b} is above the critical point p_{f-k}, so that for a given p we may have a hybrid transition for the HKC or for BPC, but not for both. Thus the giant-HKC is present everywhere in regions II and III of Fig. 9.14, the giant-BPC is present in region III, and both are absent in region I. If the value of k is increased, the locations of the hybrid transitions move toward larger values of p, and there is a limiting value of k after which these transitions disappear altogether. Our numerical calculations revealed that in the case of Erdös–Rényi graphs for both the heterogeneous k-core and bootstrap percolation, this limit is proportional to the mean degree, see also [11]. Note also that the continuous transition also moves slightly with increasing k, and in the limit $k \to \infty$, tends to the line $pf = p_c$ for both processes.

Because bootstrap percolation is an activation process, while the heterogeneous k-core is found by pruning, we can characterize them as two branches of the same process, with the difference between the curves shown in Fig. 9.13. Consider beginning from a completely inactive network (that may be damaged so that some fraction p of vertices remain). As we gradually increase f from zero, under the bootstrap percolation process, more and more vertices become active (always reaching equilibrium before further increases of f) until at a certain threshold value, f_{c1-b} a giant active component appears. As we increase f further, the size of the

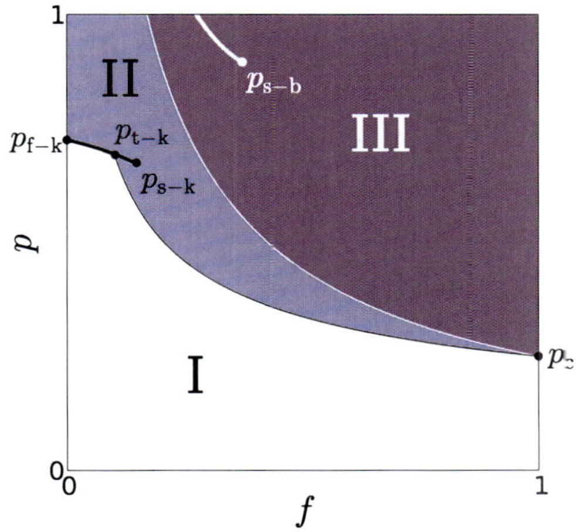

Fig. 9.14 Phase diagram for heterogeneous k-core and bootstrap percolation in the f-p plane. The giant-HKC is present in regions II and III, the giant-BPC is present in region III. The giant-HKC appears continuously at the threshold marked by the *thin black curve*. The hybrid, discontinuous transition occurs at the points marked by the *heavy black line*, beginning from p_{s-k}. Above p_{t-k} the first appearance of the giant-HKC is with a discontinuous transition. Above p_{f-k} the giant-HKC appears discontinuously for any $f > 0$. The continuous appearance of the giant-BPC is marked by the *thin white curve*, and the hybrid transition (beginning at p_{s-b} by a *heavy white curve*. This diagram is for a Bethe lattice with degree 5, and for $k = 3$, but the diagram for any network with finite second moments of the degree distribution will be qualitatively the same

giant-BPC traces the dashed curves shown in Fig. 9.13. See also [32]. The direction of this process is indicated by the arrows on these curves. Finally, at $f = 1$ all undamaged vertices are active. Now we reverse the process, beginning with a fully active network, and gradually reducing f, de-activating (equivalent to pruning) vertices that fall below their threshold k_i under the heterogeneous k-core process. As f decreases, the solid curves in Fig. 9.13 will be followed, in the direction indicated by the arrows. Notice that the size of the giant-HKC for given values of f and p is always larger than the giant-BPC.

It is clear from Figs. 9.13 and 9.14 that even though bootstrap percolation and the heterogeneous k-core described above have the same thresholds and proportions of each kind of vertex, the equilibrium size of the respective giant components is very different. The difference results from the top-down vs. bottom-up ways in which they are constructed. To find the heterogeneous k-core, we begin with the full network, and prune vertices which don't meet the criteria, until we reach equilibrium. In contrast, bootstrap percolation begins with a largely inactive network, and successively activates vertices until equilibrium is reached. To see the effect of this difference, we now describe an important concept: the subcritical clusters of bootstrap percolation.

A subcritical cluster in bootstrap percolation is a cluster of activatable vertices (i.e. not seed vertices) who each have exactly $k - 1$ active neighbors external to the cluster. Under the rules of bootstrap percolation, such clusters cannot become

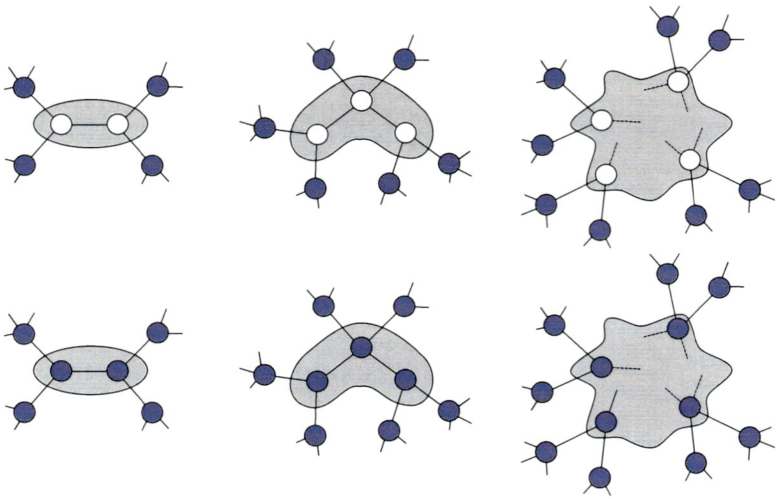

Fig. 9.15 *Top row*: Subcritical clusters of different sizes in bootstrap percolation. *Left*: Because they start in an inactive state, two connected vertices (*shaded area*) cannot become active if each has $k-1$ active neighbors. (In this example $k=3$.) The same follows for clusters of three (*center*) or more vertices (*right*). If any member of a subcritical cluster gains another active neighbor, an avalanche of activations encompasses the whole cluster. *Bottom row*: Similar clusters would be included in the heterogeneous k-core

activated. The vertices within the cluster block each other from becoming active – see Fig. 9.15. Now compare the situation for the heterogeneous k-core. Any cluster of threshold k vertices which each have $k-1$ neighbors in the core external to the cluster, is always included in the heterogeneous k-core. for the heterogeneous k-core, vertices in clusters like those in Fig. 9.12 assist one another. Thus, the exclusion of subcritical clusters from activation in bootstrap percolation accounts for the difference in sizes of the bootstrap percolation core and the heterogeneous k-core.

These results are found in a similar way to those for the heterogeneous k-core: we draw recursive diagrams for the relevant probabilities, and use them to correctly write down self-consistency equations. Activation in bootstrap percolation must spread through the network, meaning that the vertex needs k active downstream neighbors in order to become active (and thus provide an active neighbor to its upstream "parent"). For example, the probability that an arbitrarily chosen vertex is active, $\mathscr{S}_b$ is then [compare (9.15)]:

$$\mathscr{S}_b = pf \sum_{q=1}^{\infty} P(q) \sum_{l=1}^{q} \binom{q}{l} Y^l (1-Y)^{q-l} + p(1-f) \sum_{q=k}^{\infty} P(q) \sum_{l=k}^{q} \binom{q}{l} Y^l (1-Y)^{q-l} . \tag{9.29}$$

Where we define Y to be the probability (counterpart of Z) that on following an arbitrary edge, we encounter a vertex that is either a seed or has k active children. Then:

$$Y = pf + p(1-f) \sum_{q \geq k+1} \frac{qP(q)}{\langle q \rangle} \sum_{l=k}^{q-1} \binom{q-1}{l} Y^l (1-Y)^{q-1-l}$$
$$\equiv \Phi(Y, p, f). \tag{9.30}$$

Note that (9.30) differs from (9.16) because the number of children required is k not $k-1$. This is equivalent to excluding the subcritical clusters represented in Fig. 9.15. The equation for $\mathscr{S}_{gc-b}$ follows similarly.

9.10 Conclusions

We conclude that various k-core problems and bootstrap percolation on networks constitute a wide class of problems which allow a uniform treatment. We have demonstrated that the heterogeneous k-core problem and bootstrap percolation are actually complementary. In both these problems the phases transitions are determined by the divergence of some specific clusters. In the case of k-cores, these clusters are corona clusters, while for the hybrid transition in bootstrap percolation, the subcritical clusters diverge. One should note that this comparison turns out to be possible only after the formulation of the heterogeneous k-core problem, which explains the importance of this generalization.

This work was partially supported by the following projects PTDC: FIS/71551/2006, FIS/108476/2008, SAU-NEU/103904/2008, and MAT/114515/2009, and also by the SOCIALNETS EU project.

References

1. Albert, R., Barabási, A. L.: Statistical mechanics of complex networks. Rev. Mod. Phys. **74**, 47–97 (2002).
2. Dorogovtsev, S.N., Mendes, J. F. F.: Evolution of networks. Adv. Phys. **51**, 1079–1187 (2002).
3. Dorogovtsev, S.N., Mendes, J. F. F.: *Evolution of networks*. (Oxford University Press, Oxford, 2003).
4. Dorogovtsev, S. N., Goltsev, A. V., Mendes, J. F. F.: Critical phenomena in complex networks. Rev. Mod. Phys. **80**, 1275–1335 (2008).
5. Chalupa, J., Leath, P. L., Reich, G. R.: Bootstrap percolation on a Bethe lattice. J. Phys. C **12**, L31–L35 (1979).
6. B. Bollobás, in *Graph Theory and Combinatorics: Proc. Cambridge Combinatorial Conf. in honour of Paul Erdős* (B. Bollobás, ed.) (Academic Press, NY, 1984), p. 35.
7. Pazó D.: Thermodynamic limit of the first-order phase transition in the Kuramoto model. Phys. Rev. E **72**, 046211 (2005).

8. Carmi, S., Havlin, S., Kirkpatrick, S., Shavitt, Y., Shir, E.: A model of Internet topology using k-shell decomposition. PNAS **104**, 11150–11154 (2007).
9. Alvarez-Hamelin, J. I., Dall'Asta, L., Barrat, A., Vespignani, A.: Large scale networks fingerprinting and visualization using the k-core decomposition. In: Y. Weiss, B. Schölkopf, and J. Platt (eds.) Advances in Neural Information Processing Systems **18**, pp. 41–50, MIT Press, Cambridge (2006).
10. Alvarez-Hamelin, J. I., Dall'Asta, L., Barrat, A., Vespignani, A.: *k*-core decomposition of Internet graphs: Hierarchies, selfsimilarity and measurement biases. Networks and Heterogeneous Media **3**, 371 (2008).
11. Dorogovtsev, S.N., Goltsev, A.V., Mendes, J.F.F.: *k*-core organisation of complex networks. Phys. Rev. Lett. **96**, 040601 (2006).
12. Goltsev, A. V., Dorogovtsev, S. N., Mendes, J. F. F.: *k*-core (bootstrap) percolation on complex networks: Critical phenomena and nonlocal effects. Phys. Rev. E **73**, 056101 (2006).
13. Dorogovtsev, S. N., Goltsev, A. V., Mendes, J. F. F.: *k*-core architecture and *k*-core percolation on complex networks. Physica D **224**, 7–19 (2006).
14. Moukarzel, C. F.: Rigidity percolation in a field. Phys. Rev. E **68**, 056104 (2003).
15. Schwarz, J. M., Liu, A. J., Chayes, L. Q.: The onset of jamming as the sudden emergence of an infinite *k*-core cluster. Europhys. Lett. **73**, 560–566 (2006).
16. Chatterjee, N., Sinha, S.: Understanding the mind of a worm: hierarchical network structure underlying nervous system function in C. elegans. Prog. Brain Res. **168**, 145–153 (2007).
17. Schwab, D. J., Bruinsma, R. F., Feldman, J. L., Levine, A. J.: Rhythmogenic neuronal networks, emergent leaders, and *k*-cores. Phys. Rev. E **82**, 051911 (2010).
18. Klimek, P., Thurner, S., Hanel, R.: Pruning the tree of life: k-Core percolation as selection mechanism. J. Theoret. Biol. **256**, 142–146 (2009).
19. Pittel, B., Spencer, J., Wormald, N.: Sudden emergence of a giant k-core in a random graph. J. Comb. Theory B **67**, 111–151 (1996).
20. Fernholz, D., Ramachandran, V.: The giant *k*-core of a random graph with a specified degree sequence. Tech. Rep. TR04-13, University of Texas Computer Science (2004).
21. Riordan, O.: The k-core and branching processes. Combin. Probab. Comput. **17**, 111–136 (2008).
22. Corominas-Murtra, B., Mendes, J. F. F., Solé, R. V.: Nested subgraphs of complex networks. J. Phys. A **41**, 385003 (2008).
23. Kogut, P. M., Leath, P. L.: Bootstrap percolation transitions on real lattices. J. Phys. C **14**, 3187–3194 (1981).
24. Serrano, M. A, Boguna, M.: Percolation and epidemic thresholds in clustered networks. Phys. Rev. Lett. **97**, 088701 (2006).
25. Serrano, M. A, Boguna, M.: Clustering in complex networks. II. Percolation properties. Phys. Rev. E **74**, 056115 (2006).
26. Parisi, G., Rizzo, T.: *k*-core percolation in four dimensions. Phys. Rev. E **78**, 022101 (2008).
27. Gleeson, J. P., Melnik, S.: Analytical results for bond percolation and k-core sizes on clustered networks. Phys. Rev. E **80**, 046121 (2009).
28. Gleeson, J. P., Melnik, S., Hackett, A.: How clustering affects the bond percolation threshold in complex networks. Phys. Rev. E **81**, 066114 (2010).
29. Newman, M. E. J., Strogatz, S. H., and Watts, D. J.: Random graphs with arbitrary degree distributions and their applications. Phys. Rev. E **64**, 026118 (2001).
30. Melnik, S., Hackett, A., Porter, M. A., Mucha, P. J., Gleeson, J.P.: The unreasonable effectiveness of tree-based theory for networks with clustering. arXiv:1001.1439.
31. Aizenman, M., Lebowitz, J. L.: Metastability effects in bootstrap percolation. J. Phys. A **21**, 3801–3813 (1988).
32. Baxter, G.J., Dorogovtsev, S.N., Goltsev, A.V., Mendes, J.F.F.: Bootstrap percolation on complex networks. Phys. Rev. E **82**, 011103 (2010).
33. Cohen, R., ben Avraham, D., Havlin, S.: Percolation critical exponents in scale-free networks. Phys. Rev. E **66**, 036113 (2002).

Part III
Complex Networks Optimization Techniques

Chapter 10
Hardness Complexity of Optimal Substructure Problems on Power-Law Graphs

Yilin Shen, Dung T. Nguyen, and My T. Thai

Abstract The remarkable discovery of many large-scale real networks is the power-law distribution in degree sequence: the number of vertices with degree i is proportional to $i^{-\beta}$ for some constant $\beta > 1$. A lot of researchers believe that it may be easier to solve some optimization problems in power-law graphs. Unfortunately, many problems have been proved *NP*-hard even in power-law graphs as Ferrante proposed in Ferrante et al. (Theoretical Computer Science 393(1–3):220–230, 2008). Intuitively, a theoretical question is raised: Are these problems on power-law graphs still as hard as on general graphs? The chapter shows that many optimal substructure problems, such as *minimum dominating set*, *minimum vertex cover* and *maximum independent set*, are easier to solve in power-law graphs by illustrating better inapproximability factors. An optimization problem has the property of optimal substructure if its optimal solution on some given graph is essentially the union of the optimal subsolutions on all maximal connected components. In particular, the above problems and a more general problem (ρ-*minimum dominating set*) are proven to remain *APX*-hard and their constant inapproximability factors on general power-law graphs by using the cycle-based embedding technique to embed any d-bounded graphs into a power-law graph. In addition, the corresponding inapproximability factors of these problems are further proven in simple power-law graphs based on the graphic embedding technique as well as that of *maximum clique* and *minimum coloring* using the embedding technique in Ferrante et al. (Theoretical Computer Science 393 (1–3):220–230, 2008). As a result of these inapproximability factors, the belief that there exists some $(1 + o(1))$-approximation algorithm for these problems on power-law graphs is proven not always true. In addition, this chapter contains the in-depth

Y. Shen (✉) • D.T. Nguyen • M.T. Thai
Department of Computer Information Science and Engineering, University of Florida, Gainesville, FL 32611, USA
e-mail: yshen@cise.ufl.edu; dtnguyen@cise.ufl.edu; mythai@cise.ufl.edu

M.T. Thai and P.M. Pardalos (eds.), *Handbook of Optimization in Complex Networks: Theory and Applications*, Springer Optimization and Its Applications 57, DOI 10.1007/978-1-4614-0754-6_10, investigations in the relationship between the exponential factor β and constant greedy approximation algorithms. The last part includes some minor *NP*-hardness results on simple power-law graphs for small $\beta < 1$.

10.1 Introduction

A great number of large-scale networks in real life are discovered to follow a power-law distribution in their degree sequences, ranging from the Internet [15], the World-Wide Web (WWW) [4] to social networks [25]. That is, the number of vertices with degree i is proportional to $i^{-\beta}$ for some constant β in these graphs, which is called power-law graphs. The observations show that the exponential factor β ranges between 1 and 4 for most real-world networks [10], i.e., $\beta = 2.1$ in Internet and World Wide Web, $\beta = 2.3$ in social networks and $\beta = 2.5$ in protein–protein interaction networks. Erdös collaboration network in 1997 is illustrated as an example of power-law networks in Fig. 10.1. Intuitively, the following theoretical question is raised: What are the differences in terms of complexity hardness and inapproximability factor of several optimization problems between in general graphs and in power-law graphs?

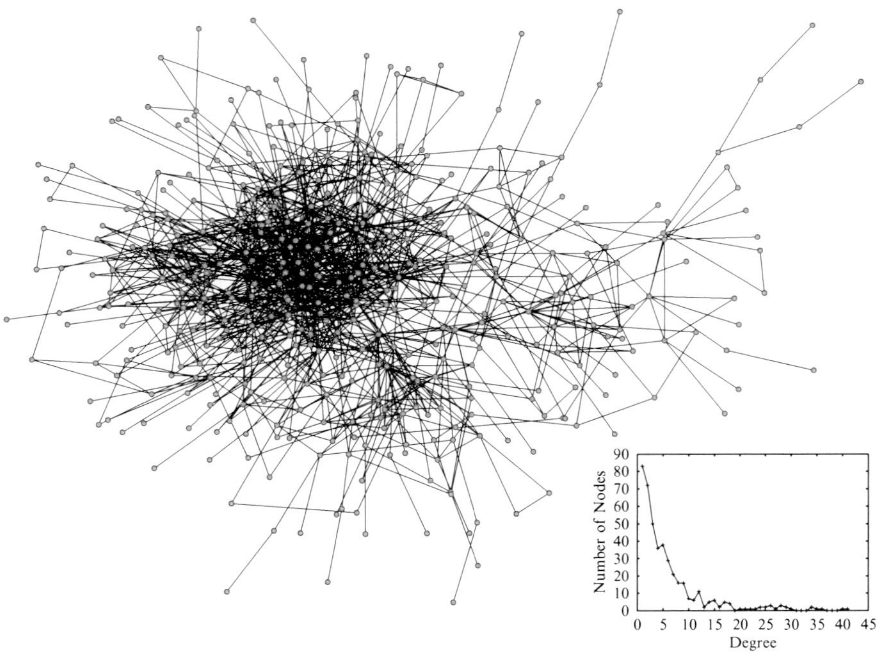

Fig. 10.1 Erdös collaboration network in 1997 [1]

Many experimental results on random power-law graphs give us a belief that the problems might be much easier to solve on power-law graphs. Eubank et al. [14] showed that a simple greedy algorithm leads to a $1 + o(1)$ approximation factor on *minimum dominating set* (MDS) and *minimum vertex cover* (MVC) on power-law graphs (without any formal proof) although MDS and MVC has been proven *NP*-hard to be approximated within $(1-\varepsilon)\log n$ and 1.366 on general graphs respectively [12]. In [24], Gopal also claimed that there exists a polynomial time algorithm that guarantees a $1 + o(1)$ approximation of the MVC problem with probability at least $1 - o(1)$. Unfortunately, there is no such formal proof for this claim either. Furthermore, several papers also have some theoretical guarantees for some problems on power-law graphs. Gkantsidis et al. [17] proved the flow through each link is at most $O(n\log^2 n)$ on power-law random graphs where the routing of $O(d_u d_v)$ units of flow between each pair of vertices u and v with degrees d_u and d_v. In [17], the authors took advantage of the property of power-law distribution by using the structural random model [2, 3] and showed the theoretical upper bound with high probability $1 - o(1)$ and the corresponding experimental results. Likewise, Janson et al. [19] gave an algorithm that approximated *maximum clique* within $1 - o(1)$ on power-law graphs with high probability on the random poisson model $G(n, \alpha)$ (i.e., the number of vertices with degree at least i decreases roughly as n^{-i}). Although these results were based on experiments and various random models, they raised an interest in investigating hardness and inapproximability of optimization problems on power-law graphs.

Recently, Ferrante et al. [16] had an initial attempt on power-law graphs to show the *NP*-hardness of *maximum clique* (CLIQUE) and *minimum graph coloring* (COLORING) ($\beta > 1$) by constructing a bipartite graph to embed a general graph into a power-law graph and *NP*-hardness of MVC, MDS and *maximum independent set* (MIS) ($\beta > 0$) based on their optimal substructure properties. Unfortunately, there is a minor flaw which makes the proof of *NP*-hardness of MIS, MVC, MDS with $\beta < 1$ no longer hold.

In this chapter, two new techniques are proposed especially for optimal substructure problems, *cycle-based embedding technique* and *graphic embedding technique*, which embed a d-bounded graph into a general power-law graph and a simple power-law graph respectively. Then we use these two techniques to further prove the *APX*-hardness and the inapproximability of MIS, MDS, and MVC on general power-law graphs and simple power-law graphs for $\beta > 1$. These inapproximability results on power-law graphs are shown in Table 10.1. Furthermore, the inapproximability results in CLIQUE and COLORING are shown by taking advantage of the reduction in [16]. The corresponding relationship is analyzed between β and constant greedy approximation algorithms for MIS and MDS. To show the minor *NP*-hardness of these problems for $\beta < 1$, we propose another *eligible embedding technique* in the last part of this chapter.

In addition, due to a lot of recent studies in online social networks on the influence propagation problem [21,22], ρ-*minimum dominating set* (ρ-MDS) is formulated and proven hard to be approximated within $2 - (2 + o_d(1))\log\log d/\log d$

Table 10.1 Inapproximability factors on power-law graphs with exponential factor $\beta > 1$

Problem	General power-law graph	Simple power-law graph
MIS	$1 + \frac{1}{140(2\zeta(\beta)3^{\beta}-1)} - \varepsilon$	$1 + \frac{1}{1120\zeta(\beta)3^{\beta}} - \varepsilon$
MDS	$1 + \frac{1}{390(2\zeta(\beta)3^{\beta}-1)}$	$1 + \frac{1}{3120\zeta(\beta)3^{\beta}}$
MVC, ρ-MDS	$1 + \frac{2\left(1-(2+o_c(1))\frac{\log\log c}{\log c}\right)}{\left(\zeta(\beta)c^{\beta}+c^{\frac{1}{\beta}}\right)(c+1)}$	$1 + \frac{2-(2+o_c(1))\frac{\log\log c}{\log c}}{2\zeta(\beta)c^{\beta}(c+1)}$
CLIQUE	–	$O\left(n^{1/(\beta+1)-\varepsilon}\right)$
COLORING	–	$O\left(n^{1/(\beta+1)-\varepsilon}\right)$

[a]Conditions: MIS and MDS: P≠NP; MVC, ρ-MDS: unique games conjecture; CLIQUE, COLORING: NP≠ZPP

[b]c is a constant which is the smallest d satisfying the condition in [6]

factor on d-bounded graphs under unique games conjecture, which further leads to the following inapproximability result on power-law graphs (shown in Table 10.1).

The rest of chapter is organized as follows. In Sect. 10.2, we introduce some problem definitions, the model of power-law graphs, and some related concepts. The inapproximability optimal substructure framework is presented in Sect. 10.3. We show the hardness and inapproximability of MIS, MDS, MVC on general power-law graphs using the cycle-based embedding technique in Sect. 10.4. More inapproximability results in simple power-law graphs are illustrated in Sect. 10.5 based on the graphic embedding technique, which implies the *APX*-hardness of these problems. Additionally, the inapproximability factor on maximum clique and minimum coloring problems are proven. In Sect. 10.6, we analyze the relationship between β and constant approximation algorithms, which further proves that the integral gap is typically small for optimization problems on power-law graphs than that on general bounded graphs. Some minor *NP*-hardness results of optimal substructure problems for $\beta < 1$ are presented in Sect. 10.7.

10.2 Preliminaries

In this section, we first recall the definition of several classical optimization problems having the optimal substructure property and formulate the new optimization problem ρ-minimum dominating set. Then the power-law model and some corresponding concepts are proposed. At last, some special graphs are introduced which will be used in the analysis throughout the whole chapter.

10.2.1 Problem Definitions

Definition 10.1 (Maximum Independent Set). Given an undirected graph $G = (V,E)$, find a subset $S \subseteq V$ with the maximum size such that no two vertices in S are adjacent.

Definition 10.2 (Minimum Vertex Cover). Given an undirected graph $G = (V,E)$, find a subset $S \subseteq V$ with the minimum size such that for each edge E at least one endpoint belongs to S.

Definition 10.3 (Minimum Dominating Set). Given an undirected graph $G = (V,E)$, find a subset $S \subseteq V$ with the minimum size such that for each vertex $v_i \in V \setminus S$, at least one neighbor of v_i belongs to S.

Definition 10.4 (Maximum Clique). Given an undirected graph $G = (V,E)$, find a clique with maximum size where a subgraph of G is called a clique if all its vertices are pairwise adjacent.

Definition 10.5 (Minimum Graph Coloring). Given an undirected graph $G = (V,E)$, label the vertices in V with minimum number of colors such that no two adjacent vertices share the same color.

The ρ-minimum dominating set is defined as general version of MDS problem. In the context of influence propagation, the ρ-MDS problem aims to find a subset of nodes with minimum size such that all nodes in the whole network can be influenced within t rounds. In particular, a node is influenced when ρ fraction of its neighbors are influenced. For simplicity, we define ρ-MDS problem in the case that $t = 1$.

Definition 10.6 (ρ-Minimum Dominating Set). Given an undirected graph $G = (V,E)$, find a subset $S \subseteq V$ with the minimum size such that for each vertex $v_i \in V \setminus S$, $|S \cap N(v_i)| \geq \rho |N(v_i)|$.

10.2.2 Power-Law Model and Some Notations

A great number of models [2, 3, 7, 8, 23] on power-law graphs are emerging in the past recent years. The analysis in this chapter is based on the general (α,β) graph model, that is, the graphs are only constrained by the power-law distribution in degree sequences. To begin with, the following two types of degree sequences are defined.

Definition 10.7 (y-Degree Sequence). Given a graph $G = (V,E)$, the y-degree sequence of G is a sequence $Y = \langle y_1, y_2, \ldots, y_\Delta \rangle$ where Δ is the maximum degree of G and $y_i = |\{u \; u \in V \wedge deg(u) = i\}|$.

Definition 10.8 (d-Degree Sequence). Given a graph $G = (V,E)$, the d-degree sequence of G is a sequence $D = \langle d_1, d_2, \ldots, d_n \rangle$ of vertex in non-increasing order of their degrees.

Note that y-degree sequence and d-degree sequence are interchangeable. Given a y-degree sequence $Y = \langle y_1, y_2, \ldots, y_\Delta \rangle$, the corresponding d-degree sequence is $D = \langle \Delta, \Delta, \ldots, \Delta - 1, \Delta - 1, \ldots, \Delta - 1, \ldots, 1, \ldots, 1 \rangle$ where the number i appears y_i times. Because of their equivalence, we may use only y-degree sequence or d-degree sequence or both without changing the meaning or validity of results. The definition of power-law graphs can be expressed via y-degree sequences as follows.

Definition 10.9 (General (α, β) Power-Law Graph Model). A graph $G = (V, E)$ is called a (α, β) power-law graph $G_{(\alpha,\beta)}$ where multiedges and self-loops are allowed if the maximum degree is $\Delta = \lfloor e^{\alpha/\beta} \rfloor$ and the number of vertices of degree i is

$$y_i = \begin{cases} \lfloor e^\alpha / i^\beta \rfloor, & \text{if } i > 1 \text{ or } \sum_{i=1}^{\Delta} \lfloor e^\alpha / i^\beta \rfloor \text{ is even} \\ \lfloor e^\alpha \rfloor + 1, & \text{otherwise.} \end{cases} \tag{10.1}$$

In simple (α, β) power-law graphs, there are no multiedges and self-loops.

Note that a power-law graph are represented by two parameters α and β. Since graphs with the same β express the same behaviors, we categorize all graphs with the same β into a β-family of graphs such that β is regarded as a constant instead of an input. In addition, we focus more on the case $\beta > 1$ because almost all real large-scale networks have $\beta > 1$. In this case, the number of vertices is:

$$\sum_{i=1}^{\Delta} \frac{e^\alpha}{i^\beta} = \zeta(\beta) e^\alpha + n^{\frac{1}{\beta}} \approx \zeta(\beta) e^\alpha,$$

where $\zeta(\beta) = \sum_{i=1}^{\infty} \frac{1}{i^\beta}$ is the *Riemann Zeta function*. Also, the d-degree sequence of any (α, β) power-law graph is continuous according to the following definition.

Definition 10.10 (Continuous Sequence). An integer sequence $\langle d_1, d_2, \ldots, d_n \rangle$, where $d_1 \geq d_2 \geq \cdots \geq d_n$, is continuous if $\forall 1 \leq i \leq n-1$, $|d_i - d_{i+1}| \leq 1$.

Definition 10.11 (Graphic Sequence). A sequence D is said to be graphic if there exists a graph such that D is its d-degree sequence.

Definition 10.12 (Degree Set). Given a graph G, let $D_i(G)$ be the set of vertices of degree i on G.

Definition 10.13 (d-Bounded Graph). Given a graph $G = (V, E)$, G is a d-bounded graph if the degree of any vertex is upper bounded by an integer constant d.

10.2.3 Special Graphs

Definition 10.14 (d-Regular Cycle $RC_n^{\mathbf{d}}$). Given a vector $\mathbf{d} = (d_1, \ldots, d_n)$, a **d**-regular cycle RC_n^d is composed of two cycles. Each cycle has n vertices and two

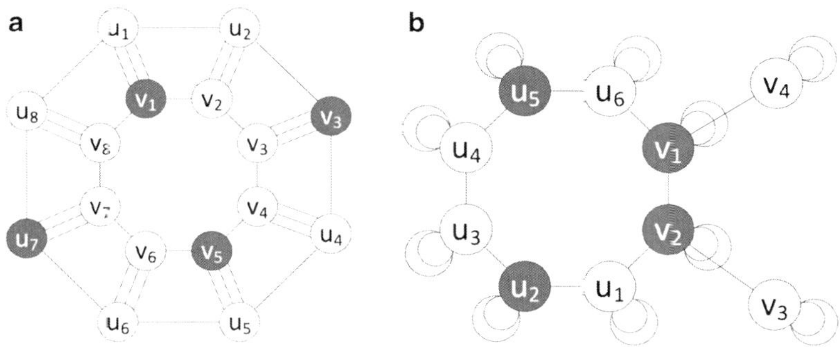

Fig. 10.2 Special graph examples: the left one is a (3,3,3,3,3,3,3,3)-regular cycle and the right one is a (3,3,3,3)-branch-(2,2,2,2,2,2)-cycle. The *grey vertices* consist of the optimal solution of MDS on these two special graphs

i^{th} vertices in each cycle are adjacent with each other by $d_i - 2$ multiedges. That is, **d**-regular cycle RC_n^d has $2n$ vertices and the two ith vertex has the same degree d_i. An example RC_8^d is shown in Fig. 10.2a.

Definition 10.15 (κ-Branch-d-Cycle κ-$BC_n^{\mathbf{d}}$). Given two vectors $\mathbf{d} = (d_1, \ldots, d_n)$ and $\kappa = (\kappa_1, \ldots, \kappa_m)$, the κ-branch-**d**-cycle is composed of a cycle with a number of vertices n such that each vertex has degree d_i as well as $|\kappa|/2$ appendant branches, where $|\kappa|$ is a even number. Note that any κ-branch-**d**-cycle has $|\kappa|$ even number of vertices with odd degrees. An example is shown in Fig. 10.2b.

10.2.4 Existing Inapproximability Results

Some inapproximability results are listed here in the literature to use later in our proofs.

1. MVC is hard to be approximated into $2-(2+o_d(1))\log\log d/\log d$ for every sufficiently large integer d in d-bounded graphs under unique games conjecture [6, 11].
2. In three-bounded graphs, MIS and MDS is NP-hard to be approximated into $\frac{140}{139} - \varepsilon$ for any $\varepsilon > 0$ and $\frac{391}{390}$ respectively [5].
3. Maximum clique and minimum coloring problem is hard to be approximated into $n^{1-\varepsilon}$ on general graphs unless NP = ZPP [18].

10.3 Inapproximability Optimal Substructure Framework in Power-Law Graphs

In this section, by taking advantage of the optimal substructure property, we introduce a framework to derive the approximation hardness of the class of problems in power-law graphs. Recall that a graph optimization problem is said to satisfy optimal substructure if its optimal solution is the union of the optimal solutions on each connected component. Therefore, when a graph G is embedded into a power-law graph G', the optimal solution in G' consists of a subset of the optimal solution in G. According to this important property, *inapproximability optimal substructure framework* is proposed to prove the inapproximability factor if there exists a *embedded-approximation-preserving reduction* that relates the approximation hardness in general graphs and power-law graphs by guaranteeing the relationship between the solutions in the original graph and the constructed graph.

Definition 10.16 (Embedded-Approximation-Preserving Reduction). Given an optimal substructure problem O, a reduction from an instance on graph $G=(V,E)$ to another instance on a power-law graph $G'=(V',E')$ is called embedded-approximation-preserving if it satisfies the following properties:

1. G is a subset of maximal connected components of G'.
2. The optimal solution of O on G', $OPT(G')$, is upper bounded by $\mathfrak{C}OPT(G)$ where $\mathfrak{C}$ is a constant correspondent to the growth of the optimal solution.

Theorem 10.1 (Inapproximability Optimal Substructure Framework). *Given an optimal substructure problem O, if there exists an embedded-approximation-preserving reduction from a graph G to another graph G', we can extract the inapproximability factor δ of O on G' using* ε-inapproximability *of O on G, where δ is lower bounded by $\frac{\varepsilon\mathfrak{C}}{(\mathfrak{C}-1)\varepsilon+1}$ and $\frac{\varepsilon+\mathfrak{C}-1}{\mathfrak{C}}$ when O is a maximum and minimum optimization problem respectively.*

Proof. Suppose that there exists an algorithm providing a solution of O on G' with size at most δ times the optimal solution. Denote A and B to be the sizes of the produced solution on G and $G' \setminus G$ and A^* and B^* to be their corresponding optimal values. Hence, we have $B^* \leq (\mathfrak{C}-1)A^*$. With the completeness that $OPT(G)=A^* \Rightarrow OPT(G')=B^*$, the soundness leads to the lower bound of δ which is dependent on the type of O, maximization or minimization problem, as follows.

Case 10.1. When O is a maximization problem, the proof of soundness is as follows

$$A^*+B^* \leq \delta(A+B) \tag{10.2}$$

$$\Leftrightarrow A^* \leq \delta A+(\delta-1)B^* \tag{10.3}$$

$$\Leftrightarrow A^* \leq \delta A+(\delta-1)(\mathfrak{C}-1)A^*, \tag{10.4}$$

where (10.3) holds since $B \leq B^*$ and (10.4) holds since $B^* \leq (\mathfrak{C}-1)A^*$.

On the other hand, it is hard to approximate O within ε on G, thus $A^* > \varepsilon A$. Replace it to the above inequality, we have:

$$A^* < A^*\delta/\varepsilon + (\delta - 1)(\mathfrak{C} - 1)A^* \Leftrightarrow \delta > \frac{\varepsilon\mathfrak{C}}{(\mathfrak{C} - 1)\varepsilon + 1}.$$

Case 10.2. When O is a minimization problem, since $B^* \leq B$, similarly

$$\begin{aligned} A + B &\leq \delta(A^* + B^*) \\ \Leftrightarrow A &\leq \delta A^* + (\delta - 1)B^* \\ \Leftrightarrow A &\leq \delta A^* + (\delta - 1)(\mathfrak{C} - 1)A^*. \end{aligned}$$

Then from $A > \varepsilon A^*$,

$$\varepsilon < \delta + (\delta - 1)(\mathfrak{C} - 1) \Leftrightarrow \delta > \frac{\varepsilon + \mathfrak{C} - 1}{\mathfrak{C}}.$$

10.4 Hardness and Inapproximability of Optimal Substructure Problems on General Power-Law Graphs

10.4.1 General Cycle-Based Embedding Technique

In this section, a *general cycle-based embedding technique* is proposed on (α, β) power-law graphs with $\beta > 1$. The basic idea is to embed an arbitrary d-bounded graph into a power-law graph using a $\mathbf{d_1}$-regular cycle, a κ-branch-$\mathbf{d_2}$-cycle, and a number of cliques K_2, where $\mathbf{d_1}$, $\mathbf{d_2}$, and κ are defined by α and β. Before discussing the main embedding technique, we first show that most optimal substructure problems can be polynomially solved in both $\mathbf{d}$-regular cycles and κ-branch-$\mathbf{d}$-cycle. In this context, the cycle-based embedding technique helps to prove the complexity of these optimal substructure problems on power-law graphs according to their corresponding complexity results on general bounded graphs.

Lemma 10.1. *MDS, MVC, and MIS are polynomially solvable on* $\mathbf{d}$*-regular cycles.*

Proof. Here, we just prove MDS problem is polynomially solvable on $\mathbf{d}$-regular cycles. The algorithm is simple. From an arbitrarily vertex, we select the vertex on the other cycle in two hops. The algorithm will terminate until all vertices are dominated. Now we will show that this gives the optimal solution. Let's take RC_8^3 as an example. As shown in Fig. 10.2a, the size of MDS is 4. Notice that each vertex can dominate exact three vertices, that is, 4 vertices can dominate exactly 12 vertices. However, in RC_8^3, there are altogether 16 vertices, which have to be dominated by at least four vertices apart from the vertices in MDS. That is, the algorithm returns an optimal solution. The proofs of MVC and MIS are similar.

Algorithm 2: Cycle Embedding Algorithm

1 $\alpha \leftarrow \max\{\ln \max_{1 \le i \le d}\{n_i \cdot i^\beta\}, \beta \ln d\}$;
2 For $\tau(1)$ vertices of degree 1, add $\lfloor \tau(1)/2 \rfloor$ number of cliques K_2;
3 For $\tau(2)$ vertices of degree 2, add a cycle with the size $\tau(2)$;
4 For all vertices of degree larger than 2 and smaller than Δ, construct a $\mathbf{d_1}$-regular cycle where $\mathbf{d_1}$ is a vector composed of $\lfloor \tau(i)/2 \rfloor$ number of elements i for all i satisfying $\tau(i) > 0$;
5 For all leftover isolated vertices L such that $\tau(i) - 2\lfloor \tau(i)/2 \rfloor = 1$, construct a $\mathbf{d_2^1}$-branch-$\mathbf{d_2^2}$-cycle, where $\mathbf{d_2^1}$ and $\mathbf{d_2^2}$ are the vectors containing odd and even elements correspondent to the vertices of odd and even degrees in L respectively.

Lemma 10.2. *MDS, MVC, and MIS is polynomially solvable on* κ*-branch-*$\mathbf{d}$*-cycles.*

Proof. Again, we show the proof of MDS. First, we select the vertices connecting both the branches and the cycle. Then by removing the branches, we will have a line graph regardless of self-loops, on which MDS is polynomially solvable. It is easy to see that the size of MDS will increase if any one vertex connecting both the branch and the cycle in MDS is replaced by some other vertices. The proof of MIS is similar. Note that the optimal solution for MVC consists of all vertices since all edges need to be covered.

Theorem 10.2 (Cycle-Based Embedding Technique). *Any d-bounded graph* G_d *can be embedded into a power-law graph* $G_{(\alpha,\beta)}$ *with* $\beta > 1$ *such that* G_d *is a maximal component and most optimal substructure problems can be polynomially solvable on* $G_{(\alpha,\beta)} \setminus G_d$.

Proof. With the given β, we choose α to be $\max\{\ln \max_{1 \le i \le d}\{n_i \cdot i^\beta\}, \beta \ln d\}$. Based on $\tau(i) = \lfloor e^\alpha / i^\beta \rfloor - n_i$ where $n_i = 0$ when $i > d$, we construct the power-law graph $G_{(\alpha,\beta)}$ as the following Algorithm 2. The last step holds since the number of vertices of odd degrees has to be even. From Step 1, we know $e^\alpha = \max\{\max_{1 \le i \le d}\{n_i \cdot i^\beta\}, d^\beta\} \le d^\beta n$, that is, the number of vertices N in graph $G_{(\alpha,\beta)}$ satisfies $N \le \zeta(\beta) d^\beta n$, which means that N/n is a constant. According to Lemma 10.1 and Lemma 10.2, since $G_{(\alpha,\beta)} \setminus G_d$ is composed of a $\mathbf{d_1}$-regular cycle and a $\mathbf{d_2^1}$-branch-$\mathbf{d_2}$-cycle, it can be polynomially solvable. Note that the number of vertices in L is at most Δ since there is at most one leftover vertex of each degree.

10.4.2 APX-*Hardness*

In this section, MIS, MDS, and MVC are proven to remain *APX*-hard even on power-law graphs.

Theorem 10.3. *MDS is* APX*-hard on power-law graphs.*

Proof. According to Theorem 10.2, we use the cycle-based embedding technique to show $\mathscr{L}$-reduction from MDS on any d-bounded graph G_d to MDS on a power-law graph $G_{(\alpha,\beta)}$ since MDS is proven *APX*-hard on d-bounded graphs [20].

Letting ϕ be a feasible solution on G_d, we can construct MDS in G' such that MDS on a K_2 is 1, $n/4$ on a **d**-regular cycle and $n/3$ on a cycle and a κ-branch-**d**-cycle. Therefore, for a solution ϕ on G_d, we have a solution φ on $G_{(\alpha,\beta)}$ to be $\varphi = \phi + n_1/2 + n_2/3 + n_3/4$, where n_1, n_2 and n_3 corresponds to $\tau(1)$, $\tau(2) \cup L$ and all leftover vertices. Hence, we have $OPT(\varphi) = OPT(\phi) + n_1/2 + n_2/3 + n_3/4$.

On one hand, for a d-bounded graph with vertices n, the optimal MDS is lower bounded by $n/(d+1)$. Thus, we know

$$\begin{aligned}
OPT(\varphi) &= OPT(\phi) + n_1/2 + n_2/3 + n_3/4 \\
&\leq OPT(\phi) + (N-n)/2 \\
&\leq OPT(\phi) + (\zeta(\beta)d^{\beta} - 1)n/2 \\
&\leq OPT(\phi) + (\zeta(\beta)d^{\beta} - 1)(d+1)OPT(\phi)/2 \\
&= \left[1 + (\zeta(\beta)d^{\beta} - 1)(d+1)/2\right] OPT(\varphi),
\end{aligned}$$

where N is the number of vertices in $G_{(\alpha,\beta)}$.

On the other hand, with $|OPT(\phi) - \phi| = |OPT(\varphi) - \varphi|$, we proved the $\mathscr{L}$-reduction with $c_1 = 1 + (\zeta(\beta)d^{\beta} - 1)(d+1)/2$ and $c_2 = 1$.

Theorem 10.4. *MVC is* APX-*hard on power-law graphs.*

Proof. In this proof, we show $\mathscr{L}$-reduction from MVC on d-bounded graph G_d to MVC on power-law graph $G_{(\alpha,\beta)}$ using cycle-based embedding technique.

Let ϕ be a feasible solution on G_d. We construct the solution $\varphi \leq \phi + (N-n)$ since the optimal solution of MVC is $n/2$ on K_2, cycle, **d**-regular cycle and n on κ-branch-**d**-cycle. Therefore, since the optimal MVC on a d-bounded graph is lower bounded by $n/(d+1)$, we have

$$OPT(\varphi) \leq \left[1 + (\zeta(\beta)d^{\beta} - 1)(d+1)\right] OPT(\phi).$$

On the other hand, with $|OPT(\phi) - \phi| = |OPT(\varphi) - \varphi|$, we proved the $\mathscr{L}$-reduction with $c_1 = 1 + (\zeta(\beta)d^{\beta} - 1)(d+1)$ and $c_2 = 1$.

Corollary 10.1. *MIS is* APX-*hard on power-law graphs.*

10.4.3 Inapproximability Factors

In this section, we show the inapproximability factors on MIS, MVC, and MDS on power-law graphs, respectively, using the results in Sect. 10.2.4.

Theorem 10.5. *For any $\varepsilon > 0$, there is no $1 + \frac{1}{140(2\zeta(\beta)3^{\beta}-1)} - \varepsilon$ approximation algorithm for maximum independent set on power-law graphs.*

Proof. In this proof, we construct the power-law graph $G_{(\alpha,\beta)}$ based on cycle-based embedding technique in Theorem 10.2 from d-bounded graph G_d. Let ϕ and φ be feasible solutions of MIS on G_d and $G_{(\alpha,\beta)}$. Then $OPT(\varphi)$ composed of $OPT(\phi)$, clique K_2, cycle, **d**-regular cycle and κ-branch-**d**-cycles are all exactly half number of vertices. Hence, we have $OPT(\varphi) = OPT(\phi) + (N-n)/2$ where n and N is the number of vertices in G_d and $G_{(\alpha,\beta)}$, respectively. Since $OPT(\phi) \geq n/(d+1)$ on d-bounded graphs for MIS and $N \leq \zeta(\beta)d^{\beta}n$, we further have $\mathfrak{C} = 1 + \frac{(\zeta(\beta)d^{\beta}-1)(d+1)}{2}$ from

$$\begin{aligned} OPT(\varphi) &= OPT(\phi) + \frac{N-n}{2} \\ &\leq OPT(\phi) + \frac{(\zeta(\beta)d^{\beta}-1)}{2}n \\ &\leq OPT(\phi) + \frac{(\zeta(\beta)d^{\beta}-1)(d+1)}{2}OPT(\phi) \\ &= \left(1 + \frac{(\zeta(\beta)d^{\beta}-1)(d+1)}{2}\right)OPT(\phi). \end{aligned}$$

Since $\varepsilon = \frac{140}{139} - \varepsilon'$ for any $\varepsilon' > 0$ on three-bounded graphs, the inapproximability factor can be derived from inapproximability optimal substructure framework as

$$\delta > \frac{\varepsilon\mathfrak{C}}{(\mathfrak{C}-1)\varepsilon + 1} > 1 + \frac{1}{140\mathfrak{C}} - \varepsilon = 1 + \frac{1}{140(2\zeta(\beta)3^{\beta}-1)} - \varepsilon,$$

where the last step follows from $d = 3$.

Theorem 10.6. *There is no $1 + \frac{1}{390(2\zeta(\beta)3^{\beta}-1)}$ approximation algorithm for minimum dominating set on power-law graphs.*

Proof. In this proof, we construct the power-law graph $G_{(\alpha,\beta)}$ based on cycle-based embedding technique in Theorem 10.2 from d-bounded graph G_d. Let ϕ and φ be feasible solutions of MDS on G_d and $G_{(\alpha,\beta)}$. The optimal MDS on $OPT(\phi)$, clique K_2, cycle, **d**-regular cycle and κ-branch-**d**-cycles are $n/2$, $n/4$ and $n/3$ respectively. Let ϕ and φ be feasible solutions of MDS on G_d and $G_{(\alpha,\beta)}$. Then we have $\mathfrak{C} = 1 + \frac{(\zeta(\beta)d^{\beta}-1)(d+1)}{2}$ similar as the proof in Theorem 10.5.

Since $\varepsilon = \frac{391}{390}$ in three-bounded graphs, the inapproximability factor can be derived from inapproximability optimal substructure framework as

$$\delta > 1 + \frac{\varepsilon - 1}{\mathfrak{C}} = 1 + \frac{1}{390(2\zeta(\beta)3^{\beta}-1)},$$

where the last step follows from $d = 3$.

Theorem 10.7. *MVC is hard to be approximated within* $1+\frac{2\left(1-(2+o_c(1))\frac{\log\log c}{\log c}\right)}{\left(\zeta(\beta)c^\beta+c^{\frac{1}{\beta}}\right)(c+1)}$ *on power-law graphs under unique games conjecture.*

Proof. By constructing the power-law graph $G_{(\alpha,\beta)}$ based on cycle-based embedding technique in Theorem 10.2 from d-bounded graph G_d, the optimal MVC on clique K_2, cycle, **d**-regular cycle are half number of vertices while the optimal MVC on κ-branch-**d**-cycles are all vertices. Thus, we have $\mathfrak{C}=1+\frac{\left(\zeta(\beta)d^\beta-1+d^{\frac{1}{\beta}}\right)(d+1)}{2}$ since

$$OPT(\varphi)\leq OPT(\phi)+\frac{N-n-\Delta}{2}+\Delta \tag{10.5}$$

$$\leq OPT(\phi)+\frac{(\zeta(\beta)d^\beta-1)n+n^{\frac{1}{\beta}}d}{2} \tag{10.6}$$

$$=OPT(\phi)+\frac{\left(\zeta(\beta)d^\beta-1+\frac{d}{n^{1-\frac{1}{\beta}}}\right)n}{2} \tag{10.7}$$

$$\leq OPT(\phi)+\frac{\left(\zeta(\beta)d^\beta-1+\frac{d}{(d+1)^{1-\frac{1}{\beta}}}\right)(d+1)}{2}OPT(\phi) \tag{10.8}$$

$$\leq\left(1+\frac{\left(\zeta(\beta)d^\beta-1+d^{\frac{1}{\beta}}\right)(d+1)}{2}\right)OPT(\phi), \tag{10.9}$$

where ϕ and φ be feasible solutions of MVC on G_d and $G_{(\alpha,\beta)}$, Δ is the maximum degree in $G_{(\alpha,\beta)}$. The inequality (10.6) holds since there are at most Δ vertices in κ-branch-**d**-cycle, i.e., $\Delta=e^{\alpha/\beta}\leq n^{1/\beta}d$; (10.8) holds since there are at least $d+1$ vertices in a d-bounded graph and the optimal MVC in a d-bounded graph is at least $n/(d+1)$.

Since $\varepsilon=2-(2+o_d(1))\log\log d/\log d$, the inapproximability factor can be derived from inapproximability optimal substructure framework as

$$\delta>1+\frac{\varepsilon-1}{\mathfrak{C}}\geq 1+\frac{2\left(1-(2+o_c(1))\frac{\log\log c}{\log c}\right)}{\left(\zeta(\beta)c^\beta+c^{\frac{1}{\beta}}\right)(c+1)},$$

where c is the smallest d satisfying the condition in [6]. The last inequality holds since function $f(x)=(1-(2+o_x(1))\log\log x/\log x)/g(x)(x+1)$ is monotonously decreasing when $f(x)>0$ for all $x>0$ when $g(x)$ is monotonously increasing.

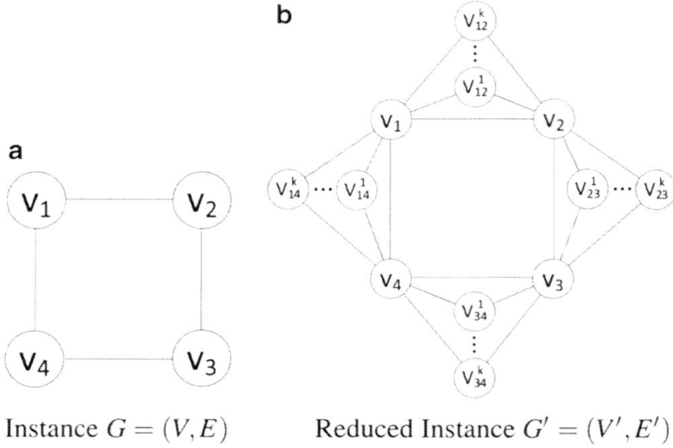

Fig. 10.3 The reduction from MVC to ρ-MDS

Theorem 10.8. *ρ-MDS is hard to be approximated into $2-(2+o_d(1))\frac{\log\log d}{\log d}$ on d-bounded graphs under unique games conjecture.*

Proof. In this proof, we show the gap-preserving from MVC on (d/ρ)-bounded graph $G=(V,E)$ to ρ-MDS on d-bounded graph $G'=(V',E')$. *w.l.o.g.*, we assume that d and d/ρ are integers. First, we construct a graph $G'=(V',E')$ by adding new vertices and edges to G as follows. For each edge $(v_i,v_j)\in E$, create k new vertices $v_{ij}^1,\ldots,v_{ij}^k$ where $1\le k\le\lfloor 1/\rho\rfloor$ and $\rho\le 1/2$. Then we add $2k$ new edges (v_{ij}^l,v_i) and (v_{ij}^l,v_j) for all $l\in[1,k]$ as shown in Fig. 10.3. Clearly, $G'=(V',E')$ is a d-bounded graph.

Let ϕ and φ be feasible solutions to MVC on G and G' respectively. We claim that $OPT(\phi)=OPT(\varphi)$.

On one hand, if $S=\{v_1,v_2,\ldots,v_j\}\in V$ is the minimum vertex cover on G. Then $\{v_1,v_2,\ldots,v_j\}$ is a ρ-MDS on G' because each vertex in V has ρ of all neighbors in MVC and every new vertex in $V'\setminus V$ has at least one of two neighbors in MVC. Thus, $OPT(\phi)\ge OPT(\varphi)$.

One the other hand, we can prove that $OPT(\varphi)$ does not contain new vertices, that is, $V'\setminus V$. Consider a vertex $v_i\in V$, if $v_i\in OPT(\varphi)$, the new vertices v_{ij}^l for all $v_j\in N(v_i)$ and all $l\in[1,k]$ are not needed to be selected. If $v_i\notin OPT(\varphi)$, it has to be dominated by ρ proportion of its all neighbors. That is, for each edge (v_i,v_j) incident to v_i, either v_j or all v_{ij}^l have to be selected since every v_{ij}^l has to be either selected or dominated. If all v_{ij}^l are selected in $OPT(\varphi)$ for some edge (v_i,v_j), v_j is still not dominated by enough vertices if there are some more edges incident to v_j and the number of vertices v_{ij}^l k is great than 1, that is, $\lfloor 1/\rho\rfloor\ge 1$. In this case, v_j will be selected to dominate all v_{ij}^l. Thus, $OPT(\varphi)$ does not contain new vertices. Since the vertices in V selected is a solution to ρ-MDS, that is, for each vertex v_i

in graph G, v_i will be selected or at least the number of neighbors of v_i will be selected. Therefore, the vertices in $OPT(\varphi)$ consist of a vertex cover in G. Thus, $OPT(\phi) \le OPT(\varphi)$. Then we show the completeness and soundness as follows.

- If $OPT(\phi) = m \Rightarrow OPT(\varphi) = m$
- If $OPT(\phi) > \left(2-(2+o_d(1))\frac{\log\log(d/2)}{\log(d/2)}\right)m \Rightarrow OPT(\varphi) > \left(2-(2+o_d(1))\frac{\log\log d}{\log d}\right)m$

$$OPT(\varphi) > \left(2-(2+o_d(1))\frac{\log\log(d/\rho)}{\log(d/\rho)}\right)m > \left(2-(2+o_d(1))\frac{\log\log d}{\log d}\right)m$$

since the function $f(x) = 2 - \log\log x/\log x$ is monotonously increasing for any $x > 0$.

Theorem 10.9. *ρ-MDS is hard to be approximated into* $1 + \frac{2\left(1-(2+o_c(1))\frac{\log\log c}{\log c}\right)}{2+(\zeta(\beta)c^{\beta}-1)(c+1)}$ *on power-law graphs under unique games conjecture.*

Proof. By constructing the power-law graph $G_{(\alpha,\beta)}$ based on cycle-based embedding technique in Theorem 10.2 from d-bounded graph G_d we have $\mathfrak{C} = 1 + \frac{(\zeta(\beta)d^{\beta}-1)(d+1)}{2}$ from the optimal MVC on $OPT(\phi)$, clique K_2, cycle, **d**-regular cycle and κ-branch-**d**-cycles as

$$\begin{aligned} OPT(\varphi) &= OPT(\phi) + n_1/2 + f(\rho)n_2 + g(\rho)n_3 \\ &\le OPT(\phi) + \frac{N-n}{2} \le \left(1 + \frac{(\zeta(\beta)d^{\beta}-1)(d+1)}{2}\right) OPT(\phi), \end{aligned}$$

where $f(\rho) = \begin{cases} \frac{1}{4}, & \rho \le \frac{1}{3} \\ \frac{1}{3}, & \frac{1}{3} < \rho \le \frac{1}{2} \end{cases}$, $g(\rho) = \frac{1}{3}$ for all $\rho \le \frac{1}{2}$ and ϕ, φ are feasible solutions of MVC on G_d and $G_{(\alpha,\beta)}$. n_1, n_2 and n_3 are correspondent to the number of vertices in cliques K_2, cycle, **d**-regular cycle and κ-branch-**d**-cycle.

Since $\varepsilon = 2 - (2 + o_d(1))\log\log d/\log d$, the inapproximability factor can be derived from inapproximability optimal substructure framework as

$$\delta > 1 + \frac{\varepsilon - 1}{\mathfrak{C}} \ge 1 + \frac{2\left(1-(2+o_c(1))\frac{\log\log c}{\log c}\right)}{2+(\zeta(\beta)c^{\beta}-1)(c+1)},$$

where c is the smallest d satisfying the condition in [6]. The last inequality holds since function $f(x) = (1-(2+o_x(1))\log\log x/\log x)/g(x)(x+1)$ is monotonously decreasing when $f(x) > 0$ for all $x > 0$ when $g(x)$ is monotonously increasing.

Algorithm 3: Graphic Sequence Construction Algorithm

```
Input  : d-degree sequence D = ⟨d_1, d_2, ..., d_n⟩ where d_1 ≥ d_2 ≥ ... ≥ d_n
Output: Graph H
while D ≠ ∅ do
    Connect vertex of d_1 to vertices of d_2, d_3, ..., d_{d_1+1};
    d_1 ← 0;
    for i = 2 to d_1 + 1 do
        d_i ← d_i − 1;
    end
    Sort D in non-increasing order;
    Remove all zero elements in D;
end
```

10.5 More Inapproximability Results on Simple Power-Law Graphs

10.5.1 General Graphic Embedding Technique

In this section, we introduce a general graphic embedding technique to embed a d bounded graph into a simple power-law graph. Before presenting the embedding technique, we first show that a graph can be constructed in polynomial time from a class of integer sequences.

Lemma 10.3. *Given a sequence of integers $D = \langle d_1, d_2, \ldots, d_n \rangle$ which is non-increasing, continuous and the number of elements is at least as twice as the largest element in D, i.e., $n \geq 2d_1$, it is possible to construct a simple graph G whose d-degree sequence is D in polynomial time $O(n^2 \log n)$.*

Proof. Starting with a set of individual vertices S of degree 0 and $|S| = n$, we iteratively connect vertices together to increase their degrees up to given degree sequence. In each step, the leftover vertex of highest degree is connected to other vertices one by one in the decreasing order of their degrees. Then the sequence D will be resorted and all zero elements will be removed. The algorithm stops until D is empty. The whole algorithm is shown as follows (Algorithm 3).

After each while loop, the new degree sequence, called D', is still continuous and its number of elements is at least as twice as its maximum element. To show this, we consider three cases: (1) If the maximum degree in D' remains the same, there are at least $d_1 + 2$ vertices in D. Since D is continuous, the number of elements in D is at least $d_1 + 2 + d_1 - 1$, that is, $2d_1 + 1$. Therefore, the number of elements in D' is $2d_1$, i.e., $n \geq 2d_1$ still holds. (2) If the maximum degree in D' is decreased by 1, there are at least two elements of degree d_1 in D. Thus, at most one element in D will become 0. Then we have $n \geq 2d_1 - 2 = 2(d_1 - 1)$. (3) If the maximum degree in D' is decreased by 2, there are at most two elements in D becoming 0. Thus, $n \geq 2d_1 - 3 > 2(d_1 - 2)$.

Algorithm 4: Graphic Embedding Algorithm

1 $\alpha \leftarrow \max\{\frac{\beta}{\beta-1}(\ln 4+\beta\ln d), \ln 2+\ln n+\beta\ln d\}$ and corresponding $G_{(\alpha,\beta)}$;
2 D be the d-degree sequence of $G_{(\alpha,\beta)} \setminus G_d$;
3 Construct $G_{(\alpha,\beta)} \setminus G_d$ using Algorithm 3.

The time complexity of the algorithm is $O(n^2 \log n)$ since there are at most n iterations and each iteration takes at most $O(n\log n)$ to sort the new sequence D.

Theorem 10.10 (Graphic Embedding Technique). *Any d-bounded graph G_d can be embedded into a simple power-law graph $G_{(\alpha,\beta)}$ with $\beta > 1$ in polynomial time such that G_d is a maximal component and the number of vertices in $G_{(\alpha,\beta)}$ can be polynomially bounded by the number of vertices in G_d.*

Proof. Given a d-bounded degree graph $G_d = (V,E)$ and $\beta > 1$, we construct a power-law graph $G_{(\alpha,\beta)}$ of exponential factor β which includes G_d as a set of maximal components. The construction is shown as Algorithm 4.

According to Lemma 10.3, the above construction is valid and finishes in polynomial time. Then we show that N is upper bounded by $\zeta(\beta)2d^{\beta}n$, where n and N are the number of vertices in G_d and $G_{(\alpha,\beta)}$ respectively. From the construction, we know either

$$\alpha \geq \frac{\beta}{\beta-1}(\ln 4+\beta\ln d) \Rightarrow \alpha \geq \ln 4+\beta\ln d+\alpha/\beta \Rightarrow \frac{e^{\alpha}}{d^{\beta}} \geq 4e^{\frac{\alpha}{\beta}}$$

or

$$\alpha \geq \ln 2+\ln n+\beta\ln d \Rightarrow \frac{e^{\alpha}}{d^{\beta}} \geq 2n.$$

Therefore, $\frac{e^{\alpha}}{d^{\beta}} \geq 2e^{\frac{\alpha}{\beta}} + n$. Note that $\left\lfloor \frac{e^{\alpha}}{d^{\beta}} \right\rfloor$ is the number of vertices of degree d. In addition, G has at most n vertices of degree d, so D is continuous degree sequence and has the number of vertices at least as twice as the maximum degree.

In addition, when n is large enough, we have $\alpha = \ln 2+\ln n+\beta\ln d$. Hence, the number of vertices N in $G_{(\alpha,\beta)}$ is bound as $N \leq \zeta(\beta)e^{\alpha} = 2\zeta(\beta)d^{\beta}n$, i.e., the number of vertices of $G_{(\alpha,\beta)}$ is polynomial bounded by the number of vertices in G_d.

10.5.2 Inapproximability of MIS, MVC and MDS

Theorem 10.11. *For any $\varepsilon > 0$, it is* NP*-hard to approximate maximum independent set within $1+\frac{1}{1120\zeta(\beta)3^{\beta}} - \varepsilon$ on simple power-law graphs.*

Proof. In this proof, we construct the simple power-law graph $G_{(\alpha,\beta)}$ based on graphic embedding technique in Theorem 10.10 from d-bounded graph G_d. Let ϕ

and φ be feasible solutions of MIS on G_d and $G_{(\alpha,\beta)}$. Since $OPT(\phi) \geq n/(d+1)$ on d-bounded graphs and $N \leq 2\zeta(\beta)d^\beta n$, we further have $\mathfrak{C} = 2\zeta(\beta)d^\beta(d+1)$ from

$$OPT(\varphi) \leq N \leq 2\zeta(\beta)d^\beta n \leq 2\zeta(\beta)d^\beta(d+1)OPT(\phi).$$

Since $\varepsilon = \frac{140}{139} - \varepsilon'$ for any $\varepsilon' > 0$ on three-bounded graphs, the inapproximability factor can be derived from inapproximability optimal substructure framework as

$$\delta > \frac{\varepsilon\mathfrak{C}}{(\mathfrak{C}-1)\varepsilon+1} = 1 + \frac{1}{140\mathfrak{C}-1} - \varepsilon > 1 + \frac{1}{1120\zeta(\beta)3^\beta} - \varepsilon.$$

Theorem 10.12. *It is* NP*-hard to approximate minimum dominating set within* $1 + \frac{1}{3120\zeta(\beta)3^\beta}$ *on simple power-law graphs.*

Proof. From the proof of Theorem 10.11, we have $\mathfrak{C} = 2\zeta(\beta)d^\beta(d+1)$. Then since $\varepsilon = \frac{391}{390}$ on three-bounded graphs, we have

$$\delta > 1 + \frac{\varepsilon - 1}{\mathfrak{C}} \geq 1 + \frac{1}{3120\zeta(\beta)3^\beta}.$$

Theorem 10.13. *There is no* $1 + \frac{2-(2+o_c(1))\frac{\log\log c}{\log c}}{2\zeta(\beta)c^\beta(c+1)}$ *approximation algorithm of Minimum Vertex Cover on simple power-law graphs under unique games conjecture.*

Proof. Similar as the proof of Theorem 10.12, we have $\mathfrak{C} = 2\zeta(\beta)d^\beta(d+1)$. Then since $\varepsilon = 2-(2+o_d(1))\log\log d/\log d$, the inapproximability factor can be derived from inapproximability optimal substructure framework as

$$\delta > 1 + \frac{\varepsilon - 1}{\mathfrak{C}} \geq 1 + \frac{2-(2+o_c(1))\frac{\log\log c}{\log c}}{2\zeta(\beta)c^\beta(c+1)},$$

where c is the smallest d satisfying the condition in [6].

Theorem 10.14. *There is no* $1 + \frac{2-(2+o_c(1))\frac{\log\log c}{\log c}}{2\zeta(\beta)c^\beta(c+1)}$ *approximation algorithm for minimum positive dominating set on simple power-law graphs.*

Proof. Similar Theorem 10.14, the proof follows from Theorem 10.8.

10.5.3 Maximum Clique, Minimum Coloring

Lemma 10.4 (Ferrante et al. [16]). *Let* $G = (V,E)$ *be a simple graph with* n *vertices,* $\beta \geq 1$ *and* $\alpha \geq \max\{4\beta, \beta\log n + \log(n+1)\}$. *Then, we can construct a graph* G_2 *such that* $G_2 = G_1 \setminus G$ *is a bipartite graph and* G_1 *is a simple* (α,β) *power-law graphs.*

Lemma 10.5. *Given a function $f(x)$ ($x \in \mathbb{Z}$, $f(x) \in \mathbb{Z}^+$) monotonously decreases, then $\sum_x f(x) \leq \int_x f(x)$.*

Corollary 10.2. $e^\alpha \sum_{i=1}^{e^{\alpha/\beta}} \left(\frac{1}{d}\right)^\beta < (e^\alpha - e^{\alpha/\beta})/(\beta - 1)$.

Theorem 10.15. *Maximum clique cannot be approximated within $O\left(n^{1/(\beta+1)-\varepsilon}\right)$ on simple large power-law graphs with $\beta > 1$ and $n > 54$ for any $\varepsilon > 0$ unless* NP=ZPP.

Proof. In [16], the authors proved the hardness of maximum clique problem on power-law graphs. Here we use the same construction. According to Lemma 10.4, $G_2 = G \setminus G_1$ is a bipartite graph when $\alpha \geq \max\{4\beta, \beta \log n + \log(n+1)\}$ for any $\beta \geq 1$. Let ϕ be a solution on general graph G and φ be a solution on power-law graph G_2. We show the completeness and soundness.

- If $OPT(\phi) = m \Rightarrow OPT(\varphi) = m$
  If $OPT(\phi) \leq 2$ on graph G, we can solve clique problem in polynomial time by iterating the edges and their endpoints one by one. However G is not a general graph in this case. w.l.o.g., assuming $OPT(\phi) > 2$, then $OPT(\varphi) = OPT(\phi) > 2$ since the maximum clique on bipartite graph is 2.
- If $OPT(\phi) \leq m/n^{1-\varepsilon} \Rightarrow OPT(\varphi) < O\left(1/(N^{1/(\beta+1)-\varepsilon'})\right) m$
  In this case, we consider the case that $4\beta < \beta \log n + \log(n+1)$, that is, $n > 54$. According to Lemma 10.4, let $\alpha = \beta \log n + \log(n+1)$. From Corollary 10.2, we have

$$N = e^\alpha \sum_{i=1}^{\Delta} \left(\frac{1}{i}\right)^\beta < \frac{e^\alpha - e^{\alpha/\beta}}{\beta - 1} = \frac{n^\beta(n+1) - n(n+1)^{1/\beta}}{\beta - 1} < \frac{2n^{\beta+1} - n}{\beta - 1}.$$

  Therefore, $OPT(\varphi) = OPT(\phi) \leq m/n^{1-\varepsilon} < O\left(m/\left(N^{1/(\beta+1)-\varepsilon'}\right)\right)$.

Corollary 10.3. *Minimum coloring problem cannot be approximated within $O\left(n^{1/(\beta+1)-\varepsilon}\right)$ on simple large power-law graphs with $\beta > 1$ and $n > 54$ for any $\varepsilon > 0$ unless* NP=ZPP.

10.6 Relationship Between β and Approximation Hardness

As shown in previous sections, many hardness and inapproximability results are dependent on β. In this section, we analyze the hardness of some optimal substructure problems based on β by showing that trivial greedy algorithms can achieve constant guarantee factors for MIS and MDS.

Lemma 10.6. *When $\beta > 2$, the size of MDS of a power-law graph is greater than Cn where n is the number of vertices, C is some constant only dependent on β.*

Proof. Let $S = (v_1, v_2, \ldots, v_t)$ of degrees $d_1, d_2, \ldots, d_t$ be the MDS of power-law graph $G_{(\alpha,\beta)}$. Observing that the total degrees of vertices in dominating set must be at least the number of vertices outside the dominating set, we have $\sum_{i=1}^{i=t} d_i \geq |V \setminus S|$. With a given total degree, a set of vertices has minimum size when it includes the vertices of highest degrees. Since the function $\zeta(\beta-1) = \sum_{i=1}^{\infty} \frac{1}{i^{\beta-1}}$ converges when $\beta > 2$, there exists a constant $t_0 = t_0(\beta)$ such that

$$\sum_{i=t_0}^{\Delta} i \left\lfloor \frac{e^\alpha}{i^\beta} \right\rfloor \geq \sum_{i=1}^{t_0} \left\lfloor \frac{e^\alpha}{i^\beta} \right\rfloor,$$

where α is any large enough constant. Thus, the size of MDS is at least

$$\sum_{i=t_0}^{\Delta} \left\lfloor \frac{e^\alpha}{i^\beta} \right\rfloor \approx \left(\zeta(\beta) - \sum_{i=1}^{t_0-1} \frac{1}{i^\beta} \right) e^\alpha \approx C|V|,$$

where $C = (\zeta(\beta) - \sum_{i=1}^{t_0} \frac{1}{i^\beta})/(\zeta(\beta))$.

Consider the greedy algorithm which selects from the vertices of the highest degree to the lowest. In the worst case, it selects all vertices with degree greater than 1 and a half of vertices with degree 1 to form a dominating set. The approximation factor of this simple algorithm is a constant.

Corollary 10.4. *Given a power-law graph with $\beta > 2$, the greedy algorithm that selects vertices in decreasing order of degrees provides a dominating set of size at most $\sum_{i=2}^{\Delta} \lfloor e^\alpha / i^\beta \rfloor + \frac{1}{2} e^\alpha \approx (\zeta(\beta) - 1/2) e^\alpha$. Thus the approximation ratio is $(\zeta(\beta) - \frac{1}{2})/(\zeta(\beta) - \sum_{i=1}^{t_0} 1/i^\beta)$.*

Let us consider another maximization problem MIS, we propose a greedy algorithm Power-law-Greedy-MIS as follows. We sort the vertices in non-increasing order of degrees and start checking from the vertex of the lowest degree. If the vertex is not adjacent to any selected vertex, it is selected. The set of selected vertices forms an independent set with the size at least a half the number of vertices of degree 1 which is $e^\alpha/2$. The size of MIS is at most a half of number of vertices. Thus, the following lemma holds.

Lemma 10.7. *Power-law-Greedy-MIS has factor $1/(2\zeta(\beta))$ on power-law graphs with $\beta > 1$.*

10.7 Minor *NP*-Hardness on Simple Power-Law Graphs for $\beta < 1$

In the section, we show some minor *NP*-hardness of optimal substructure problems on simple power-law graphs for small $\beta < 1$.

Definition 10.17 (Eligible Sequences). A sequence of integers $S = \langle s_1, \ldots, s_n \rangle$ is eligible if $s_1 \geq s_2 \geq \ldots \geq s_n$ and $f_S(k) \geq 0$ for all $k \in [n]$, where

$$f_S(k) = k(k-1) + \sum_{i=k+1}^{n} \min\{k, s_i\} - \sum_{i=1}^{k} s_i.$$

Erdős and Gallai [13] showed that an integer sequence is graphic – d-degree sequence of an graph, if and only if it is eligible and the total of all elements is even. Then Havel and Hakimi [9] gave an algorithm to construct a simple graph from a degree sequence. We now prove the following eligible embedding technique based on this result.

Theorem 10.16 (Eligible Embedding Technique). *Given an undirected simple graph $G = (V, E)$ and $0 < \beta < 1$, there exists polynomial time algorithm to construct a power-law graph $G' = (V', E')$ of exponential factor β such that G is a set of maximal components of G'.*

Proof. To construct G', we choose $\alpha = \max\{\beta \ln(n-1) + \ln(n+2), 3\ln 2\}$. Then $\lfloor e^\alpha/((n-1)^\beta) \rfloor > n+2$, i.e., there are at least two vertices of degree d in $G' \setminus G$ if there are a least two vertices of degree d in G'. According to the definition, the total degrees of all vertices in G' and G are even. Therefore, the lemma will follow if we prove that the degree sequence D of $G' \setminus G$ is eligible.

In D, the maximum degree is $\lfloor e^{\alpha/\beta} \rfloor$. There is only one vertex of degree i if $1 \leq e^\alpha / i^\beta < 2$, i.e., $e^{\alpha/\beta} \geq i > (e^\alpha/2)^{1/\beta}$.

Let us consider $f_D(k)$ in two cases:

1. ***Case***: $k \leq \lfloor e^{\alpha/\beta}/2 \rfloor$

$$\begin{aligned} f_D(k) &= k(k-1) + \sum_{i=k+1}^{n} min\{k, d_i\} - \sum_{i=1}^{k} d_i \\ &> k(k-1) + \sum_{i=k}^{T-k} k + \sum_{i=B}^{k-1} i + \sum_{i=1}^{B-1} 2 - \sum_{i=1}^{k} (T-k+1) \\ &= k(T-k) + (k-B)(k-1+B)/2 + B(B-1) - k(2T-k+1)/2 \\ &= (B^2 - B)/2 - k, \end{aligned}$$

where $T = \lfloor e^{\alpha/\beta} \rfloor$ and $B = \lfloor (e^\alpha/2)^{1/\beta} \rfloor + 1$. Note that $\alpha/\beta > \ln 2\,(2/\beta + 1)$ since $\alpha > 3\ln 2$ and $0 < \beta < 1$. Hence $\left(\lfloor (e^\alpha/2)^{1/\beta} \rfloor + 1\right)\left(\lfloor (e^\alpha/2)^{1/\beta} \rfloor\right) > \lfloor e^{\alpha/\beta} \rfloor \geq 2k$, that is, $f_D(k) > 0$.

2. ***Case***: $k > \lfloor e^{\alpha/\beta}/2 \rfloor$

$$f_D(k+1) \geq f_D(k) + 2k - 2d_{k+1} \geq f_D(k) \geq \ldots \geq f_D(\lfloor e^{\alpha/\beta}/2 \rfloor) > 0.$$

Corollary 10.5. *An optimal substructure problem is also* NP*-hard on power-law graphs for all* $0 < \beta < 1$ *if it is* NP*-hard on simple general graphs.*

Proof. According to Theorem 10.16, we can embed an undirected graph $G = (V, E)$ into a power-law graph G' of β lying in $(0, 1)$ and of vertices polynomial time in the size of G. Since the optimization problem has optimal substructure property and G is a set of maximal connected components of G', its optimum solution for the graph G can be computed easily from an optimal solution for G'. This completes the proof of NP-hardness.

10.8 Conclusion

This chapter focuses on the analysis of approximation hardness and inapproximability for optimal substructure problems on power-law graphs. These problems are only illustrated not be able to approximated into some constant factors on both general and simple power-law graphs although they remain APX-hard. However, we also notice that the gap between inapproximability factor and the simple constant approximation ratio of these problems is still not small enough and the hardness on power-law graph is weaker than that on degree bounded graphs. Is there any efficient reduction which is not from bounded graph will improve the hardness results on power-law graphs? Can we obtain stronger hardness results based on some specific power-law models? For example, if the number of vertices only follow power-law distribution when degree is larger than some constant i_0, we can reduce from graph of degree bounded by i_0 and get better results.

On the contrary, we also show that maximum clique and minimum coloring are still very hard to be approximated since the optimal solutions to these problems are dependent on the structure of local graph components rather than global graph. In other words, the power-law distribution in degree sequence does not help much for such optimization problems without optimal substructure property.

Acknowledgement This work is partially supported by NSF Career Award ♯ 0953284, DTRA; Young Investigator Award, Basic Research Program ♯ HDTRA1-09-1-0061; and DTRA ♯ HDTRA1-08-10.

References

1. http://vlado.fmf.uni-lj.si/pub/networks/data/erdos/erdos971.net. Data of Erdös Collaboration Network in 1997
2. Aiello, W., Chung, F., Lu, L.: A random graph model for massive graphs. In: STOC '00, pp. 171–180. ACM, New York, NY, USA (2000)
3. Aiello, W., Chung, F., Lu, L.: A random graph model for power law graphs. Experimental Math **10**, 53–66 (2000)

4. Albert, R., Jeong, H., Barabasi, A.L.: The diameter of the world wide web. Nature **401**, 130–131 (1999)
5. Alimonti, P., Kann, V.: Hardness of approximating problems on cubic graphs. In: CIAC '97, pp. 288–298. Springer-Verlag, London, UK (1997)
6. Austrin, P., Khot, S., Safra, M.: Inapproximability of vertex cover and independent set in bounded degree graphs. In: CCC '09, pp. 74–80 (2009)
7. Barabási, A.L., Albert, R.: Emergence of scaling in random networks. Science **286**, 509–512 (1999)
8. Bianconi, G., Barabási, A.L.: Bose-einstein condensation in complex networks (2000)
9. Bondy, J., Murty, U.: Graph theory with applications. MacMillan London (1976)
10. Bornholdt, S., Schuster, H.G. (eds.): Handbook of Graphs and Networks: From the Genome to the Internet. John Wiley & Sons, Inc., New York, NY, USA (2003)
11. Chlebík, M., Chlebíková, J.: Approximation hardness of dominating set problems in bounded degree graphs. Inf. Comput. **206**(11), 1264–1275 (2008)
12. Dinur, I., Safra, S.: On the hardness of approximating minimum vertex cover. Annals of Mathematics **162**, 2005 (2004)
13. Erdos, P., Gallai, T.: Graphs with prescribed degrees of vertices. Mat. Lapok **11**, 264–274 (1960)
14. Eubank, S., Kumar, V.S.A., Marathe, M.V., Srinivasan, A., Wang, N.: Structural and algorithmic aspects of massive social networks. In: SODA '04, pp. 718–727. Society for Industrial and Applied Mathematics, Philadelphia, PA, USA (2004)
15. Faloutsos, M., Faloutsos, P., Faloutsos, C.: On power-law relationships of the internet topology. In: Proceedings of the conference on Applications, technologies, architectures, and protocols for computer communication, SIGCOMM '99, pp. 251–262. ACM, New York, NY, USA (1999)
16. Ferrante, A., Pandurangan, G., Park, K.: On the hardness of optimization in power-law graphs. Theoretical Computer Science **393**(1-3), 220–230 (2008)
17. Gkantsidis, C., Mihail, M., Saberi, A.: Conductance and congestion in power law graphs. SIGMETRICS Perform. Eval. Rev. **31**(1), 148–159 (2003)
18. Hastad, J.: Clique is hard to approximate within $n^{1-\varepsilon}$. In: FOCS '96, p. 627. IEEE Computer Society, Washington, DC, USA (1996)
19. Janson, S., Luczak, T., Norros, I.: Large cliques in a power-law random graph (2009)
20. Kann, V.: On the Approximability of NP-complete Optimization Problems. Ph.D. thesis, Royal Institute of Technology Stockholm (1992)
21. Kempe, D., Kleinberg, J., 07va Tardos: Influential nodes in a diffusion model for social networks. In: IN ICALP, pp. 1127–1138. Springer Verlag (2005)
22. Kempe, D., Kleinberg, J., Tardos, E.: Maximizing the spread of influence through a social network. In: In KDD, pp. 137–146. ACM Press (2003)
23. Norros, I., Reittu, H.: On a conditionally poissonian graph process. Advances in Applied Probability pp. 38–59 (2006)
24. Pandurangan, G.: http://www.cs.purdue.edu/homes/gopal/powerlawtalk.pdf (2006)
25. Redner, S.: How popular is your paper? An empirical study of the citation distribution. The European Physical Journal B - Condensed Matter and Complex Systems **4**(2), 131–134 (1998)

Chapter 11
Path Problems in Complex Networks

Pavel Ghosh and Arun Sen

Abstract In this chapter network path problems arising in several different domains have been discussed. Based on the different characteristics of the paths typical to the nature of the application domain, a general classification of the problems has been made. The goal of path computation may be finding a single path or multiple paths between a source-destination node pair in a network. In case of multiple path computation, one may seek to find totally disjoint or partially disjoint paths. In this chapter, two problems corresponding to a single path scenario and two corresponding to a multiple path scenario have been discussed in four different subsections. In these subsections, the problems have been formally defined first, followed by discussion on the proposed algorithms for solution of the problems, complexity analysis and experimental results.

11.1 Introduction

One of the most complex networks ever created by the human beings is the Internet.Since its inception in the 1960, it has grown be a vital part of everyday life of nearly a third of the human race. It is estimated that 1.9 billion people are using the Internet today with 5 million terabytes of data available on it. There are 193 million registered domain names and according to the Internet Systems Consortium (www.isc.org) as of July 2008 there were nearly 600 million end systems attached to the Internet. This number does not include cell phones, laptops and other devices that are only intermittently connected to the Internet. Given the proliferation of the smart phones in the last few years, the number of end systems today could very

P. Ghosh • A. Sen (✉)
School of Computing, Informatics and Decision Systems Engineering,
Arizona State University, Tempe, AZ, USA
e-mail: pavel.ghosh@asu.edu; asen@asu.edu

M.T. Thai and P.M. Pardalos (eds.), *Handbook of Optimization in Complex Networks: Theory and Applications*, Springer Optimization and Its Applications 57,
DOI 10.1007/978-1-4614-0754-6_11,

well be above a billion. In the studies undertaken in the 1990s, it has been shown that the Internet is a scale-free network that with a power-law degree distribution. The focus of this chapter is to study path problems in this complex network known as the Internet. Path problems in networks have been extensively studied by many researchers in Mathematics, Computer Science, and Operations Research because of their application in problems in many different domains. In most of these problems, one weight is associated with a link representing, among other things, the *cost, delay* or the *reliability* of that link. The objective most often is to find a least weighted path between a specified source–destination pair. The classical shortest path algorithms due to Dijkstra [7] and Bellman–Ford [1, 8] can be utilized for this purpose. However, there can be many different variations of the path problems. Some of the variants are listed below.

- One might be interested in finding a *single path* between a source–destination node pair, or *multiple paths* between them.
- In case of multiple paths, one might be interested in finding *completely disjoint paths* or *partially (maximally) disjoint paths*.
- Both in case of single and multiple paths, there may be only a *single weight* associated with each link or *multiple weights* associated with each link.
- In case of multiple weights, there can be *multiple path lengths* associated with each path (computed using each weight independently) or a *single path length* associated with each path (computed by some combination of multiple link weights into a single super-weight).
- The weights associated with each link may be either *time varying* or *time invariant*.
- The contribution of a link weight on the path length computation may be an *identity function* ($f(x) = x$) or a *nonidentity function* ($f(x) \neq x$).
- The contribution of a link weight on the path length computation may be *independent* of the links traversed before traversing this link or *dependent* on it. Accordingly, path length computation process can be viewed as *without-memory (Markovian)* or *with-memory (non-Markovian)*.
- Weights on various links of a path can be combined using an *additive* function or a *nonadditive* function.

From the above discussion, it is clear that based on the desired objective of different applications, path problems can be classified into several different fashions. Various ways of classifying the path problems has been shown in Fig. 11.1. In order to enhance the clarity of the presentation, several similar subtrees in different branches has been omitted.

It is clear from the above classification that there can be many different variations of the path problems. In this chapter, we focus only four of them – two single path problems and two multiple path problems.

The first single path problem studied here relates to scenarios where the contribution of a link to the path length computation depends not only on the weight of that link but also the weights of the links already traversed. This is significantly different from most of the other path problems where the contribution of a link to the

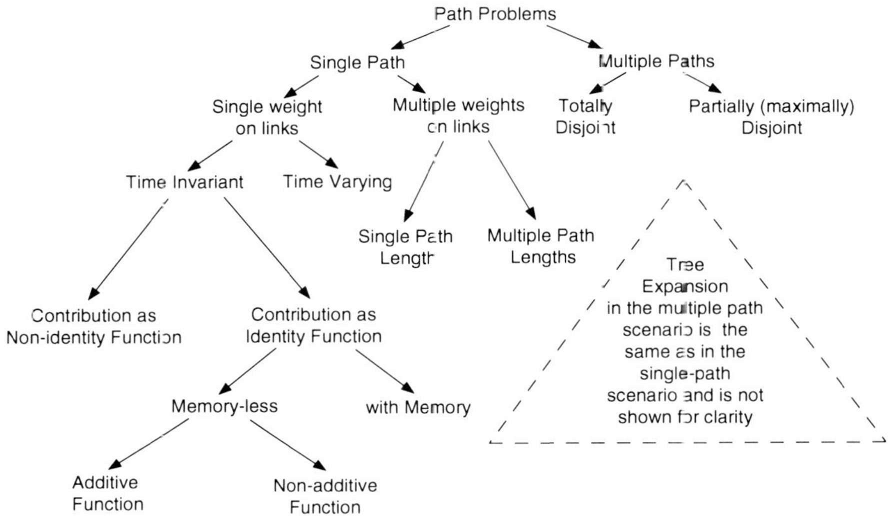

Fig. 11.1 Variation of the path problems

path length computation depends only on the weight of that link and is independent of the weights of the links already traversed. This condition is similar to a Markov chain where the next state is dependent only on the current state and is independent of the past states. This class of path problems may be viewed as "non-Markovian" as the contribution of a link towards the path length depends on the current link as well as the links traversed in the past on the path from the source to the destination. We present results of such path problems in Sect. 11.2.1.

The second single path problem studied here relates to the growth of bulk data and fast distribution of voluminous data. The recent emergence of bulk data is visible both in the entertainment, business and scientific world. In entertainment, very large content such as digital video programming or VOD movies need to be distributed from central server to edge servers for subsequent on-demand streaming to recipients. In network assisted storage (NAS) appliances, large pieces of data files need to be updated, replicated and reconciled at remote storage locations inter-connected over IP network. Large software companies and enterprises require updating their mirror sites with their software products. In e-science collaboration, large experimental data are moved among geographically distributed location using underlying IP network. To that end, in an interview in ACM queue [21], Turing award winner Jim Grey indicated for his work with astronomers, he had to exchange 2 terabytes of data regularly with his collaborators and the data size is expected to grow to 40 terabytes soon. In contrast to streaming or related real-time applications, bulk data applications are elastic and do not care about packet delay or rate/flow level guarantees. Instead, the only performance metric of importance to a bulk data is the net transfer time. In Sect. 11.2.2, we focus on exactly the same metric and try to explore a relevant routing problem in bulk data transfer. Specifically, we consider a realistic case of a network where the link bandwidth varies over time. Such a scenario is of practical importance where we focus on large-scale data files requiring hours of transfer time. For example, assuming 10 megabytes/s of sustained data transfer over a period of time, it may take more than 2 days to transfer a file of size 2 terabytes. One cannot assume that link bandwidth will remain stationary for that entire duration.

The first multiple path problem studied here relates to computation of *widest pair of disjoint paths*. As multiple path routing between a source–destination node pair offers significant advantages over traditional single path routing, the topic has received considerable attention in the networking research community in the last few years. The advantages of multipath routing include better *network utilization, reduced congestion, increased probability for survival in case of failure* and *better traffic engineering*. These advantages can be exploited to its fullest extent if the paths are (node or link) disjoint. For this reason, routing using disjoint or nearly disjoint paths have received special attention from the researchers [32–34, 41–48]. Suurballe [46,47] was one of the earliest researchers to explore disjoint path routing. He presented polynomial time algorithms for computation of a pair of disjoint paths such that the sum of the path lengths is minimum. In [42, 44] it has been shown that the problem of finding a disjoint pair of paths such that the length of the longer path of the pair is shortest among all such pairs is NP-complete. In all these papers, the metric used for computation of the length of a path is additive [19]. An example of such a metric is the *delay* associated with each link of the network. Although *bandwidth* as a metric is just as important as the delay for quality-of-service routing, results comparable to [46, 47] is not known for nonadditive metrics like bandwidth.

The bandwidth of a path is equal to the bandwidth of the link with the smallest capacity used by the path. A path from a source node s to a destination node t is known as the *widest path* if the bandwidth of this path is largest among all paths between s to t [19]. Algorithms for computation of a widest path between a source–destination node pair is well studied [19]. In Sect. 11.3.1, we study problems related to disjoint path routing with a nonadditive metric like bandwidth and present results related to the computation of *widest pair of disjoint paths*. We consider two versions of the problem. In the first version, we address the problem of finding a pair of disjoint paths between a source–destination node pair, such that the combined bandwidth of this path-pair is maximum over all such path-pairs. In the second version, we want to find a pair of disjoint paths, such that the bandwidth of the first path is at least X_1 and the bandwidth of the second path is at least X_2, for some prespecified values X_1 and X_2. We prove that both versions of the problem are NP-complete. We provide exact and approximate solutions for both versions of the problem.

The second multiple path problem studied here relates to routing in the Internet using *transit hubs*. The current Internet routers select only a single path for a given destination. The choice of this default IP path is not left to the end hosts instead on the administrative system (AS) domain operator or on the BGP level policies. Relying just on the single default path may lead to various end-to-end performance bottlenecks. Empirical studies reported by authors in [62] shows that about 50% of the bandwidth bottlenecks are not on access links to users but within the ISP carrier networks. These studies suggest that just by upgrading the last mile access link, one cannot assure a better level of end-to-end performance. It, thus, makes a clear case for the need of alternate path to avoid bottleneck at core network. Such a case was validated by experimental studies conducted by authors of DETOUR [63] showing that in many cases alternate overlay paths have better latency and throughput characteristics than direct default IP paths. For example, in specific experiments the authors in [63] have shown that in 30% of direct IP paths have better alternate paths with respect to the round trip time metric and 70% of direct IP paths have better alternate paths with respect to the loss metric.

However, alternate paths cannot be directly created using existing routers as they do not provide any rerouting flexibilities. Circumventing this problem, it was proposed in [20, 63, 64] to deploy intermediate nodes that can provide a level of indirection in creating alternate path. These nodes often referred as overlay node, rendezvous point [64], transit hubs have the capability of forwarding packets and can realize an end-to-end alternate path by chaining a set of default IP paths. Forwarding and routing of packets along the alternate path can be done at application layer by forming an overlay network as [20]. Another approach is to use IP-in-IP encapsulation [65]. In this case, an IP header indicating the final destination is encapsulated in the payload of another IP header. The latter header contains in its destination address field the IP address of an intermediate router.

Formally, the *K-transit hub routing problem* can be stated as follows. Suppose the path P_{v_i,v_j} between every pair of nodes v_i and v_j is known. Given the source and the destination nodes s and t respectively, is it possible to establish an alternate path from s to d, disjoint from the primary path $P_{s,d}$, by concatenating at most K other paths P_{v_i,v_j}? In other words, is it possible to find a set of paths $\{P_{s,v_1}, P_{v_1,v_2}, P_{v_2,v_3}, P_{v_3,v_4}, \ldots, P_{v_{k-1}\,v_k}, P_{v_k,d}\}$ such that the concatenation of these paths will produce a path from s to d subject to the constraints that (1) no two paths in this set share an edge and (2) no paths in this set share an edge with the primary path $P_{s,d}$? If such paths exist, it is possible to establish two disjoint paths between the nodes s and d and utilize the bandwidths of both the paths for data transfer.

11.2 Single Path Problems

11.2.1 Shorest Path Computation with Non-Markovain Link Contribution to Path Length

In this subsection, we consider two specific problems that can be classified as non-Markovian. One of them is encountered in the multimedia data transmission domain and the other one is encountered in the VLSI circuit design domain. We consider these problems under different conditions and develop algorithms for the problems under these conditions. The shortest path problem in multimedia data transmission environment can be solved in $O(n^2)$ or $O(n^3)$ computational time. We also provide mathematical programming solutions for the multimedia data transmission problem as well as the VLSI circuit design problem.

11.2.1.1 Problem Formulation

In the classical path problem, each edge $e_i \in E$ of the graph $G = (V,E)$ has a weight w_i associated with it and if there is a path P from the node v_0 to v_k in the graph $G = (V,E)$

$$v_0 \stackrel{w_1}{\rightarrow} v_1 \stackrel{w_2}{\rightarrow} v_2 \stackrel{w_3}{\rightarrow} \ldots \stackrel{w_k}{\rightarrow} v_k,$$

then the *path length* or the *distance* between the nodes v_0 and v_k is given by

$$PL(v_0, v_k) = w_1 + w_2 + \cdots + w_k.$$

This model is valid as long as the weights on the links represents the *cost* or the *delay* associated with the link. However, if the weight represents the *reliability* or the *bandwidth* associated with the link, then *addition* of the link weights on the path is not meaningful. In case the weights represent the reliability, the calculation once again becomes meaningful if the addition operator is replaced by a *multiplication* operator. In case the weight represents the bandwidth, the calculation becomes meaningful if a *minimum* operator replaces the addition operator. Thus a generalization of the path length will be

$$PL(v_0, v_k) = w_1 \oplus w_2 \oplus w_3 \oplus \cdots \oplus w_k,$$

where $\oplus$ is a suitable operator for the particular application. In [19], the authors consider three different types of operators and call them *additive, multiplicative* and *concave metrics*, respectively.

In the next level of generalization, the path length computation is based on not the link weight itself but a function of the link weight. In this case the path length is given by

$$PL(v_0, v_k) = f(w_1) \oplus f(w_2) \oplus f(w_3) \oplus \cdots \oplus f(w_k),$$

where $f(w_i)$ can be any function of the link weight w_i, appropriate for the particular application.

In the next higher level of generalization each link has multiple weights associated with it. This model realistically captures the data transmission environment where the *Quality of Service (QoS)* issues are of paramount importance. The various weights associated with a link may represent among other things, the *delay*, the *cost*, the *jitter*, the *cell loss rate*, etc. In this case the path length computation is carried out in one of the following two ways:

Case I: In this case, each path has multiple *path lengths* associated with it. If $(w_{i,1}, w_{i,2}, \ldots, w_{i,m})$ are m different link weights associated with the link e_i, then there are m different path lengths associated with a *path* between a given source node v_0 and a given destination node v_k and they are given as the following vector:

$$\begin{bmatrix} PL_1(v_0, v_k) \\ PL_2(v_0, v_k) \\ \vdots \\ PL_m(v_0, v_k) \end{bmatrix} = \begin{bmatrix} f(w_{1,1}) \oplus f(w_{2,1}) \oplus \cdots \oplus f(w_{k,1}) \\ f(w_{1,2}) \oplus f(w_{2,2}) \oplus \cdots \oplus f(w_{k,2}) \\ \vdots \\ f(w_{1,m}) \oplus f(w_{2,m}) \oplus \cdots \oplus f(w_{k,m}) \end{bmatrix}.$$

This class of problems is known as the *multicriteria optimization problems* and is studied in [9, 10, 12, 17–19].

Case II: In this case, each path has a single path length associated it and it is given by:

$$PL(v_0, v_k) = f(w_{1,1}, \ldots, w_{1,m}) \oplus f(w_{2,1}, \ldots, w_{2,m}) \oplus \cdots \oplus f(w_{k,1}, \ldots, w_{k,m}),$$

where $(w_{i,1}, w_{i,2}, \ldots, w_{i,m})$ are m different link weights associated with the link e_i, appropriate for the particular application. It may be noted that this formulation gives rise to a *single criterion optimization problem* as opposed to the *multiple criteria optimization problem* formulated in Case I.

Both Case I and Case II of the previous level can be further generalized at the next higher level. As in the previous case, each link has multiple weights associated with them. In this level of generalization, the contribution of a link in the path length computation depends not only on the weights associated with that link but also on the weights of the links already traversed. In this case, the path length for Case I is

$$\begin{bmatrix} PL_1(v_0, v_k) \\ PL_2(v_0, v_k) \\ \vdots \\ PL_m(v_0, v_k) \end{bmatrix} = \begin{bmatrix} f(w_{1,1}) \oplus \cdots \oplus f(w_{1,1}, \cdots, w_{k,1}) \\ f(w_{1,2}) \oplus \cdots \oplus f(w_{1,2}, \cdots, w_{k,2}) \\ \vdots \\ f(w_{1,m}) \oplus \cdots \oplus f(w_{1,m}, \cdots, w_{k,m}) \end{bmatrix}.$$

At this level of generalization, the path length for Case II is

$$\begin{aligned} PL(v_0, v_k) = {} & f(w_{1,1}, \ldots, w_{1,m}) \oplus f(w_{1,1}, \ldots, w_{1,m}, w_{2,1}, \ldots, w_{2,m}) \oplus \cdots \\ & \oplus f(w_{1,1}, \ldots, w_{1,m}, \ldots, w_{k,1}, \ldots, w_{k,m}). \end{aligned}$$

The two path problems discussed here belong to this last category.

1. Path Problem in Multimedia Data Transmission

In the variant of the path problem for the multimedia data transmission, each edge e_i has two weights, δ_i and s_i associated with it, $\delta_i \geq 0$ and $s_i \geq 0$. These two weights are referred to as (1) the *per unit delay factor* and (2) the *size factor*, respectively. If P is a path from the node v_0 to v_k,

$$v_0 \stackrel{\delta_1, s_1}{\rightarrow} v_1 \stackrel{\delta_2, s_2}{\rightarrow} v_2 \stackrel{\delta_3, s_3}{\rightarrow} \dots \stackrel{\delta_k, s_k}{\rightarrow} v_k,$$

then the *path length* or the *total delay* between the nodes v_0 and v_k denoted $PL(v_0, v_k)$ is given by

$$\begin{aligned} PL(v_0, v_k) &= \delta_1 + s_1\delta_2 + s_1 s_2 \delta_3 + \dots + s_1 \dots s_{k-1}\delta_k \\ &= \sum_{i=1}^{k} \delta_i \prod_{j=1}^{i-1} s_j \text{ with} \prod_{j=1}^{0} s_j = 1. \end{aligned}$$

It is clear that the path length in this case fits into the most general case discussed in the previous paragraph with $m = 2, w_{i,1} = \delta_i, w_{i,2} = s_i$ and $f(w_{1,1}, w_{1,2}, w_{2,1}, w_{2,2}, \dots, w_{i,1}, w_{i,2}) = s_1 s_2 \dots s_{i-1}\delta_i$, for all i, $1 \leq i \leq k$, and $s_0 = 1$.

The physical significance of the parameters δ_i and s_i are as follows: the transmission delay is clearly proportional to the size of the multimedia data file being transmitted. Therefore, we consider the *per unit delay factor* δ_i and to compute the total delay, we multiply δ_i with the size of the file being transmitted. As a multimedia data file travels through different nodes in a network on its journey from the source to the destination, it passes through some *compression* or *decompression* algorithms. As a result the size of the multimedia data file may change. The size factor s_i captures this aspect of multimedia data transmission. The expression for the path length (or total delay) given above is obtained on the assumption that *one unit of data* is being transmitted from the source to the destination.

2. Path Problem in VLSI Circuit Design

In the VLSI circuit design domain, one often encounters a resistor–capacitor (RC) circuit of the form shown in Fig. 11.2. For the purpose of computing the signal propagation delay between the points S and T, the RC-circuit is modeled as a graph shown in Fig. 11.3. Each link e_i of this graph has two weights r_i and c_i associated with it and the delay between the points S and T is given by [11].

$$r_1(c_1 + c_2 + c_3 + c_4) + r_2(c_2 + c_3 + c_4) + r_3(c_3 + c_4) + r_4(c_4).$$

In general, if the RC-circuit contains k resistors and k capacitors, the delay or the *path length* between the source node v_0 and the destination node v_k is given by

$$PL(v_0, v_k) = r_1(c_1 + \dots + c_k) + r_2(c_2 + \dots + c_k) + \dots + r_k(c_k)$$

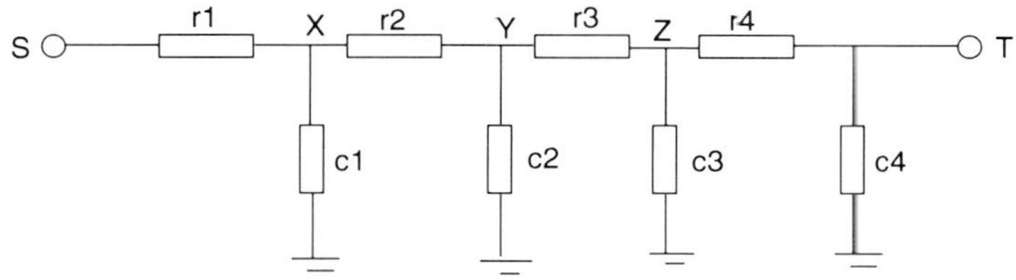

Fig. 11.2 RC circuit

Fig. 11.3 Delay model of RC circuit

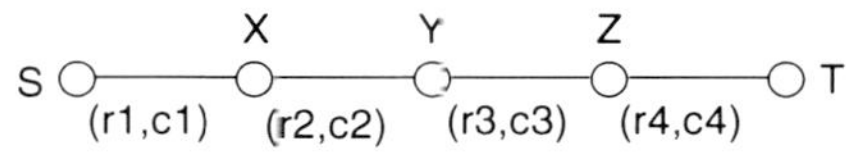

or it can be rewritten as

$$PL(v_0, v_k) = c_1(r_1) + c_2(r_1 + r_2) + \ldots + c_k(r_1 + \ldots + r_k)$$
$$= \sum_{i=1}^{k} c_i \left(\sum_{j=1}^{i} r_j \right).$$

Thus, it can be seen that the delay computation in this model also fits into the most general case of path length computation discussed earlier in this section with $m = 2, w_{i,1} = c_i, w_{i,2} = r_i$ and

$$f(w_{1,1}, w_{1,2}, w_{2,1}, w_{2,2}, \ldots, w_{i,1}, w_{i,2}) = c_i \left(\sum_{j=1}^{i} r_j \right), \text{ for all } i, 1 \le i \le k.$$

11.2.1.2 Path Problem in Multimedia Data Transmission

The length of a path $P(v_0, v_k) : v_0 \rightarrow v_1 \rightarrow v_2 \rightarrow \ldots \rightarrow v_k$, in the multimedia data transmission problem, is given by $PL(v_0, v_k) = \delta_1 + s_1\delta_2 + s_1 s_2 \delta_3 + \ldots + s_1 \ldots s_{k-1}\delta_k$. The traditional shortest path algorithms such as the *Dijkstra's algorithm* make the observation that *"subpaths of shortest paths are shortest paths"* [6] and exploits it to develop the shortest path algorithm. The main lemma on which the Dijkstra's algorithm depends is given below. The proof of the lemma can be found in [6].

Lemma 11.1. *Given a weighted directed graph $G = (V, E)$ with weight function w_i associated with each link e_i, $(e_i \in E, 1 \le i \le |E|)$, and the length of a path is computed as the sum of the weights of the links on that path. Let $P(v_0, v_k)$: $v_0 \rightarrow v_1 \rightarrow v_2 \rightarrow \ldots \rightarrow v_k$ be a shortest path from vertex v_0 to vertex v_k and for any i and j such that $0 \le i \le j \le k$, let $P(v_i, v_j) : v_i \rightarrow v_{i+1} \rightarrow \ldots \rightarrow v_j$ be a subpath of P from vertex v_i to vertex v_j. Then, $P(v_i, v_j)$ is the shortest path from v_i to v_j [6].*

Corollary 11.1. *If the shortest path from the source node s to the destination node d is given by $s = v_0 \rightarrow v_1 \rightarrow v_2 \rightarrow \ldots \rightarrow v_k = d$ then $s = v_0 \rightarrow v_1 \rightarrow v_2 \rightarrow \ldots \rightarrow v_i$ is the shortest path from s to v_i for all i, $1 \le i \le k-1$.*

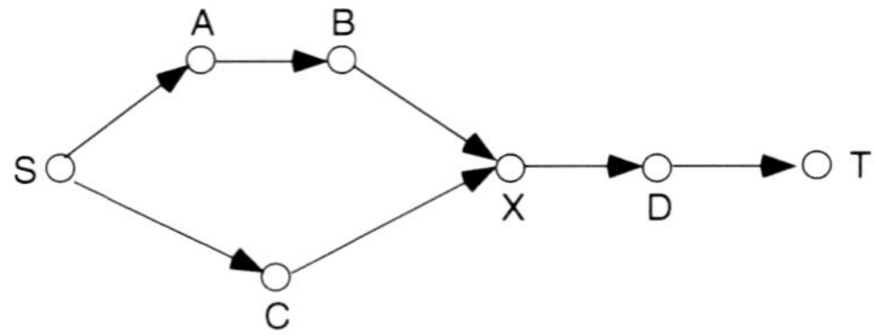

Fig. 11.4 Shortest path for MMD transmission, Example 1

Table 11.1 Delay factor δ and the size factor s of the graph in Fig. 11.4

Delay factor δ	Size factor s
$\delta_{S,A} = 1$	$s_{S,A} = 1$
$\delta_{S,C} = 1$	$s_{S,C} = 3$
$\delta_{A,B} = 1$	$s_{A,B} = 1$
$\delta_{B,X} = 1$	$s_{B,X} = 4$
$\delta_{C,X} = 1$	$s_{C,X} = 1$
$\delta_{X,D} = 1$	$s_{X,D} = 1$
$\delta_{D,T} = 1$	$s_{D,T} = 1$

In other words, to get to the destination from the source using the shortest path, the intermediate nodes must be visited using the shortest path from the source to the intermediate nodes. This is true because, the *path length* in this case is computed as the sum of weights on the links that make up the path. In case of multimedia data transmission problem, where *the path length is not computed as the sum of the links weights, this is no longer true*. This is demonstrated with an example shown in Fig. 11.4. The δ_i and s_i values associated with the links of this graph is given in Table 11.1.

With this data set the length of the path, $S \to C \to X \to D \to T$, is $\delta_{S,C} + s_{S,C}\delta_{C,X} + s_{S,C}s_{C,X}\delta_{X,D} + s_{S,C}s_{C,X}s_{X,D}\delta_{D,T} = 1 + 3.1 + 3.1.1 + 3.1.1.1 = 1 + 3 + 3 + 3 = 10$ whereas the length of the path $S \to A \to B \to X \to D \to T$ is $\delta_{S,A} + s_{S,A}\delta_{A,B} + s_{S,A}s_{A,B}\delta_{B,X} + s_{S,A}s_{A,B}s_{B,X}\delta_{X,D} + s_{S,A}s_{A,B}s_{B,X}s_{X,D}\delta_{D,T} = 1 + 1.1 + 1.1.1 + 1.1.4.1 + 1.1.4.1.1 = 1 + 1 + 1 + 4 + 4 = 11$. Thus, the path $S \to C \to X \to D \to T$ is shorter than the path $S \to A \to B \to X \to D \to T$ in the example. However, in this example the length of the path $S \to C \to X$ is $1 + 3.1 = 4$, which is *greater* than the length of the path $S \to A \to B \to X$, $1 + 1.1 + 1.1.1 = 3$.

As noted earlier, the length of a path $P(v_0, v_k) : v_0 \to v_1 \to \ldots \to v_k$, in the multimedia data transmission problem, is given by $PL(v_0, v_k) = \delta_1 + s_1\delta_2 + s_1s_2\delta_3 + \ldots + s_1 \ldots s_{k-1}\delta_k$. This path length function has an interesting property and this property is utilized to establish the following lemma.

Lemma 11.2. *Given a weighted directed graph $G = (V, E)$ with weight functions (δ_i, s_i), associated with each link e_i, $(e_i \in E, 1 \le i \le |E|)$, and the length of a path $P(v_0, v_k) : v_0 \to v_1 \to \ldots \to v_k$ computed as $PL(v_0, v_k) = \delta_1 + s_1\delta_2 + s_1s_2\delta_3 + \ldots + s_1 \ldots s_{k-1}\delta_k$. Let $P(v_0, v_k) : v_0 \to v_1 \to \ldots \to v_k$ be a shortest path from vertex v_0 to vertex v_k and for any i, $1 \le i \le k-1$, let $P(v_i, v_k) : v_i \to v_{i+1} \to \ldots \to v_k$ be a subpath of P from vertex v_i to vertex v_k. Then, $P(v_i, v_k)$ is the shortest path from v_i to v_k, $1 \le i \le k$.*

Proof. The length function $PL(v_0, v_k) = \delta_1 + s_1\delta_2 + s_1s_2\delta_3 + \ldots + s_1 \ldots s_{k-1}\delta_k$ can be rewritten as $PL(v_0, v_k) = \delta_1 + s_1(\delta_2 + s_2(\delta_3 + s_3(\delta_4 - \ldots + s_{k-2}(\delta_{k-1} + s_{k-1}\delta_k)\ldots)$. Consider the path $P(v_o, v_k) : v_0 \rightarrow v_1 \rightarrow \ldots \rightarrow v_{i-1} \rightarrow v_i \rightarrow v_{i+1} \rightarrow \ldots \rightarrow v_k$. The length of this path is $PL(v_0, v_k) = \delta_1 + s_1(\delta_2 \ldots + s_{i-1}(\delta_i + s_i(\delta_{i+1} + \ldots + s_{k-2}(\delta_{k-1} + s_{k-1}\delta_k)\ldots)$. Thus, $PL(v_0, v_k) = \delta_1 + s_1(\delta_2 \ldots + s_{i-1}(PL(v_i, v_k)\ldots)$. If $P(v_i, v_k)$ is not the shortest path between the nodes v_i and v_k with a path length $PL(v_i, v_k)$, then it can be replaced by a shorter path between the nodes v_i and v_k, reducing the total path length $PL(v_0, v_k)$ and contradicting the assumption that $P(v_0, v_k)$ is the shortest path between the nodes v_0 and v_k. □

Path Problem in Multimedia Data Transmission with No Reduction in Size

In this subsection, we consider the case where the size factor, s_i, associated with a link e_i is greater than or equal to unity for all links. This implies that the data size will never reduce from its original size while passing through a link. The more general case where the size factor, s_i, does not have any such restriction (i.e., s_i is allowed to be less than unity) will be considered in the next subsection.

Because of lemma 11.2 and the fact $s_i \geq 1, \delta_i \geq 0$, we can apply a modified version of Dijkstra's algorithm to solve the shortest path problem in the multimedia data transmission environment. The traditional version of the algorithm starts from the source node and computes the shortest path to other nodes until it finds the shortest path to the destination. In this modified version, we start from the destination node and compute the shortest path from other nodes to the destination nodes until it finds the shortest path from the source to the destination node. The algorithm is given below:

Shortest Path Algorithm for Multimedia Data Transmission Environment

Input: The directed graph $G = (V, E)$, $(V = \{1, 2, \ldots, n\})$, two $n \times n$ matrices δ and s, the (i, j)-th entry of the matrices stores the delay factor δ and the size factor s of the link from the node i to node j. If there is no link from the node i to j, both $\delta_{i,j}$ and $s_{i,j}$ is taken to be ∞. Without any loss of generality, we assume that the node 1 is the source node and node n is the destination node.

Output: Array $D(1, \ldots, n)$, that stores the shortest path length from node i to the destination node n for all i, $1 \leq i \leq n$.

Comments: The algorithm starts from the destination node and in each iteration finds the shortest path from a node i in the graph to the destination node n, $1 \leq i \leq n-1$.

Theorem 11.1. *If $\forall i, j$ the delay factor $\delta(i, j) \geq 0$ and the size factor $s(i, j) \geq 1$, then the above algorithm correctly computes the shortest path from any node $i, 1 \leq i \leq n-1$ to the destination node n.*

Algorithm 5 Shortest Path for Multimedia Data Transfer

```
C = {1,2,...,n−1};
for i = n−1 downto 1 do do
    D[i] = δ[i,n]
end for
repeat
    v = i ∈ C such that D[i] has the minimum value;
    C = C \ {v};
    for each w ∈ C do
        D[w] = min(D[w], δ[w,v] + s[w,v]D[v]);
    end for
until n − 2 times
```

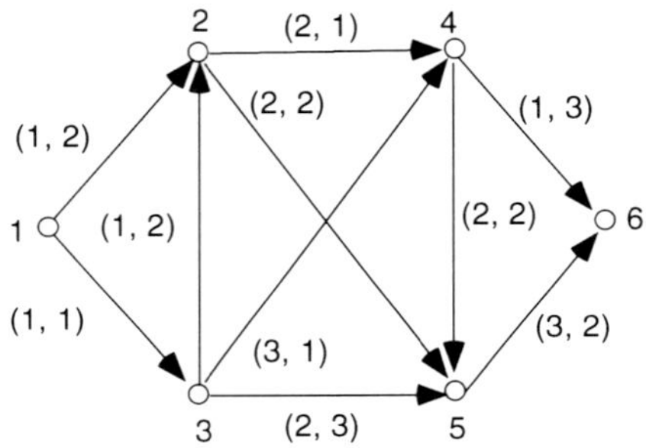

Fig. 11.5 Shortest path for MMD transmission, Example 2

Proof. In each iteration of the repeat loop, the node i that has the smallest value of $D[i]$ among the set of nodes in the set C, is removed from the set C. Suppose that during the p-th iteration of the loop some node v was removed from the set C and the corresponding $D[v]$ never changed till the termination of the algorithm. Suppose, if possible, that $D[v]$ is not the shortest path length from node v to the destination node n. Suppose that there exists a path from v to n through some node $w \in C$, and the length of this path is shorter than $D[v]$. The length of such a path from v to n is $\delta(v,w) + s[v,w]D[w]$. If $\delta(v,w) + s[v,w]D[w] < D[v]$, when $\delta(i,j) \geq 0$ and $s(i,j) \geq 1$ then $D[w] < D[v]$. However, if $D[w]$ is indeed smaller than $D[v]$, during the p-th iteration, the node w would have been removed from the set C, instead of the node v, contradicting the assumption that the $D[v]$ is not the shortest path length from the node v to the destination node n. □

Theorem 11.2. *The complexity of the Algorithm 5 is $O(n^2)$.*

Proof. The algorithm is a slight modification of the classical shortest path algorithm due to Dijkstra. Following similar arguments as in the complexity analysis of Dijkstra's algorithm [6], the complexity of this algorithm can be shown to be $O(n^2)$. □

An example of the result of execution of the algorithm on the graph in Fig. 11.5 is shown in Table 11.2. In this graph, the source node is 1 and the destination

Table 11.2 Shortest path computation for the graph in Fig. 11.5

Iteration number	Nodes of the graph					
	1	2	3	4	5	6
1	∞	∞	∞	1*	3	0
2	∞	3	4	1	3*	0
3	∞	3*	4	1	3	0
4	7	3	4*	1	3	0
5	5*	3	4	1	3	0

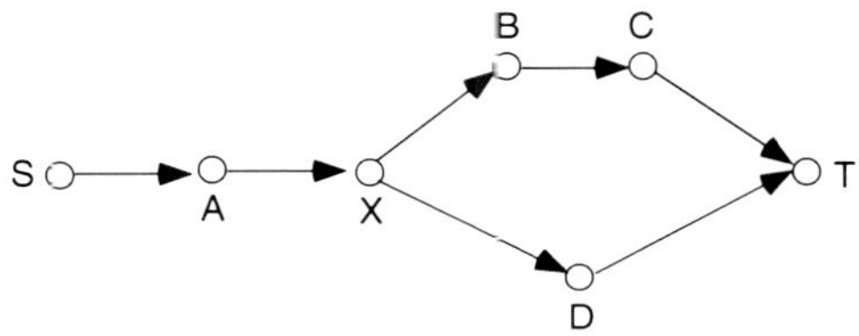

Fig. 11.6 Shortest path for MMD transmission, Example 3

Table 11.3 Delay factor δ and the size factor s of the graph in Fig. 11.6

Delay factor δ	Size factor s
$\delta_{S,A} = 1$	$s_{S,A} = 1$
$\delta_{A,X} = 2$	$s_{A,X} = 1$
$\delta_{X,B} =$ [illegible]	$s_{X,B} = 0.25$
$\delta_{X,D} = 2$	$s_{X,D} = 1$
$\delta_{B,C} = 2$	$s_{B,C} = 1$
$\delta_{C,T} = 2$	$s_{C,T} = 1$
$\delta_{D,T} =$ [illegible]	$s_{D,T} = 1$

node 6. The shortest path length from node 1 to node 6 is 5 and the path is $v_1 \rightarrow v_3 \rightarrow v_4 \rightarrow v_6$.

It is well known that if the path length is measured as the sum of the weights on the links, Dijkstra's algorithm fails to compute the shortest path betwen the source–destination nodes, in case some of the link weights are *negative*. For exactly the same reason, our modified version of the Dijkstra's algorithm fails to compute the shortest path in case $s_{i,j} < 1$. An example of the case where the above algorithm fails to compute the shortest path is shown in Fig. 11.6. The δ_i and s_i values associated with the links of this graph is given in Table 11.3. The result of the execution of the modified Dijkstra algorithm on this graph is shown in Table 11.4.

At termination of the algorithm, the shortest path length between the source node S and the destination node T is given as 6 and the path is $S \rightarrow A \rightarrow X \rightarrow D \rightarrow T$ ($\delta_{S,A} + s_{S,A}\delta_{A,X} + s_{S,A}s_{A,X}\delta_{X,D} + s_{S,A}s_{A,X}s_{X,D}\delta D, T = 1 + 1.2 + 1.1.2 + 1.1.1.1 = 6$). However, this result is *incorrect* because the length of the path $S \rightarrow A \rightarrow X \rightarrow B \rightarrow C \rightarrow T$ is $\delta_{S,A} + s_{S,A}\delta_{A,X} + s_{S,A}s_{A,X}\delta_{X,B} + s_{S,A}s_{A,X}s_{X,B}\delta B, C + s_{S,A}s_{A,X}s_{X,B}s_{B,C}\delta C, T = 1 + 1.2 + 1.1.1 - 1.1.(0.25).2 + 1.1.(0.25).1.2 = 5$. In this case, the algorithm computes the shortest path length incorrectly, because one of the size factors, $s_{X,B} < 1$ ($s_{X,B} = 0.25$).

Table 11.4 Shortest path computation for the graph in Fig. 11.6

Iteration number	Nodes of the graph						
	S	A	X	B	C	D	T
1	∞	∞	∞	∞	2	1*	0
2	∞	∞	3	∞	2*	1	0
3	∞	∞	3*	4	2	1	0
4	∞	5	3	4*	2	1	0
5	∞	5*	3		2	1	0
6	6*	5	3	4	2	1	o

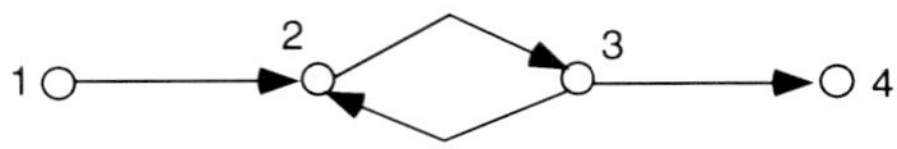

Fig. 11.7 Shortest path with negative weighted cycle

Path Problem in Multimedia Data Transmission with Reduction in Size

It was mentioned in the previous section that in this path problem if some size factor $s_i < 1$, it has the same effect as a negative weighted link in a traditional shortest path problem. The example given earlier, shows that our version of the Dijkstra's algorithm fails to correctly compute the shortest path from the source to the destination in this situation. In the traditional shortest path problem, where the path length is computed as the sum of the weights on the links of a path, there is a notion of a *negative weighted cycle*. A cycle is referred to as a *negative weighted cycle* if sum of the weights on the links making up the cycle is a negative number. The multimedia data transmission problem that is presently under consideration, both the weights (the delay factor δ_i and size factor s_i) associated with a link e_i, are nonnegative. However, in this problem path length is computed in a different way. In this problem also we have a notion of a negative weighted cycle. Suppose the links $e_1, e_2, \ldots, e_p$ makes a cycle in the graph. This cycle will be referred to a *negative weighted cycle*, if $\delta_i \prod_{j=1}^{p} s_j < 1$.

The implication of such a negative weighted cycle is that the data size decreases every time it goes around such a cycle. As the source to destination transmission delay is dependent on the size of the data, this delay is also reduced every time the data goes around such a cycle. An example of such a phenomenon is given next.

Consider the graph in Fig. 11.7 with the nodes 1 and 4 being the source and the destination respectively. The nodes 2 and 3 form a loop, as shown in the figure. The delay and the size factor on the links is given in the Table 11.5 below.

Consider the path $1 \rightarrow 2 \rightarrow 3 \rightarrow 4$. This is a *no loop* path between the nodes 1 and 4. The path length of this path is $d_1 + 1.\varepsilon + 1.(1/p).d_2 = d_1 + \varepsilon + d_2/p$. The length of the path from 1 to 4, if it passes through the loop $1, 2, \ldots, k$ times is given in Table 11.6.

Thus the path length is $d_1 + (1/(p-1))(p - 1/p^{2k})\varepsilon + (1/p^{2k+1})d_2$, if the path loops through k times and $d_1 + (1/(p-1))(p - 1/p^{2k+2})\varepsilon + (1/p^{2k+3})d_2$ if the path loops through $k+1$ times. Thus the path length increases by $((p+1)/p^{2k+2})\varepsilon$ and

Table 11.5 Delay factor δ and the size factor s of the graph in Fig. 11.7, (p is an integer, ε is a very small number)

Delay factor δ	Size factor s
$\delta_{1,2} = d_1$	$s_{1,2} = 1$
$\delta_{2,3} = \varepsilon$	$s_{2,3} = 1/p$
$\delta_{3,2} = \varepsilon$	$s_{3,2} = 1/p$
$\delta_{3,4} = d_2$	$s_{3,4} = 1$

Table 11.6 Delay after going through the loops

Loop count	Path length
1	$d_1 + (1 + 1/p + 1/p^2)\varepsilon + d_2/p^3$
2	$d_1 + (1 + 1/p + \ldots + 1/p^4)\varepsilon + d_2/p^5$
3	$d_1 + (1 + 1/p + \ldots + 1/p^6)\varepsilon + d_2/p^7$
...	...
k	$d_1 + (1 + 1/p + \ldots + 1/p^{2k})\varepsilon + d_2/p^{2k+1}$

decreases by $((p^2 - 1)/p^{2k+3})d_2$ if the number of times the path goes through the loop increases from k to $k + 1$. If d_2 is much larger than ε, the total decrease is much larger than the total increase and as a result if the path goes through the loop one more time, the path length decreases. The situation is similar to the *negative weighted cycle* problem in the traditional shortest path length.

In the path problem for multimedia data transmission environment, we can compute the shortest path between a specified source–destination pair, even with "negative" weights (i.e., with size factor $s_i < 1$) on the links, as long as there is no negative weighted cycles in the graph. We use a modified version of the Bellman–Ford algorithm for this purpose.

Just like the traditional Bellman–Ford algorithm, we find the shortest path lengths subject to the constraint that paths contain at most one link, then relax the condition on the length of the path and find the shortest path length subject to the constraint that paths contain at most two links and so on. Using the same terminology and notations as in [2] we call the shortest path that uses at most h links as the *shortest* ($\leq h$) *path*.

Suppose that $D^h{}_i$ denotes the shortest ($\leq h$) path length from node i to the destination node n, ($1 \leq i \leq n-1$). $D^h{}_n = 0$ for all h. The algorithm works as follows.

Shortest Path Algorithm for Multimedia Data Transmission Environment

Input: The directed graph $G = (V, E)$, ($V = \{1, 2, \ldots, n\}$), two $n \times n$ matrices δ and s, the (i, j)-th entry of the matrices stores the delay factor δ and the size factor s of the link from the node i to node j. If there is no link from the node i to j, both $\delta_{i,j}$ and $s_{i,j}$ is taken to be ∞. Without any loss of generality, we assume that the node 1 is the source node and node n is the destination node.

Output: The shortest path length from every node in the graph to the destination node n.

Comments: The algorithm starts from the destination node and in each iteration finds the shortest ($\leq h$) path from a node i in the graph to the destination node n, $1 \leq i, h \leq n-1$.

Algorithm 6 Shortest Path for Multimedia Data Transfer with Reduction in Size

for $i = 1$ to $n-1$ **do**
  $D^0{}_i = \infty$;
end for
for $i = 1$ to $n-1$ **do**
  $D^1{}_i = \delta(i,n)$;
end for
for $h = 1$ to $n-2$ **do**
  for $i = 1$ to $n-1$ **do**
    $D^{h+1}{}_i = \min_{1 \leq j \leq n-1}[s(i,j)D^h{}_j + \delta(i,j)]$;
  end for
end for

Theorem 11.3. *If the graph $G = (V,E)$ does not contain any negative weighted cycle, then the above algorithm correctly computes the shortest path length from any node $i, 1 \leq i \leq n-1$ to the destination node n, even when some of the size factors s_i associated with a link e_i is less than 1.*

Proof. The proof follows along the same lines as the correctness proof of the traditional Bellman–Ford algorithm. We follow the notations, terminologies and the proof technique given in [2].

From the algorithm, $D^1{}_i := \delta_{i,n}$ for all $i, 1 \leq i \leq n-1$. Clearly, these are the correct shortest (≤ 1) path lengths from the node i to destination node n. Induction can be used on h to prove the correctness of the algorithm. We show that if $D^h{}_i$ is the shortest ($\leq h$) path length from node i to n for all $i, 1 \leq i \leq n-1$, then $D^{h+1}{}_i$ is the shortest ($\leq h+1$) path length from node i to n for all $i, 1 \leq i \leq n-1$.

First, we show that $D^{h+1}{}_i \geq \min_{1 \leq j \leq n-1}[s_{i,j}D^h{}_j + \delta_{i,j}]$. Suppose that $(i,k,\ldots,n)$ is a shortest ($\leq h+1$) path from i to n. Then, its path length is the sum of path length from i to k and path length from k to n. The shortest ($\leq h$) path length from k to n is D^h_k. The path length from i to k is $\delta_{i,k}$. Taking the size factor $s_{i,k}$ into account, we can say that $D^{h+1}{}_i \geq \min_{1 \leq j \leq n-1}[s_{i,j}D^h{}_j + \delta_{i,j}]$.

Next, we show that $D^{h+1}{}_i \leq \min_{1 \leq j \leq n-1}[s_{i,j}D^h{}_j + \delta_{i,j}]$. Suppose that some k finds the minimum of $s_{i,j}D^h{}_j + \delta_{i,j}, 1 \leq j \leq n-1$ and that $(k,\ldots,n$ is the shortest ($\leq h$) path from k to n. By our inductive hypothesis, this length is $D^h{}_k$. Taking the size factor $\delta_{i,j}$ in to account, the length of $(i,k,\ldots,n)$ path is $s_{i,k}D^h{}_k + \delta_{i,k}$. Clearly, $s_{i,k}D^h{}_k + \delta_{i,k} = \min(s_{i,j}D^h{}_j + \delta_{i,j}), 1 \leq j \leq n-1$. If $D^{h+1}{}_i$ is the shortest path length from i to n, then $D^{h+1}{}_i \leq s_{i,k}D^h{}_k + \delta_{i,k}$. Therefore, $D^{h+1}{}_i \leq \min_{1 \leq j \leq n-1}[s_{i,j}D^h{}_j + \delta_{i,j}]$.

Therefore, $D^{h+1}{}_i = \min_{1 \leq j \leq n-1}[s_{i,j}D^h{}_j + \delta_{i,j}]$, and this is computed in the last line of the algorithm. □

Table 11.7 Shortest path computation for the graph in Fig. 11.6 using modified Bellman–Ford algorithm

Iteration number	Nodes of the graph						
	S	A	X	B	C	D	T
1	∞	∞	∞	∞	2	1	0
2	∞	∞	3	4	2	1	0
3	∞	∞	2	4	2	1	0
4	∞	4	2	4	2	1	0
5	5	4	2	4	2	1	0
6	5	4	2	4	2	1	0

Theorem 11.4. *The Complexity of the Algorithm 6 is $O(n^3)$.*

Proof. The algorithm is a slight modification of the classical shortest path algorithm due to Bellman and Ford. Following similar arguments as in the complexity analysis of the Bellman–Ford algorithm [2], the complexity of this algorithm can be shown to be $O(n^3)$. □

An example of the result of execution of the algorithm on the graph in Fig. 11.6 is shown in Table 11.7.

Mathematical Programming Solution to the Path Problem in Multimedia Data Transmission

In this subsection, we show that the shortest path problem for the multimedia data transmission problem can also be solved using mathematical programming techniques.

Given a graph $G = (V,E)$ with weights δ_i and s_i associated with each link $e_i \in E$ and two specified vertices s and t, the problem is to find the shortest (or the least weighted) path from s to t.

In the mathematical programming formulation of the problem, we associate a binary indicator variable $x_{i,j}$ with each link (i,j) of the directed graph $G = (V,E)$. By assigning a zero or a one to the variable $x_{i,j}$ the solution indicates whether or not the link (i,j) is a part of the shortest path from the source to the destination. We also introduce two other variables $y_{i,j}$ and $z_{i,j}$, $(1 \leq i,j \leq |V|)$.

Minimize $\sum_{(i,j)\in E} \delta_{i,j} z_{i,j}$

Subject to the following constraints:

1 $$\sum_{\{j|(i,j)\in E\}} x_{i,j} - \sum_{\{j|(j,i)\in E\}} x_{j,i} = r_i, \forall i \in V$$

$r_i = 1$, if i is the source node, $r_i = -1$ if i is the destination node and $r_i = 0$ if i is any other node.

2 $$\sum_{\{j|(i,j)\in E\}} y_{i,j} - \sum_{\{j|(j,i)\in E\}} z_{j,i} = 0, \forall i \in V - \{s,t\},$$

3 $$z_{i,j} = s_{i,j} y_{i,j}, \forall (i,j) \in E,$$

4 $$z_{i,j} \leq K x_{i,j}, \forall (i,j) \in E.$$

where K is a large constant.

The first constraint establishes a path from the source to the destination. As data passes through a link (i,j), its size changes by a factor $s_{i,j}$. This is ensured by the constraint (3). The delay keeps accumulating as the data file passes through various links on its journey from the source to the destination. This aspect is captured by the constraint (2). The purpose of constraint (4) is to ensure that the variable $z_{i,j}$ does not have a nonzero value when $x_{i,j} = 0$, i.e., when the the link (i,j) is not part of the path from the source to the destination. The contribution of the delay factor $\delta_{i,j}$ associated with the link (i,j) is taken into account in the objective function of the formulation.

11.2.1.3 Path Problem in VLSI Circuit Design

The shortest path problem in the VLSI circuit design environment was introduced in Sect. 11.3. The path length in such an environment is carried out in the following way: If the RC-circuit shown in Fig. 11.2 and modeled in Fig. 11.3 contains k resistors and k capacitors, the delay or the *path length* between the source node v_0 and the destination node v_k is given by

$$PL(v_0, v_k) = r_1(c_1 + \ldots + c_k) + r_2(c_2 + \ldots + c_k) + \ldots + r_k(c_k)$$

or it can be rewritten as

$$\begin{aligned} PL(v_0, v_k) &= c_1(r_1) + c_2(r_1 + r_2) + \ldots + c_k(r_1 + \ldots + r_k) \\ &= \sum_{i=1}^{k} c_i \left(\sum_{j=1}^{i} r_j \right). \end{aligned}$$

It may be interesting to note that neither lemma 11.1 (based on which the traditional shortest path algorithm of Dijkstra is developed) nor lemma 11.2 (based on which our modified Dijkstra and Bellman–Ford algorithms for multimedia data transmission are developed) *is true* for this path length function. We demonstrate this with the following two examples.

Consider the graph shown Fig. 11.8. The first of the two parameters associated with a link represents resistance r and the second represents the capacitance c.

With this data set the length of the path, $A \to B \to D$, is $c_{A,B} r_{A,B} + c_{B,D}(r_{A,B} + r_{B,D}) = 1.1 + 1.(1+2) = 4$ whereas the length of the path $A \to C \to D$ is $c_{A,C} r_{A,C} + c_{C,D}(r_{A,C} + r_{C,D}) = 3.1 + 1.(1+1) = 5$. Thus, the path $A \to B \to D$ is shorter than

Fig. 11.8 Shortest path for VLSI circuit, Example 1

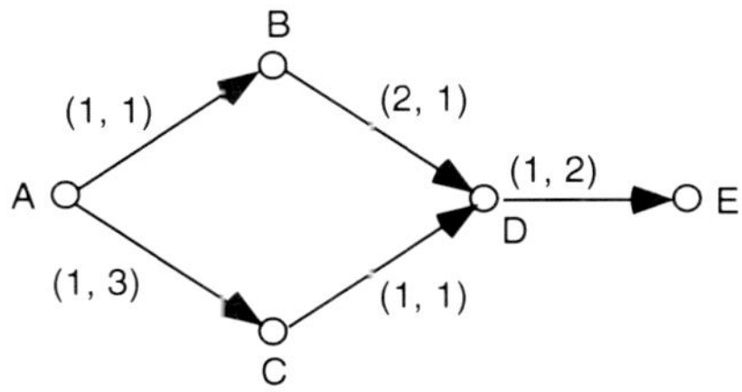

Fig. 11.9 Shortest path for VLSI circuit, Example 2

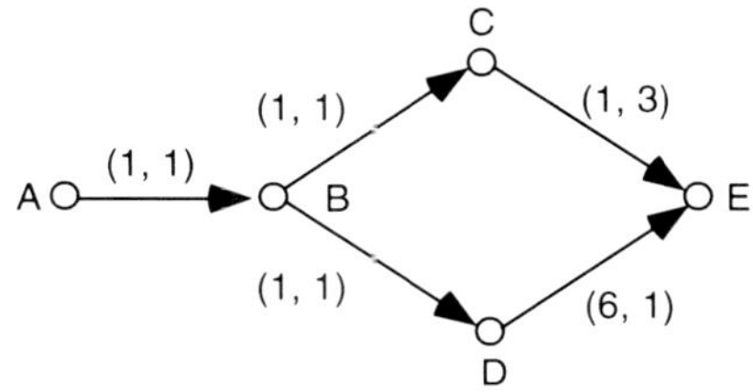

the path $A \to C \to D$. However, in this example the length of the path $A \to B \to D \to E$ is $c_{A,B}r_{A,B} + c_{B,D}(r_{A,B} + r_{B,D}) + c_{D,E}(r_{A,B} + r_{B,D} + r_{D,E}) = 1.1 + 1.(1+2) + 2(1+2+\ 1) = 12$, which is *greater* than the length of the path $A \to C \to D \to E$ is $c_{A,C}r_{A,C} + c_{C,D}(r_{A,C} + r_{C,D}) + c_{D,E}(r_{A,C} + r_{C,D} + r_{D,E}) = 3.1 + 1.(1+1) + 2(1+1+\ 1) = 11$. This example shows that lemma 11.1 does not hold for this problem.

Next, we show that even lemma 11.2 does not hold for this problem. Consider the graph shown in Fig. 11.9.

With this data set the length of the path, $B \to C \to E$, is $c_{B,C}r_{B,C} + c_{C,E}(r_{B,C} + r_{C,E}) = 1.1 + 3.(1+1) = 7$ whereas the length of the path $B \to D \to E$ is $c_{B,D}r_{B,D} + c_{D,E}(r_{B,D} + r_{D,E}) = 1.1 + 1.(1+6) = 8$. Thus, the path $B \to C \to E$ is shorter than the path $B \to D \to E$. However, in this example the length of the path $A \to B \to C \to E$ is $c_{A,B}r_{A,B} + c_{B,C}(r_{A,B} + r_{B,C}) + c_{C,E}(r_{A,E} + r_{B,C} + r_{C,E}) = 1.1 + 1.(1+1) + 3(1+1+1) = 12$, which is *greater* than the length of the path $A \to B \to D \to E$ is $c_{A,B}r_{A,B} + c_{B,D}(r_{A,B} + r_{B,D}) + c_{D,E}(r_{A,B} + r_{B,D} + r_{D,E}) = 1.1 + 1.(1+1) - 1(1+1+6) = 11$. This example shows that lemma 11.2 does not hold for this problem.

Although neither lemma 11.1 nor lemma 11.2 hold for this path problem, we can still solve the problem using mathematical programming techniques.

Mathematical Programming Solution to the Path Problem in VLSI Circuit Design

We will use the variables $x_{i,j}, y_{i,j}$, and $z_{i,j}$ associated with the links, just like the ones in the multimedia data transmission problem. The source and the destination of the path is denoted by s and t respectively.

Minimize $\sum_{(i,j)\in E} c_{i,j}z_{i,j}$

Subject to the following constraints:

1 $$\sum_{\{j|(i,j)\in E\}} x_{i,j} - \sum_{\{j|(j,i)\in E\}} x_{j,i} = r_i, \forall i \in V,$$

$r_i = 1$ if i is the source node, $r_i = -1$ if i is the destination node and $r_i = 0$ if i is any other node.

2 $$\sum_{\{j|(i,j)\in E\}} y_{i,j} - \sum_{\{j|(j,i)\in E\}} z_{j,i} = 0, \forall i \in V - \{s,t\},$$

3 $$z_{i,j} = y_{i,j} + r_{i,j}x_{i,j}, \forall (i,j) \in \{E - \alpha(s)\},$$

where $\alpha(s) = \{s,v\} \in E, v \in V - \{s\}$

4 $$z_{i,j} = r_{i,j}x_{i,j}, \forall (i,j) \in \alpha(s),$$

5 $$z_{i,j} \leq Kx_{i,j}, \forall (i,j) \in E,$$

where K is a large constant.

The first constraint establishes a path from the source to the destination. As data passes through a link (i,j), its size changes by a factor $r_{i,j}$. This is ensured by the constraints (3) and (4). The delay keeps accumulating as the data passes through various links on its journey from the source to the destination. This aspect is captured by the constraint (2). The purpose of constraint (5) is to ensure that the variable $z_{i,j}$ does not have a nonzero value when $x_{i,j} = 0$, i.e., when the the link (i,j) is not part of the path from the source to the destination. The contribution of the capacitance $c_{i,j}$ associated with the link (i,j) is taken into account in the objective function of the formulation.

11.2.2 Fast Transfer of Bulk Data Files in Time-Varying Networks

Focusing on this important issue of time-varying link bandwidths, we address two relevant problems in this subsection: (1) how to find an optimal path from source to destination in a graph with time-varying bandwidth that minimizes the total data transfer time and (2) if we adapt our path with time, how do we minimize the number of path shifts. One must observe that in a network with stationary bandwidth, the solution is trivial and corresponds to computing a shortest widest path by modifying Dijkstra's algorithm. However, in time-varying bandwidth scenario, wideness of a given path varies with time and therefore the challenge is in finding a path that is optimal over a specified duration of time. The second problem that we address is relevant to the scalability issues if we allow shifting of path based on the bandwidth variation.

Application of our proposed solution can be used in the context of overlay networks or private networks. In this work, we assume that knowledge of the bandwidth variation is available for routing purpose. In a shared network like Internet, bandwidth prediction on a daily or hourly basis is possible up to certain level of accuracy based on measurement (a relevant study of this issue is available in [30]). Further, in a advanced reservation framework as was proposed in [28], the available bandwidth at different time can be known directly based on reservation states at each link. Corollary of our solution can be directly applied to advanced reservation framework for bulk-data transfer where the amount of bandwidth to be reserved at different slots can be computed.

11.2.2.1 Related Work

Various point-to-point routing schemes were proposed in the past mostly targeting real-time applications. These real-time applications have their bandwidth demand either fixed or is adaptive over a specified range and requires packet level delay/jitter guarantees. Routing of such applications in an advanced reservation framework as discussed in [28] although considers time variation of link bandwidth but significantly differs in the following aspect. Bulk data applications unlike, for example, real-time streaming applications do not have a fixed bandwidth requirement and can use as much bandwidth as provided to it at any given time towards minimizing the transfer time. The above goal differentiates the underlying approach in routing bulk data.

Previous work in bulk data transfer has looked into the data transport mechanisms and concentrated on how to provide a fast and reliable transfer of bulk data by employing erasure correcting codes [22, 23].

In the context of overlay network, several schemes were proposed that through coordination among overlay nodes results in efficient transfer of bulk data or content. In [24], authors propose establishing parallel connections from multiple sources with complementary data sets to increase the total download speed. Similar work is also proposed in [26], where authors proposed efficient scheduling of bulk data upload from multiple source sites where these sites can coordinate among themselves. In [25], the authors propose a coordinated approach where multiple receivers can exchange mutually exclusive data sets in addition to relying on the actual data source to minimize total download time. In another recent work presented in [27], authors propose multiple multicast trees to minimize the replication time of bulk content to multiple sites. Most of these aforementioned works require strong coordination among servers and are only applications to overlay networks. Also, these work focus more on coordinating the content distribution and not the routing. In contrast, we here consider point-to-point optimal routing of content or bulk data.

The closest work with similar goals as ours is presented in [20], where optimal single point-to-point path is created based on network congestion or path failures. However, optimality of the path is based on present network condition and not conditioned over the entire duration of transfer. Further, the above solution does not address the scalability issue when switching of paths are used to better adapt to network bandwidth variability. Frequent switching of path can introduce significant signaling in setting up new path and can also lead to route instability and flapping. In this work, we address the issue of how to minimize the switching of paths with a limited increase in the data transfer time. Shortest path problems in time dependent networks have been studied in [4, 13–15, 29].

11.2.2.2 Problem Formulation

When the file size is so large that the data transmission is likely to continue over an extended period of time, the *residual bandwidth* of each link of the network is likely to change during this period. Suppose that the data transfer begins at 9 AM on Monday and continues up to 9 AM on Tuesday. At 9 AM on Monday, there exists some residual bandwidth on each link of the network. Since we want the data to be transferred to the destination as early as possible, we may use the *widest path*[1], i.e., the path with the largest residual bandwidth for data transfer (maximum bandwidth path). The problem with this approach is, since the residual bandwidth changes with time, the widest path between the source and the destination may change with time. We give a concrete example to illustrate this point. Suppose a large file needs to be transferred from A to B starting at 9 AM. The widest path from A to B at 9 AM may be through the intermediate node C. At 10 AM, the widest path from A to B may be through the intermediate node D and at 11 AM through the node E. If we want the file transfer to be completed at the earliest possible time, we should use A–C–B path at 9 AM, A–D–B path at 10 AM, A–E–B path at 11 AM and so on. The first question is then which is the single optimal path that we can use for the entire duration of the file transfer. If we even allow path changes whenever there is a change in the residual bandwidth, the signaling overhead in setting up new path maybe significant. We next formally state the above problems in details.

To capture the notion of time varying residual bandwidth on each link, we divide the time into slots of some fixed duration. We consider a graph $G = (V, E)$, representing the network and time slots 1, 2, 3, ... of equal size. The availability of network resources such as the residual bandwidth is tracked at the granularity of a slot. The elements $b_l[i]$ of the vector $b_l = \{b_l[1], b_l[2], ...,\}$ associated with each link $l \in E$, represents residual bandwidth of link l at time slot i.

[1] A widest path from the source to the destination with respect to bandwidth b_l, associated with link l, is a path P^* that maximizes the value of $min_{l \in P} b_l$, over all path P i.e., $P^* = argmax_P \; min_{l \in P} w_l$.

- **Routing problem:** Given the above definition of graph G with b_l for each link $l \in E$, a file size F, source s and destination d, compute a path from s to d that minimizes the number of slots required to transfer F.
- **Path switching problem:** Given the above definition of graph G with b_l for each link $l \in E$, a file size F, source s, destination d and specified slots T_s, find the minimum number of path switching and associated switching positions (in slots) such that the given file can be transferred in T_s number of slots.

In the context of the above problems, we define $T_{\min}$ as the minimum number of slots necessary to transfer a file of size F from the source to the destination. $T_{\min}$ corresponds to the case where widest path in each time slot is used to transfer the file. Therefore, in our first problem, number of slots corresponding to the optimal path is always greater than or equal to $T_{\min}$. Similarly, $T_s \geq T_{\min}$ for a valid solution to the path switching problem.

In discussion of the algorithms, we define bandwidth for a link in terms of number of bytes/slot since we assume bandwidth does not vary during the slot duration. In that case, link bandwidth b also means the amount of data that can be transferred in a slot which we refer here as throughput. Therefore, we use link bandwidth, file size in a given slot and throughput interchangeably throughout the text. In the next two sections, we provide the solutions to the above two problems.

11.2.2.3 Optimal Routing Algorithm

The algorithm takes as input a directed graph $G = (V,E)$ and $n \times n \times T$ matrix B. (i, j, k)-th entry of the matrix stores the residual bandwidth of the link from the node i to node j during time slot k, $(1 \leq k \leq T)$. The specified source and destination nodes are s and d, respectively. Based on the inputs, we refer this algorithm as *Route(G,B)*. We use the term *Throughput* on a given time interval in slots to be size of the data that can be transferred in that time interval.

Algorithm *Route(G,B)* uses an algorithm, called *FileRoute(G,B,F,p,q)*, which finds if there is a route in the slots $[p:q]$ [2] where file with size F can be transferred, to find out the minimum number of time slots needed to send F, in the following way. The maximum number of slots $T_{\max}$ can be easily found by taking the minimum residual bandwidth among all slots for each link in the graph and finding the widest path. Therefore, we can safely say that the minimum number of slots using the optimal path will belong in the range $[T_{\min}, T_{\max}]$. We next perform a binary search on this range to find the exact number of slots and the corresponding path.

Before presenting the algorithm *FileRoute(G,B,F,p,q)*, we first define certain notations and operations for ease of explanations. Let $E_G(p,q)$ denote the subset of the edges of G, consisting of edges whose residual bandwidth for time slot p

[2] The notation $[t_1 : t_2]$ will be used to denote the range starting with time slot t_1 and ending with time slot t_2.

Algorithm 7 FileRoute

```
Initialize_Q(p, f, G): {NOTE: a subroutine called by algorithm FileRoute}
   For all slot r; i ≤ r ≤ j
   Q(p) ← enqueue distinct bandwidth values (less than or equal to f) for slot p in G in
   descending order
Initialize_Q(i, F, G)
if (i == j) then
    β = throughput of widest path P from s → d in G′
    if (β > F) then
        return yes and path P
    else
        return no
    end if
end if
while (Q(i)! = ∅) do
    α = dequeue (Q(p))
    G′ = G − E_G(p, α)
    if (∃ path P from s → d in G′) then
        F = F − α
    end if
    if (F ≤ 0) then
        return yes and path P
    end if
    test = FileRoute(G′, B, F − α, i + 1, j)
    if (test = yes) then
        return yes and path P {NOTE: P is the path found by the previous recursive call.}
    end if
end while
```

is less than α. Let $G' = G - E_G(p,\alpha)$ be the operation of deleting the edges in $E_G(p,\alpha)$ from graph G. Let $Q(p)$ denotes a *Queue* for a given slot p with enqueue and dequeue operations defined with the additional property that all elements in Q are in sorted decreasing order.

Now, we present the algorithm $FileRoute(G,B,F,i,j)$ where G is a capacitated graph, i,j are slots where $i \leq j$ and F is the file size. The output of the algorithm is a decision: yes or no based on if a single path exists such that file size F can be transferred in the time range of slots $[i:j]$ or not. If the decision is *yes*the algorithm also provides the optimal path. Let us now formally present the algorithm with G', i, j, F initialized to G, p, q, f.

The algorithm starts with the p-th slot. The algorithm recursively calls itself by modifying the graph, advancing to the next slot and adjusting the size of the file to be transmitted. The idea is as follows: During the first recursive call, it asks if the slots $i+1$ through j will be able to transfer a file of size $F - \alpha$. If the answer to this question is *yes,* then it implies that total file of size F can be completely transferred in the slot range $[i:j]$. This is true because the minimum bandwidth associated with a link in the graph at this time is α since G' has no edges less than value α and there

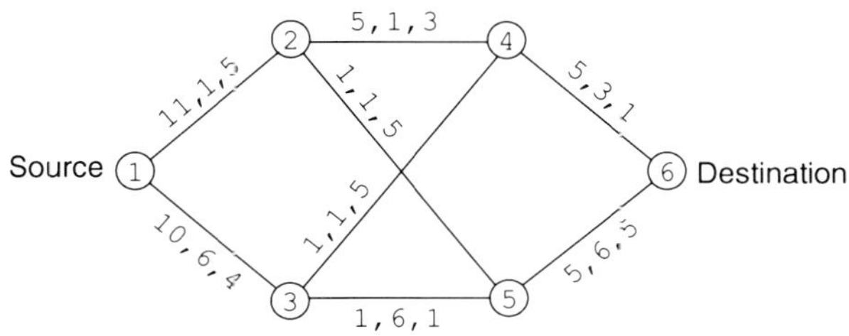

Fig. 11.10 Graph1: G' of recursion#0

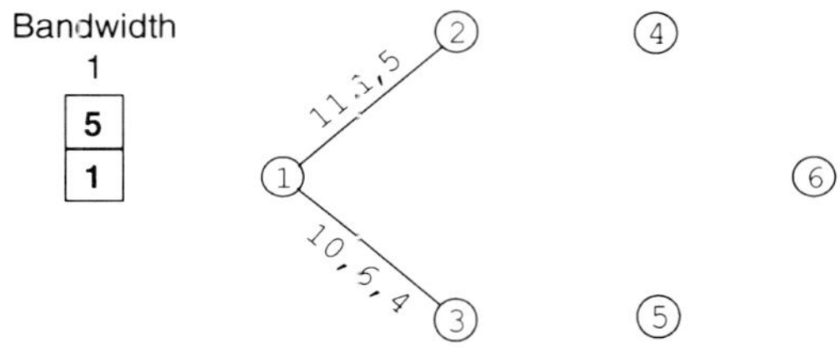

Fig. 11.11 Graph2: G' of Recursion#0

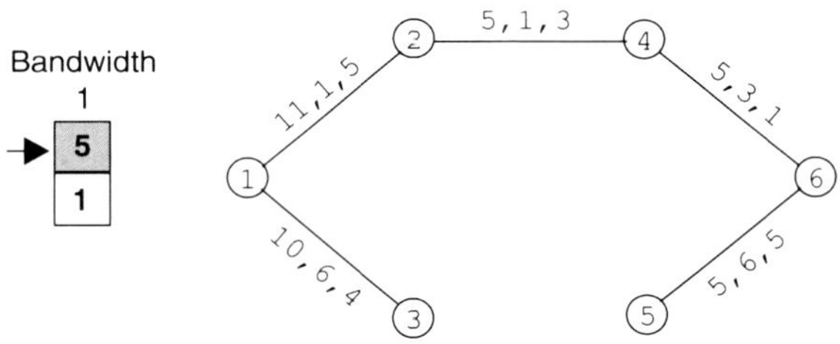

Fig. 11.12 Graph1: G' of recursion#0

exist a path from $s \rightarrow d$ in G'. Accordingly, the first slot will be able to transfer a file of size α along with $F - \alpha$ sent over the slots $i+1$ through j.

The terminating condition is invoked when the recursion reaches the last slot j or when the queue gets empty any slot. At the last slot, if widest path is found with throughput $\beta > F$ (F here is the remaining part of file that needs to be transferred), then it returns yes.

An example Suppose we have a 6-node network, Graph0, as in the Fig. 11.10 We want to know if it can transfer a file of size 8, from node 1 to node 6, within three time slots by calling the function $FileRoute(Graph0, B, 8, 1, 3)$.
Iteration 1: ($G' = Graph0, F = 8, p = 1, q = 3$) Recursion: 0
Get $G' = G'(1) - 10$ by deleting all edges less than 10 from $G'(1)$. (shown in Fig. 11.11).

Iteration 2: ($G = Graph0, F = 8, p = 1, q = 3$) Recursion :0
Since there is no path from node 1 to 6 in G', dequeue $Q(1)$ gives 5 and G' is updated accordingly as showing in Fig. 11.12. Being there is a path from 1 to 6 with

throughput 5 resulting in $F = 8 - 5 = 3$. Next *FileRoute*$(G', B, F = 3, p = 2, q = 3)$ is called.

Iteration 3: $(G = Graph2, F = 3, p = 2, q = 3)$ Dequeue $Q(2)$ giving 3 and resultant graph G' after updating using operation $G' - 3$ is shown in Fig. 11.13.

Iteration 4: $(G = Graph2, F = 3, p = 2, q = 3)$ Recursion :1
There is no path from node 1 to 6 in G' (Graph3), therefore, next dequeue $Q(2)$ gives 1 and corresponding updating $G' = G' - 1$ gives the graph as shown in Fig. 11.14. Since there is a path from 1 to 6 in G' with throughput 1, F gets updated to $F = 3 - 1 = 2$ and *FileRoute*$(G', B, F = 2, p = 3, q = 3)$ is called.

Iteration 5: $(G = Graph4, F = 2, p = 3, q = 3)$ Recursion :2
Since $p = q$, throughput can be computed using the widest path algorithm (WidestPath). WidestPath() algorithm in this case returns 1. Hence, return NO to the upper level of recursion, Recursion#1. Reset $F = 3$.

Iteration 6: $(G = Graph2, F = 3, p = 2, q = 3)$ Recursion :1
Since $Q(2)$ is empty, return NO to the upper level recursion, Recursion 0. Reset $F = 8$.

Iteration 7: $(G = Graph0, F = 8, p = 1, q = 3)$ Recursion :0
Dequeue $Q(1)$ giving 1 and G' is updated as shown in Fig. 11.15 Update $F = 8 - 1 = 7$. Recursive call to $FileRoute(G', B, 7, 2, 3)$.

Iteration 8: $(G = Graph5, F = 7, p = 2, q = 3)$ Recursion :1
Dequeue $Q(2)$ gives 7 which makes no 1–6 path in the updated graph shown in Fig. 11.16.

Iteration 9: $(G = Graph5, F = 7, p = 2, q = 3)$ Recursion :1
Further dequeue $Q(2)$ gives 6 which results in a path in updated G' show in in Fig. 11.17. F is updated to $F = 7 - 6 = 1$ and $FileRoute(G', B, 1, 3, 3)$ is called.

Iteration 10: $(G = Graph7, F = 1, p = 3, q = 3)$ Recursion :2
Since $p = q$, compute the throughput in slot3 by using the widest path algorithm which returns 1 that is equal to F, hence return *yes*.

Solution:
Path: 1-3-5-6, found by widest path as the terminating stage of the algorithm, is the solution. Thus the contribution for 1,2 and 3 slots in terms of throughput is 1, 6 and 1 respectively adding to 8 (input file size).

11.2.2.4 Path Switching Algorithm

We first describe the idea of our algorithms before their formal presentation. The algorithms compute the paths for data transfer from the source to the destination, such that a file of size F can be transferred from the source to the destination with fewest number of path switching and the number of time slots used for data transfer will at most be $T_{\min} + \delta$, where δ is a prespecified threshold value.

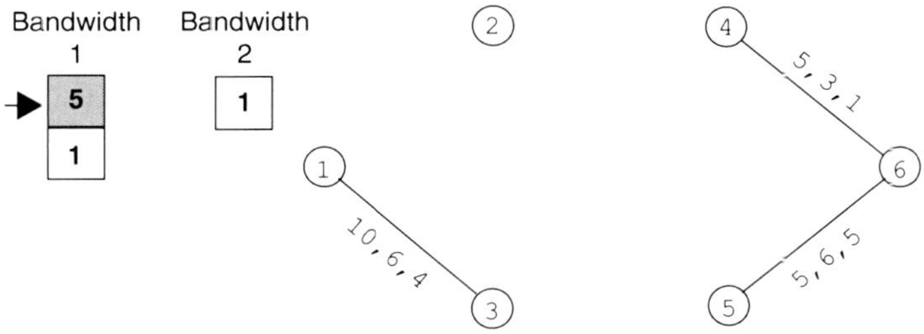

Fig. 11.13 Graph2: G' of recursion#0

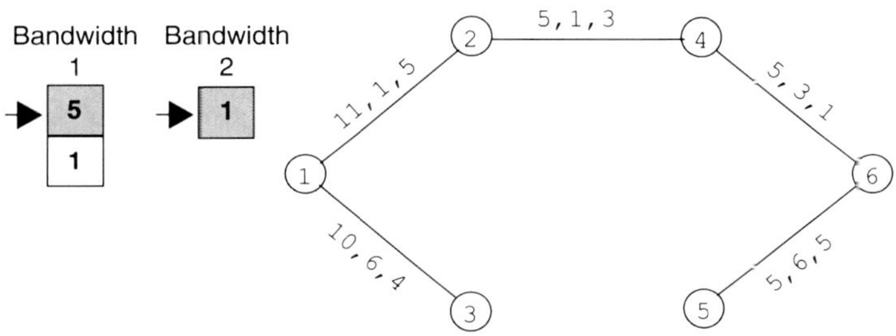

Fig. 11.14 Graph1: G' of recursion#0

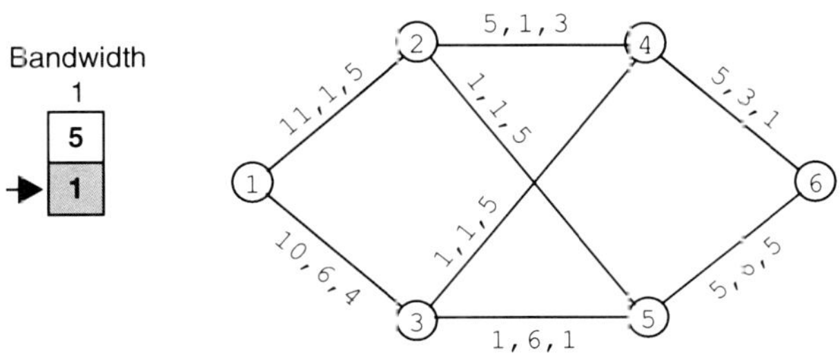

Fig. 11.15 Graph2: G' of recursion#0

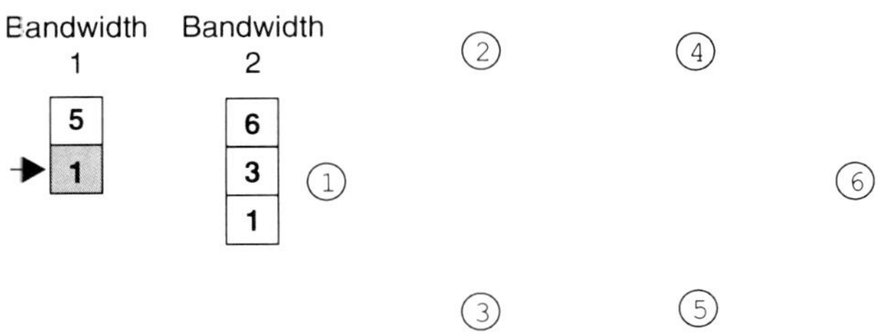

Fig. 11.16 Graph1: G' of recursion#0

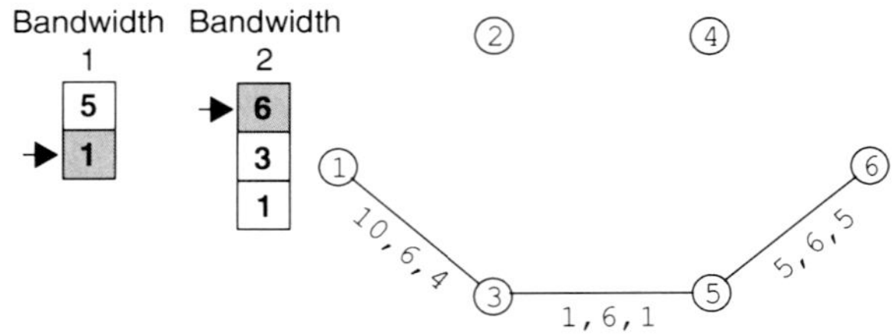

Fig. 11.17 Graph2: G' of recursion#0

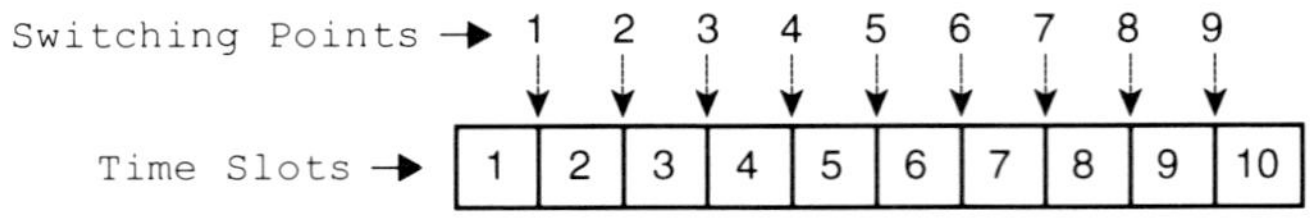

Fig. 11.18 Slots and switching points

In Fig. 11.18, we show an example with ten time slots. The number of points on the time line, where the paths can change are known as the *switching points*. In this example, there are nine switching points. In general, there will be $n-1$ switching points corresponding to n slots.

It may be noted that the throughput may be different if path switching is allowed. We use the matrix *Max_Thp[p, q, r]* in our algorithms. The (p,q,r)-th entry in the *Max_Thp* matrix indicates the maximum size file that can be transferred between the time slots p and q (both inclusive) with r path switching.

First, we describe the *MaxThroughput* algorithm which we will subsequently use in the main algorithm that finds the minimum number of paths. The algorithm *MaxThroughput* computes the maximum file size that can be transferred fromthe source to the destination using one path (i.e., without path switching) within the specified time slots by repeated calls to the *FileRoute(G,B,F,p,q)*. We note that the maximum file size $F_{\max}$ that can be sent over the range $[p:q]$ can be obtained by summing the throughput obtained by running widest path algorithm for each time slot in $[p:q]$. Let us now define G' which is derived from G with same nodes and edges, but capacity of edge e in G' is given by $c(e) = \min_{r\in[p:q]} b_e(r)$, where $b_e(r)$ is the capacity of the edge in original graph G for time slot r. Let $F_{\min}$ be the file size that can be sent over G' using widest path algorithm. Next, we perform a binary search within in the range $[F_{\min} : F_{\max}]$ using the algorithm *FileRoute(G,B,F,p,q),* to find maximum F for which *FileRoute* gives a *yes* decision.

As noted earlier, we will use $Max\_Thp[p,q,r]$ to indicate the maximum size file that can be transferred between the time slots p and q (both inclusive) with r path switching. In addition, we will use a matrix *switch_on* such that the (p,q,r)-th entry in the *switch_on* matrix will indicate the position of the first switching point where path switching takes place for maximum data transfer between the slots p through q with r path switching. $Max\_Thp[p,q,r]$ can be computed in the following way.

$$Max\_Thp[p,q,r] = \max_{p\leq s\leq q-r} [Max\_Thp[p,s,0] + Max\_Thp[s+1,q,r-1]].$$

Algorithm 8 Path Switching

Using widest path algorithm compute $Max\_Thp[i,i,0] for 1 \leq i \leq T$;
$TotalThroughput = \sum_{i=1}^{T} Max\_Thp[i,i,0]$;
if $(TotalThroughput < F)$ **then**
    return {"File F cannot be transferred in T slots"}; *EXIT*;
else
    Call *Algorithm_MaxThroughput* to compute $Max\_Thp[1,T,0]$
    if $(Max\_Thp[1,T,0] \geq F)$ **then**
        return {"File F can be transferred in T slots without
    else
        Call *Algorithm_MaxThroughput* to compute $Max\_Thp[i,T,0] for 1 \leq i \leq T-1$;
        for $r := 1$ to $T-1$ **do**
            for $p := 1$ to $T-r$ **do**
                $Max\_Thp[p,T,r] = \max_{p \leq s \leq T-r}[Max\_Thp[p,s,0] + Max\_Thp[s+1,T,r-1]]$;
                s^* = the value of s that maximizes $[Max\_Thp[p,s,0]+Max\_Thp[s+1,T,r-1]]$;
                $switch\_on[p,T,r] = s^*$;
                if $(Max\_thp[p,T,r] \geq F)$ **then**
                    return $switch\_on[p,T,r]$; EXIT;
                end if
            end for
        end for
    end if
end if

The algorithm presented next, Algorithm 8, uses dynamic programming technique and computes the value of $Max\_Thp[p,T,r]$, by varying r, the number of switches used and p, the index of the starting slot for file transfer. It may be noted that once, r and p is fixed, there may be several switching positions in the slot interval p to T, where path switching can take place. s^* stores the position of first path switching for realization of maximum data transfer between the slots p and T with r path switchings.

Example of Path Switching Algorithm

We consider the network shown in Fig. 11.10. Suppose that we want to transfer a file of size 10. In this graph, $Max\_Thp[1,1,0] = 5$, $Max\_Thp[2,2,0] = 6$, and $Max\_Thp[3,3,0] = 5$. Since *TotalThroughput* (16) is greater than the file size (10), algorithm *MaxThroughput* is called to compute $Max\_Thp[1,3,0]$. In this example $Max\_Thp[1,3,0] = 8$. Since $Max\_Thp[1,3,0]$ is less than the file size, path switching will be necessary in this case. $Max\_Thp[1,3,0] = 8$, $Max\_Thp[2,3,0] = 7$.

$$\begin{aligned}Max\_Thp[1,3,1] &= max\{(Max\_Thp[1,1,0] + Max\_Thp[2,3,0]),\\ &\quad (Max\_Thp[1,2,0] + Max\_Thp[3,3,0])\}\\ &= max\{(5+7),(7+5)\} = 12\end{aligned}$$

and *switch_on*$[1,3,1] = 1$. This means two paths, one only over the slot 1 and the other over the slots 2 and 3 will be used in this case.

Computational Complexity

Finally, *Algorithm_PathSwitching* uses the *Algorithm_MaxThroughput* for computing the points where the paths have to be switched. If m is the number of edges of the network and T is the total number of slots, then the complexity of the *Algorithm_ThroughputTest* is $O(m^T)$. Since the *Algorithm_MaxThroughput* executes *Algorithm_ThroughputTest*, $O(log\ \mathscr{B})$ time, where $\mathscr{B}$ is the sum of the maximum throughput in each time slot, the complexity of the *Algorithm_MaxThroughput* is $O(m^T log\ \mathscr{B})$. The complexity of the *Algorithnm_PathSwitching* is $O(m^T + n^2 + T^3)$.

11.3 Multiple Path Problems

11.3.1 Widest Pair of Disjoint Paths

As indicated earlier, we consider two versions of the problem in this section. In the first version, we address the problem of finding a pair of disjoint paths between a source–destination node pair, such that the combined bandwidth of this path-pair is maximum over all such path-pairs. In the second version, we want to find a pair of disjoint paths, such that the bandwidth of the first path is at least X_1 and the bandwidth of the second path is at least X_2, for some prespecified values X_1 and X_2. We prove that both versions of the problem are NP-complete. We provide exact and approximate solutions for both versions of the problem. In our extensive experimentations, the heuristics produced optimal or near-optimal solutions for almost all the instances of the problem.

Kishimoto in [55] introduced the concept of *maximum multiroute flows* in a network, which at a first glance may appear to be same as the *widest pair of disjoint paths* problem being introduced here. However, on closer examination one will find that the *maximum 2-route flow* problem [31, 39, 55] is completely different from the problem under consideration here. *Maximum 2-route flow* is defined in terms of *elementary 2-flow*. An *elementary 2-flow* is defined to be a flow of one unit along a pair of link-disjoint paths. A 2-route flow is any flow such that it can be expressed as a nonnegative linear sum of elementary 2-flows. The *maximum 2-route flow* between two nodes is a 2-route flow with the maximum value that can be sent between the two nodes, without exceeding the link capacities [31, 39, 55]. To see the difference between maximum 2-route flow and widest pair of disjoint paths, consider the example shown in Fig. 11.19. The maximum 2-route flow between the nodes s and t is 30, whereas the flow using the widest pair of disjoint paths will only be 25. It may be noted that, whereas widest pair of disjoint paths can use only two

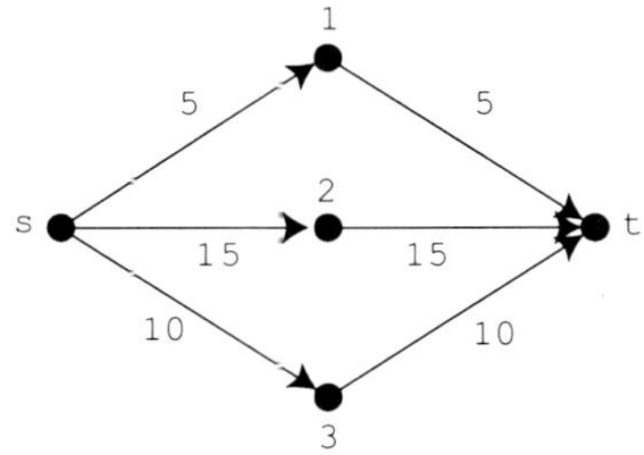

Fig. 11.19 An example of maximum 2-route flow

paths from the source to the destination for data transmission maximum 2-route flow is not restricted to use only two paths. In the example of Fig. 11.19, maximum 2-route flow of 30 units is achieved by using three paths from s to t, $s \longrightarrow 1 \longrightarrow t$, $s \longrightarrow 2 \longrightarrow t$ and $s \longrightarrow 3 \longrightarrow t$.

Taft-Plotkin et al. in [48] used maximally disjoint paths for QoS routing. They present *Maximally Disjoint Shortest and Widest Paths (MADSWIP)* algorithm, which can compute *maximum bandwidth disjoint or maximally disjoint paths* between a source–destination node pair. Although the objective of MADSWIP appears to be same as the widest pair of disjoint path problem, closer examination reveals significant difference between the two. If $BW(P_1)$ and $BW(P_2)$ are the bandwidths associated with disjoint paths P_1 and P_2, in our formulation of the problem, the bandwidth of path-pair (P_1,P_2), $BW(P_1,P_2) = BW(P_1) + BW(P_2)$. In contrast, in [48] bandwidth of path pair (P_1,P_2) is defined to be $BW(P_1,P_2) = min\{BW(P_1),BW(P_2)\}$. This difference in the definition has a significant impact in the computational complexity of the two problems in that, whereas the MADSWIP can be computed in $O(|E|log\ |V|)$ time, we prove that the widest pair of disjoint path problem is NP-complete.

11.3.1.1 Model and Problem Formulation

Like most of the researchers [31, 38, 43, 46–48, 55], we model the network as a directed graph $G = (V,E)$ with a capacity $c(e)$ associated with each arc $e \in E$ [35]. We will use the notation (u,v) to indicate an arc $e \in E$, where the direction of the arc is from u to v, $(u \longrightarrow v)$. In addition, a source node s and a destination node t are specified. If P is a path from the s to t, then the *bandwidth* of the path P, denoted by $BW(P)$ is equal to the minimum of the capacities of the arcs that make up the path P. We consider two different versions of the *Widest Pair of Disjoint Paths* (WPDP) problem. In one version, we want the combined bandwidths of two disjoint paths to be at least a specified value X, and in the other version we want the bandwidths of the two disjoint paths individually be greater than X_1 and X_2, respectively. We refer to the first version as the *Widest Pair of Disjoint Paths (Coupled)* (WPDPC) and the second version as the *Widest Pair of Disjoint Paths (Decoupled))* (WPDPD) problem. The two problems are defined formally next.

Widest Pair of Disjoint Paths (Coupled)

Instance: Given a graph $G = (V, E)$, a capacity $c(e)$ associated with each arc $e \in E$, a source node s, a destination node t and a specified number X.

Question: Are there disjoint paths P_1 and P_2 from s to t, such that the sum of the bandwidths of the paths P_1 and P_2 is greater than or equal to X?

Widest Pair of Disjoint Paths (Decoupled)

Instance: Given a graph $G = (V, E)$, a capacity $c(e)$ associated with each arc $e \in E$, a source node s, a destination node t and specified numbers X_1 and X_2.

Question: Are there disjoint paths P_1 and P_2 from s to t, such that the bandwidth of the path P_1 is greater than or equal to X_1 and the bandwidth of of the path P_2 is greater than or equal to X_2?

It may be noted that the paths P_1 and P_2 may be either node-disjoint or arc-disjoint [35]. The NP-completeness proofs presented here are for the arc-disjoint version of the problems. However, the results are true for the node disjoint versions as well. For the sake of brevity, the proofs related to node-disjoint version of the problem are not presented here. Without loss of generality, we will assume that $X_1 > X_2$. It may be noted that the WPDPD problem is solvable in polynomial time if $X_1 = X_2$, [46, 47]. WPDPD can be viewed as a path finding problem where each arc is colored either red or blue, and the objective is to find a pair of disjoint paths such that at least one of the paths uses the red arcs only. Formally, the disjoint path problem in a graph with red and blue colored arcs can be stated as follows.

Disjoint Path Problem with Red and Blue Arcs (DPPRB)

Instance: A graph $G = (V, E)$, where each arc $e \in E$ is colored either red or blue; a source node s and a destination node t.

Question: Is it possible to establish two disjoint paths from s to t, such that at least one of the paths uses the red arcs only?

From an instance of the WPDPD problem, an instance of the DPPRB can be constructed in the following way. From the graph $G = (V, E)$ of the instance of the WPDPD problem, delete all the arcs with capacity $c(e)$ *less* than X_2. Call this graph $G' = (V, E')$. If the capacity of an arc in G' is at least X_1, then color the arc red, otherwise color it blue. It can be easily verified that it is possible to find disjoint paths P_1 and P_2 from the source node s to the destination node t in the graph G with $BW(P_1) \geq X_1$ and $BW(P_2) \geq X_2$, if and only if, it is possible to find two disjoint paths in G' from s to t such that at least one of the paths use red arcs only.

11.3.1.2 Complexity Analysis

In this section we first prove that the DPPRB problem is NP-complete. Any path P from s to t in the graph G, an instance of the DPPRB problem, will be called a *red–blue* path. A path from s to t will be called a *red* path if all the arcs in the path are red. It may be noted from the definition that any red path is a red–blue path, while the reverse is not true. The proof technique used here is similar to the one used in one of our earlier papers [32].

Theorem 11.5. *The DPPRB Problem is NP-Complete.*

Proof. Clearly, DPPRB is in NP, as one can easily verify if the two paths are arc-disjoint and at least one of them uses red arcs only.

We give a polynomial time transformation from 3SAT to DPPRB. From a instance ϕ of the 3SAT problem, we generate an instance of the DPPRB problem (G,s,t), so that ϕ is satisfiable if and only if G contains a red s–t path p^R and a red–blue s–t path p^{RB} such that p^R and p^{RB} are arc-disjoint in G.

Suppose the 3SAT instance ϕ is made up of clauses $C = \{C_1,\ldots,C_k\}$ and variables $\{x_1,x_2,\ldots,x_l\}$. Let $L = \{x_1,\bar{x}_1,\ldots x_l,\bar{x}_l\}$ be the set of literals made from the variables $\{x_1,x_2,\ldots,x_l\}$. It may be noted that the number of clauses is k and the number of variables is l in ϕ.

The construction of the graph G, the instance of the DPPRB problem, takes place in two steps. In the first step, we construct G_1, a subgraph of G and in the second step augment G_1 with additional nodes and arcs to construct G.

Step 1. We define a graph G_1 in the following way: Create vertices s and t. For each clause-literal pair (C_i,x_j), create two vertices $u_{i,j}$ and $v_{i,j}$ and a red arc connecting them. For each clause-literal pair $(C_i,\bar{x}_j)$, create two vertices $u'_{i,j}$ and $v'_{i,j}$ and a red arc connecting them. For each variable x_j, create two vertices u_j and v_j and blue arcs $(u_j,u_{1,j})$, $(u_j,u'_{1,j})$, $(v_{k,j},v_j)$, $(v'_{k,j},v_j)$, as well as blue arcs $(v_{i,j},u_{i+1,j})$ and $(v'_{i,j},u'_{i+1,j})$ for $i = 1,2,\ldots,k-1$. Finally we add blue arcs (s,u_1), (v_l,t) and blue arcs (v_j,u_{j+1}) for $j = 1,2,\ldots,l-1$.

If the instance ϕ of the 3SAT problem is given by $\phi = (\bar{x}_1 \vee \bar{x}_2 \vee \bar{x}_3) \wedge (\bar{x}_1 \vee x_2 \vee x_3) \wedge (x_1 \vee x_2 \vee x_3) \wedge (\bar{x}_1 \vee x_2 \vee \bar{x}_3)$, then the graph G_1 corresponding to ϕ is shown in Fig. 11.20.

Step 2. Next we add $2k+2$ vertices and $7k+3$ red arcs to G_1 in the following way to obtain the graph G: Add vertices $w_{0,1},w_{0,2},w_{1,1},w_{1,2}\ldots,w_{k,1},w_{k,2}$ and two red arcs $(s,w_{0,1})$ and $(w_{k,2},t)$.

Let C_i be the ith clause and α_i be one of the three literals in C_i. If α_i is x_j, we add two red arcs $(w_{i-1,2},u_{i,j})$ and $(v_{i,j},w_{i,1})$; if α_i is $\bar{x}_j$, we add two red arcs $(w_{i-1,2},u'_{i,j})$ and $(v'_{i,j},w_{i,1})$. $6k$ red arcs are added in this way. In addition $k+1$ red arcs $(w_{i,1},w_{i,2}), 0 \le i \le k$ are added.

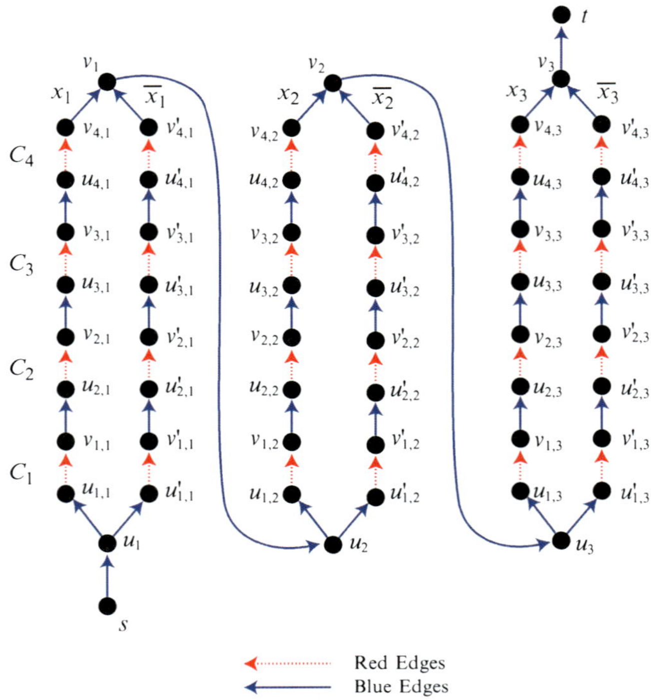

Fig. 11.20 G_1 corresponding to the sample 3SAT instance

The graph G_1 (shown in Fig. 11.20) after augmentation with these additional nodes and arcs is shown in Fig. 11.21. This is graph G, the instance of the DPPRB problem generated from the instance of the 3SAT problem.

It is clear that graph G can be constructed from the instance of the 3SAT problem ϕ in polynomial time, since G contains $4kl + 2l + 2k + 4$ nodes, $2kl + 6k + 4$ red arcs, and $2kl + 3l + 1$ blue arcs.

From the construction of G as shown in Fig. 11.21 we can see the following facts: Any red path must use the arcs $(w_{i,1}, w_{i,2}), 0 \leq i \leq k$, and the red–blue path, if it has to be arc-disjoint from the red path, cannot use these arcs. This implies p^{RB}can only use arcs from G_1. Let p^R be any red s–t path in G and p^{RB} be any red–blue s–t path in G which is arc-disjoint with p^R. For any variable x_j, if p^R goes through any of the vertices corresponding to literal x_j, then p^{RB} must go through all of the vertices corresponding to literal $\bar{x}_j$. Similarly, if p^R goes through any of the vertices corresponding to literal $\bar{x}_j$, then p^{RB} must go through all of the vertices corresponding to literal x_j. If p^R goes through some of the vertices corresponding to literal x_j as well as some of the vertices corresponding to literal $\bar{x}_j$, then p^{RB} cannot be arc-disjoint with p^R.

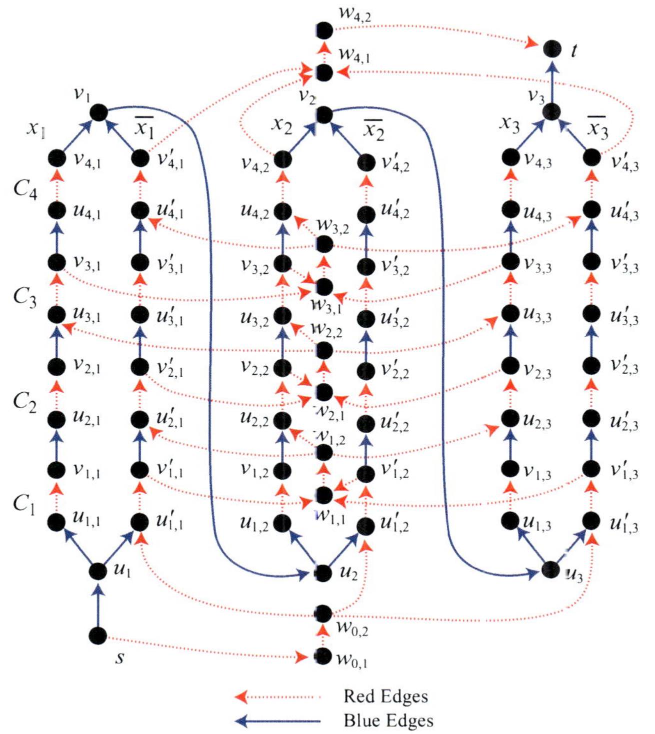

Fig. 11.21 G corresponding to the sample 3SAT instance

We now prove the desired result that (G,s,t) contains a pair of arc-disjoint red and red–blue s–t paths if and only if ϕ is satisfiable. Let us first assume that (G,s,t) contains arc-disjoint paths p^R and p^{RB}, where p^R goes through only the red arcs and p^{RB} goes through red and blue arcs. We will define a truth assignment of the variables $f : \{x_1,x_2,\ldots,x_l\} \mapsto \{TRUE, FALSE\}$ so that ϕ is true under this assignment.

Let x_j be any literal in ϕ. If p^R goes through any of the vertices corresponding to literal x_j, then p^{RB} must go through all of the vertices corresponding to literal $\bar{x}_j$. In this case, we assign $f(x_j) = TRUE$.

If p^R goes through any of the vertices corresponding to literal $\bar{x}_j$, then p^{RB} must go through all of the vertices corresponding to literal x_j. In this case, we assign $f(x_j) = FALSE$.

If p^R does not go through any of the vertices corresponding to literal x_j or literal $\bar{x}_j$, then p^{RB} goes through either all of vertices corresponding to literal x_j or all the vertices corresponding to literal $\bar{x}_j$. In the former case, we set $f(x_j) = FALSE$. In the latter case, we set $f(x_j) = TRUE$.

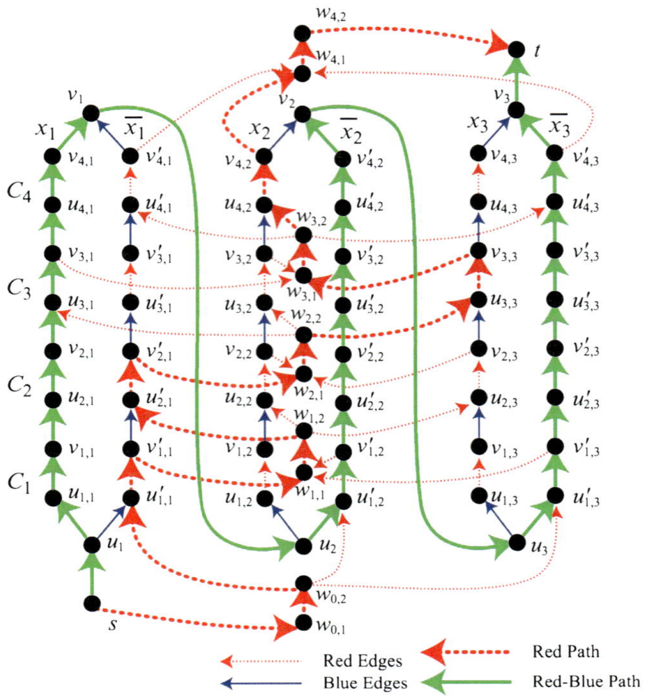

Fig. 11.22 Arc-disjoint R/B paths to truth assignment satisfying ϕ

In short, p^R goes through the arcs $(w_{i,1}, w_{i,2}), 0 \le i \le k$, and the arcs $(u_{i,j}, v_{i,j})$ (with $f(x_j) = TRUE$) and arcs $(u'_{i,j}, v'_{i,j})$ (with $f(x_j) = FALSE$) in the truth assignment. Since the red path p^R passes through the $(w_{i,1}, w_{i,2}), 0 \le i \le k$ in that order, ϕ is satisfied with the above truth assignment.

Figure 11.22 illustrates the two arc disjoint paths from s to t. The red path p^R: $s \to w_{0,1} \to w_{0,2} \to u'_{1,1} \to v'_{1,1} \to w_{1,1} \to w_{1,2} \to u'_{2,1} \to v'_{2,1} \to w_{2,1} \to w_{2,2} \to u_{3,3} \to v_{3,3} \to w_{3,1} \to w_{3,2} \to u_{4,2} \to v_{4,2} \to w_{4,1} \to w_{4,2} \to t$ and the red–blue path p^{RB}: $s \to u_1 \to u_{1,1} \to v_{1,1} \to u_{2,1} \to v_{2,1} \to u_{3,1} \to v_{3,1} \to u_{4,1} \to v_{4,1} \to v_1 \to u_2 \to u'_{1,2} \to v'_{1,2} \to u'_{2,2} \to v'_{2,2} \to u'_{3,2} \to v'_{3,2} \to u'_{4,2} \to v'_{4,2} \to v_2 \to u_3 \to u'_{1,3} \to v'_{1,3} \to u'_{2,3} \to v'_{2,3} \to u'_{3,3} \to v'_{3,3} \to u'_{4,3} \to v'_{4,3} \to v_3 \to t$.

Since p^R goes through the vertices $u'_{1,1}$, $v'_{1,1}$, $u'_{2,1}$, $v'_{2,1}$ which correspond to the literal $\bar{x}_1$, we set $f(x_1) = FALSE$ in the truth assignment. Since p^R goes through the vertices $u_{3,3}$, $v_{3,3}$, which correspond to the literal x_3, we set $f(x_3) = TRUE$ in the truth assignment. Since p^R goes through the vertices $u_{4,2}$, $v_{4,2}$ which correspond to the literal x_2, we set $f(x_2) = TRUE$ in the truth assignment. One can easily verify that ϕ evaluates to $TRUE$ with this truth assignment.

To show the converse, assume ϕ is satisfiable and let f be a truth assignment that satisfies ϕ. We will show that there exist a red s–t path p^R and a red–blue s–t path

p^{RB} which are arc-disjoint in G. Note that any red–blue s–t path will contain the arcs (s,u_1), (v_l,t), (v_{j-1},u_j) for $j=2,3,\ldots,l$, and a red–blue path from u_j to v_j for $j=1,2,\ldots,l$.

If $f(x_j)=FALSE$, we define the segment of p^{RB} from u_j–v_j to be u_j-$u_{1,j}$-$v_{1,j}$-$u_{2,j}$-$v_{2,j}$-$\cdots$-$u_{k,j}$-$v_{k,j}$-v_j. Otherwise, we define the segment of p^{RB} from u_j–v_j to be u_j-$u'_{1,j}$-$v'_{1,j}$-$u'_{2,j}$-$v'_{2,j}$-$\cdots$-$u'_{k,j}$-$v'_{k,j}$-v_j.

We define the red s–t path p^R to contain the red arcs $(s,w_{0,1})$, $(w_{c,2},t)$. Let C_i be the ith clause. Since C_i is $TRUE$ under truth assignment f, at least one of the three literals of C_i is $TRUE$. Let x_j ($\bar{x}_j$, respectively) be one such literal. The red path p^R contains the red arcs $(w_{i-1,2},u_{i,j})$, $(u_{i,j},v_{i,j})$, $(v_{i,j},w_{i,1})$ (the red arcs $(w_{i-1,2},u'_{i,j})$, $(u'_{i,j},v'_{i,j})$, $(v'_{i,j},w_{i,1})$, respectively). In addition, the red path p^R must contain all the red arcs $(w_{i,1},w_{i,2}), 0\le i\le k$.

Since p^R does not go through any vertices corresponding to literal whose value is $FALSE$ and the u_i–v_i segment of p^{RB} goes through only those vertices corresponding to a literal whose value is not $TRUE$, p^B and p^R are arc-disjoint. Therefore the red path p^R and the red–blue path p^{RB} thus constructed are arc disjoint.

Consider the sample 3SAT instance ϕ again. The truth assignment $f(x_1)=FALSE$, $f(x_2)=TRUE$, $f(x_3)=TRUE$ satisfies ϕ. The corresponding red and red–blue paths are shown in Fig. 11.22. □

Theorem 11.6. *The WPDPD problem is NP-complete.*

Proof. Since there exists one-to-one correspondence between the WPDPD and the DPPRB problem, and the later problem is NP-complete, so is the former.

Theorem 11.7. *The WPDPC problem is NP-complete.*

Proof. WPDPC is in NP since we can verify in polynomial time whether a given solution has two arc disjoint paths from the source to the destination and their combined bandwidth is greater than the specified value X. Let $\langle G=(V,E),c,s,t,X_1,X_2\rangle$ be an instance of the WPDPD problem, where $X_1>X_2$. We construct an instance $\langle G'=(V',E'),c',s',t',X\rangle$ of WPDPC problem from the instance of WPDPD in polynomial time as follows:

1. Construct V' by adding two new nodes s' and v'. $V'=V\cup\{s',v'\}$.
2. Construct a new arc (s',s) with capacity X_2, and two new arcs (s',v') and (v',s) both with capacity M, where M is at least as large as the largest capacity in G.
3. Let $E'=E''\cup\{(s',s),(s',v'),(v',s)\}$, where $E''\subseteq E$ is the set of arcs whose capacities are at least X_2 in G. We keep those arcs in E'' having same capacities in G'.
4. Let $t'=t$ be the destination.
5. Let $X=X_1+X_2$ be the required combined bandwidths of the disjoint paths.

Suppose that P_1 and P_2 are two arc disjoint paths in G from s to t with $BW(P_1)\ge X_1$ and $BW(P_2)\ge X_2$. By the construction of G', all arcs in P_1 and P_2 must be in E'. We now construct two arc disjoint paths P'_1 and P'_2 in G' from s' to t' to be

Maximize $y_1 + y_2$
Subject to

$$\sum_{(i,j)\in E} x_{i,j}^k - \sum_{(j,i)\in E} x_{i,j}^k = \begin{cases} 1 & \text{if } i = s, \\ -1 & \text{if } i = t, \\ 0 & \text{otherwise.} \end{cases} \quad \text{for } \forall i \in V,\ k \in \{1,2\} \quad (1)$$

$$x_{i,j}^1 + x_{i,j}^2 \leqslant 1 \quad \text{for } \forall (i,j) \in E \quad (2)$$

$$y_k \leqslant B_{i,j} x_{i,j}^k + M(1 - x_{i,j}^k) \quad \text{for } \forall (i,j) \in E,\ k \in \{1,2\} \quad (3)$$

$$x_{i,j}^k = 0/1 \quad \text{for } \forall (i,j) \in E,\ k \in \{1,2\} \quad (4)$$

$$y_k \geqslant 0 \quad \text{for } k \in \{1,2\} \quad (5)$$

Fig. 11.23 ILP formulation of the WPDPC problem

$\{(s',v'),(v',s)\} \cup P_1$ and $\{(s',s)\} \cup P_2$, respectively. Obviously, $BW(P_1') \geq X_1$ and $BW(P_2') = X_2$. We have $BW(P_1') + BW(P_2') \geq X_1 + X_2 \geq X$. Suppose that P_1' and P_2' are two arc disjoint paths in G' from s' to t' with combined bandwidth at least X. Without lost of generality, assume that $\{(s',v'),(v',s)\} \subseteq P_1'$ and $\{(s',s)\} \subseteq P_2'$. By the construction of G', the capacity of any arc in G' is at least X_2. This implies that $BW(P_2')$ must be X_2, since (s',s) is a bottleneck of P_2' and its capacity is X_2. Since $BW(P_1') + BW(P_2') \geq X$ and $BW(P_2') = X_2$, $BW(P_1') \geq X - X_2 = X_1$. We now construct path P_1 from $P_1' \backslash \{(s',v'),(v',s)\}$ and its arc disjoint path P_2 from $P_2' \backslash \{(s',s)\}$. It is easy to verify that removing those arcs does not reduce the bandwidths of the paths. We have $BW(P_1) \geq X_1$ and $BW(P_1) \geq X_2$. □

11.3.1.3 Exact Solution Using Integer Linear Programming

In this section, we provide the integer linear programming formulations of both WPDPC and WPDPD problems. The formulations for the WPDPC and WPDPD are shown in Figs. 11.23 and 11.24 respectively. The formulations of the two problems have considerable similarity. The variable y_k represents the bandwidth of path k. The binary variable $x_{i,j}^k$ indicates that the arc (i,j) is in path k if and only if it is equal to 1. The objective is to maximize the combined bandwidth. Constraint (1) is for flow conservation. Constraint (2) ensures that the paths are arc disjoint. Constraint (3) determines the capacity of a bottleneck arc for each path. M in that constraint is any large constant.

In the WPDPD problem, the target values for the bandwidths of the paths P_1 and P_2, X_1 and X_2 respectively, are provided as part of the input. In the ILP formulation of the WPDPD problem given in Fig. 11.24, we use a variable ρ to determine if the goal of establishing the disjoint paths P_1 and P_2 with bandwidths X_1 and X_2 respectively can be realized. The objective of the ILP is to maximize ρ. If the value of ρ after execution of the ILP is at least 1, it implies the goal can be realized. Otherwise, it provides information about proximity of the realizable goal to the desired goal.

Maximize ρ
Subject to

$$\sum_{(i,j)\in E} x_{i,j}^k - \sum_{(j,i)\in E} x_{i,j}^k = \begin{cases} 1 & \text{if } i = s, \\ -1 & \text{if } i = t, \\ 0 & \text{otherwise.} \end{cases} \quad \text{for } \forall i \in V, k \in \{1,2\} \quad (1)$$

$$x_{i,j}^1 + x_{i,j}^2 \leqslant 1 \quad \text{for } \forall (i,j) \in E \quad (2)$$

$$\rho X_k \leqslant B_{i,j} x_{i,j}^k + M(1 - x_{i,j}^k) \quad \text{for } \forall (i,j) \in E, k \in \{1,2\} \quad (3)$$

$$x_{i,j}^k = 0/1 \quad \text{for } \forall (i,j) \in E, k \in \{1,2\} \quad (4)$$

$$\rho \geqslant 0 \quad \text{for } k \in \{1,2\} \quad (5)$$

Fig. 11.24 ILP formulation of the WPDPD problem

11.3.1.4 Heuristic Solution Using Relaxation and Randomization

In this section, we suggest two heuristic techniques for the solution of the widest pair of disjoint paths problem. The heuristics are given for the coupled version of the problem, i.e., WPDPC problem. However, same ideas can be used for the decoupled version, i.e., WPDPD problem.

Deterministic Heuristic Algorithm

The algorithm is executed in two phases. In the first phase the relaxed version of the ILP given in Fig. 11.23 is executed (i.e., the constraint number 4 is changed to $x_{i,j}^k \geq 0$ from $x_{i,j}^k = 0/1$). Phase 2 of this heuristic algorithm uses the same technique as in [46, 47].

Algorithm: Deterministic heuristic algorithm (DHA)
Phase 1

Step 1 Find M, the bandwidth of the widest path in the network from s to t.
Step 2 Relax the integrality constraint (4 in Fig. 11.23) and solve the LP, using the value of M obtained in the previous step. If the LP returns a solution with integral values for the variables $x_{i,j}^k$, then stop. Otherwise, go to Phase 2.

Phase 2

Step 1 Let the arc sets $P = \emptyset$ and $F = \emptyset$.
Step 2 For each arc (i,j) in the network, replace its capacity by the value of variable $x_{i,j}^1$ obtained in the LP solution. Find a widest path from s to t with the new capacities on the arcs and add the arcs that make up the widest path to the set P.
Step 3 For each arc (i,j) of P, add arc (j,i) to the network if the network already does not have it. Add those new arcs to the set F.
Step 4 Remove all the arcs in the set P from the network.

Step 5 For each arc (i, j) in the network, replace its capacity by the value of variable $x^2_{i,j}$ obtained in the LP solution. Find a widest path from s to t with the new capacities on the arcs. If the arcs that make up this widest path is not in the set F, then add them to the set P.

Step 6 Construct two arc disjoint paths from the set of arcs in P.

Randomized Heuristic Algorithm

The randomized heuristic algorithm (RHA) also executed in two phases and its first phase is identical to that of the DHA.

Algorithm: RHA
Phase 1 Same as Phase 1 of DHA.
Phase 2

Step 1 Construct two networks G_1 and G_2 from the LP solution obtained in Phase 1. Both G_1 and G_2 has the same set of nodes as in the original network. An edge (i, j) in the network is in G_1 if and only if $x^1_{i,j} > 0$ and its capacity is set equal to $x^1_{i,j}$. Similarly, An edge (i, j) in the network is in G_2 if and only if $x^2_{i,j} > 0$ and its capacity is set equal to $x^2_{i,j}$.

Step 2 Two disjoint paths P_1 and P_2 are constructed by random walks on G_1 and G_2. Both paths start at the source node and attempt to reach the destination. They use *Random_Pick_Arc* function to determine the next node on their paths from the current node u. Once an arc is traversed by a path, it is no longer available for the other path. The paths get a chance to call *Random_Pick_Arc* alternatively using its current node u until they reach the destination.

Random_Pick_Arc(u_i, G_i, G)
u_i: current node of path P_i.

Step 1 Let A_{u_i} be the set of arcs leaving $u_i$in G_i. Let Sum_i be the sum of the capacities of the set of arcs A_{u_i}. If $A_{u_i} = \emptyset$, then goto Step 2. Otherwise, goto Step 3.

Step 2 Divide the interval $[0, Sum_i)$ into several left-open-right-close subintervals, where each subinterval corresponds to an arc a in A_{u_i} and the size of the subinterval is equal to the capacity of a. Randomly generate a number R between 0 and *Sum*. Clearly, R will fall into one of the subintervals. Return the arc corresponding to the subinterval and then Stop.

Step 3 Let A'_{u_i} be the set of arcs leaving u_i in G. Let Num_i be the number of all these arcs. Divide the interval $[0, Num_i)$ into Num_i left-open-right-close subintervals, where each subinterval corresponds to an arc a' in A'_{u_i} and the size of the subinterval is equal to 1. Randomly generate a number R between 0 and Num_i. Clearly, R will fall into one of the subintervals. Return the arc corresponding to the subinterval and then Stop.

Table 11.8 Simulation results for WPDPC problem with ARPANET topology

Source destination	Capacity set 1			Capacity set 2			Capacity set 3		
	OPT	DHA	RHA	OPT	DHA	RHA	OPT	DHA	RHA
0,19	12	12	12	225	219	219	945	928	880
2,17	12	12	12	226	215	215	1139	1049	1119
4,15	11	11	11	223	219	219	1139	1053	1084
6,13	11	10	10	229	220	218	1139	1119	1119
8,11	9	8	8	230	230	230	1062	1062	1012
10,9	11	10	10	253	253	253	1277	1105	1265
12,7	10	10	10	276	271	276	1251	1251	1230
14,5	8	7	7	219	217	217	1083	970	970
16,3	8	4	4	221	221	221	1072	976	816
18,1	9	8	8	221	215	215	1117	1084	1084

11.3.1.5 Experimental Results

In this section, we compare the performance of our heuristic algorithms against the optimal solution obtained by ILP. The simulation results are provided for both WPDPC and WPDPD. Simulation experiments were carried out on the ARPANET topology, with 20 nodes and 32 links. For each version, we tested ten different source–destination pairs, and three different sets of capacities on the arcs, i.e., Capacity Set 1, 2, and 3. In order to get a better result for RHA, we repeat the algorithm for 800 times and pick the best solution. In Table 11.8, results for coupled version are presented. For each capacity set, the combined bandwidth obtained by ILP, DHA and RHA are listed. In Table 11.9, results for decoupled version are presented. For each capacity set, the two bandwidths of the pair of paths obtained by ILP, DHA and RHA are listed. It may be observed that for the coupled version, in about one third of the instances, the heuristic algorithms produced optimal solutions. For the decoupled version, more half of the instances, the optimal solution is achieved by the heuristic algorithms. In the cases where the heuristics failed to find the optimal solution, they were within 50% of the optimal solution. From these experimental results, we conclude that the quality of heuristics are high and they produce optimal or near optimal solutions most of the time.

11.3.2 Multipath Routing Using Transit Hubs

The practical benefits of having intermediate nodes for alternate path routing have given rise to a new direction of research efforts focussing on computing alternate paths to optimize various end-to-end performance objectives. An example is a recent work proposed in [50] where authors studies the problem of computing alternate paths that minimizes the maximum load on a link. In the context of the overlay networks, significant amount of works is proposed as well in maximizing end-to-end

Table 11.9 Simulation results for WPDPD problem with ARPANET topology ($X_1 = 3$, $X_2 = 4$)

Source,	Capacity set 1			Capacity set 2			Capacity set 3		
destination	OPT	DHA	RHA	OPT	DHA	RHA	OPT	DHA	RHA
0,19	7,5	7,2	7,5	115,108	115,110	115,110	502,378	502,378	502,378
2,17	7,5	7,3	7,3	119,107	119,107	119,107	637,492	637,432	637,432
4,15	6,5	6,3	6,3	115,108	115,108	115,108	637,492	637,477	637,477
6,13	7,4	7,3	7,3	119,104	119,104	119,104	637,502	637,477	637,477
8,11	5,4	5,4	5,4	119,111	119,111	119,104	617,425	637,375	637,425
10,9	6,4	7,3	7,4	134,101	134,101	134,101	617,458	819,458	819,409
12,7	6,4	6,4	6,4	146,111	146,111	146,111	640,502	640,426	640,590
14,5	4,3	5,2	5,2	115,102	115,104	115,104	601,443	640,409	640,409
16,3	5,3	5,3	5,3	117,104	117,101	117,101	611,458	640,359	640,359
18,1	4,3	6,2	6,2	117,101	117,104	117,104	659,458	659,332	659,457

bandwidth, minimizing loss and delay. In this subsection, we focus on computing alternate disjoint path that can provide resilience to end-to-end path outages – an event whose occurrence has become quite frequent in present Internet.

Resilience to Path Failure

Path failures or outages can happen due to various reasons such as physical link disconnection, software errors, router misconfigurations. Due to path outages, the end hosts experience broken connection, packets loss or congestion. In many cases, the recovery phase by BGP takes many minutes before converging into a stable path. Empirical studies done on Internet as reported in [66] shows up to ten path failures per day for a given end host. The same study done over various hosts over a long period of time shows that the outage time is greater than 30 min for 5% of the cases and around 4–5 min for 71% of the cases. The path failures are observed to happen with equal probability at both Inter-AS and Intra-AS level that path failure is widespread and affect any end host.

In trying to alleviate the above path failure problem, the authors in [20] proposed resilient overlay network (RON). RON presented the protocols and system architecture for deploying alternate path using overlay nodes between end hosts. Such an alternate backup path can be created either reactive to a path failure or can co-exist with the default path.

The goal of this work is the design of algorithm for the computation of an alternate path P such that P does not share any underlying IP links with the default path. The understanding is that by having disjoint path, we can avoid any possible correlation between the failure probability of default and the back up path. In our application scenario, given the transit nodes, one can easily establish the underlying IP path in terms of the routers using traceroute and router resolution tool Rocketfuel [67].

In a traditional disjoint path computation problem, a disjoint path is defined a set of link concatenated to find a path from source to destination. However, in the case considered here, the disjoint path consists of overlay links between transit nodes $TN1 \to TN2$ that maps to a prespecified shortest path between $TN1$ and $TN2$.

We consider an application scenario where there exist a set of dedicated transit nodes which can be placed at diverse locations on the Internet. To ensure a good performance, we assume that the transit nodes have very good access link bandwidth and is not a bottleneck in terms of forwarding load.

In this scenario, we consider the problem of disjoint path routing using transit nodes where the default path between the transit nodes are prespecified. The objective is to minimize the number of transit nodes required to find the disjoint path. Such a minimizing criteria is required for various reasons. For example, one may want to minimize the deployment cost of transit nodes. In case of VPN network, one may want to minimize the number of leased connections.

Our first contribution is to present the decision version of the above problem where given a number of k transit nodes, the question is to find if there exist a path which is disjoint to the default IP path. We prove that the solution to the above decision version is NP complete.

Our second contribution is to take the next step and study the design of an exact algorithm. The solution to the exact problem leads to solving the maximal independent sets in a graph. In many practical cases if the maximum number of independent sets is small, such an exact algorithm can come to use.

Our third contribution is to present a heuristic based solution for the aforementioned problem. And we present some results on various topologies including the abelene network.

11.3.2.1 Transit Hub Routing in IP Networks

In this paper, we study the *K-transit hub routing problem*. The problem is defined as follows: Suppose the path P_{v_i,v_j} between every pair of nodes v_i and v_j is known. Given the source and the destination nodes s and t respectively, is it possible to establish an alternate path from s to d, disjoint from the primary path $P_{s,d}$, by concatenating at most K other paths P_{v_i,v_j}? In other words, is it possible to find a set of paths $\{P_{s,v_1}, P_{v_1,v_2}, P_{v_2,v_3}, P_{v_3,v_4}, \ldots, P_{v_{k-1},v_k}, P_{v_k,d}\}$ such that the concatenation of these paths will produce a path from s to d subject to the constraint that (1) no two paths in this set share an edge and (2) no paths in this set share an edge with the primary path $P_{s,d}$. If such paths exists, it is possible to establish two disjoint paths between the nodes s and d and utilize the bandwidths of both the paths for data transfer.

11.3.2.2 Problem Formulation and Complexity Analysis

As indicated earlier, the input to the K-transit hub routing problem is (1) an undirected network graph $G = (V, E)$, (2) a set of $n * (n-1)$ paths ($|V| = n$) between every source–destination node pair (the path from node i to j is not necessarily the same as the path from j to i), and (3) specified source and destination nodes s and d respectively. The objective of the K-transit hub routing problem is to find out if it is possible to construct a path from s to d by concatenating at most K paths (from the set of $n * (n-1)$ paths) so that (1) each of the K paths is edge disjoint with the original s to d path and (2) the K paths are mutually edge disjoint.

In order to find an answer to this question, we first remove from the graph $G = (V, E)$, all the edges used by the path from s to d. Let $\mathscr{P}$ be the set of all $n * (n-1)$ paths given as the input. After removal of the edges belonging to the s to d path, many of the paths in $\mathscr{P}$ may become disconnected. We will refer to such paths as "unavailable" and will denote by P_{unav}. The other paths in the set $\mathscr{P}$ are the "available" and will be denoted by P_{av}

Definitions and Notations

Definition 11.1. *Intersection set of paths:* The intersection set of two paths P_i and P_j is the set of edges common between the paths and is denoted by $P_i \cap P_j$.

Definition 11.2. *Compatible Paths*: Two paths P_i and P_j are said to be *compatible* if their intersection set is empty.

Definition 11.3. *Concatenation of Paths*: If P_i is a path from s_i to d_i and P_j is a path from s_j to d_j, they can be *concatenated* if $d_i = s_j$ and the result of the concatenation operation is a path from s_i to d_j.

K-Transit Hub Routing Problem

Instance: Given an undirected graph $G = (V,E)$, a set of triples $(s_i, d_i, P_i), 1 \leq i \leq r$, where s_i is a source node, d_i is destination node and P_i is a path from s_i to d_i, specified source/destination nodes s and d respectively and an integer K.

Question: Suppose $\mathscr{P}_{av} = \{P_1, \ldots, P_r\}$. Is there a subset $\mathscr{P}'_{av} \subseteq \mathscr{P}_{av}$ such that
(1) $\mid \mathscr{P}'_{av} \mid \leq K$, and
(2) The paths in $\mathscr{P}'_{av}$ are mutually compatible, i.e., if $P_i, P_j \in \mathscr{P}'_{av}$ then $P_i \cap P_j = \emptyset, \forall i \neq j$, and
(3) A path from s to d can be constructed by concatenating the paths in $\mathscr{P}'_{av}$.
Complexity Analysis Theorem. The K-transit hub routing problem is NP-complete.

Proof. It is not difficult to verify that the K-transit hub routing problem is in NP. We show that the K-transit hub routing problem is NP-complete by a polynomial transformation from the 3SAT problem. From a given instance of the 3SAT problem, specified by a set of variables $\mathscr{X} = \{\S_\infty, \ldots, \S_\backslash\}$ and a set of clauses $\mathscr{C} = \{\mathscr{C}_\infty, \ldots, \mathscr{C}_\Updownarrow\}$, we construct an instance of the K-transit hub routing problem in the following way.

For the sake of convenience in the proof, we define two types of edges in the graph $G = (V,E)$.

Definition. An edge $(uv) \in E$, is called a *path-edge*, if in the set of paths given as the input of the K-transit hub routing problem, the path from the node u to v uses this edge. Otherwise, the edge is known as a *non-path-edge*.

All the edges are classified into one of these two classes (path-edge and non-path edge). It may be noted that both of these two types of edges can be used by paths whose source/destination nodes are different from the end-points of the edge.

The instance of the K-transit hub routing problem can be generated from the instance of the 3-SAT problem in the following way. $\forall x_i \in \mathscr{X}$ and $\forall c_j \in \mathscr{C}$, construct 4-vertex subgraph, with vertex set $\{u_{i,j}, u'_{i,j}, v_{i,j}, v'_{i,j}\}$, and edge set $\{(u_{i,j}, u'_{i,j}), (v_{i,j}, v'_{i,j})\}$, where both edges are path-edges.

For each x_i and $\forall j = 1, \ldots, m-1$, we connect the subgraph for x_i, C_j, and the one for x_i, C_{j+1} with two non-path-edges, $(u'_{i,j}, u_{i,j+1})$ and $(v'_{i,j}, v_{i,j+1})$. Then, for each x_i,

Fig. 11.25 Subgraph constructed for x_i

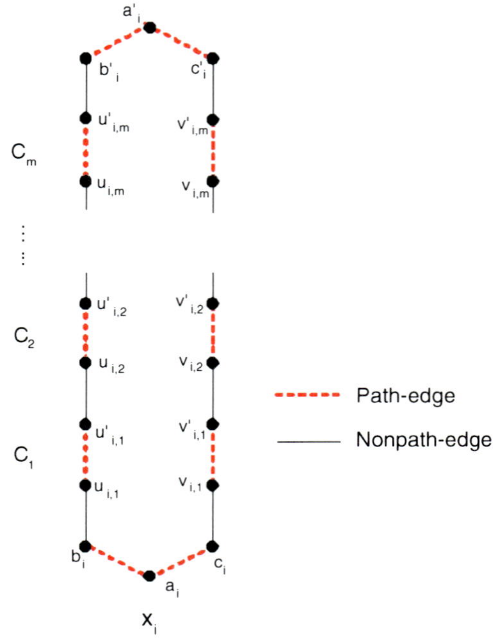

we add six more vertices: $a_i, b_i, c_i, a'_i, b'_i$, and c'_i. We connect a_i, b_i, c_i with path-edges (a_i, b_i) and (a_i, c_i) and connect a'_i, b'_i, c'_i with path-edges (a'_i, b'_i) and (a'_i, c'_i). For each x_i, we add four more non-path-edges: $(b_i, u_{i,1}), (c_i, v_{i,1}), (b'_i, u'_{i,m})$, and $(c'_i, v'_{i,m})$. Now, for each x_i, we have a subgraph as shown in Fig. 11.25.

In the next step, we add a set of vertices, $s, w_0, w_1, \ldots, w_m, d'$ and two path-edges $(s, w_0), (w_m, d')$. Furthermore, we add a set of path-edges between $w'_j s$ and the subgraphs corresponding to x_i, C_j for all i, j, in the following way: $\forall i = 1, \ldots, n, \forall j = 1, \ldots, m,$, if $x_i \in C_j$, then add $(w_{j-1}, u_{i,j})$ and $(u'_{i,j}, w_j)$; if $\bar{x}_i \in C_j$, then add $(w_{j-1}, v_{i,j})$ and $(v'_{i,j}, w_j)$.

Now, we add the destination node d, and a set of path-edges: $(d', a_1), (a'_1, a_2), (a'2, a_3), \ldots, (a'_{n-1}, a_n)$, and (a'_n, d). The construction of the graph G is now complete and is shown in Fig. 11.26.

Now, we specify the paths as part of the input of transit hub routing problem. First of all, every path-edge in G represents a path between its endpoints. Besides these one-edge paths, we specify a set of longer paths: $\forall i = 1, \ldots, n$, a path between b_i and b'_i, $P_{b_i} = b_i - u_{i,1} - u'_{i,1} \ldots - u'_{i,m} - b'_i$, and a path between c_i and c'_i, $P_{c_i} = c_i - v_{i,1} - v'_{i,1} \ldots - v'_{i,m} - c'_i$.

Set s, d to be the source node and destination node respectively and set K to be infinity. Construction of an instance of the K-transit hub problem is now complete.

Claim. There exists a truth assignment satisfying the instance of the 3SAT problem, if and only if a path from s to d can be constructed in the generated instance of the K-transit hub routing problem by concatenating at most K mutually compatible paths.

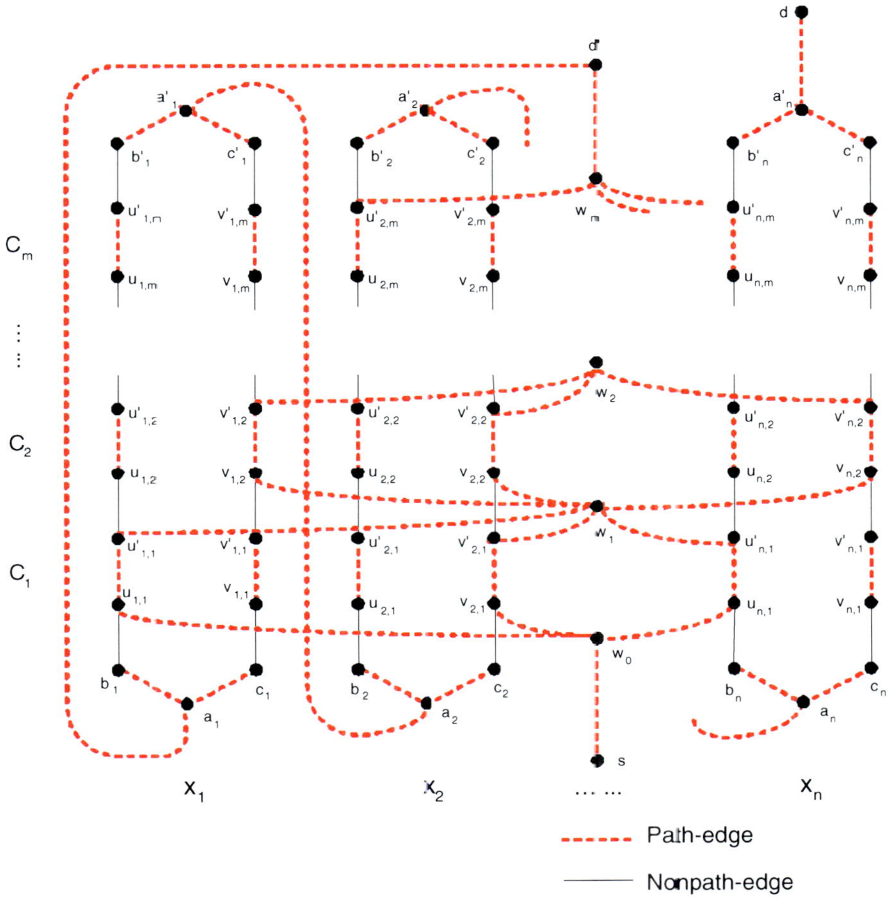

Fig. 11.26 The graph G constructed for NP-completeness proof

Proof of the claim: Suppose there is a truth assignment satisfying the instance of the 3SAT problem. We can construct a path from s to d by concatenating a subset of paths in the following way: (1) go from s to w_0 following the path-edge between them, (2) Each C_j, $j = 1, \ldots, m$, has at least one literal, z that has been assigned "true" by the truth assignment. This means we can go from w_{j-1} to w_j using the corresponding path-edges (i.e., $w_{j-1} - u_{i,j} - u'_{i,j} - w_j$ or $w_{j-1} - v_{i,j} - v'_{i,j} - w_j$), (3) go from w_m to d' using the path-edge between them, (4) go from d' to a_1 using the path-edge between them. (5) If $x_1 =$ "true," then no edge on the path from c_1 to c'_1 has been used so far; otherwise, if $x_1 =$ "false," then no edge on the path from b_1 to b'_1 has been used. Hence we can get to a'_1 from a_1 following the unused path. (6) Go from a'_i to a_{i+1}, for $1 \leq i \leq n-1$, using the path-edge between them. (7) At last, go from a'_n to d following the path-edge between them.

Thus, we find a $s-d$ path, which is a concatenation of a sequence of mutually compatible paths.

Now to prove the reverse, suppose we can go from s to d by concatenating a sequence of mutually compatible paths. It is not hard see that we must go from s to d' by following the path-edges connecting the $w'_j s$ first. Then, from d', we have to go through each subgraph for the $x'_i s$, from a_i to a'_i. For each $i = 1, \ldots, n$, if the path from b_i to b'_i is used, then assign x_i to be "false"; if the path from c_i to c'_i is used, then assign x_i to be "true." It is easy to check this assignment satisfies the corresponding 3SAT problem.

This proves that the K-transit hub routing problem is NP-complete.

11.3.2.3 Exact Solution for the K-Transit Hub Routing Problem

In this section, we provide an exact algorithm for the solution of the K-transit hub routing problem as shown in Algorithm 9. Recall from Sect. 11.3 that the set $\mathscr{P}_{av}$ represents the set of paths available for the construction of an alternate path from s to d, disjoint from the original s to d path in the path set $\mathscr{P}$. Because of the way of construction of the set $\mathscr{P}_{av}$ from the set $\mathscr{P}$, a path between s and d, constructed by concatenating a subset of the path in $\mathscr{P}_{av}$, will automatically be edge disjoint from the original s to d path. Thus, requirement (2) of the K-transit hub problem will automatically be satisfied. Therefore, we only need to make sure that requirement (1) is satisfied, i.e., the paths from the set P_{av} concatenated to generate a path from s to d must be mutually edge disjoint.

In Sect. 11.3, two paths P_i and P_j were defined to be *compatible* if they do not share an edge. Therefore, it is clear that when we attempt to construct the alternate s to d path, it must be done with only the compatible paths in the set P_{av}. As a first step in that direction, we first construct a *path intersection graph (PIG)*.

Definition 11.4. A *path intersection graph* is the intersection graph of paths in the set $\mathscr{P}_{av}$. This is a graph $G_{\text{pig}} = (V_{\text{pig}}, E_{\text{pig}})$, where each node represents a path in the set $\mathscr{P}_{av}$ and two nodes have an edge between them, if the corresponding paths have any common edge.

Definition 11.5. An independent set (or a stable set) in a graph $G = (V, E)$ is a subset $V' \subseteq V$, such that no two nodes in V' are adjacent to each other in the graph $G = (V, E)$.

Definition 11.6. An independent set in a graph $G = (V, E)$ is called a *maximal independent set* if it is not a proper subset of any other independent set in the graph.

As a second step towards construction of the alternate s to d path, we compute all the maximal independent sets of the path intersection graph. The maximal independent sets of the path intersection graph will correspond to the sets of maximal compatible paths in $\mathscr{P}_{av}$. Suppose $\{MIS_1, MIS_2, \ldots\}$ represent the set of maximal independent sets of the path intersection graph.

Algorithm 9 K-Transit Hub Routing Algorithm $(G, \mathscr{P}_{av}, s, d, K)$

setp_1 Compute the path intersection graph, $G_{\text{pig}} = (V_{\text{pig}}, E_{\text{pig}})$ corresponding to the paths in $\mathscr{P}_{av}$.
setp_2 Compute all Maximal Independent Sets of G_{pig}, $\mathscr{MIS} = \{MIS_1, MIS_2, \ldots, MIS_t\}$.
setp_3 Compute a subset $\mathscr{MIS}' \subseteq \mathscr{MIS}$, such that all elements of $\mathscr{MIS}'$, contain at least one path whose terminating point is s and another path whose terminating point is d.
setp_4 Repeat steps 5–7 for each elements MIS_i of $\mathscr{MIS}'$.
setp_5 Compute the path construction graph $G_{\text{pcg}}(i)$ corresponding to MIS_i.
setp_6 Let $V_{i,s}$ be the set of nodes in $G_{\text{pcg}}(i)$ that corresponds to those paths whose one terminating point is s and $V_{i,d}$ be the set of nodes in $G_{\text{pcg}}(i)$ that corresponds to those paths whose one terminating point is s. Repeat step 5 for each element $v_{i,s} \in V_{i,s}$ and for each element $v_{i,d} \in V_{i,d}$.
setp_7 Compute the shortest path from $v_{i,s}$ to $v_{i,d}$. If the shortest path length is at most K then an alternate path from s to d using compatible paths from the set $\mathscr{P}_{av}$ exists. EXIT from the loop.
setp_8 If no path of length at most K can be found in any of the combinations of $v_{i,s}$ and $v_{i,d}$, then an alternate path from s to d using compatible paths from the set $\mathscr{P}_{av}$ does not exist.
setp_9 EXIT.

As a third step in the process to construct an alternate s to d path, we construct a *path construction graph* corresponding to each maximal independent set $MIS_1, MIS_2, \ldots, MIS_t$, computed in the previous step.

Definition 11.7. Each node in a *path construction graph* corresponding to a $MIS_i, 1 \leq i \leq t$, $G_{\text{pcg}}(i) = (V_{\text{pcg}}(i), E_{\text{pcg}}(i))$, corresponds to a path in MIS_i and two nodes have an edge between them if the corresponding paths have a common terminating point, i.e., if the terminating points of a path is v_i and v_j and the terminating points of another path is v_k and v_j, then the nodes corresponding to these two paths will have an edge between them in the graph $G_{\text{pcg}}(i)$.

Some nodes in the graph $G_{\text{pcg}}(i)$ will correspond to paths whose one terminating point is the designated source node s. Similarly, there will be a set of nodes in the graph $G_{\text{pcg}}(i)$ that will correspond to paths whose one terminating point is the designated destination node d. Let $V_{\text{pcg}}(i,s) = \{v_{s,1}, v_{s,2}, \ldots, v_{s,p}\}$ denote the set of nodes that correspond to paths whose one terminating point is the designated source node s. Similarly, let $V_{\text{pcg}}(i,d) = \{v_{d,1}, v_{d,2}, \ldots, v_{d,q}\}$ denote the set of nodes that correspond to paths whose one terminating point is the designated destination node d. Now in the graph $G_{\text{pcg}}(i)$, we compute the shortest path between the nodes $v_{s,j}, 1 \leq j \leq p$ and $v_{d,k}, 1 \leq k \leq q$. If any of these paths have length at most K, then it is *possible* to construct an alternate path from s to d, disjoint from the original path $P_{s,d}$ in the graph $G = (V, E)$, by concatenating compatible paths in the set $\mathscr{P}_{av}$. This process of building a path construction graph $G_{\text{pcg}}(i)$ from MIS_i and then computation of shortest path needs to be repeated $\forall i, 1 \leq i \leq t$. If a shortest path of length at most K cannot be found in any one of these graphs $G_{\text{pcg}}(i), 1 \leq i \leq t$, then it is *impossible* to construct an alternate path from s to d, disjoint from the original path $P_{s,d}$ in the graph $G = (V, E)$, byconcatenating compatible paths in the set $\mathscr{P}_{av}$.

We present the algorithm in pseudocode below:

11.3.2.4 Algorithm Analysis

The algorithm first computes the path intersection graph of the set of available paths $\mathscr{P}_{av}$ and then computes all maximal independent sets of this graph. The maximal independent sets gives the set of compatible paths that can be concatenated for constructing the path from the source s to destination d. In step 5 of the algorithm the path construction graph is constructed and in step 7, the shortest path between a $v_{i,s}$ and $v_{i,d}$ is computed. Since the process is repeated for all maximal independent sets that contains a $v_{i,s}$ and $v_{i,d}$ and for all $v_{i,s}$ and $v_{i,d}$, if a path between s to d can be obtained by concatenating at most K compatible paths in the set $\mathscr{P}_{av}$, this process will find it. This ensures the correctness of the algorithm.

Since in Sect. 11.3 we proved that the K-transit hub routing problem is NP-complete, it is highly unlikely that an exact solution to the problem can be found in polynomial time. The complexity of the algorithm presented in this section is exponential and it is due to the fact that the number of maximal independent sets in a graph can be an exponential function of the number of nodes in the graph. An upper bound on the number of maximal independent sets in a graph was established by Moon and Moser [58] in 1965. They proved that the number of maximal independent sets in any graph is at most $3^{n/3}$ where n is number of nodes in the graph. Recently, Eppstein in [51] and Nielsen in [59] have improved the bound Nielsen has shown in [59] that the number of maximal independent sets of size exactly k in any graph of size n is $n/k^{k-(n \ mod \ k)}(n/k+1)n \ mod \ k$. It was also shown in [59] that for maximal independent sets of size at most k the same bound holds for $k \leq n/3$ and for $k > n/3$ the bound is $3^{n/3}$.

For generating all maximal independent sets of a graph, algorithms such as the ones presented in [60] and [54] can be used. Both the algorithms produce the maximal independent sets one after another in such a way that that the delay between generation of two consecutive maximal independent sets is bounded by a polynomial function of the input size. The computation complexity of the algorithm in [60] is $O(n*m*\alpha)$ [56] and the algorithm in [54] is $O(n^3*\alpha)$ [54] where n, m and α represents the number of nodes, edges and the maximal independent sets of the graph respectively. We use the algorithm in [54] for generating all maximal independent sets in step 2 of the K-transit hub routing algorithm.

The worst case computational complexity of the step 1 of the algorithm is $O(\beta^2)$, step 2 is $O(\beta^3*\alpha)$ and step 3 is $O(\beta^2*\alpha)$. The combined complexity of the steps 4, 5, 6 and 7 is $O(\beta^4*\alpha)$ where α, β represents the number maximal independent sets of the path intersection graph and the paths (i.e., $|\mathscr{P}_{av}|$), respectively. Thus, the overall complexity of the algorithm is $O(\alpha*\beta^4)$.

Special Cases The above analysis is based on the assumption that the path intersection graph does not have any special structure. However, if the path intersection graph has a special structure than the properties of that structure can

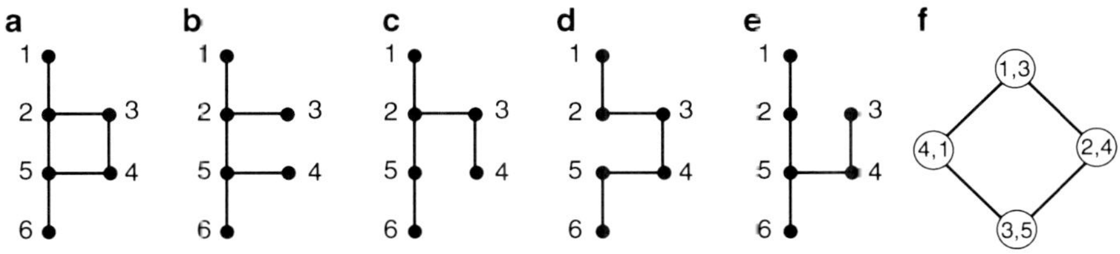

Fig. 11.27 Network topology (**a**), spanning trees generated by the nodes 1–4 (**b–e**) and the cycle of size 4 in the path intersection graph

be exploited to design more efficient algorithms. During the routing process, each node establishes a path to every other node of the graph. These paths essentially establish a spanning tree of the underlying graph rooted at this particular node. This is shown in Fig. 11.29. The spanning trees rooted at nodes 1 through 4 is shown in Fig. 11.29b–e.

Definition 11.8. An undirected graph G is called a *chordal* or *triangulated* graph if every cycle of length strictly greater than 3 possesses a chord, that is, an edge joining two nonconsecutive nodes of the cycle [53]. Equivalently G does not contain any induced subgraph isomorphic to a cycle of size 3 or greater

Intersection graphs of paths in a tree was studied by Gavril in [52] and [57]. It has been shown that [52] that the intersection graphs of paths in a tree are chordal. Efficient algorithms for the generation of all maximal independent sets of a chordal graph is presented in [56]. The complexity of generating all maximal independent sets of a chordal graph is $O(n+m)*\alpha$ whereas for general graphs it is $O(n*m*\alpha)$ where n, m, α are the number of nodes, edges and maximal independent sets, respectively.

Since each node of the network generates a spanning tree rooted at that node, the path intersection graph produced by the paths starting at the root node will be chordal. Figure 11.29a shows a network topology. The Fig. 11.29a–d shows the spanning trees rooted at nodes 1 through 4, respectively. The path intersection graph produced from each of the individual spanning trees will be chordal [52]. Unfortunately, when the path intersection graph produced by the spanning trees rooted at different nodes are combined, they are no longer chordal if the underlying spanning trees are different. The Fig. 11.29e shows such a phenomenon. The node (1,3) represents the path from 1 to 3 obtained from the spanning tree rooted at node 1 (Fig. 11.29b); the node (2, 4) represents the path from 2 to 4 obtained from the spanning tree rooted at node 2 (Fig. 11.29c); the node (3, 5) represents the path from 3 to 5 obtained from the spanning tree rooted at node 3 (Fig. 11.29d); the node (4, 1) represents the path from 4 to 1 obtained from the spanning tree rooted at node 4 (Fig. 11.29e). It can be seen from Fig. 11.29e, the path intersection graph in this example will have a cycle of size 4 as a subgraph. Accordingly, the path intersection graph cannot be chordal. Such a situation is encountered because the spanning trees produced by nodes 1, 2, 3 and 4 are different and when they are combined, it is no

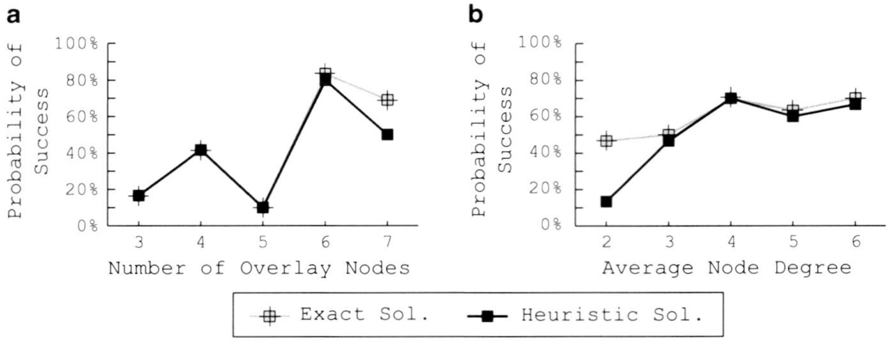

Fig. 11.28 Performance of the heuristic solution, (**a**) probability of success vs. number of overlay nodes; (**b**) probability of success vs. average node degree

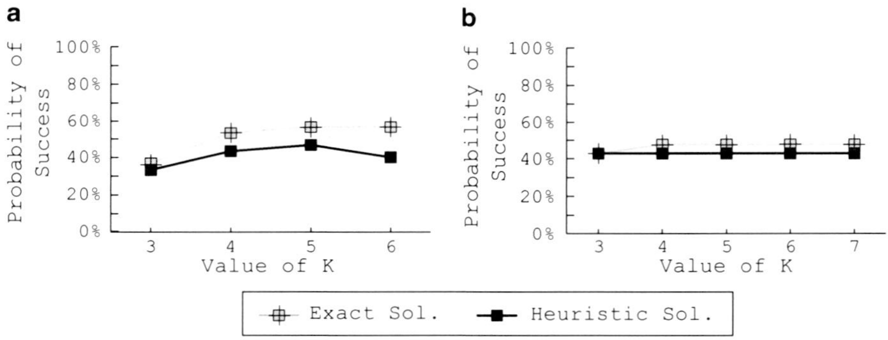

Fig. 11.29 Performance of the heuristic solution, (**a**) probability of success vs. value of K for instance 1; (**b**) Probability of success vs. value of K for instance 2

longer a tree. However, in a special case if the spanning trees produced by different nodes are identical, after combination of all the spanning trees, it will still be a tree. In this case, the path intersection graph will be chordal and the efficient algorithm for generating all maximal independent sets, given in [56], can be utilized to reduce overall computational complexity.

11.3.2.5 Heuristic Solution for the *K*-Transit Hub Routing Problem

The main overhead involved in the exact algorithm is in the computation of all the Maximal Independent Sets of the path intersection graph. In this section, we present a heuristic solution using randomization technique for the K-transit hub routing problem that produces a solution with high probability as shown in Algorithm 10. The complexity of the solution is bounded by a polynomial function of the number of nodes in the overlay network.

Algorithm 10 Heuristic Solution for the K-Transit Hub Routing Problem $(G, \mathscr{P}_{av}, s, d, K)$

1. Compute the path intersection graph $G_{\text{pig}} = (V_{\text{pig}}, E_{\text{pig}})$ corresponding to the paths in P_{av}.
2. Determine $V_s \subseteq V_{\text{pig}}$ as the set of nodes that correspond to the paths whose one terminating point is s and $V_d \in V_{\text{pig}}$ as the set of nodes that correspond to the paths whose one termination point is d.
3. Repeat steps 4 through 7for every node-pair $(v_s, v_d) \in V_s \times V_d$.
4. Construct a maximal independent set with two nodes v_s and v_d, $\mathscr{MIS} = \{v_s, v_d\}$.
5. Let $\mathscr{NNS}(S)$, the "Non-Neighborhood Set" of S be defined as $\mathscr{NNS}(S) = V_{\text{pig}} \setminus (\mathscr{N}(S) \bigcup S)$, where $\mathscr{N}(S)$ represents the neighborhood set of S. Select with equal probability a node $v \in \mathscr{NNS}(S)$. Augment the maximal independent set, $\mathscr{MIS} = \mathscr{MIS} \bigcup \{v\}$.
6. If $\mathscr{MIS}$ is not a maximal independent set, go back to step_5 Otherwise, form the path construction graph G_{pcg}. Compute $V_{i,s}, V_{i,d} \in V(G_{\text{pcg}})$ as the set of nodes corresponding to paths having one terminating point in s, d respectively.
7. Compute the shortest path between every $v_{i,s} \in V_{i,s}$ and $v_{i,d} \in V_{i,d}$. there exists a shortest path of length at most K between any $v_{i,s} \in V_{i,s}$ and $v_{i,d} \in V_{i,d}$, then an alternate path from s to d using compatible paths from the set $\mathscr{P}$ exists. EXIT from the loop.
8. If none of the combinations of v_s and v_d report a path of length at most K, then there is no alternate path from s to d using compatible paths from the set $\mathscr{P}_{av}$.
9. EXIT.

Complexity Analysis As in the exact algorithm, the heuristic solution starts by determining the path intersection graph of the available paths $\mathscr{P}_{av}$. However, instead of finding all maximal independent sets involving the two nodes(paths) v_s and v_d that terminate in s and d respectively, the algorithm randomly generates a maximal independent set for each pair-wise combination of v_s and v_d. The random generation procedure first includes the two nodes v_s and v_d into a working set $\mathscr{MIS}$ of independent nodes. It then randomly selects a node from all the remaining non-neighboring nodes of $\mathscr{MIS}$ in the path intersection graph and includes it into $\mathscr{MIS}$. This process is continued until $\mathscr{MIS}$ is maximally independent.

Step 1 of the algorithm has the worst case computational complexity $O(\beta^2)$, where β is the number of nodes in the path intersection graph. Steps 5 and 6 perform $O(\beta^2)$ operations in the worst case to compute a Maximal Independent Set. Step 7 of the algorithm performs $O(\beta^2)$ operations in the worst case to check if there exists compatible paths in the path construction graph between the source node and the destination node. Thus, the overall complexity of the algorithm is $O(\beta^4)$.

Performance of the Heuristic Solution

In order to evaluate the performance of our proposed exact and heuristic solutions, we conducted experiments for the K-transit hub routing problem, on randomly generated topologies.

The problem instances are generated in three steps.

Step 1. GT-ITM (Georgia-Tech Internet Topology Model) [61] topology generator was used to generate the physical layer topologies. Several random graphs consisting of 30 nodes and for different values of the average node degree were generated for the simulation experiments.

Step 2. A subset of these nodes were randomly chosen, with uniform distribution, as the set of overlay nodes.

Step 3. Shortest Paths using Dijkstra's algorithm between every pair of the overlay nodes were computed.

One of the metrics used for the evaluation of the performance of the heuristic technique is the *success ratio*. The success ratio is defined as the ratio of the number of source–destination pairs for which a path was found by the algorithm to the total number of source–destination pairs.

Three sets of experiments were conducted to study the performance of the heuristic solutions. In the first set of experiments, the number of overlay nodes were varied from 3 to 7 and success ratio of both the exact and the heuristic solutions were measured for all source–destination pairs. The value of K was chosen to be greater than the number of overlay nodes.

In the second set of experiments, different physical topologies consisting of 30 nodes were chosen with varying average node degrees. In each case, six nodes were chosen to be overlay nodes and the success ratio of the exact and heuristic algorithms were measured for all the 30 source–destination pairs. The aim of this experiment was to study the impact of the average node degree on the performance of the algorithm.

The third set of experiments were conducted with two data sets. For various values of K, the success ratio of both the algorithms were recorded. The physical topology had 30 nodes with an average node-degree of 4 and the overlay structure had seven nodes.

In most of the cases, the success ratio of the heuristic was close to the exact algorithm. From the first set of results, among the five cases, the success ratio of the heuristic algorithm is within 4% deviation from the success ratio of the exact algorithm in four cases.

In the second set of results, increasing the average node-degree in the physical topology has a positive effect on finding alternate paths in the overlay. The success ratio for both the heuristic and the exact algorithms increase with increased average node-degree. The success ratio of the heuristic algorithm follows closely with that of exact algorithm. Both the algorithms benefit from increased path availability.

The results of the third simulation indicate that the performance of the heuristic solution is not significantly dependent on the value of K, the number of paths that are allowed to be concatenated to construct the source to destination path.

In all of these cases, the performance of the heuristic technique was close to the exact solution. We noted the execution timings of both the heuristic and the exact solution. In almost all the cases, the computational time for the exact solution was significantly higher than that of the heuristic solution. In many instances, the computational time for the exact solution was about 100 times more than that for the heuristic solution. We thus conclude that our heuristic technique almost always produces a very high quality solution in a fraction of time needed to find the exact solution.

References

1. R. Bellman,"On a Routing Problem", *Quarterly of Applied Mathematics*,vol. 16, no. 1, pp. 87–90, 1958.
2. D. Bertsekas and R. Gallager, *Data Networks*, Prentice Hall, 1987.
3. G. Brassard and P. Bratley, *Fundamentals of Algorithmics*, Prentice-Hall, 1996.
4. X. Cai, T. Kloks and C.K. Wong,"Time-Varying Shortest Path Problems with Constraints", *Networks*, vol. 29, pp. 141–149, 1997.
5. C.T. Chou,"Linear Broadcast Routing", *Journal of Algorithms*, vol. 10, no. 4, pp. 490–517, 1989.
6. T.H. Cormen, C.E. Leiserson and R.L. Rivest, *Introduction to Algorithms*, McGraw-Hill Book Company, 1990.
7. E.W. Dijkstra, "A Note on Two Problems in Connection with Graphs", *Numerische Mathematik*, vol. 1, pp. 269–271, 1959.
8. L.E. Ford and S.M. Johnson, "A Tournament Problem", *The American Mathematical Monthly*, vol. 66, pp. 387–389, 1959.
9. G.Y. Handler and I. Zang, "A Dual Algorithm for the Constrained Shortest Path Problem", *Networks*, vol. 10, pp. 293–310, 1980.
10. J.M. Jaffe, "Algorithms for Finding Paths with Multiple Constraints", *Networks*, vol. 14, pp. 95–116, 1984.
11. J. Lillis, "Timing Optimization for Multi-Source Nets: Characterization and Optimal Repeater Insertion", *Proc. of Design Automation Conference*, pp. 214–219, 1997.
12. E.Q.V. Martins, "On a Multicriteria Shortest Path Problem", *European Journal of Operations Research*, vol. 16, pp. 236–245, 1984.
13. A. Orda and R. Rom, "Shortest-Path and Minimum-Delay Algorithms in Networks with Time-Dependent Edge-Lengths", *Journal of the Association for Computing Machinery*, vol. 37, no. 3, pp. 605–625, 1990.
14. A. Orda and R. Rom, "Minimum Weight Paths with Time-Dependent Networks", *Networks*, vol. 21, pp. 295–319, 1991.
15. A. Orda and R. Rom, "Distributed Shortest-Path Protocols for Time-Dependent Networks", *Distributed Computing*, vol. 10, pp. 49–62, 1996.
16. J.B. Rosen, S. Z. Sun and G.L. Xue, "Algorithms for the Quickest Path Problem and the Enumeration of Quickest Paths", *Computers and Operation Research*, vol. 18, no. 6, pp. 579–584, 1991.
17. H.F. Salama, D.S. Reeves and Y. Viniotis, "A Distributed Algorithm for Delay-Constrained Unicast Routing", *Proceedings of IEEE INFOCOM'97*, pp. 1c2.1–1c.2.8, 1997.
18. C.C. Skiscim and B.L. Golden, "Solving k-Shortest and Constrained Shortest Path Problems Efficiently" *Annals of Operation Research*, vol. 20, pp. 249–282, 1989.
19. Z. Wang and J. Crowcroft, "Quality-of-Service Routing for Supporting Multimedia Applications", *IEEE Journal on Selected Areas of Communications*, vol. 14, no. 7, pp. 1228–1234, 1996.
20. D. Anderson, H. Balakrishnan, M. Kaashoek and R. Morris, "Resilient overlay networks", *Proc. 18th ACM SOSP conference,* Oct 2001.
21. D. Patterson, "A Conversation with Jim Gray", ACM Queue, vol. 1, no. 4, June 2003.

22. Y. Birk and D. Crupnicoff, "A multicast transmission schedule for scalable multi-rate distribution of bulk data using non-scalable erasure-correcting codes", *Proc. of IEEE Infocom 2003,* San Francisco.
23. J.W. Byers, M. Luby, M. Mitzenmacher and A. Rege, "A digital fountain approach to reliable distribution of bulk data", *Proc. of ACM SIGCOMM,* Vancouver, Sept 1998.
24. J. W. Byers, M. Luby and M. Mitzenmacher, "Accessing multiple mirror sites in parallel: using tornado codes to speed up downloads", *Proc. IEEE Infocom* 1999.
25. J.W. Byers, J. Considine, M. Mitzenmacher and S. Rost, "Informed content delivery across adaptive overlay networks", *Proc. ACM SIGCOM* 2002.
26. W. Cheng, C. Chou, L. Golubchik, S. Khuller and Y. Wan,"Large-scale data collection: a coordinated approach", *Proc. of IEEE Infocom 2003,* San Francisco.
27. L. Cherkasova and J. Lee, "FastReplica: Efficient Large File Distribution within Content Delivery Networks", *Proc. of the 4-th USENIX Symposium on Internet Technologies and Systems*, Seattle, Washington, March 26–28, 2003.
28. R. A. Guerin and A. Orda "Networks with Advance Reservations: The Routing Perspective", Proc. of IEEE Infocom, April, 2000.
29. I. Ioachim, S. Gelinas, F. Soiumis, J. Desrosiers, "A Dynamic Programming Algorithm for the Shortest Path Problem with Time Windows and Linear Node Costs", Vol. 31, pp. 193–204, John Wiley & Sons Inc., 1998.
30. M. Roughan, A. Greenberg, C. Kalmanek, M. Rumsewicz, J. Yates and Y. Zhang, "Experience in measuring internet backbone traffic variability: models, metrics, measurements and meaning", Proc. ITC 2003, *Berlin.*
31. C. Aggarwal and J. Orlin, "On Multi-route Maximum Flows in Networks", *Networks*, vol. 39, no. 1, pp. 43–52, 2002.
32. R. Anderson, F. Chung, A. Sen and G. Xue, "On Disjoint Path Pairs with Wavelength Continuity Constraint in WDM Networks", In Proceedings of the *IEEE Infocom*, 2004.
33. Y. Bejerano, Y. Breitbart, A. Orda, R. Rastogi and A. Sprintson, "Algorithms for Computing QoS paths with Restoration", *Proc. of IEEE Infocom*, 2002.
34. R. Bhandari, "Optimal Diverse Routing in Tele-communication Fiber Networks", *Proc. of IEEE Infocom*, 1994.
35. J.A. Bondy and U.S.R. Murthy, "Graph Theory with Applications", North Holland, 1976.
36. I. Cidon, R. Rom and Y. Shavitt,"Analysis of Multi-path Routing", *IEEE/ACM Trans. on Networking*, vol. 7, no. 6, pp. 885–896, 1999.
37. M.R. Garey and D.S. Johnson, "Computers and Intractability: A Guide to the Theory of NP-Completeness", W.H. Freeman Publishers, San Francisco, 1978.
38. Y. Jia, I. Nikolaidis and P. Gburzynski, "Multiple Path Routing in Networks with Inaccurate Link State Information, *Proc. of the IEEE ICC 2001*", Helsinki, Finland, 11–14 June 2001, pp. 2583–2587.
39. K. Kar, M. Kodialam and T.V. Lakshman, "Routing Restorable Bandwidth Guaranteed Connections using Maximum 2-Route Flows", *Proceedings of IEEE Infocom*, 2002.
40. W. Kishimoto, "A Method for Obtaining the Maximum Multi-Route Flows in a Network", *Networks*, vol. 27, no. 4, pp. 279–291, 1996.
41. S. Lee and M. Gerla, "Split Multipath Routing with Maximally Disjoint Paths in Ad-hoc Networks", *Proc. of IEEE ICC* 2001.
42. C.L. Li, S.T. McCormick and D. Simchi-Levi, "The Complexity of Finding Two Disjoint Paths with Min-Max Objective Functions", *Discrete Applied Mathematics*, vol. 26, pp. 105–115, 1990.
43. R. Ogier, V. Ruttenburg and N. Shacham, Distributed Algorithms for Computing Shortest Pair of Disjoint Paths, *IEEE Trans. on Information Theory*, vol. 39, no. 2, 1993.
44. A. Sen, B. Shen and S. Bandyopadhyay, "Survivability of Lightwave Networks – Path Lengths in WDM Protection Scheme", *Journal of High Speed Networks*, vol. 10, pp. 303–315, 2001.
45. D. Sidhu, R. Nair and S. Abdallah, "Finding Disjoint Paths in Networks", *Proc. of ACM Sigcomm*, 1999.
46. S. Suurballe, "Disjoint Paths in Networks", *Networks*, vol. 4, pp. 125–145, 1974.

47. S. Suurballe and R. Tarjan,"A Quick Method for Finding Shortest Pair of Disjoint Paths", *Networks*, vol. 14, pp. 325–336, 1984.
48. N. Taft-Plotkin, B. Bellur and R.G. Ogier, "Quality-of-Service Routing Using Maximally Disjoint Paths", *Proc. of IEEE/IFIP IQQoS*, 1999.
49. S. Vutukury, "Multipath Routing Mechanisms for Traffic Engineering and Quality of Service in the Internet, PhD Dissertation, University of California, Santa Cruz, March 2001.
50. R. Cohen and G. Nakibli, "On the computational complexity and effectiveness of "N-hub shortest path routing"", *Proc. of IEEE Infocom*, 2004.
51. D. Eppstein, "Small maximal independent sets and faster exact graph coloring". *Proc. 7th Workshop on Algorithms and Data Structures*, vol. 2125, Lecture Notes in Computer Science, pp. 462–470, Springer-Verlag, 2001.
52. F. Gavril, "The Intersection Graphs of Subtrees in Trees Are Exactly the Chordal Graphs", *Journal of Combinatorial Theory (B)*, vol. 16, pp. 47–56, 1974.
53. M.C. Golumbic, "Algorithmic Graph Theory and Perfect Graphs", Second Edition, Elsevier Publishers, 2004.
54. D.S. Johnson, M. Yannakakis and C.H. Papadimitriou, "On Generating All Maximal Independent Sets", *Information Processing Letters*, vol. 27, pp. 119–123, 1988
55. W. Kishimoto, "A Method for Obtaining the Maximum Multi-Route Flows in a Network", *Networks*, vol. 27, no. 4, pp. 279–291, 1996.
56. J.Y.T. Leung, "Fast Algorithms for Generating All Maximal Independent Sets of Interval", Circular Arc and Chordal Graphs, *Journal of Algorithms*, vol. 5, pp. 22–35, 1984.
57. C.L. Monma and V.K. Wei, "Intersection Graph of Paths in a Tree", *Journal of the Combinatorial Theory*, Series B, vol. 41, pp. 141–181, 1986.
58. J.W. Moon and L. Moser, "On Cliques in Graphs", *Israel Journal of Mathematics*, vol. 3, pp. 23–28, 1965.
59. J.M. Nielson, "On the Number of Maximal Independent Sets in Graphs", BRICS Report Series RS-02-15, Department of Computer Science, University of Aarhus, Denmark, ISSN 0909-0878; http://www.brics.dk
60. S. Tsukiyama, M. Ide, H. Ariyoshi and I. Shirakawa, "A New Algorithm for Generating All Maximal Independent Sets", *SIAM Journal of Computing*, vol. 6, pp. 505–517, 1977.
61. E.W. Zegura, K.L. Calvert and S. Bhattacharjee, "How to model an internetwork", *Proc. IEEE Infocom*, pp. 595–602, 1996.
62. A. Akella, S. Seshan and A. Shaikh, "An Empirical Evaluation of Wide-Area Internet Bottlenecks", *In Proc. of IMC*, San Diego, CA, 2003.
63. S. Savage, T. Anderson, A. Aggarwal, D. Becker, N. Cardwell, A. Collins, E. Hoffman, J. Snell, A. Vahdat, G. Voelker and J. Zahorjan, "Detour: a case for informed Internet routing and transport", *IEEE Micro,* vol(19), Jan 1999.
64. S.Z.S.S.I. Stoica, D. Adkins and S. Surana, "Internet indirection infrastructure," *in Proceedings of ACM SIGCOMM 2002,* August 2002.
65. C. Perkins, "IP encapsulation within IP," *IETF RFC 2003,* October 1996.
66. N. Feamster, D. Anderson, H. Balakrishnan and M. Kaashoek, "Measuring the Effects of Internet Path Faults on Reactive Routing," *In Proc. of ACM SIGMETRICS,* San Diego, CA, June 2003.
67. N. Spring, R. Mahajan, and D. Wetherall, "Measuring ISP topologies with Rocketfuel," *In Proc. ACM SIGCOMM,* Aug. 2002.

Chapter 12
Optimized Design of Large-Scale Social Welfare Supporting Systems on Complex Networks

Jaroslav Janáček, Ľudmila Jánošíková, and Ľuboš Buzna

Abstract Our contemporary societies are supported by several systems of high importance providing large-scale services substantial for citizens everyday life. Typically, these systems are built or rely on various types of complex networks such as road networks, railway networks, electricity networks, communication networks etc. Examples of such systems are a set of emergency medical stations, fire or police stations covering the area of a state, social or administration infrastructure. The problem of how to design these systems efficiently, fairly, and reliably is still timely and it brings along many new research challenges. This book chapter presents a brief survey of optimization models and approaches applicable to the problem. We pay special attention to the methods based on the branch and bound principle and show how their computational properties can be improved. Furthermore, we discuss how some of these models can be rearranged in order to allow using the existing solving techniques as approximative methods. The presented numerical experiments are conducted on realistic data describing the topology of the Slovak road network. On the one hand, we hope that this chapter can come handy to researchers working in the area of complex networks, as it presents efficient methods to design public service systems on the networks. On the other hand, we can picture the benefits potentially resulting from the knowledge of the network properties and possibly being utilized in the algorithms design.

Keywords Public service systems • Large-scale emergency systems

---

J. Janáček • Ľ. Jánošíková • Ľ. Buzna (✉)
Faculty of Management Science and Informatics, Department of Transportation Networks, University of Žilina, Univerzitná 8215/1, 010 26 Žilina, Slovak Republic
e-mail: Jaroslav.Janacek@fri.uniza.sk; janosik@frdsa.uniza.sk; buzna@frdsa.uniza.sk

M.T. Thai and P.M. Pardalos (eds.), *Handbook of Optimization in Complex Networks: Theory and Applications*, Springer Optimization and Its Applications 57,
DOI 10.1007/978-1-4614-0754-6_12,

12.1 Introduction to Public Service Systems

The topology of many real-world systems can be described as a network. In some cases, the system organization and the network topology result from some long-term evolution processes directed by billions of interactions [15, 28, 39]. In other cases this is given by human-controlled decision-making activities, supported by sophisticated optimization methods [1, 18, 42]. In this book chapter, we revise some of the techniques that can be applied to the problem of designing the public service systems operating on all sorts of transportation networks. We focus mainly on road networks, where the typical applications are locating emergency medical centers, police or fire stations [38, 41] or other types of social and administration infrastructures [19, 27]. Nevertheless, these approaches can also be applied to many other man-made systems such as railway networks, telecommunication networks [37], power grid networks or gas networks [34, 40].

The road network is the example of the spatial complex network featuring many non-trivial topological properties [31]. So far, it is relatively little known how these properties influence the efficiency of optimization algorithms. For location problems it is, for example, known that the corresponding solving methods perform much more efficiently on real-world networks than on their random counterparts [30]. The explanation is in the spatial structure of the underlying network, which generates suitable diversity in cost coefficients. This influences the structure of lower and upper bounds and finally results in shorter computational time.

The road network is a substrate for the public service systems. These systems provide various kinds of services to inhabitants of a certain geographical area. The services can include goods delivery, presence of some necessary facilities or they can be some kind of civil services, such as medical care, fire-fighting or house maintenance. Contrary to the private service systems, none of the demands listed above can be ignored. Hereafter, a serviced person or a group of serviced people, characterized by their demand and located within the same municipality, will be referred to as a customer. The customer's demand will be characterized by its weight. If a customer is a group of people, the weight can be proportional to the number of the people in the given group.

Even if the serviced population is concentrated in municipalities, the number of municipalities is usually too large to have a source of the services at each customer's location. That is why placing the sources needs to be really thought through. Hereby, we assume that there is only a finite set of possible service center locations within the serviced area. Nevertheless, the set can be very large, e.g. it can be composed of all municipalities. Therefore, if an appropriate number of service center locations is to be found, a rather large combinatorial problem arises.

When a public service system is designed, two essential types of objectives, namely the economic efficiency and the fairness, should be taken into account [9]. The economic efficiency can be measured through how much the system and its operation cost. The costs of a service system usually consist of fixed expenses paid for opening and operating service centers, and of the cost of transportation related to the serving the customer [13].

This scheme applies only if the system provider serves customers at their locations. However, some systems can provide services only at the service centers. In this case, a customer has to travel to the service center, and obviously, the travel cost is not included in the system costs. The servicing costs are often considered as being proportional to the network distance between the customer and the service center.

Fairness or welfare in the public systems design is related to the notion of justice perceived by the customers resulting from the distribution of limited resources. This topic has been extensively studied, notably in social sciences, welfare economics, and engineering. The overview of basic concepts of fairness can be found in the reference [4]. The customer's welfare is difficult to quantify. Therefore, one possibility is to convert it into a customer's discomfort, expressing the accessibility of a service. The discomfort of an individual customer can be, for example, estimated by considering the distance or travel time required to get from the customer's residence to the nearest service center. When we use this approach, several possibilities emerge. The discomfort can be expressed directly as the distance or only as a part of the distance which exceeds a given accepted real valued threshold $D_{\max}$. The simplest option is to utilize the information whether the distance is longer than $D_{\max}$. If the distance is longer or equal to $D_{\max}$, then the customer's discomfort is considered to be affected.

Next, two extreme cases, the average customer discomfort or the worst case of customer's discomfort, can be considered as possible criteria expressing the quality of the proposed design. In the first case, the estimations of individual discomforts are summed up, and the resulting design minimizes the total discomfort. The second criteria (sometimes denoted as Rawlsian criterion) measures the quality of the design by the discomfort perceived by the worst positioned customer [36]. This approach can be extended by repeating the minimization for the second, third, fourth, etc., worst positioned customer. This extension has been studied throughly in the context of flow networks, where it is known as the max-min fair allocation of flows [33], however, only little attention has been paid to it in the context of location problems [10]. Another possibility, known as the proportional fairness, is based on the minimization of the utility function, which is the sum of logarithms of individual discomforts [35].

If the system costs and the customer's welfare are defined, it is necessary to decide which type of the public service system design should be preferred. Is it the cost-oriented design or the customer's welfare-oriented design? The cost-oriented design searches for the system optimum, minimizing the system costs, while assuring the desired level of the welfare. The welfare-oriented design, on the other hand, minimizes the customer's discomfort subject to a constraint keeping the system costs under a given limit. Both approaches can be combined with an arbitrary evaluation criteria for either cost or discomfort.

In Sect. 12.2, we show some basic examples of how to formulate costs and discomfort mathematically. With each individual case we demonstrate how to build mathematical models which can be then solved by optimization tools. In Sect. 12.3, as examples of solving methods, we briefly review the available universal solvers and exact algorithms to solve the uncapacitated facility location problem. To conclude, in Sect. 12.4, we describe a case study where we examine the network of emergency medical service (EMS) centers operating in the Slovak Republic.

12.2 Modeling Approaches to Designing Public Service Systems

Let us assume that the serviced area consists of the municipalities located at the nodes of the graph $G(N,E)$, where N is a set of nodes and E is a set of edges. The municipalities forming the finite set $J \subseteq N$ are considered as customers. The demand or the number of inhabitants living in the node $j \in J$ is denoted as b_j. The service system design problem can be reduced to the task to decide where to locate the centers within the large finite set I of possible locations, so that the value of the chosen criterion is minimal.

The shortest path length between the nodes $i \in N$ and $j \in N$ is denoted as d_{ij}, and the associated travel time as t_{ij}. The fixed charge f_i is paid to locate the service center at the node i, whereas the costs to satisfy the j-th customer from the service center i can be expressed as $(ed_{ij} + g_i)b_j$, where g_i are the costs spent to satisfy one unit of the demanded volume, and e are the travel costs per one unit of volume and one unit of distance.

Let $s(j)$ be the index of the center which is the closest to the node j, considering either the time or the distance, and which belongs to the set $I_1 \subseteq I$ of nodes where a service center is located. Then, the total system costs can be expressed as:

$$\sum_{i \in I_1} f_i + \sum_{j \in J} (ed_{s(j),j} + g_{s(j)})b_j. \tag{12.1}$$

Here, we ask the question: How could we estimate the discomfort of the customers? Let j be the customer's location and $s(j)$ be the nearest service center as defined above. If the individual customer's discomfort $u_{s(j),j}$ can be expressed as being linearly proportional to the distance between the customer j and the nearest service center, then $u_{s(j),j}$ is equal to $d_{s(j),j}b_j$, considering b_j as the weight. In the case, when only the part of the distance exceeding the given threshold $D_{\max}$ is considered as being proportional to the discomfort perceived by the customer, then $u_{s(j),j}$ is equal to $(d_{s(j),j} - D_{\max})b_j$ subject to $d_{s(j),j} \geq D_{\max}$, and otherwise the discomfort $u_{s(j),j}$ is equal to zero. Whenever only the information whether the distance is longer than $D_{\max}$ is used, the customer's discomfort $u_{s(j),j}$ equals b_j subject to $d_{s(j),j} \geq D_{\max}$; otherwise, the discomfort $u_{s(j),j}$ is equal to zero.

Equivalent definitions of customer's discomfort can be used for travel time $t_{s(j),j}$ and the threshold value $T_{\max}$.

If the average discomfort is considered to be an appropriate measure, then the objective function expressing the discomfort experienced by the population can be expressed by the formula (12.2). Alternatively, if the worst customer's discomfort is some more characteristic quantity, then the expression (12.3) is used instead, to describe the general discomfort affecting the serviced population.

$$\sum_{j \in J} u_{s(j),j} \tag{12.2}$$

$$\max\{u_{s(j),j} : j \in J\} \tag{12.3}$$

To formulate the mathematical model, both the set $I_1 \subseteq I$ and assignments $s(j)$ are expressed as decision variables. Each possible location $i \in I$ is subject to the decision whether to provide a service center or not. This decision can be modeled by the variable y_i, which takes the value of one if a service center is located at the node $i \in I$ and the value of zero otherwise. The assignment $s(j)$ representing the center i to be assigned to the customer j is expressed by the zero-one decision variable z_{ij}. The decision variable z_{ij} takes the value of one, if $i = s(j)$.

Using the variables y_i and z_{ij} and the substitution $c_{ij} = (ed_{ij} + g_i)b_j$, expressions (12.1), (12.2), and (12.3) can be rewritten as expressions (12.4), (12.5), and (12.6), respectively:

$$\sum_{i \in I} f_i y_i + \sum_{i \in I} \sum_{j \in J} c_{ij} z_{ij} \tag{12.4}$$

$$\sum_{i \in I} \sum_{j \in J} u_{ij} z_{ij} \tag{12.5}$$

$$\max\left\{\sum_{i \in I} u_{ij} z_{ij} : j \in J\right\} \tag{12.6}$$

The non-linearity of the expression (12.6) can be removed by adding a new variable which is minimized and by adding a set of constraints ensuring that this variable takes the values greater or equal to the discomfort perceived by each customer.

In this subsection, we revised the basic forms of the objective functions used when designing public service systems. For more systematic overviews we refer the reader to the references [6, 16, 22].

12.2.1 Cost-Oriented Service System Design

The cost-oriented design leads to the system, which minimizes the system cost, assuring a certain level of the welfare. The structure of the resulting system is described by the above-introduced variables y_i, which determine the nodes, at which the service centers are located. The quality criterion corresponds, for example, to the expression (12.4). In addition, some consistency constraints must be now imposed on the decision variables. With each customer j, only a single allocation variable

z_{ij} is allowed to take value of one, and, furthermore, if the variable z_{ij} takes the value of one, then the associated variable y_i must take the value of one as well. If no other restriction on the customers' welfare is put except being served from a service center, then the model of the cost-oriented system design can be stated as follows:

$$\text{Minimize} \sum_{i \in I} f_i y_i + \sum_{i \in I} \sum_{j \in J} c_{ij} z_{ij} \tag{12.7}$$

$$\text{subject to} \sum_{i \in I} z_{ij} = 1 \quad \text{for } j \in J \tag{12.8}$$

$$z_{ij} \leq y_i \quad \text{for } i \in I,\ j \in J \tag{12.9}$$

$$y_i,\ z_{ij} \in \{0,\ 1\} \quad \text{for } i \in I,\ j \in J \tag{12.10}$$

The constraints (12.8) ensure that each municipality (customer) is assigned to exactly one location. Whenever the customer j is assigned to the location i, the constraints (12.9) (so called linking constraints) force the variable y_i to take the value of one. This problem, introduced in [2], is known as the uncapacitated facility location problem (UFLP), or the simple plant location problem. This problem is NP-hard [21] and its properties and solving techniques are broadly discussed, for example, in [3, 32].

If a certain level of welfare is supposed to be preserved, the model (12.7) – (12.10) is supplemented with the conditions that keep the discomfort described by the expression (12.6) below a given threshold $U_{\max}$. Thus, for this purpose, either the set of constraints (12.11) or (12.12) is added to the model (12.7)–(12.10).

$$\sum_{i \in I} u_{ij} z_{ij} \leq U_{\max} \quad \text{for } j \in J \tag{12.11}$$

$$u_{ij} z_{ij} \leq U_{\max} \quad \text{for } i \in I,\ j \in J \tag{12.12}$$

Note that the extended model can be easily turned into the original uncapacitated facility location problem by redefining the cost coefficients c_{ij}. It can be done for example by setting c_{ij} to sufficiently high value, whenever the inequality $u_{ij} \leq U_{\max}$ does not hold. In case the total costs of the designed system do not include the costs c_{ij}, then much simpler model can be formulated instead. As the relation between the customer j and the center location i becomes irrelevant, with the exception of the constraints (12.12), we can even avoid using the allocation variables z_{ij}. We define the coefficients a_{ij} so that a_{ij} takes the value of one if and only if the inequality $u_{ij} \leq U_{\max}$ holds, and otherwise it takes the value of zero. We use the zero-one location variables y_i as before, and formulate the following model:

$$\text{Minimize} \sum_{i \in I} f_i y_i \tag{12.13}$$

$$\text{subject to} \sum_{i \in I} a_{ij} y_i \geq 1 \quad \text{for } j \in J \tag{12.14}$$

$$y_i \in \{0,\ 1\} \quad \text{for } i \in I \tag{12.15}$$

The constraints (12.14) are satisfied, if at least one service center is located in the neighborhood of the customer j, so that the discomfort constraint $u_{ij} \leq U_{\max}$ is met. The problem (12.13)–(12.15), known as the set covering problem, was for the first time formulated in [41]. The problem is NP-hard [29] and its solving algorithms are over-viewed, for instance, in [11].

12.2.2 Customer's Welfare-Oriented Service System Design

The welfare-oriented design aims at minimizing the customer's discomfort, provided the system costs do not exceed the given level $C_{\max}$. Also, in this case, the structure of the designed system is described by the location variables y_i and allocation variables z_{ij}. The quality criterion expressing the average customer's welfare could correspond to the expression (12.5). The consistency constraints imposed on the decision variables y_i and z_{ij} are the same as before. A simple model illustrating the system design oriented at the average customer's welfare can be stated as follows:

$$\text{Minimize} \sum_{i \in I} \sum_{j \in J} u_{ij} z_{ij} \tag{12.16}$$

$$\text{subject to} \sum_{i \in I} z_{ij} = 1 \qquad \text{for } j \in J \tag{12.17}$$

$$z_{ij} \leq y_i \qquad \text{for } i \in I,\ j \in J \tag{12.18}$$

$$\sum_{i \in I} f_i y_i + \sum_{i \in I} \sum_{j \in J} c_{ij} z_{ij} \leq C_{\max} \tag{12.19}$$

$$y_i, z_{ij} \in \{0, 1\} \qquad \text{for } i \in I,\ j \in J \tag{12.20}$$

This problem can be solved directly, using a universal optimization tool. However, if the sets I and J are too large, then the problem can be rearranged as the uncapacitated facility location problem, using the Lagrangian relaxation. If the positive Lagrangian multiplier λ is introduced, and we apply the relaxation to the constraint (12.19), we obtain a new objective function:

$$\begin{aligned} &\sum_{i \in I} \sum_{j \in J} u_{ij} z_{ij} + \lambda \left(\sum_{i \in I} f_i y_i + \sum_{i \in I} \sum_{j \in J} c_{ij} z_{ij} - C_{\max} \right) \\ &= \sum_{i \in I} \lambda f_i y_i + \sum_{i \in I} \sum_{j \in J} (u_{ij} + \lambda c_{ij}) z_{ij} - \lambda C_{\max} \end{aligned} \tag{12.21}$$

For the given value of λ we can minimize the expression (12.21) subject to the constraints (12.17), (12.18), and (12.20) as an instance of the uncapacitated facility location problem. Then the optimal or the near-optimal solution of the original problem (12.16)–(12.20) can be obtained by iterative process, where the bisection method is used to find such value of λ that the resulting solution $(\mathbf{y}, \mathbf{z})$ fulfills the constraint (12.19) as the equality with an arbitrary precision.

The same approach can be applied when the total costs do not include the costs c_{ij}. Thus, the constraint (12.19) is replaced by simpler constraint (12.22). If the parameter p limits the number of the located service centers instead of constraining the costs, then the constraint (12.19) can be replaced by the constraint (12.23).

$$\sum_{i \in I} f_i y_i \leq C_{\max} \tag{12.22}$$

$$\sum_{i \in I} y_i \leq p \tag{12.23}$$

If the discomfort is expressed as $u_{ij} = b_j$ for $d_{ij} > D_{\max}$ and $u_{ij} = 0$ otherwise, then the instance of the set covering model can be formulated. The coefficients a_{ij} are defined as above, i.e. a_{ij} takes the value of one, if and only if the inequality $d_{ij} \leq D_{\max}$ holds, otherwise it takes the value of zero. Also, the location variables y_i are defined as before and, in addition, the auxiliary variables x_j are introduced for each customer $j \in J$. It is expected that the variable x_j takes the value of one, if there is no located service center within the radius $D_{\max}$ from the customer j. Then the following set covering model describes the welfare-based system design problem:

$$\text{Minimize} \sum_{j \in J} b_j x_j \tag{12.24}$$

$$\text{subject to} \sum_{i \in I} a_{ij} y_i \geq 1 - x_j \quad \text{for } j \in J \tag{12.25}$$

$$\sum_{i \in I} f_i y_i \leq C_{\max} \tag{12.26}$$

$$y_i \in \{0,\ 1\} \quad \text{for } i \in I \tag{12.27}$$

$$x_j \in \{0,\ 1\} \quad \text{for } j \in J \tag{12.28}$$

The term (12.24) expresses the overall volume of uncovered demands which is minimized. The constraints (12.25) ensure that the variables x_j take the value of one, if and only if there is no service center located within the radius $D_{\max}$ from the customers location j. The constraint (12.26) puts the limit $C_{\max}$ on the system costs. More complicated situation arises if the welfare-oriented system is designed and the discomfort of the worst positioned customer is used as the quality criterion. The model of this type assumes the following form:

$$\text{Minimize } h \tag{12.29}$$

$$\text{subject to} \sum_{i \in I} u_{ij} z_{ij} \leq h \quad \text{for } j \in J \tag{12.30}$$

$$\sum_{i \in I} z_{ij} = 1 \quad \text{for } j \in J \tag{12.31}$$

$$z_{ij} \leq y_i \quad \text{for } i \in I, j \in J \tag{12.32}$$

$$\sum_{i\in I} f_i y_i + \sum_{i\in I}\sum_{j\in J} c_{ij} z_{ij} \leq C_{\max} \tag{12.33}$$

$$y_i, z_{ij} \in \{0,\ 1\} \quad \text{for } i \in I, j \in J \tag{12.34}$$

The transformation of the model (12.29)–(12.34) either into the uncapacitated facility location problem or into the set covering problem is possible, although only under very strong and simplifying assumptions on both customers' discomforts and cost coefficients.

The family of the above reported models can comprise more complex service system design problems than we have explained. Thus, it is often necessary to add some additional constraints which can capture restrictions such as limited capacity of service centers, capacity and time constraints on the communication links (roads) etc. Another issue is how to find a reasonable trade-off between various objectives. For example, what should be preferably minimized the total discomfort of all customers or the worst individual discomfort? As shown above, each of these criteria leads to a specific objective function. The criterion of the total discomfort is described by the term (12.16), and the criterion of the worst situated customer discomfort is described by the term (12.29). To a certain extent, these two approaches can be combined in one optimization process as the following example shows. Let us introduce the real valued threshold $U_{\max}$, and define the penalty P as

$$P = \sum_{j\in J} \max\{u_{ij} : i \in I\}. \tag{12.35}$$

Then we define the new disutility $\underline{u_{ij}}$ so that $\underline{u_{ij}} = u_{ij}$ if $\underline{u_{ij}} \leq U_{\max}$ and $u_{ij} = P$ otherwise. The optimal solution of the problem described by (12.16)–(12.20) with the new coefficients $\underline{u_{ij}}$ does not allow to assign a customer to a service center, if it should cause bigger discomfort than $U_{\max}$, under the assumption that such a solution exists. Using this construction, an optimization process can be formulated. The process starts with a big value of the threshold $U_{\max}$, and repeatedly solves the problem (12.16)–(12.20) for smaller and smaller $U_{\max}$ until the value of (12.16) exceeds penalty P. The last but one solution is a good compromise between these two approaches. Another option is to combine different objectives into one model whereby a multi-objective problem is created.

12.3 Solving the Uncapacitated Facility Location Problem to Optimality

As we have demonstrated in the previous section, the UFLP can be often either directly or indirectly used to tackle the public service system design problem. For this reason we describe in this section the methods enabling to solve this problem efficiently. There are usually two options of solving the optimization problem derived from the real-world situation. We might either choose a specialized algorithm restricted to the problem, which almost always requires the implementation of the algorithm using a convenient programming language, or, if the problem can be seen as the instance of a standard class of optimization problems, we can use a universal optimization tool.

12.3.1 Universal Optimization Tools

A universal optimization tool is a software package designed to solve standard classes of optimization problems, such as linear programming problems, linear integer programming problems or quadratic programming problems. Typically, such a tool consists of a modeler which allows to separate the mathematical model from the data, describing the particular instance of the optimization problem. Thus, the formulation of the mathematical model is independent on parameter values, which makes it easier to maintain both, the model and the data. Nowadays, the most popular tools are CPLEX (http://www.cplex.com), XPRESS-MP (http://www.fico.com) or MOSEK (http://www.mosek.com).

Modern optimization packages offer programming languages to simplify the work with mathematical models, debugging environments and interfaces for graphical output. Moreover, they can be embedded into development environments, such as AIMMS (http://www.aimms.com). To supplement the results provided by specialized algorithms, presented in the next section, we compare them with the results obtained by XPRESS-MP.

12.3.2 Specialized Algorithms

As we have already mentioned in Sect. 12.2, the uncapacitated facility location problem is broadly applicable in the design of public service systems. The corresponding solving technique can be used not only to design the cost-optimal two-levels distribution system [24] but it can be extended in order to solve more complex location problems. As it was discussed in the reference [20], it is possible to turn the maximum distance problem, the maximum covering problem and the p-median problem into the form of the UFLP. Also, the capacitated version of the UFLP or the p-center problem can be approximatively solved using the solving algorithms for UFLP [26].

Many scholars have dealt with UFLP [14]. Nevertheless, as far as the exact algorithms are concerned, the now seminal procedure DualLoc proposed by Donald Erlenkotter [17] is still one of the most efficient methods, and it enables to find an optimal solution in tractable computational times [12]. Inspired by this approach in [30], Manfred Koerkel proposed several successful modifications, which speed up the solving process and the resulting algorithm was named PDLoc.

In [25], we have shown how both algorithms, DualLoc and PDLoc, can be accelerated by modifying the famous procedure of dual ascent (DA) introduced in [5]. We implemented both algorithms using the integrated programming development environment Delphi. In our implementation, we restricted the values of coefficients f_i and c_{ij} to integers. For historical reasons, from now on we will refer to our implementation of the DualLoc algorithm as to the BBDual algorithm [23].

12.3.2.1 Algorithm BBDual

If the variables y_i are known, the optimal values of the variables z_{ij} can be found easily. It is sufficient to assign the customer j to the facility i, for which the value of coefficient c_{ij} is minimal. Thus, the most difficult problem is to determine the setting of the variables y_i. The BBDual algorithm is based on the branch and bound method, in which two subproblems emerge by fixing the variables y_i either to zeros or ones. The algorithm uses of the depth-first strategy. To decide if a given subproblem should be processed or excluded from the searching process, a lower bound of high quality is needed. Such lower bound can be obtained by successively performing the dual ascent (DA) and the dual adjustment algorithms (DAD). The former, the DA procedure, starts from an arbitrary feasible solution of the dual problem, and it subsequently increases the value of its objective function. The latter procedure enables a further improvement of the dual solution by searching for a configuration in which a small decrease of the objective function will allow larger increase. These two procedures provide dual feasible solution and the corresponding value of the objective function serves as the lower bound. Furthermore, a corresponding primal feasible solution is generated. This is done by the PRIMA procedure, which follows the complementary conditions holding between an LP relaxation of the problem (12.8)–(12.10) and its dual. The best-found primal solution is stored and its objective function value constitutes the upper bound for the optimal solution.

12.3.2.2 Algorithm PDLoc

Algorithm PDLoc comprehends a number of effective modifications and improvements of procedures originally proposed by Erlekotter, and in addition, it is enhanced by several new procedures. Similarly to the BBDual algorithm, the PDLoc employs the branch and bound method to determine the optimal solution, but in contrast to the BBDual, the strategy of the lowest lower bound is used to process the searching tree. Varying the order of customers in the PRIMA procedure enables to open new locations, and to explore a broader spectrum of primal solutions. This leads to faster decrease of the upper bound and to faster termination of the searching process.

The number of steps in the incremental build-up process that is used to construct a dual solution by the DA procedure depends on the gap between the f_i and c_{ij} values. Therefore, more rapid incrementation leads to considerable improvements in the cases when the fixed charges f_i are considerably higher than the allocation costs c_{ij} are. This is ensured by using the dual multi-ascent procedure (MDA) instead of the DA procedure.

Another improvement results from applying the simple exchange heuristic right after the first primal solution is found. Moreover, if a large difference between the upper and the lower bound occurs, it is reduced by the modified dual adjustment procedure. This procedure consists of two phases. The first of them is called the primal-dual multi adjustment (PDMAdj) and the second is the primal-dual adjustment (PDAdj). In a loop, both procedures alternatively decrease and increase the reached lower bound in order to find a combination of operations which would allow to increase the lower bound. The difference between those two is in the scheme that is used to decide which and how many variables are modified in one step.

The used branch and bound searching scheme allows to fix the selected locations to be permanently open ($y_i = 1$) or closed ($y_i = 0$) and thereby to reduce the size of the solved problem. To fix a variable, special conditions have to be met. The evaluation of the conditions is time consuming, especially, if the searching process is nested deeply in the searching tree. Therefore, fixing variables is preferably done on the top of the searching tree (pre-processing). If the processed branch is far down from the root, the variables are fixed only if there is a chance to fix a couple of them simultaneously.

12.3.2.3 Benchmarks

Benchmarks that we used to test our implementation of algorithms were derived from the real-world network. The set G700 consists of 700 problems derived from the road network of the Slovak Republic. This set includes ten subgroups the sizes of which range from $100 \times 2{,}906$, $200 \times 2{,}906$, to as large as $1{,}000 \times 2{,}906$. The first number is the number of candidates for a facility location ($|I|$), and the second number refers to the size of the customer set ($|J|$). Each subgroup contains 70 benchmarks. For each size of the benchmark ten different random subgraphs of the road network graph of corresponding size were generated. Each subgraph was used as a base for creating seven benchmarks by modifying the coefficients c_{ij} and f_i to cover evenly the whole spectrum of the centers located by the optimal solution.

For instance, for the problem of size $100 \times 2{,}906$ the optimal numbers of located facilities were 1, 17, 33, 50, 66, 83 and 100, respectively. We will provide the source code of the algorithms upon request. Our benchmarks were uploaded onto the supplementary Web page http://frdsa.uniza.sk/~buzna/supplement.

12.3.2.4 Preliminary Experiments

We solved all problems to get the frequency in which the particular procedures are executed. These numerical experiments were performed on a PC equipped with Intel 2.4 GHz processor and 256 MB RAM. The average computational time distributed among the inner procedures is shown in Fig.12.1.

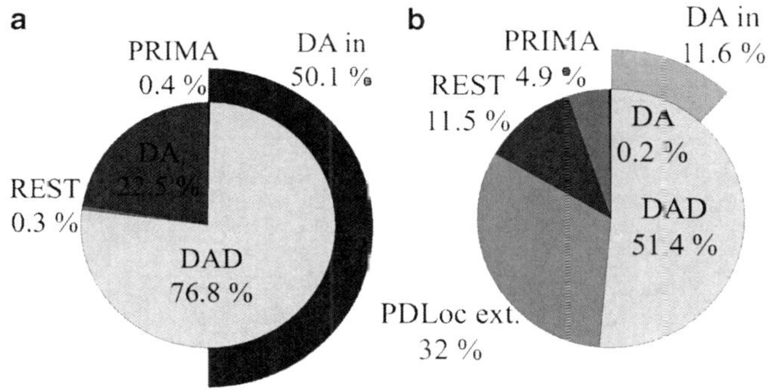

Fig. 12.1 The average distribution of computational time among the procedures of the BBDual algorithm in (**a**) and the PDLoc algorithm in (**b**) (these results were obtained for the set G700)

The abbreviations "DA," "DAD," and "PRIMA" denote the relative average time taken by the corresponding procedure. The "REST" includes the time consumed by the branch and bound method, lower bound computation, as well as necessary memory operations. "PDLOC ext." represents the time spent on running the procedures which are exclusively included only in the PDLOC algorithm. The label "DA in" stands for the time taken by the procedure DA, which was called from other procedures (e.g. the DA procedure is called inside the DAD procedure). This also explains why we plotted this value outside the pie graph. The results clearly show that the BBDual algorithm took in average 72.6% by performing the DA procedure, while the PDLOC algorithm devoted only 11.8% of the computational time to this procedure. The time consumed by the DAD procedure is approximately equal with both algorithms, although its distribution among the time consuming activities differs considerably.

In the case of the BBDual algorithm, the DAD procedure took 50.1% of the time. On the contrary, the PDLoc algorithm needs only a small portion of the time to perform the DA procedure nested in the DAD procedure. This implies that the PDLoc algorithm focuses more on intensive searching for improving operations. More detailed comparison of the computational performance can be found in Table 12.1.

12.3.2.5 Amendments of Algorithms BBDual and PDLoc

Amendments of algorithms BBDual and PDLoc were inspired by preliminary experiments, which showed that the DA procedure consumes large portion of the computational time. In order to describe the procedure DA, we need to consider the following form of the dual problem [30], derived from the LP relaxation of the primal problem (12.7)–(12.10).

$$\text{Maximize } z_D = \sum_{j \in J} v_j \tag{12.36}$$

Table 12.1 Average time in seconds and corresponding standard deviation obtained for benchmarks G700

	BBDual		BBDual*		PDLoc		PDLoc*	
Size of problems	t[s]	Std D	t[s]	Std D	t[s]	Std D	t[s]	Std D
100 × 2,906	5.343	12.64	0.28	0.29	1.44	0.93	0.77	0.34
200 × 2,906	27.16	68.33	0.41	0.39	1.80	0.99	0.87	0.22
300 × 2,906	52.95	143.11	0.74	0.64	2.40	1.48	1.20	0.90
400 × 2,906	127.06	337.58	1.01	0.47	2.74	1.82	1.46	0.88
500 × 2,906	134.17	340.52	1.75	1.05	5.29	6.52	2.83	0.90
600 × 2,906	277.59	700.73	2.54	1.78	5.21	5.54	3.64	2.88
700 × 2,906	277.70	704.57	3.90	2.77	6.12	5.45	4.63	4.38
800 × 2,906	497.26	1,248.42	5.07	4.23	8.56	8.87	6.45	6.35
900 × 2,906	640.44	1,652.65	7.24	6.31	11.45	11.25	8.89	8.83
1,000 × 2,906	644.88	1,595.72	7.07	5.89	10.60	11.19	7.47	8.08

Procedure DA(J^+)
While $|J^+| > 0$ do:
    Get next $j \in J^+$.
    Set $d_1 = min_i(s_i)$ with $i \in \{i : c_{ij} \leq v_j\}$.
    Set $d_2 = min_i(c_{ij} - v_j)$ with $i \in \{i : c_{ij} > v_j\}$.
    If $d_1 > d_2$, set $d = d_2$. Else, delete j from J^+ and set $d = d_1$.
    If $d > 0$, then:
        For each $i \in \{i : c_{ij} \leq v_j\}$, set $s_i = s_i - d$.
        Set $v_j = v_j + d$ and $z_D = z_D + d$.
Terminate.

Fig. 12.2 The original procedure DA as it was described in [17] or in [30]

$$\text{subject to} \sum_{j \in J} \max\{0, v_j - c_{ij}\} + s_i = f_i, \quad \text{for } i \in I. \tag{12.37}$$

$$s_i \geq 0 \quad \text{for } i \in I. \tag{12.38}$$

The original DA procedure (see Fig. 12.2) solves the problem (12.36)–(12.38) by processing the set of relevant customers J^+, customer by customer, in the order which follows the unordered input sequence. At each step the variable v_j corresponding to the customer $j \in J^+$ is incremented by the value d, whereas the value d is determined as the maximal value, which satisfies the constraints (12.6). Hence, it becomes obvious, that this variable cannot be increased in the followings steps, and the customer j can be excluded from the set J^+. This procedure is repeated until J^+ is emptied.

As we will demonstrate, the performance of this procedure depends on the ordering in which the set of relevant customers J^+ is processed. This drawback is illustrated by the example with two possible locations, i_1 and i_2, and three customers j_1, j_2 and j_3, as it is depicted in Fig. 12.3. The locations i_1 and i_2 are associated with two slack variables, s_{i1} and s_{i2}, respectively. The edge connecting the customer j

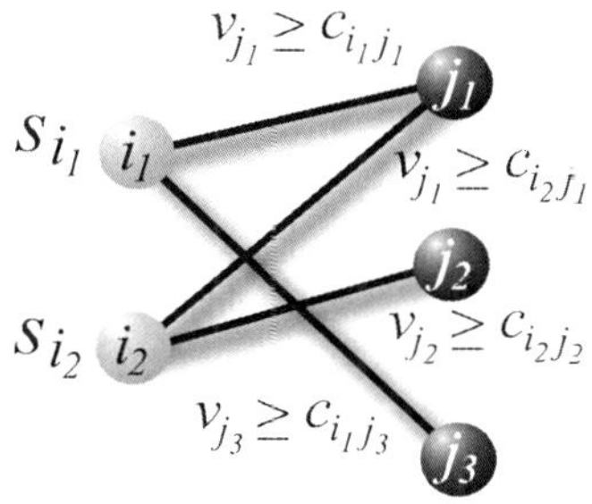

Fig. 12.3 The scenario where the ordering of customers improves the efficiency of the DA procedure. *White-colored* nodes represent the candidates for the facility location and *black* nodes are customers. The variables s_{i1} and s_{i2} are slack variables (see the constraints (12.37)) and v_j are dual variables corresponding to customers. The location i is connected with the customer j by the edge only if the inequality $v_j \geq c_{ij}$ holds

Procedure DA*(J$^+$)
Order the set J^* of relevant customers into the sequence $j_1, j_2, \ldots, j_k, \ldots, j_n$ ascendingly with respect to the cardinalities K_j.
For $k = 1$ to $|J^+|$ do:
  Set $j = j_k$.
  $d = \min\{\min\{s_i : i \in I, c_{ij} \leq v_j\}, \{c_{ij} - v_j : i \in I, c_{ij} > v_j\}\}$.
  If $d > 0$, then:
    Update s_i for $i \in I, c_{ij} \leq v_j$ by $s_i := s_i - d$. Set $v_j := v_j + d$ and $z_D = z_D + d$.
    If the cardinality of K_j has increased, then reorder the sequence of customers $j_k, \ldots, j_n$.
Terminate.

Fig. 12.4 Modified DA* procedure

with the location i symbolizes that the inequality $v_j \geq c_{ij}$ holds. By the symbol K_j we denote the cardinality of the set $\{i \in I : c_{ij} \leq v_j\}$ for the customer j. From Fig. 12.3 we get $K_{j_1} = 2$, $K_{j_2} = 1$ and $K_{j_3} = 1$.

The procedure DA (see Fig. 12.2) processes the set of customers J^+ in the order which is given by the sequence in which they enter the procedure. Thus, the first to be processed is the customer j_1 followed by j_2 and j_3. If the processing of the customer j_1 enables to increase the variable v_{j_1} by a value β, then the lower bound z_D is increased exactly by β. The maximal theoretical increase of the lower bound z_D is thus given by the sum of variables s_{i1} and s_{i2}. In the example from Fig. 12.3, the increment β has to be subtracted from both slack variables s_{i1} and s_{i2}, in order to meet the constraints (12.37). In summary, the theoretical capacity $s_{i1} + s_{i2}$ is reduced by 2β, in order to increase the lower bound z_D by β.

Considering the theoretical capacity and its possible decrease by modifying the variables v_j, we formulate a new DA* procedure [25] (see Fig. 12.4). This procedure exploits better the potential of increasing the lower bound. This approach is based on prioritizing those customers, which preserve better chances for the increase of the lower bound z_D in the following steps. We order the relevant customers J^+ ascendingly according to the cardinalities K_j. The benefit of this modification is demonstrated by the example in Fig. 12.3.

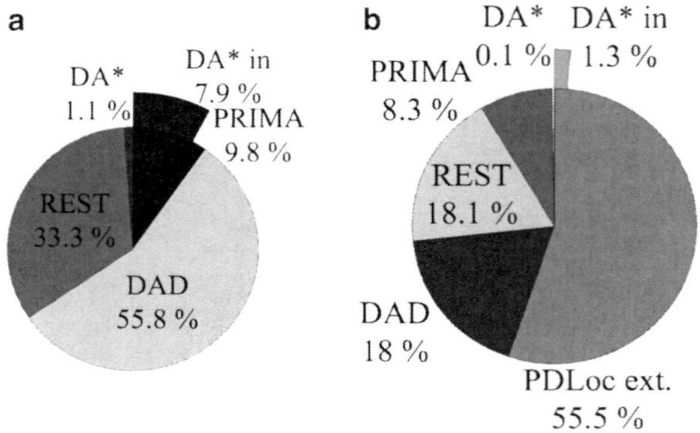

Fig. 12.5 The average distribution of computational time among the procedures of the BBDual* algorithm in (**a**) and the PDLoc* algorithm in (**b**) (these results were obtained for the set G700)

Having applied the above mentioned ordering of customers, they will be processed in the order j_2, j_3 and j_1. If processing the customer j_2 enables to increase the variable v_{j2} by β, then, considering the constraints (12.37), only slack variable s_{i2} has to be reduced by β. This way the theoretical capacity is reduced by the value β, and the lower bound z_D increases by β. Compared to the unordered case, the ordering may reduce the sum of slack variables s_i less than the original DA procedure does, while preserving the same growth rate for the variables v_j.

12.3.2.6 Verification of the Proposed Amendments

Verification of the proposed amendments was performed with the same set of benchmarks as described previously. We applied the new DA* procedure to both algorithms. We inserted the MDA procedure into the BBDual algorithm, as it had proved itself to be very efficient with the PDLoc algorithm when dealing with the cases in which the costs f_i are considerably higher than the costs c_{ij}. We also amended the evaluation of subproblems in the BBDual algorithm. Both incoming subproblems are solved simultaneously, and the most perspective subproblem is processed as the first. We would like to point out that we have tested all these modifications separately [25]. However, none of them brought any remarkable improvements compared to when applied together. To distinguish the original and the new versions of the algorithms BBDual and PDLoc, the modified version is denoted with the superscript "*."

The effects of the proposed modifications were extensively examined by numerical experiments [25]. Similarly, as in the preliminary experiments, we evaluated the average computational time and its distribution among the inner procedures. Figure 12.5 gives the evidence of the significant change in the time distribution of the inner procedures. The total usage of the DA procedure was reduced from

72.6% to 9% for the BBDual* algorithm and from 11.8% to 1.4% for the PDLoc* algorithm. These results also suggest that the new DA* procedure has a significant influence on the performance of both algorithms. Table 12.1 compares the results achieved by the original and modified versions of the algorithms. The results indicate that the proposed modification can save time considerably. Please note that using the MDA procedure in the BBDual algorithm contributed to this significant reduction, especially for the benchmarks where $f_i \gg c_{ij}$. However, the improvement brought by the new DA* is perceptible in the whole range of the parameter values. More details on this computational study can be found in the reference [25].

12.4 Case Study

In this section, we show how the choice of the particular criteria can influence the resulting design of the public service system. As an illustration example we use the system of EMS. More precisely, we present the optimized location of ambulance stations in the area of the Slovak Republic. Another purpose of the following text is to introduce improving modifications of the mathematical model resulting in better quality of the final design. When we were carrying out this study, the deployment of EMS stations was defined by the regulations of the Ministry of Health of the Slovak Republic Nos. 30/2006 and 365/2006. In accordance with these regulations, there were 264 EMS stations located in 223 cities and villages, including urban districts of the capital Bratislava and of the second largest Slovakian city Košice (see Fig. 12.6). They altogether served 2,916 municipalities populated with 5,379,455 inhabitants.

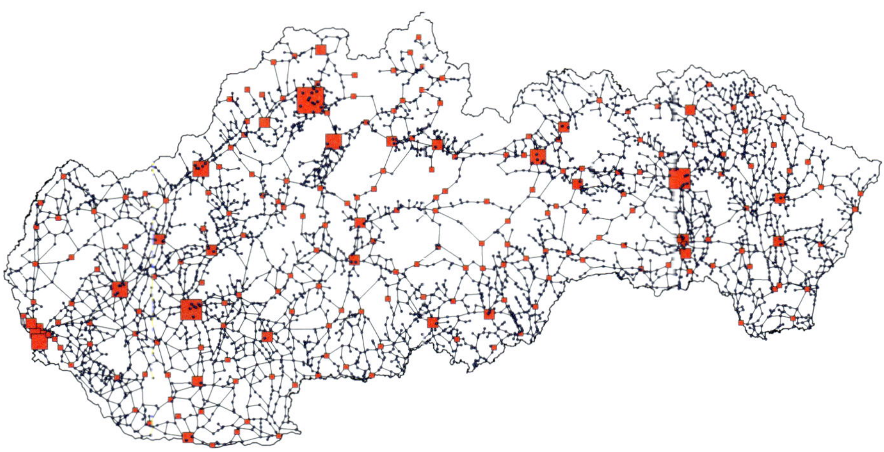

Fig. 12.6 Road network of the Slovak Republic. Locations of the existing EMS stations are marked with *squares*. The size of the *squares* is proportional to the number of the stations located in the given municipality

We evaluate four different criteria:

$$C1 : \sum_{j \in J} b_j t_{s(j),j}, \tag{12.39}$$

expressing the total traveling time from all customers to the closest EMS location weighted by the number of inhabitants b_j,

$$C2 : \sum_{\substack{j \in J \\ t_{s(j),j} > T_{\max}}} b_j (t_{s(j),j} - T_{\max}), \tag{12.40}$$

summing up the travel time exceeding the threshold value $T_{\max}$ weighted by the number of inhabitants b_j,

$$C3 : \sum_{\substack{j \in J \\ t_{s(j),j} > T_{\max}}} b_j, \tag{12.41}$$

representing the number of inhabitants which are further from the closest EMS than the threshold $T_{\max}$ states, and

$$C4 : \sum_{\substack{j \in J \\ t_{m(j),j} > T_{\max}}} b_j, \tag{12.42}$$

expressing the number of inhabitants which are covered by fewer than two EMS stations, where we define $m(j)$ as the index of the second closest EMS station to the customer j.

We combine the criterion $C1$ with the location–allocation type of model (12.7)–(12.10) and the criteria $C3$ and $C4$ with the set covering type of model (12.13)–(12.15). In all three cases we propose the deployment of 264 stations, which is optimal with respect to the particular criteria.

The set I of the candidate stations consists of the existing EMS stations defined by the official regulations, and of the municipalities with at least 300 inhabitants. Altogether we found 2,283 cities and villages meeting the conditions. Following the recommendation of the Ministry of Health to deliver medical care within 15 min from an emergency call, we regard $T_{\max} = 15$ min as the threshold value.

As it was indicated in Sect. 12.2, different criteria lead to different kinds of mathematical models, and therefore we also apply different solution techniques. The Location–allocation model was solved using the modified algorithm BBDual*, described in Sect. 12.3. The size and structure of the covering models are simpler, and they allow to use the universal optimization tool Xpress-MP (http://www.fico.com). The experiments were performed on a personal computer equipped with the Intel Core 26,700 processor with the following parameters: 2.66 GHz and 3 GB RAM. The computational time for the BBDual* algorithm was about 116 min. Xpress-MP managed to solve the covering models in 1.5 s and 0.3 s, respectively.

Table 12.2 Values of criteria calculated for three solutions obtained by the optimization (Location–allocation model, Set covering model, Double covering model) and for the existing design of EMS stations (Man-made)

	Criterion C1	Criterion C2	Criterion C3	Criterion C4	Number of changed EMS stations
Location–allocation model (C1)	13,677,537	35,109	15,613	285,324	104
Set covering model (C3)	23,613,635	180	91	188,937	207
Double covering model (C4)	25,597,594	182	91	11,684	199
Man-made	16,378,985	92,032	31,672	431,415	0

The last column shows the number of changed EMS locations proposed by the optimization compared to the existing design

Table 12.2 summarizes the values of the criteria (12.39)–(12.42) for the optimal solutions obtained by three selected models and for the current deployment. The results for the Location–allocation model with the criterion (12.39) are in the first row, the results for the Set covering model with criterion (12.41) are in the second row, and the results for the Double covering model (12.42) are listed in the third row. The Man-made row in the table corresponds to the current distribution of EMS stations over the area of the Slovak Republic.

Comparing the results of mathematical modeling to the current situation, we can observe that the first model, the Location–allocation model, is able to improve all of the defined criteria. Although this model minimizes the total travel time from ambulance locations to all potential patients, the best improvement is achieved by criterion C3, reflecting the discomfort of uncovered people. It means that the solution obtained by this model is more social, than the current system offers, in terms of customer's equity in the access to the provided service. As expected, the highest level of equity is assured for the set covering models They were able to locate the ambulances in such a way that almost all municipalities were covered. The only uncovered village was Runina with 91 inhabitants. This small village in the most eastern part of Slovakia has no candidate location within the radius of 15 min.

To achieve a more fair design, the covering models change substantially the current deployment of EMS stations (see the last column in Table 12.2). To implement these solutions would thus require to restructure the existing infrastructure considerably what might be too costly. If this issue is of importance, it can be practical to reduce the number of changed EMS stations by including the following constraint

$$\sum_{i \in I_0} (1 - y_i) \leq r, \tag{12.43}$$

where I_0 denotes the set of current center locations and r is the upper bound for the acceptable number of changes.

Let us look more closely at the results obtained by the Double covering model. In the solution presented in Table 12.2, there are all candidates for EMS stations equivalent regardless their population or the distance from a hospital. As a consequence of such simplification the model deploys, for example, only two stations in the capital Bratislava and none in the county capital Žilina. At first sight, this was not a reasonable design, which was also later confirmed by a computer simulation. To get a more realistic solution, the model needs to be modified. We add constraint $y_i \geq 1$ for the selected cities, to ensure that at least one station is located there. In further experiments, we apply this constraint to those cities that have a hospital.

The existing situation and the solution of the Double covering model were evaluated using computer simulations. The goal of the simulations is to assess how efficiently the system operates in real conditions, and to evaluate the parameters and properties that are not explicitly captured by mathematical models. From a customer's (patient's) point of view, the simulation should give answers to the following questions (evaluated separately for every municipality, as well as for the entire region):

1. What is the real coverage rate, i.e. what is the percentage of the calls processed within the required time limit?
2. What are the average and maximal waiting times for an ambulance arrival?

The analysis combining mathematical optimization with simple Monte Carlo simulations allows to verify the arrangements improving the system performance, such as the number of the ambulances allocated to a station. Simulating the EMS system can be viewed as a queuing system with the Poisson arrival of events and exponentially distributed service time [8]. Supposing that every station is equipped with one ambulance only, the system has as many service lines as the number of the stations is. Random events are the emergency calls that come from the populated areas. Every municipality has a specific arrival rate λ_j (calls per time unit). The statistics describing the number of calls for particular municipalities were not available to us. Therefore we used aggregated statistics for the Slovak Republic mapping the year 2001 (http://www.kezachranka.sk), reporting that the overall number of patients was 232,904. We calculated the rate λ per one inhabitant, and used it to estimate $\lambda_j = \lambda b_j$, where b_j is the number of inhabitants registered in the municipality j. Processing a call requires to take the following steps:

1. Call handling (a dispatcher has to obtain the address, and assess how serious the call is, decides which ambulance to assign to it, and contacts the ambulance crew, then the crew has to reach the vehicle);
2. Driving the ambulance from the station to the patients location;
3. Treating the patient by the ambulance crew;
4. Transporting the patient to the nearest hospital;
5. Passing the patient to the hospital staff;
6. Driving the ambulance back to the station.

Table 12.3 Comparison of the real-world situation from the year 2006 with the deployment of the stations proposed by the mathematical models

	Man-made	Double covering	Double covering with EMS located at hospitals
Number of municipalities with at least one EMS station	223	217	245
Multiply covered inhabitants [%]	91.90	99.77	99.49
Calls not serviced within 15 min [%]	1.22	1.25	0.25
Average waiting time [min]	3.4	5.4	3.9
Maximal waiting time [min]	43.5	30	28.4

When modeling the service time, we neglect the call handling and we assume deterministic travel times (defined by the distance and the average speed). According to [8], the treatment time can be considered as to be exponentially distributed with the mean value 10 min. Hence, as we did not have the relevant statistical data on the EMS operation, we supposed that every patient was transported to the nearest hospital, and the time the ambulance spent waiting at the hospital was constant (10 min).

There are many possible dispatching policies for allocating the calls to EMS stations. We always assign a call to the nearest station. If the nearest ambulance is busy, the used policy differs with each situation. The call either waits until the ambulance returns back to the station or it can be reallocated to the next nearest station with an idle ambulance. A simulation model allows to experiment with dispatching policies and consequently to choose the best one. With the experiments, reported in Table 12.3, we suppose that we know in advance the time when the ambulance is returning back to the station. The call is then allocated to the station by which the patient gets served the earliest.

Table 12.3 compares the characteristics of the EMS system given by the official regulations from the year 2006 (column Man-made), with the results obtained by the Double covering model and with the design proposed by the modified Double covering model with stations located at hospitals. In all three cases there are located 264 stations. In Table 12.3, we can see that the solutions proposed by mathematical modeling increase the occurrence of multiply covered inhabitants from 91.9% to almost 100%. The last three rows list the performance characteristics of the system obtained by the computer simulations.

Comparing the solution of the Double covering model with the current deployment shows that the rate of calls not serviced within 15 min remains almost identical (approximately 1.2%). The average waiting time for an ambulance to arrive increases (by 59%), while the maximal waiting time decreases (by 31%). Comparing the solution of the modified Double covering model (with stations located at hospitals) with the current deployment we can observe significant improvement in both quantities: The rate of calls not serviced within 15 min declines to one fifth, and the maximal waiting time is shortened by one third.

a

b

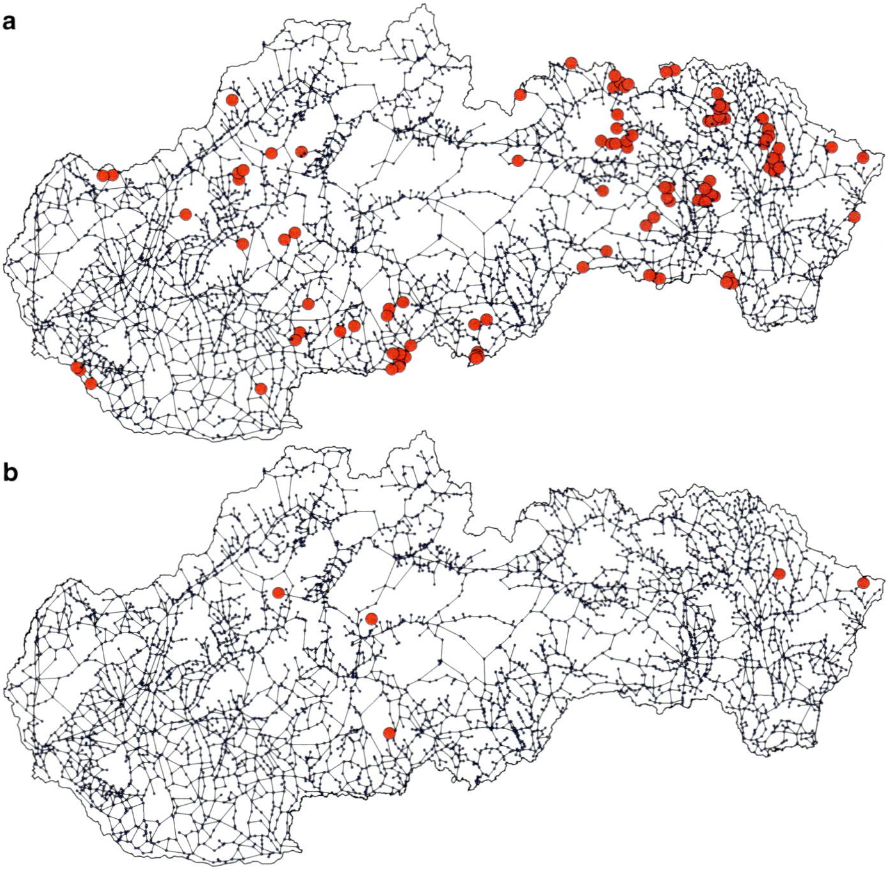

Fig. 12.7 Municipalities with the average waiting time for an ambulance arrival longer than 15 min. In (**a**) we show the situation for the existing EMS stations in the year 2006, and (**b**) depicts the results we reached when we performed Monte Carlo simulations for the solution obtained by the Double covering model imposing the stations to be located close to hospitals

Although the average waiting time remains the same, Fig. 12.7a, b confirm that the number of municipalities where the arrival time exceeds 15 min decreases dramatically.

In conclusion, the statistics presented in Table 12.3 indicate that the design proposed by mathematical modeling is better than the current solution with respect to the service availability to patients. The rate of calls not serviced within 15 min declines to one fifth, the maximal waiting time for an ambulance arrival shortens by one third. The number of municipalities with the average waiting time longer than 15 min falls from 102 to 5.

12.5 Conclusions

In this chapter, we have discussed some issues arising when designing public service systems on complex networks. We focused mainly on effective solving methods, and we searched for a desirable compromise between fairness and economic efficiency, which can be attributed to the resulting design. As possible directions for further research we consider:

- The analysis of the behavior with focus on customers' preferences based on real-world data, and utilization of the results in the modeling of public service systems,
- More detailed theoretical analysis of the price of fairness in the context of location problems,
- and last but not least, possible utilization of centrality measures and other network properties [7] as early indicators of promising solutions in the design of optimization algorithms.

Acknowledgements The authors are grateful for the financial support provided by the Ministry of Education of the Slovak Republic (project VEGA 1/0361/10) and thank J. Slavík, P. Tarábek, M. Koháni, and two anonymous reviewers for thorough reading of the manuscript and the valuable comments.

References

1. Ahuja, R.K., Magnanti, T., Orlin, J.: Network flows: Theory, algorithms, and applications, Prentice Hall, (1993).
2. Balinski, M.: Integer programming: methods, uses computation, Management Science, 12, pp. 254–313. (1965).
3. Beltran-Royo, C., Vial J.P., Alonso-Ayuso, A.: Semi-Lagrangian relaxation applied to the uncapacitated facility location problem, Computational Optimization and Applications, Online First, DOI: 10.1007/s10589-010-9338-2, (2010).
4. Bertsimas, D., Farias, V.F., Trichakis, N.: The price of fairness, Operations Research, (forthcoming).
5. Bilde, O., Krarup, J.: Sharp lower bounds and efficient algorithms for the simple plant location problem. Annals of Discrete Mathematics, 1, pp. 77–97, (1977).
6. Brandeau, M.L., Chiu, S.S., Kumar, S., Grossmann, T.A.: Location with Market Externalities, In: Z., Drezner, ed., Facility Location: A Survey of Applications and Methods, Springer Verlag, New York, (1995).
7. Brandes, U., Erlenbach, T.: Network Analysis: Methodological Foundations, Springer-Verlag Berlin Heidelberg, (2005).
8. Budge, S., Erkut, E., Ingolfsson, A.: Optimal Ambulance Location with Random Delays and Travel Times. Working paper, University of Alberta School of Business, (2005).
9. Butler, M., Wiliams, H.P.: Fairness versus efficiency in charging for the use of common facilities. Journal of Operational Research Society, 53, pp. 1324–1329 (2002).
10. Butler, M., Wiliams, H., P.: The allocation of shared fixed costs, European Journal of Operational Research 170, pp. 391–397, (2006).

11. Caprara A., Toth, P., Fischetti, M.: Algorithms for the set covering problem, Annals of Operations Research 98, pp. 353–371, (2000).
12. Crainic, T.G.: Long-haul freight transportation. In Handbook of transportation science, New York: Springer, (2003).
13. Daganzo, C.F.: Logistics System Analysis, Springer Verlag Berlin, (1996).
14. Daskin, M.S.: Network and Discrete Location. Models, Algorithms, and Applications. John Wiley & Sons, New York, NY, (1995).
15. Dorogovtsev, S., Mendes, J.: Evolution of networks: From biological nets to the Internet and WWW, Oxford University Press, Oxford (2003).
16. Eiselt, H.A., Laporte, G.: Objectives in Location Problems, In: Drezner, Z., ed., Facility Location: A Survey of Applications and Methods, Springer Verlag, New York, (1995).
17. Erlenkotter, D.: A dual-based procedure for uncapacitated facility location. Operations Research, **26**, 6, pp. 992–1009, (1978).
18. Frank, H., Frisch, I.T.: Communication, Transmission and Transportation Networks, Addison Wesley, Series in Electrical Engineering, (1971).
19. Galvao, R.D., Nascimento, E.M.: The location of benefit-distributing facilities in the Brazilian social security system. Operational Research'90 (Proceedings of the IFORS 1990 Conference), pp. 433–443, (1990).
20. Galvao, R.D.: Uncapacitated facility location problems: contributions, Pesqui. Oper. 2004, **24**, 1, pp. 7–38 . ISSN 0101–7438, (2004).
21. Garey, M.R., Johnson, D.S.: Computers and Intractability: A guide to the theory of NP Completeness, San Francisco, Freeman and Co., (1979).
22. Hakimi, S.L.: Optimum location of switching centers and the absolute centers and medians of a graph, Operations Research, 12, pp. 450–459, (1964).
23. Janáček, J., Kováčiková, J.: Exact solution techniques for large location problems. In Proceedings of the International Conference: Mathematical Methods in Economics, pp. 80–84, Ostrava, (1997).
24. Janáček, J., Buzna, Ľ.: A comparison of continuous approximation with mathematical programming approach to location problems, Central European Journal of Operation Research, **12**, 3, pp. 295–305, (2004).
25. Janáček, J., Buzna, Ľ.: An acceleration of Erlenkotter-Koerkels algorithms for the uncapacitated facility location problem. Ann. Oper. Res. **164**, pp. 97–109, (2008).
26. Janáček, J., Gábrišová, L.: A two-phase method for the capacitated facility problem of compact customer subsets. Transport, **24**, 4, pp. 274–282, (2009).
27. J. Janáček, B., Linda, E., Ritschelová: Optimization of municipalities with extended competence selection, Prague economic papers: quarterly journal of economic theory and policy, **19**, 1 pp. 21–34, (2010).
28. Jeong, H., Tombor, B., Albert, R., Oltvai, Z.N., Barabási, A.L.: The large-scale organization of metabolic networks, Nature, **407**, (2000).
29. Karp, R.M.: Reducibility among Combinatorial Problems: In Complexity of Computer Computations, Miller, E.W., Tatcher J.W. eds., New York, Plenum Press, pp. 85–104, (1972)
30. Koerkel, M.: On the exact solution of large-scale simple plant location problems. Eur. J. Oper. Res. **39**, 2, pp. 157–173, (1989).
31. Lämmer, S., Gehlsen, B., Helbing, D.:Scaling laws in the spatial structure of urban road networks, Physica A, **363**, 1, pp. 89–65, (2006).
32. Mladenovic, N., Brimberg, J., Hansen, P.: A note on duality gap in the simple plant location problem. European Journal of Operational Research, **174**, pp. 11–22, (2006).
33. Nace, D., Doan, L.N., Klopfenstein, O., Bashllari, A.: Max-min fairness in multi-commodity flows, Computers & Operations Research **35**, pp. 557–573. (2008).
34. Nogueirai, T., Mendes, R., Valie, Z., Cardoso, J. C.: Optimal Location of Natural Gas Sources in the Iberian System, Proceedings of the 6th WSEAS International Conference on Power Systems, Lisbon, Portugal, September 22–24, (2006).
35. Pioro, M., Medhi, D.: Routing, flow, and capacity design in communication and computer networks, Morgan Kaufmann Publishers, (2004).

36. Rawls, J.: A Theory of justice. Harward University Press, (1971).
37. Resende, M.G.C., Pardalos, P.M.: Handbook of optimization in telecommunications, Springer Science+Business Media, Inc., (2006).
38. Silva, F., Serra, D.: Locating Emergency Services with Different Priorities: The Priority Queuing Covering Location Problem, The Journal of the Operational Research Society, **59**, 9, pp. 1229–1238, (2008).
39. Srinivasan, U.T., Dunne, J.A., Harte, J., Martinez, N.D.: Response of complex food webs to realistic extinction sequences, Ecology **88**, 517, (2007).
40. Sripetch, A., Saengudomlert, P.: Topology Design of Optical Networks Based on Existing Power Grids, Fifth Annual Conference on Communication Networks and Services Research(CNSR'07), (2007).
41. Toregas, C., Swain, R., ReVelle C., Bergman, L.: The location of emergency service facilities, Operations Research, **19**, pp. 1363–1373, (1973).
42. Whittle, P., Networks optimization and evolution, Cambridge University Press, (2007).

Chapter 13
Optimal Flows in Dynamic Networks and Algorithms for their Finding

Maria Fonoberova

Abstract The minimum cost flow problem and the maximum flow problem on networks with time-dependent characteristics and nonlinear cost functions on arcs are considered. The algorithms for determining optimal solutions of the single-commodity and multicommodity network flow problems based on the time-expanded network method are elaborated. Some applications of the optimal flow problems are provided.

13.1 Introduction

The chapter considers the minimum cost flow problem and the maximum flow problem on dynamic networks with different forms of restrictions by parameters of network and time. The classical optimal flow problems on networks are extended and generalized for the cases of nonlinear cost functions on arcs, multicommodity flows, and time- and flow-dependent transactions on arcs of the network. The time-varying flow models which capture the essential properties of flows arising in real-life complex systems are studied. The main goal of this chapter consists in providing methods and algorithms for finding the minimum cost flow and the maximum flow in single-commodity and multicommodity dynamic networks.

Recently network theory has become an important part in the study and analysis of complex systems. Dynamic flows can be used in modelling of processes from different complex systems such as transport networks, social interactions, the Internet. Communication, transportation, economic planning, road traffic assignment, evacuation planning, scheduling planning, cash flow, and management problems can be formulated and solved as single-commodity or multicommodity flow problems

M. Fonoberova (✉)
Aimdyn, Inc., 1919 State St., Santa Barbara, CA 93101, USA
e-mail: mfonoberova@aimdyn.com

M.T. Thai and P.M. Pardalos (eds.), *Handbook of Optimization in Complex Networks: Theory and Applications*, Springer Optimization and Its Applications 57,
DOI 10.1007/978-1-4614-0754-6_13,

[1, 3, 4, 47]. The maximum and the minimum cost flow problems have large implementation for many practical problems in complex systems of electronic communication, transportation, production, and distribution. The considered problems also have theoretical importance for investigation and solving of various optimization problems on graphs. For example, the minimum cost flow problem can be used for research and solving of the distribution problem, the synthesis problem of communication networks, or the allocation problem.

The field of network flows blossomed in the 1940s and 1950s with interest in transportation planning and has developed rapidly since then. There is a significant body of literature devoted to this subject (see, for example, [1, 4, 12, 31, 46]). However, it has largely ignored a crucial aspect of transportation: transportation occurs over time. In the 1960s, Ford and Fulkerson [20, 21] introduced flows over time to include time in the network model.

The following two aspects of dynamic flows distinguish them from the traditional model. Firstly, the flow value on an arc may change over time. This feature is important in applications, where the supplies and demands are not given as fixed measures; instead, they change over time. Naturally, the flow value on each arc should adjust to these changes. Secondly, there is a transit time on every arc which specifies the amount of time flow units need to traverse the arc.

In the Ford and Fulkerson model transit times and arc capacities are fixed. Given a network with capacities and transit times on arcs, they study the problem of sending a maximum amount of flow from a source node s to a sink node t within a pre-specified time horizon T. Ford and Fulkerson show that this problem can be solved by one minimum cost static flow computation, where transit times on arcs are interpreted as cost coefficients. They prove that an optimal solution of this minimum cost flow problem can be turned into a maximum flow over time by first decomposing it into flows on $s - t$ paths. The corresponding flow over time starts to send flow on each path at time zero, and repeats each so long as there is enough time left in the T time units for the flow along the path to arrive at the sink.

Subsequently, linear models of optimal dynamic flows have been studied by Cai, Carey, Fleischer, Glockner, Hoppe, Klinz, Lozovanu, Ma, Nemhauser, Subrahmanian, Tardos, Woeginger and others in [6, 7, 16, 17, 23, 30, 32, 33, 39–41]. In this chapter, the minimum cost and the maximum dynamic flow problems are formulated and studied for nonlinear and multicommodity models. Dynamic networks are considered with time-varying capacities of arcs. For the minimum cost flow problem it is assumed that cost functions, defined on arcs, are nonlinear and depend on time and flow, but the demand-supply function depends on time. It is also considered the dynamic model with transit time functions that depend on flow and time. The game-theoretic formulation of the multiobjective multicommodity flow problem is proposed.

In Sect. 13.2, the main results related to static optimal flow problems on networks are presented. The basic methods and algorithms for solving the minimum cost flow problem and the maximum flow problem are examined. The obtained results are extended for the nonlinear and multicommodity models.

In Sect. 13.3, the minimum cost flow problem is consider on dynamic networks with nonlinear cost functions, that depend on flow and time, and demand-supply functions and capacity functions, that depend on time. The maximum flow problem is considered on dynamic networks with time-varying capacities of arcs. To solve the optimal dynamic flow problems methods and algorithms based on the reduction of dynamic problems to static ones on auxiliary networks are proposed. Dynamic problems with transit time functions that depend on flow and time are analyzed and algorithms for solving such problems are elaborated.

In Sect. 13.4, the minimum cost and the maximum multicommodity flow problems on dynamic networks are considered. The multicommodity flow problem consists of shipping several different commodities from their respective sources to their sinks through a given network satisfying certain objectives in such a way that the total flow going through arcs does not exceed their capacities. No commodity ever transforms into another commodity, so that each one has its own flow conservation constraints, but they compete for the resources of the common network. In this section, the minimum cost multicommodity flow problems are considered on dynamic networks with time-varying capacities of arcs and transit times on arcs that depend on sort of commodity entering them. The cost functions, defined on arcs, are assumed to be nonlinear and depend on time and flow, and demand-supply functions depend on time. For solving the considered problems algorithms based on the modification of the time-expanded network method are proposed. The dynamic problems with transit time functions that depend on flow and time are also considered and algorithms for their solving are proposed.

In Sect. 13.5, the game-theoretic formulation of the multiobjective multicommodity flow problem is considered. If we associate to each commodity a player, we can regard this problem as a game problem, where players interact between them and the choices of one player influence the choices of the others.

13.2 Static Flow Models and Determining Optimal Flows in Networks

In this section we present the most important results related to static flow problems and systemize some of the relevant methods and algorithms for solving the maximum and minimum cost flow problems. These results are extended for the nonlinear and multicommodity cases of the considered models.

13.2.1 Static Flows: Definitions and some Properties

13.2.1.1 Single-Commodity Flow

Let be given a directed graph $G = (V, E)$ with a vertex set V and an arc set E (see [12, 21]). This graph becomes a network $N = (V, E, u, d)$ if we assign to each vertex

$v \in V$ the demand-supply d_v and to each arc $e \in E$ the non-negative capacity u_e, which represents the maximum amount of flow that can be carried on the arc e. By using the demand-supply function d we distinct nodes as net generators of flow, net absorbers of flow, or neither. Nodes $v \in V$ with $d_v > 0$ are called sources, nodes $v \in V$ with $d_v < 0$ are called sinks and nodes $v \in V$ with $d_v = 0$ are called intermediate nodes. A source node or source has the property that the flow out of the node exceeds the flow into the node. The inverse case is a sink node or sink, where the flow into the node exceeds the flow out of node. An intermediate node or transshipment node has the property that flow in equals flow out.

The considered network admits flow, if there are such values $x_e \geq 0$, $\forall e \in E$, which satisfy the following conditions:

$$\sum_{e \in E^-(v)} x_e - \sum_{e \in E^+(v)} x_e = d_v, \; \forall v \in V, \tag{13.1}$$

$$0 \leq x_e \leq u_e, \forall e \in E, \tag{13.2}$$

where $E^-(v) = \{(v,z)\,|\,(v,z) \in E\}$, $E^+(v) = \{(z,v)\,|\,(z,v) \in E\}$.

The function x, defined on E and satisfying conditions (13.1)–(13.2), is called the feasible flow in a network. The value x_e is the quantity of flow on arc e. Condition (13.1) is the flow conservation condition and condition (13.2) is the condition of flow feasibility.

In [14], it is proved that in order for the feasible flow to exist in network N it is necessary and sufficient that for every set of vertices $A \subseteq V$ the following conditions are true:

$$d_A \leq u_{(A,\overline{A})},$$

$$d_V = 0,$$

where $d_A = \sum_{v \in A} d_v$; $\overline{A} = V \setminus A$; $u_{(Z,Y)} = \sum_{z \in Z} \sum_{y \in Y} u_{(z,y)}$, $\forall Z, Y \subseteq V$.

In the case when instead of condition (13.2) the following condition holds:

$$u'_e \leq x_e \leq u''_e, \forall e \in E,$$

where u'_e, $u''_e \geq 0$ are lower and upper capacities of arc e, respectively, then in order for the feasible flow to exist in such a network it is necessary and sufficient that for every set of vertices $A \subseteq V$ the following conditions are true:

$$d_A \leq u''_{(A,\overline{A})} - u'_{(\overline{A},A)},$$

$$d_V = 0.$$

For the first time this result was proved by Hoffman in [29]. In [14], it is shown that it can be easily obtained from the previous affirmation.

If there are no restrictions on arc capacities, i.e. $u''_e = +\infty$, $u'_e = -\infty$, $\forall e \in E$, then in order for the feasible flow to exist in such a network it is necessary and sufficient that the following condition holds:

$$d_V = 0.$$

13.2.1.2 Multicommodity Flow

We consider the following network $N = (V, E, K, u, d)$, where V is a set of vertices, E is a set of arcs, $K = \{1, 2, \dots, q\}$ is a set of commodities, u: $E \to R_+$ is a capacity function, d: $V \times K \to R$ is a demand-supply function. A flow x assigns to every arc $e \in E$ for each commodity $k \in K$ a flow $x_e^k \geq 0$ such that the following flow conservation constraints are satisfied:

$$\sum_{e \in E^-(v)} x_e^k - \sum_{e \in E^+(v)} x_e^k = d_v^k, \; \forall v \in V, \; \forall k \in K. \tag{13.3}$$

The multicommodity flow x is called feasible if it obeys the following capacity constraints:

$$\sum_{k \in K} x_e^k \leq u_e, \; \forall e \in E. \tag{13.4}$$

The value x_e^k is the quantity of the flow of commodity k on the arc e. In fact, the considered above single-commodity flow is the multicommodity flow with $q = 1$.

In order for the feasible multicommodity flow to exist it is required that $\sum_{v \in V} d_v^k = 0$, $\forall k \in K$. Nodes $v \in V$ with $d_v^k > 0$, $k \in K$, are called sources for commodity k, nodes $v \in V$ with $d_v^k < 0$, $k \in K$, are called sinks for commodity k and nodes $v \in V$ with $d_v^k = 0$, $k \in K$, are called intermediate nodes for commodity k.

13.2.2 The Maximum Flow Problem

Let us consider the single-commodity maximum flow problem on a network with only one source s and one sink t. In the case of many sources and sinks, the maximum flow problem can be reduced to the standard one by introducing one additional artificial source and one additional artificial sink as well as arcs leading from the new source to initial sources and from initial sinks to the new sink. The capacities of arcs connecting the artificial source with the initial sources are bounded by the supplies of these sources; the capacities of arcs connecting the initial sinks with the artificial sink are bounded by the demands of these sinks.

Let us consider the flow $x \geq 0$ which satisfies condition (13.2) and the following conditions:

$$\sum_{e \in E^-(v)} x_e - \sum_{e \in E^+(v)} x_e = \begin{cases} h, \; v = s, \\ 0, \; v \neq s, t, \\ -h, \; v = t. \end{cases} \tag{13.5}$$

The object of the maximum flow problem is to send a maximum amount of flow from the source to the sink such that arc capacities are not exceeded.

The maximum flow problem is formulated as follows:

to maximize the objective function

$$F = h$$

subject to (13.2), (13.5).

In the following, we introduce a notion of the network cutset, which separates s from t and is defined by every set $R \subseteq V$ such that $s \in R$, $t \notin R$. So, if $s \in R$, $t \notin R$, then the network cutset separating s from t is a set of arcs $(R, \overline{R}) = \{(v,z) \in E | v \in R, z \in \overline{R}\}$. The quantity $u_{(R,\overline{R})} = \sum_{v \in R} \sum_{z \in \overline{R}} u_{(v,z)}$ is called the capacity of the cutset $(R, \overline{R})$. The cutset, which has the minimum capacity, is called the minimum cutset.

The most important result related to the maximum flow in a network is formulated by Ford and Fulkerson in [21] and consists in the fact that for every network the maximum value of a flow from s to t is equal to the minimum capacity of a cutset separating s and t.

We refer to a path from s to t as an augmenting path for the flow x if $x < u$ for all its forward arcs and $x > 0$ for all its backward arcs. We say that arc (v,z) is saturated with flow x if $x_{(v,z)} = u_{(v,z)}$, and we say that arc (v,z) is free from flow x if $x_{(v,z)} = 0$.

Using these definitions and the Ford–Fulkerson theorem about the maximum flow and the minimum cutset the following affirmations can be obtained.

The flow x is maximum if and only if there is no augmenting path for the flow x.

The cutset $(R, \overline{R})$ is minimum if and only if every maximum flow x saturates all arcs of the cutset $(R, \overline{R})$ and leaves free all arcs that belong to $(\overline{R}, R)$.

The Ford–Fulkerson theorem and these affirmations lead to a simple and effective algorithm for finding the maximum flow, which is named the Ford–Fulkerson algorithm. This algorithm is proposed and argued by Ford and Fulkerson in [21]. To guarantee the end of the process it is required that arc capacities are integer. This assumption is not essential from the computational point of view, because in the case of rational capacities the problem can be reduced to the initial one by introducing a new quantity such that all existing fractions are its integral multiplies.

13.2.3 The Minimum Cost Flow Problem

13.2.3.1 The Linear Case

The minimum cost flow problem is the problem of sending flows in a network from supply nodes to demand nodes at minimum total cost such that arc capacities are not exceeded. The linear minimum cost flow problem consists in finding a feasible flow, satisfying (13.1)–(13.2), that minimizes the following objective function:

$$F(x) = \sum_{e \in E} c_e x_e,$$

where c: $E \rightarrow R_+$ is a cost function.

There are different approaches for solving this problem. One of the most propagated methods is the potential method for constructing an optimal flow in the case when the initial flow is known and its base graph, i.e. a partial graph on arcs of which there is a positive feasible flow, is a connected graph. If the last condition is not true, the perturbation method can be applied. The following affirmation, proved in [14], justifies the potential method.

The flow x is an optimal one if and only if for every vertex $v \in V$ there exists a number P_v such that

$$P_z - P_v \leq c_{(v,z)}, \text{ if } x_{(v,z)} = 0;$$

$$P_z - P_v = c_{(v,z)}, \text{ if } 0 < x_{(v,z)} < u_{(v,z)};$$

$$P_z - P_v \geq c_{(v,z)}, \text{ if } x_{(v,z)} = u_{(v,z)},$$

where $z \in V$, $(v,z) \in E$. The numbers P_v may be considered non-negative and are called potentials or node numbers.

If $c_e, u_e, d_v, \forall e \in E, \forall v \in V$, are integer or real numbers, then the potential method is a finite method, i.e. the solution of the problem using this method can be obtained after a finite number of steps. Besides that, if the initial flow is integer-valued, then the optimal flow will be also integer-valued.

13.2.3.2 The Non-linear Case

The nonlinear minimum cost flow problem is a problem of determining a feasible flow, that satisfies (13.1)–(13.2) and minimizes the following objective function:

$$F(x) = \sum_{e \in E} \varphi_e(x_e),$$

where φ: $E \times R_+ \to R_+$ is a nonlinear cost function.

Let us consider that the cost functions $\varphi_e(x_e)$, $\forall e \in E$, are convex downwards functions, i.e. for arbitrary nonnegative quantities $x_e^{(1)}$, $x_e^{(2)}$ the following inequality is true:

$$\varphi_e\left(\lambda x_e^{(1)} + (1-\lambda)x_e^{(2)}\right) \leq \lambda \varphi_e\left(x_e^{(1)}\right) + (1-\lambda)\varphi_e\left(x_e^{(2)}\right),\ 0 \leq \lambda \leq 1.$$

We denote by $\varphi_e^+(x_e)$ the right derivative of the function $\varphi_e(x_e)$ and by $\varphi_e^-(x_e)$ the left derivative of the function $\varphi_e(x_e)$.

In [14] it is proved that the flow x is an optimal flow in the network without restrictions on arc capacities if and only if for every vertex $v \in V$ there exists a number P_v such that

$$P_z - P_v \leq \varphi^+_{(v,z)}(x_{(v,z)}), \text{ if } x_{(v,z)} = 0;$$

$$\varphi^-_{(v,z)}(x_{(v,z)}) \leq P_z - P_v \leq \varphi^+_{(v,z)}(x_{(v,z)}), \text{ if } x_{(v,z)} > 0,$$

where $z \in V$, $(v,z) \in E$.

In the case when the functions $\varphi_{(v,z)}(x_{(v,z)})$ are everywhere differentiable, i.e. if

$$\varphi^{+}_{(v,z)}(x_{(v,z)}) = \varphi^{-}_{(v,z)}(x_{(v,z)}) = \varphi'_{(v,z)}(x_{(v,z)}),$$

and there are no restrictions on arc capacities, the flow x is an optimal one if and only if for every vertex $v \in V$ there exists a number P_v such that

$$P_z - P_v \leq \varphi'_{(v,z)}(x_{(v,z)}), \text{ if } x_{(v,z)} = 0;$$
$$P_z - P_v = \varphi'_{(v,z)}(x_{(v,z)}), \text{ if } x_{(v,z)} > 0,$$

where $z \in V$, $(v,z) \in E$.

If the functions $\varphi_{(v,z)}(x_{(v,z)})$ are differentiable functions and there exist restrictions (13.2) on arc capacities, the flow x is an optimal one if and only if for every vertex $v \in V$ there exists a number P_v such that

$$P_z - P_v \leq \varphi'_{(v,z)}(x_{(v,z)}), \text{ if } x_{(v,z)} = 0;$$
$$P_z - P_v = \varphi'_{(v,z)}(x_{(v,z)}), \text{ if } 0 < x_{(v,z)} < u_{(v,z)};$$
$$P_z - P_v \geq \varphi'_{(v,z)}(x_{(v,z)}), \text{ if } x_{(v,z)} = u_{(v,z)},$$

where $z \in V$, $(v,z) \in E$.

In the case when the functions $\varphi_{(v,z)}(x_{(v,z)})$ are non-differentiable, the flow x is an optimal flow in the network with restrictions (13.2) on arc capacities if and only if for every vertex $v \in V$ there exists a number P_v and for each saturated arc $(v,z) \in E$ (for which $x_{(v,z)} = u_{(v,z)}$) there exists a nonnegative arc number $\gamma_{(v,z)}$, such that

$$P_z - P_v \leq \varphi^{+}_{(v,z)}(x_{(v,z)}), \text{ if } x_{(v,z)} = 0;$$
$$\varphi^{-}_{(v,z)}(x_{(v,z)}) \leq P_z - P_v \leq \varphi^{+}_{(v,z)}(x_{(v,z)}), 0 < x_{(v,z)} < u_{(v,z)};$$
$$\varphi^{-}_{(v,z)}(x_{(v,z)}) + \gamma_{(v,z)} \leq P_z - P_v \leq \varphi^{+}_{(v,z)}(x_{(v,z)}) + \gamma_{(v,z)}, x_{(v,z)} = u_{(v,z)},$$

where $z \in V$.

We'd like to mention that if the functions $\varphi_e(x_e)$, $\forall e \in E$, are not assumed to be convex downwards, then in the considered above cases we can speak about only the local extremum.

To solve the non-linear minimum cost flow problem the potential method can be used, which generalizes the potential method for the linear minimum cost flow problem. To apply this method we consider that the base graph of the initial flow and every intermediary flow is a connected graph. If this condition is not true, we apply the perturbation method.

13.2.4 The Minimum Cost Multicommodity Flow Problem

13.2.4.1 The Continuously Differentiable and Convex Objective Function

The minimum cost multicommodity flow problem is formulated as follows:
to minimize the objective function

$$F(\mathrm{x}) = \sum_{e\in E} \varphi_e(x_e^1, x_e^2, \dots, x_e^q) \tag{13.6}$$

subject to (13.3)–(13.4),

where φ: $E \times R_+^q \to R_+$ is a cost function.

We consider that functions $\varphi_e(\mathrm{x})$ from (13.6) are convex downwards with regard to vector $\mathrm{x} = (x_e^1, x_e^2, \dots, x_e^q) \geq 0$, i.e. for every $0 \leq \lambda \leq 1$ the following inequality is true:

$$\varphi_e(\lambda \mathrm{x}^{(1)} + (1-\lambda)\mathrm{x}^{(2)}) \leq \lambda \varphi_e(\mathrm{x}^{(1)}) + (1-\lambda)\varphi_e(\mathrm{x}^{(2)}),$$

where $\mathrm{x}^{(1)} \geq 0$, $\mathrm{x}^{(2)} \geq 0$. Moreover, let us consider that these functions are continuously differentiable. We denote by $\varphi_e^{'(k)}(\mathrm{x})$ the partial derivative of the function $\varphi_e(\mathrm{x})$ by the k-th component of the vector $\mathrm{x} = (x_e^1, x_e^2, \dots, x_e^q)$.

In [14], it is proved that the multicommodity flow $\mathrm{x} = (x_e^1, x_e^2, \dots, x_e^q)$ is an optimal flow if and only if for every vertex $v \in V$ there exists a vector of potentials $P_v = (P_v^1, P_v^2, \dots, P_v^q)$ and for every arc $(v,z) \in E$, for which $\sum_{k=1}^{q} x_{(v,z)}^k = u_{(v,z)}$, there exists an arc number $\gamma_{(v,z)} \geq 0$, such that

$$P_z^k - P_v^k \leq \varphi_{(v,z)}^{'(k)}(\mathrm{x}_{(v,z)}), \text{ if } x_{(v,z)}^k = 0, \ \sum_{k=1}^{q} x_{(v,z)}^k < u_{(v,z)};$$

$$P_z^k - P_v^k = \varphi_{(v,z)}^{'(k)}(\mathrm{x}_{(v,z)}), \text{ if } x_{(v,z)}^k > 0, \ \sum_{k=1}^{q} x_{(v,z)}^k < u_{(v,z)};$$

$$P_z^k - P_v^k \leq \varphi_{(v,z)}^{'(k)}(\mathrm{x}_{(v,z)}) + \gamma_{(v,z)}, \text{ if } x_{(v,z)}^k = 0, \ \sum_{k=1}^{q} x_{(v,z)}^k = u_{(v,z)};$$

$$P_z^k - P_v^k = \varphi_{(v,z)}^{'(k)}(\mathrm{x}_{(v,z)}) + \gamma_{(v,z)}, \text{ if } x_{(v,z)}^k > 0, \ \sum_{k=1}^{q} x_{(v,z)}^k = u_{(v,z)}, \tag{13.7}$$

where $z \in V$, $k \in K$.

If we assume that $u_{(v,z)} = \infty$, then conditions (13.7) are reduced to the following conditions:

$$P_z^k - P_v^k \leq \varphi_{(v,z)}^{'(k)}(\mathrm{x}_{(v,z)}), \text{ if } x_{(v,z)}^k = 0,$$

$$P_z^k - P_v^k = \varphi_{(v,z)}^{'(k)}(\mathrm{x}_{(v,z)}), \text{ if } x_{(v,z)}^k > 0.$$

13.2.4.2 The Separable Objective Function

In many practical problems, the objective function has the following form:

$$F(\mathbf{x}) = \sum_{k \in K} \sum_{e \in E} \varphi_e^k(x_e^k), \tag{13.8}$$

where $\varphi_e^k \colon R_+ \to R_+$ is a cost function of commodity $k \in K$ for arc $e \in E$.

We consider that functions $\varphi_e^k(x_e^k)$ from (13.8) are convex downwards if $x_e^k \geq 0$ and are continuous in $x_e^k = 0$. By $\varphi_e^{-k}(x_e^k)$ and $\varphi_e^{+k}(x_e^k)$ we denote the left and the right derivatives of the function $\varphi_e^k(x_e^k)$, respectively.

In [14] it is proved that the multicommodity flow $\mathbf{x} = (x_e^1, x_e^2, \ldots, x_e^q)$ is an optimal one if and only if for every vertex $v \in V$ there exists a vector of potentials $P_v = (P_v^1, P_v^2, \ldots, P_v^q)$ and for every arc $(v,z) \in E$, for which $\sum_{k=1}^{q} x_{(v,z)}^k = u_{(v,z)}$, there exists an arc number $\gamma_{(v,z)} \geq 0$, such that

$$P_z^k - P_v^k \leq \varphi_{(v,z)}^{+k}(x_{(v,z)}^k), \text{ if } x_{(v,z)}^k = 0, \ \sum_{k=1}^{q} x_{(v,z)}^k < u_{(v,z)};$$

$$\varphi_{(v,z)}^{-k}(x_{(v,z)}^k) \leq P_z^k - P_v^k \leq \varphi_{(v,z)}^{+k}(x_{(v,z)}^k), \ \text{ if } x_{(v,z)}^k > 0, \ \sum_{k=1}^{q} x_{(v,z)}^k < u_{(v,z)};$$

$$P_z^k - P_v^k \leq \varphi_{(v,z)}^{+k}(x_{(v,z)}^k) + \gamma_{(v,z)}, \text{ if } x_{(v,z)}^k = 0, \ \sum_{k=1}^{q} x_{(v,z)}^k = u_{(v,z)};$$

$$\varphi_{(v,z)}^{-k}(x_{(v,z)}^k) + \gamma_{(v,z)} \leq P_z^k - P_v^k \leq \varphi_{(v,z)}^{+k}(x_{(v,z)}^k) + \gamma_{(v,z)},$$
$$\text{if } x_{(v,z)}^k > 0, \ \textstyle\sum_{k=1}^{q} x_{(v,z)}^k = u_{(v,z)},$$

where $z \in V$, $k \in K$.

To solve the minimum cost multicommodity flow problem we can apply the potential method, the linearization method, the decomposition method and some other methods [14].

13.3 Optimal Single-Commodity Flow Problems on Dynamic Networks and Methods for their Solving

In this section, we consider the minimum cost flow problem and the maximum flow problem on dynamic networks. For the minimum cost flow problem we assume that demand–supply and capacity functions depend on time, and cost functions, defined on arcs, are nonlinear and depend both on time and on flow. The maximum flow problem is considered on dynamic networks with time-varying capacities of arcs. We also study a dynamic model with transit time functions that depend on the amount of flow and the entering time-moment of flow in the arc. Methods and algorithms for solving the formulated problems are proposed.

13.3.1 The Minimum Cost Dynamic Flow Problem

A dynamic network $N = (V, E, \tau, d, u, \varphi)$ is determined by directed graph $G = (V, E)$ with set of vertices V, $|V| = n$, and set of arcs E, $|E| = m$, transit time function $\tau\colon E \to R_+$, demand-supply function $d\colon V \times T \to R$, capacity function $u\colon E \times T \to R_+$, and cost function $\varphi\colon E \times R_+ \times T \to R_+$. We consider the discrete time model, in which all times are integral and bounded by horizon T. The time horizon is the time until which the flow can travel in the network and it defines the set $T = \{0, 1, \ldots, T\}$ of the considered time moments. Time is measured in discrete steps, so that if one unit of flow leaves vertex z at time t on arc $e = (z, v)$, one unit of flow arrives at vertex v at time $t + \tau_e$, where τ_e is the transit time on arc e. The continuous flow model formulations can be found in [15, 16, 18].

In order for the flow to exist it is required that $\sum_{t\in T}\sum_{v\in V} d_v(t) = 0$. If for an arbitrary node $v \in V$ at a moment of time $t \in T$ the condition $d_v(t) > 0$ holds, then this node v at the time-moment t is treated as a source. If at a moment of time $t \in T$ the condition $d_v(t) < 0$ holds, then the node v at the time-moment t is regarded as a sink. In the case $d_v(t) = 0$ at a moment of time $t \in T$, the node v at the time-moment t is considered as an intermediate node.

Without losing generality we consider that the set of vertices V is divided into three disjoint subsets V_+, V_-, V_*, such that:

V_+ consists of nodes $v \in V$, for which $d_v(t) \geq 0$ for $t \in T$ and there exists at least one moment of time $t_0 \in T$ such that $d_v(t_0) > 0$;

V_- consists of nodes $v \in V$, for which $d_v(t) \leq 0$ for $t \in T$ and there exists at least one moment of time $t_0 \in T$ such that $d_v(t_0) < 0$;

V_* consists of nodes $v \in V$, for which $d_v(t) = 0$ for every $t \in T$.

So, V_+ is a set of sources, V_- is a set of sinks and V_* is a set of intermediate nodes of the network N.

A feasible dynamic flow in network N is a function $x\colon E \times T \to R_+$ that satisfies the following conditions:

$$\sum_{e\in E^-(v)} x_e(t) - \sum_{\substack{e\in E^+(v)\\ t-\tau_e\geq 0}} x_e(t-\tau_e) = d_v(t),\ \forall t \in T,\ \forall v \in V; \tag{13.9}$$

$$0 \leq x_e(t) \leq u_e(t),\ \ \forall t \in T,\ \forall e \in E; \tag{13.10}$$

$$x_e(t) = 0,\ \forall e \in E,\ t = \overline{T - \tau_e + 1, T}, \tag{13.11}$$

where $E^-(v) = \{(v,z) \,|\, (v,z) \in E\},\ \ E^+(v) = \{(z,v) \,|\, (z,v) \in E\}$.

Here the function x defines the value $x_e(t)$ of flow entering arc e at time t. The flow does not enter arc e at time t if it has to leave the arc after time T; this is ensured by condition (13.11). Restrictions (13.10) are capacity constraints. Conditions (13.9) represent flow conservation constraints.

To model transit costs, which may change over time, we define the cost function $\varphi_e(x_e(t),t)$ with the meaning that flow of value $\rho = x_e(t)$ entering arc e at time t will incur a transit cost of $\varphi_e(\rho,t)$. It is assumed that $\varphi_e(0,t) = 0$ for all $e \in E$ and $t \in T$.

The total cost of the dynamic flow x in the network N is defined as follows:

$$F(x) = \sum_{t \in T} \sum_{e \in E} \varphi_e(x_e(t),t). \tag{13.12}$$

The minimum cost dynamic flow problem consists in finding a feasible dynamic flow that minimizes the objective function (13.12).

13.3.2 The Method for Solving the Minimum Cost Dynamic Flow Problem

We propose an approach based on the reduction of the dynamic problem to a corresponding static problem to solve the formulated above problem. We show that the minimum cost dynamic flow problem on network $N = (V,E,\tau,d,u,\varphi)$ can be reduced to a minimum cost static flow problem on an auxiliary time-expanded network $N^{\mathrm{T}} = (V^{\mathrm{T}},E^{\mathrm{T}},d^{\mathrm{T}},u^{\mathrm{T}},\varphi^{\mathrm{T}})$. The advantage of such an approach is that it turns the problem of determining an optimal flow over time into a classical network flow problem.

The essence of the time-expanded network is that it contains a copy of the vertex set of the dynamic network for each moment of time $t \in T$, and the transit times and flows are implicit in arcs linking those copies. The network N^{T} is defined as follows:

1. $V^{\mathrm{T}} := \{v(t) \,|\, v \in V,\ t \in T\}$;
2. $E^{\mathrm{T}} := \{e(t) = (v(t), z(t+\tau_e)) \,|\, e \in E,\ 0 \le t \le T - \tau_e\}$;
3. ${d_{v(t)}}^{\mathrm{T}} := d_v(t)$ for $v(t) \in V^{\mathrm{T}}$;
4. ${u_{e(t)}}^{\mathrm{T}} := u_e(t)$ for $e(t) \in E^{\mathrm{T}}$;
5. ${\varphi_{e(t)}}^{\mathrm{T}}({x_{e(t)}}^{\mathrm{T}}) := \varphi_e(x_e(t),t)$ for $e(t) \in E^{\mathrm{T}}$.

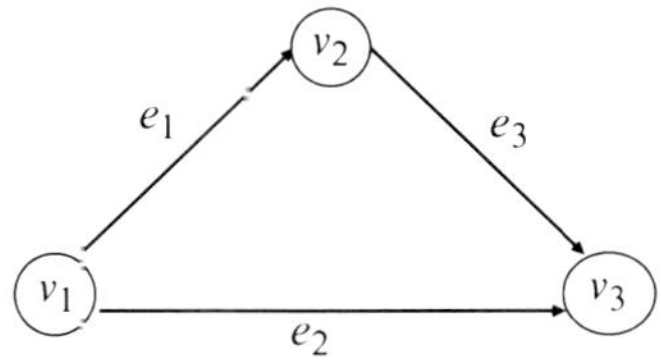

Fig. 13.1 The dynamic network N

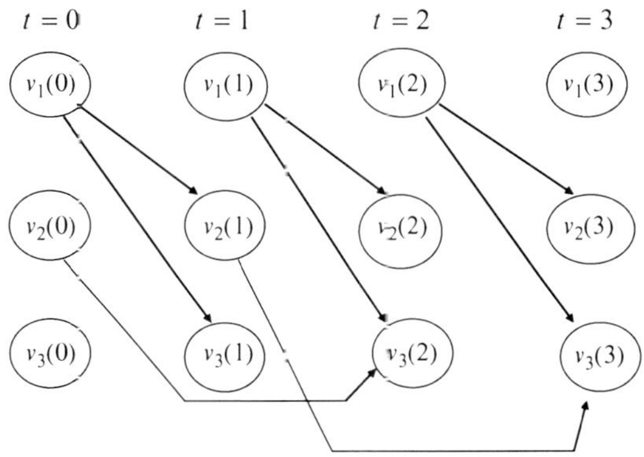

Fig. 13.2 The time-expanded network N^{T}

In the following, we construct the time-expanded network N^{T} for the dynamic network N given in Fig. 13.1. The set of time moments is $T = \{0, 1, 2, 3\}$. The transit times on each arc are as follows: $\tau_{e_1} = 1$, $\tau_{e_2} = 1$, $\tau_{e_3} = 2$. The capacity, demand–supply and cost functions are considered to be given. The constructed time-expanded network is presented in Fig. 13.2.

The correspondence between feasible flows in the dynamic network N and feasible flows in the time-expanded network N^{T} is stated in the following way. Let $x_e(t)$ be a flow in the dynamic network N, then the function x^{T} defined as follows:

$$x_{e(t)}{}^{\mathrm{T}} = x_e(t), \ \forall e(t) \in E^{\mathrm{T}}, \tag{13.13}$$

represents a flow in the time-expanded network N^{T}.

Lemma 13.1. *The correspondence (13.13) is a bijection from the set of feasible flows in the dynamic network N onto the set of feasible flows in the time-expanded network* N^{T}.

Proof. It is obvious that the correspondence (13.13) is a bijection from the set of T-horizon functions in the dynamic network N onto the set of functions in the time-expanded network N^{T}. In the following we have to show that each dynamic flow in the dynamic network N is put into the correspondence with a static flow in the time-expanded network N^{T} and vice-versa.

Let $x_e(t)$ be a dynamic flow in N, and let ${x_{e(t)}}^{\mathrm{T}}$ be a corresponding function in N^{T}. Let's prove that ${x_{e(t)}}^{\mathrm{T}}$ satisfies the conservation constraints in the static network N^{T}. Let $v \in V$ be an arbitrary vertex in N and t, $0 \le t \le T - \tau_e$, an arbitrary moment of time:

$$d_v(t) \overset{(i)}{=} \sum_{e \in E^-(v)} x_e(t) - \sum_{\substack{e \in E^+(v) \\ t-\tau_e \ge 0}} x_e(t-\tau_e)$$

$$= \sum_{e(t) \in E^-(v(t))} {x_{e(t)}}^{\mathrm{T}} - \sum_{e(t-\tau_e) \in E^+(v(t))} {x_{e(t-\tau_e)}}^{\mathrm{T}} \overset{(ii)}{=} {d_{v(t)}}^{\mathrm{T}}. \quad (13.14)$$

Note that according to the definition of the time-expanded network the set of arcs $\{e(t-\tau_e) | e(t-\tau_e) \in E^+(v(t))\}$ consists of all arcs that enter $v(t)$, while the set of arcs $\{e(t) | e(t) \in E^-(v(t))\}$ consists of all arcs that originate from $v(t)$. Therefore, all necessary conditions are satisfied for each vertex $v(t) \in V^{\mathrm{T}}$. Hence, ${x_{e(t)}}^{\mathrm{T}}$ is a flow in the time-expanded network N^{T}.

Let ${x_{e(t)}}^{\mathrm{T}}$ be a static flow in the time-expanded network N^{T}, and let $x_e(t)$ be a corresponding function in the dynamic network N. Let $v(t) \in V^{\mathrm{T}}$ be an arbitrary vertex in N^{T}. The conservation constraints for this vertex in the static network are expressed by equality (ii) from (13.14), which holds for all $v(t) \in V^{\mathrm{T}}$ at all times t, $0 \le t \le T - \tau_e$. Therefore, equality (i) holds for all $v \in V$ at all moments of time t, $0 \le t \le T - \tau_e$. In such a way $x_e(t)$ is a flow in the dynamic network N.

It is easy to verify that a feasible flow in the dynamic network N is a feasible flow in the time-expanded network N^{T} and vice-versa. Indeed,

$$0 \le {x_{e(t)}}^{\mathrm{T}} = x_e(t) \le u_e(t) = {u_{e(t)}}^{\mathrm{T}}.$$

The lemma is proved. □

Theorem 13.1. *If x is a flow in the dynamic network N and x^{T} is a corresponding flow in the time-expanded network N^{T}, then*

$$F(x) = F^{\mathrm{T}}(x^{\mathrm{T}}),$$

where

$$F^{\mathrm{T}}(x^{\mathrm{T}}) = \sum_{t \in T} \sum_{e(t) \in E^{\mathrm{T}}} {\varphi_{e(t)}}^{\mathrm{T}}({x_{e(t)}}^{\mathrm{T}})$$

is the total cost of the static flow x^{T} in the time-expanded network N^{T}.

Moreover, for each minimum cost flow x^* in the dynamic network N there is a corresponding minimum cost flow $x^{*\mathrm{T}}$ in the static network N^{T} such that

$$F(x^*) = F^{\mathrm{T}}(x^{*\mathrm{T}})$$

and vice-versa.

Proof. Let $x : E \times T \to R_+$ be an arbitrary dynamic flow in the dynamic network N. Then according to Lemma 13.1 the unique flow x^{T} in N^{T} corresponds to the flow x in N, and therefore we have:

$$F(x) = \sum_{t \in T} \sum_{e \in E} \varphi_e(x_e(t), t) = \sum_{t \in T} \sum_{e(t) \in E^{\mathrm{T}}} \varphi_{e(t)}{}^{\mathrm{T}}(x_{e(t)}{}^{\mathrm{T}}) = F^{\mathrm{T}}(x^{\mathrm{T}}).$$

So, the first part of the theorem is proved.

To prove the second part of the theorem we again use Lemma 13.1. Let $x^* : E \times T \to R_+$ be the optimal dynamic flow in N and $x^{*\mathrm{T}}$ be the corresponding optimal flow in N^{T}. Then

$$F^{\mathrm{T}}(x^{*\mathrm{T}}) = \sum_{t \in T} \sum_{e(t) \in E^{\mathrm{T}}} \varphi_{e(t)}{}^{\mathrm{T}}(x^*_{e(t)}{}^{\mathrm{T}}) = \sum_{t \in T} \sum_{e \in E} \varphi_e(x^*_e(t), t) = F(x^*).$$

The converse proposition is proved in an analogous way. □

The following algorithm for solving the minimum cost dynamic flow problem can be proposed.

1. To build the time-expanded network N^{T} for the dynamic network N.
2. To solve the classical minimum cost flow problem on the static network N^{T} [1,5, 14,25,26,31,46].
3. To reconstruct the solution of the static problem on the network N^{T} to the dynamic problem on the network N.

Building the time-expanded network and reconstructing the solution of the minimum cost static flow problem to the dynamic one has complexity $O(nT + mT)$. The complexity of step 2 depends on the complexity of the algorithm used for the minimum cost flow problem on static networks. If such an algorithm has complexity $O(f(n', m'))$, where n' is a number of vertices and m' is a number of arcs in the network, then the complexity of solving the minimum cost flow problem on the time-expanded network employing the same algorithm is $O(f(nT, mT))$.

Some specific algorithms are proposed in [37] to minimize the size of the auxiliary static network. In the case of uncapacitated dynamic networks with cost functions that are concave with regard to flow value and do not change over time, the problem can be reduced to the minimum cost flow problem on a static network of equal size, not the time-expanded network.

13.3.3 The Dynamic Model with Flow Storage at Nodes

The previous mathematical model can be extended for the case with flow storage at nodes if we associate a transit time τ_v to each node $v \in V$ which means that the flow passage through this node takes τ_v units of time. If in addition we associate the capacity function $u_v(t)$ and the cost function $\varphi_v(x_v(t),t)$ to each node v, a more general model can be obtained. In this case, the problem can be reduced to the previous one by simple transformation of the network where each node v is changed by a couple of vertices v' and v'' connected with directed arc $e_v = (v', v'')$. Here, v' preserves all entering arcs and v'' preserves all leaving arcs of the previous network. The transit time $\tau_{e_v} = \tau_v$, the cost function $\varphi_{e_v}(x_{e_v}(t),t) = \varphi_v(x_v(t),t)$ and the capacity function $u_{e_v}(t) = u_v(t)$ are associated to arc e_v.

An important particular case of the minimum cost dynamic flow problem is the one when all amount of flow is dumped into the network from sources $v \in V_+$ at the time-moment $t = 0$ and it arrives at sinks $v \in V_-$ at the time-moment $t = T$. This means that the supply–demand function $d : V \times T \to R$ satisfies the conditions:

(a) $d_v(0) > 0$, $d_v(t) = 0$, $t = 1, 2, \ldots, T$, for $v \in V_+$;

(b) $d_v(T) < 0$, $d_v(t) = 0$, $t = 0, 1, 2, \ldots, T-1$, for $v \in V_-$.

So, let us consider the minimum cost flow problem on the dynamic network with flow storage at nodes and integral constant demand–supply functions. Let be given a dynamic network $N = (V, E, \tau, d, u, \varphi)$, where the demand–supply function $d : V \to R$ does not depend on time. Without losing generality, we assume that no arcs enter sources or exit sinks. In order for a flow to exist supply must equal demand: $\sum_{v \in V} d_v = 0$.

The mathematical model of the minimum cost flow problem on this dynamic network is the following:

$$\sum_{e \in E^-(v)} \sum_{t=0}^{T} x_e(t) - \sum_{e \in E^+(v)} \sum_{t=\tau_e}^{T} x_e(t - \tau_e) = d_v, \ \forall v \in V; \tag{13.15}$$

$$\sum_{e \in E^-(v)} \sum_{t=0}^{\theta} x_e(t) - \sum_{e \in E^+(v)} \sum_{t=\tau_e}^{\theta} x_e(t - \tau_e) \leq 0, \ \forall v \in V_*, \ \forall \theta \in T; \tag{13.16}$$

$$0 \leq x_e(t) \leq u_e(t), \ \ \forall t \in T, \ \forall e \in E; \tag{13.17}$$

$$x_e(t) = 0, \ \forall e \in E, \ t = \overline{T - \tau_e + 1, T}. \tag{13.18}$$

Condition (13.18) ensures that there is no flow in the network after time horizon T. Conditions (13.17) are capacity constraints. As flow travels through the network, unlimited flow storage at the nodes is allowed, but any deficit is prohibited by constraint (13.16). Finally, all demands must be met, flow must not remain in the network after time T, and each source must not exceed its supply. This is ensured by constraint (13.15).

As above we seek for a feasible dynamic flow x that minimizes the total cost:

$$F(x) = \sum_{t \in T} \sum_{e \in E} \varphi_e(x_e(t), t).$$

We'd like to mention that the more general model can be obtained if we define the cost function as also dependent on the flow storage at nodes. In this case, the problem can be solved by using the similar approach.

To solve the formulated above minimum cost dynamic flow problem we use the modified time-expanded network method. The auxiliary static network N^{T} is constructed as follows:

1. $V^{\mathrm{T}} := \{v(t) | v \in V, t \in T\}$;
2. $V_+^{\mathrm{T}} := \{v(0) | v \in V_+\}$ and $V_-^{\mathrm{T}} := \{v(T) | v \in V_-\}$;
3. $E^{\mathrm{T}} := \{(v(t), z(t+\tau_e)) \mid e = (v,z) \in E,\ 0 \leq t \leq T - \tau_e\} \cup$
   $\{v(t), v(t+1) \mid v \in V,\ 0 \leq t < T\}$;
4. ${d_{v(t)}}^{\mathrm{T}} := d_v$ for $v(t) \in V_+^{\mathrm{T}} \cup V_-^{\mathrm{T}}$; otherwise ${d_{v(t)}}^{\mathrm{T}} := 0$;
5. ${u_{(v(t), z(t+\tau_{(v,z)}))}}^{\mathrm{T}} := u_{(v,z)}(t)$ for $(v(t), z(t+\tau_{(v,z)})) \in E^{\mathrm{T}}$;
   ${u_{(v(t), v(t+1))}}^{\mathrm{T}} := \infty$ for $(v(t), v(t+1)) \in E^{\mathrm{T}}$;
6. ${\varphi_{(v(t), z(t+\tau_{(v,z)}))}}^{\mathrm{T}}({x_{(v(t), z(t+\tau_{(v,z)}))}}^{\mathrm{T}}) := \varphi_{(v,z)}(x_{(v,z)}(t), t)$
   for $(v(t), z(t+\tau_{(v,z)})) \in E^{\mathrm{T}}$;
   ${\varphi_{(v(t), v(t+1))}}^{\mathrm{T}}({x_{(v(t), v(t+1))}}^{\mathrm{T}}) := 0$ for $(v(t), v(t+1)) \in E^{\mathrm{T}}$.

If the flow correspondence is the following: ${x_{e(t)}}^{\mathrm{T}} := x_e(t)$, where ${x_{(v(t), v(t+1))}}^{\mathrm{T}}$ in N^{T} corresponds to the flow in N stored at node v at period of time from t to $t+1$, the minimum cost flow problem on dynamic networks can be solved by solving the minimum cost static flow problem on the time-expanded network.

13.3.4 The Minimum Cost Dynamic Flow Problems with Different Types of Cost Functions on Arcs and some Generalizations

If the cost function $\varphi_e(x_e(t), t)$ is linear with regard to $x_e(t)$, then the cost function of the time-expanded network is linear. In this case, we can apply well-established methods for minimum cost flow problems, including linear programming algorithms [28], combinatorial algorithms, as well as other developments, like [24].

If the cost function $\varphi_e(x_e(t), t)$ is convex with regard to $x_e(t)$, then the cost function of the time-expanded network is convex. To solve the obtained static problem we can apply methods from convex programming as well as the specialization of such methods for the minimum cost flow problem.

If there is exactly one source, and the cost function $\varphi_e(x_e(t),t)$ is concave with regard to $x_e(t)$, then the cost function of the time-expanded network is concave. If the dynamic network is acyclic, then the time-expanded network is acyclic and finite [39, 40]. Therefore we can solve the static problem with polynomial algorithms for the minimum cost flow problem on acyclic networks with concave cost functions [36].

The same reasoning to solve the minimum cost dynamic flow problem can be held in the case when, instead of condition (13.10) in the definition of the feasible dynamic flow, the following condition takes place:

$$u'_e(t) \leq x_e(t) \leq u''_e(t), \quad \forall t \in \mathbb{T}, \; \forall e \in E,$$

where $u'_e(t) \geq 0$ and $u''_e(t) \geq 0$ are lower and upper boundaries of the capacity of arc e, respectively.

13.3.5 Dynamic Networks with Transit Time Functions that Depend on Flow and Time

In the above dynamic models, the transit time functions are assumed to be constant on each arc of the network. In this setting, the time it takes to traverse an arc does not depend on the current flow situation on the arc and the moment of time. Intuitively, it is clear that in many applications the amount of time needed to traverse an arc of the network increases as the arc becomes more congested and it also depends on the entering time-moment of flow in the arc. If these assumptions are taking into account, a more realistic model can be obtained. In this model, we assume that the transit time function $\tau_e(x_e(t),t)$ is a non-negative non-decreasing left-continuous step function with respect to the amount of flow $x_e(t)$ for every fixed time-moment $t \in T$ and an arbitrary given arc $e \in E$. We also consider two-side restrictions on arc capacities $u'_e(t) \leq x_e(t) \leq u''_e(t), \forall t \in T, \forall e \in E$, where $u', u''\colon E \times T \to R_+$ are lower and upper capacities, respectively.

It is shown [19] that the minimum cost flow problem on dynamic network with transit time functions that depend on the amount of flow and the entering time-moment of flow in the arc can be reduced to a static problem on a special time-expanded network $N^{\mathrm{T}} = (V^{\mathrm{T}}, E^{\mathrm{T}}, d^{\mathrm{T}}, u'^{\mathrm{T}}, u''^{\mathrm{T}}, \varphi^{\mathrm{T}})$, which is defined as follows:

1. $\overline{V}^{\mathrm{T}}\colon = \{v(t) \,|\, v \in V, \; t \in T\}$;
2. $\widetilde{V}^{\mathrm{T}}\colon = \{e(v(t)) \,|\, v(t) \in \overline{V}^{\mathrm{T}}, \; e \in E^-(v), \; t \in T \setminus \{T\}\}$;
3. $V^{\mathrm{T}}\colon = \overline{V}^{\mathrm{T}} \cup \widetilde{V}^{\mathrm{T}}$;
4. $\widetilde{E}^{\mathrm{T}}\colon = \{\widetilde{e}(t) = (v(t), e(v(t))) \,|\, v(t) \in \overline{V}^{\mathrm{T}}$ and corresponding $e(v(t)) \in \widetilde{V}^{\mathrm{T}}, \; t \in T \setminus \{T\}\}$;

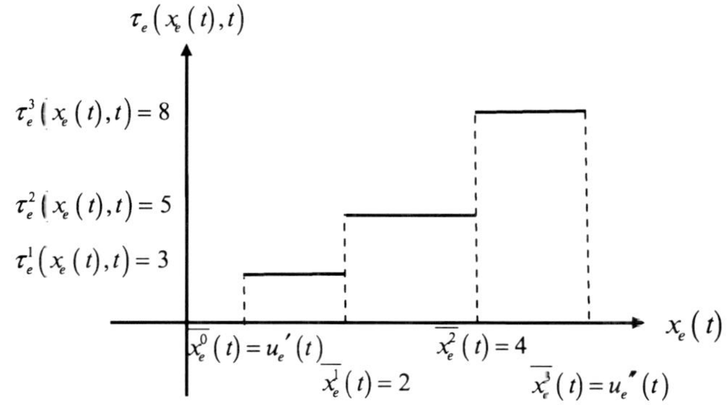

Fig. 13.3 The transit time function for the fixed moment of time t and the given arc $e=(v,z)$

5. $\overline{E}^{\mathrm{T}} := \{e^p(t) = (e(v(t)), z(t+\tau_e^p(x_e(t),t))) \mid e(v(t)) \in \widetilde{V}^{\mathrm{T}},\ z(t+\tau_e^p(x_e(t),t)) \in \overline{V}^{\mathrm{T}},\ e=(v,z) \in E,\ 0 \le t \le T - \tau_e^p(x_e(t),t),\ p \in P_{e,t}$ - set of numbers of steps of the transit time function $\tau_e^p(x_e(t)\ t)\}$;
6. $E^{\mathrm{T}} := \overline{E}^{\mathrm{T}} \cup \widetilde{E}^{\mathrm{T}}$;
7. ${d_{v(t)}}^{\mathrm{T}} := d_v(t)$ for $v(t) \in \overline{V}^{\mathrm{T}}$;
   ${d_{e(v(t))}}^{\mathrm{T}} := 0$ for $e(v(t)) \in \widetilde{V}^{\mathrm{T}}$;
8. ${u'_{\widetilde{e}(t)}}^{\mathrm{T}} := u'_e(t)$ for $\widetilde{e}(t) \in \widetilde{E}^{\mathrm{T}}$;
   ${u''_{\widetilde{e}(t)}}^{\mathrm{T}} := u''_e(t)$ for $\widetilde{e}(t) \in \widetilde{E}^{\mathrm{T}}$;
   ${u'_{e^p(t)}}^{\mathrm{T}} := \overline{x_e^{p-1}}(t)$ for $e^p(t) \in \overline{E}^{\mathrm{T}}$, where $\overline{x_e^0}(t) = u'_e(t)$;
   ${u''_{e^p(t)}}^{\mathrm{T}} := \overline{x_e^p}(t)$ for $e^p(t) \in \overline{E}^{\mathrm{T}}$;
9. ${\varphi_{\widetilde{e}(t)}}^{\mathrm{T}}({x_{\widetilde{e}(t)}}^{\mathrm{T}}) := \varphi_e(x_e(t),t)$ for $\widetilde{e}(t) \in \widetilde{E}^{\mathrm{T}}$;
   ${\varphi_{e^p(t)}}^{\mathrm{T}}({x_{e^p(t)}}^{\mathrm{T}}) := \varepsilon_p$ for $e^p(t) \in \overline{E}^{\mathrm{T}}$, where $\varepsilon_1 < \varepsilon_2 < \cdots < \varepsilon_{|P_{e,t}|}$ are small numbers.

Let us consider, for example, the transit time function $\tau_e = \tau_e(x_e(t),t)$, graphic of which for the fixed moment of time t and the given arc e is presented in Fig. 13.3. Here $P_{e,t} = \{1,2,3\}$. So, for the fixed moment of time t on the given arc e the transit time is equal to 3 if the value of flow belongs to interval $[u'_e(t),2]$; the transit time is equal to 5 if the value of flow belongs to interval $(2,4]$; the transit time is equal to 8 if the value of flow belongs to interval $(4,u''_e(t)]$.

In Fig. 13.4 a part of the obtained time-expanded network is presented for the fixed moment of time t for the given arc $e=(v,z)$ with the transit time function in Fig. 13.3. Lower and upper capacities of arcs are written above each arc and costs are written below each arc.

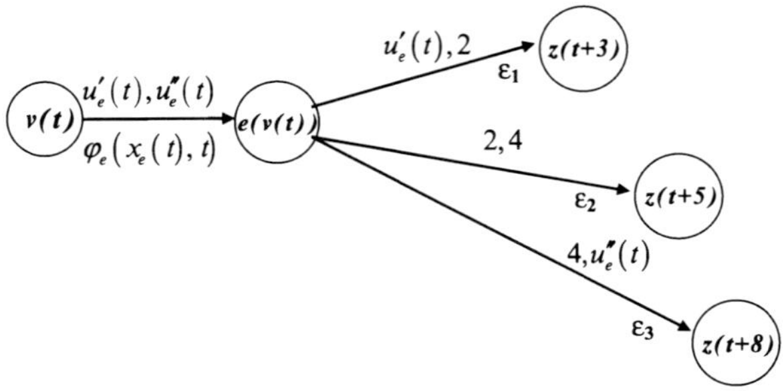

Fig. 13.4 The part of the constructed time-expanded network N^{T} for the fixed moment of time t for the arc $e = (v, z)$

The solution of the dynamic problem can be found on the basis of the following results.

Lemma 13.2. *Let x^{T}: $E^{\mathrm{T}} \to R_+$ be a flow in the static network N^{T}. Then the function x: $E \times T \to R_+$ defined as follows:*

$$x_e(t) = {x_{\widetilde{e}(t)}}^{\mathrm{T}} = {x_{e^p(t)}}^{\mathrm{T}}$$

$$\text{for } e = (v,z) \in E,\ \widetilde{e}(t) = (v(t), e(v(t))) \in \widetilde{E}^{\mathrm{T}},$$

$$e^p(t) = (e(v(t)), z(t + \tau_e^p(x_e(t), t))) \in \overline{E}^{\mathrm{T}},$$

$$p \in P_{e,t} \text{ is such that } {x_{\widetilde{e}(t)}}^{\mathrm{T}} \in \left(\overline{x_e^{p-1}}(t), \overline{x_e^p}(t)\right],\ t \in T,$$

represents a flow in the dynamic network N.

Let x: $E \times T \to R_+$ be a flow in the dynamic network N. Then the function x^{T}: $E^{\mathrm{T}} \to R_+$ defined as follows:

$${x_{\widetilde{e}(t)}}^{\mathrm{T}} = x_e(t) \text{ for } \widetilde{e}(t) = (v(t), e(v(t))) \in \widetilde{E}^{\mathrm{T}},\ e = (v,z) \in E,\ t \in T;$$

$${x_{e^p(t)}}^{\mathrm{T}} = x_e(t) \text{ for such } p \in P_{e,t} \text{ that } x_e(t) \in \left(\overline{x_e^{p-1}}(t), \overline{x_e^p}(t)\right]$$

$$\text{and } {x_{e^p(t)}}^{\mathrm{T}} = 0 \text{ for all other } p \in P_{e,t}$$

$$\text{for } e^p(t) = (e(v(t)), z(t + \tau_e^p(x_e(t), t))) \in \overline{E}^{\mathrm{T}},\ e = (v,z) \in E,\ t \in T,$$

represents a flow in the static network N^{T}.

Theorem 13.2. *If ${x^*}^{\mathrm{T}}$ is a static minimum cost flow in the static network N^{T}, then the corresponding according to Lemma 13.2 dynamic flow x^* in the dynamic network N is also a minimum cost flow and vice-versa.*

The proofs of the above lemma and theorem can be obtained by using the arguments similar to the ones in the proofs of Lemma 13.1 and Theorem 13.1.

13.3.6 The Maximum Flow Problem on Dynamic Network

Let be given a dynamic network N, determined by a directed graph $G = (V,E)$, where V is a set of vertices and E is a set of arcs. As above, we consider the discrete time model, in which all times are integral and bounded by horizon T. The object of the dynamic problem is to find a maximum flow over time in the network N within makespan $\mathbb{T} = \{0,1,2,\ldots,T\}$ while respecting the following restrictions. Each arc $e \in E$ has a nonnegative time-varying capacity $u_e(t)$ which bounds the amount of flow allowed on each arc at every moment of time. Moreover, each arc e has an associated nonnegative transit time τ_e which determines the amount of time it takes for flow to travel through the arc.

A feasible dynamic flow in the network N is a function $x: E \times \mathbb{T} \to R_+$ that satisfies conditions (13.10)–(13.11) and the following conditions:

$$\sum_{e\in E^-(v)} x_e(t) - \sum_{\substack{e\in E^+(v)\\ t-\tau_e\geq 0}} x_e(t-\tau_e) = \begin{cases} y_v(t),\ v\in V_+, \\ 0,\ v\in V_*, \\ -y_v(t),\ v\in V_-, \end{cases} \qquad \forall t\in\mathbb{T},\ \forall v\in V;$$

$$y_v(t) \geq 0,\ \forall t\in\mathbb{T},\ \forall v\in V.$$

The total value of the dynamic flow x in the network N is defined as follows:

$$|x| = \sum_{t\in\mathbb{T}}\sum_{v\in V_+} y_v(t).$$

The maximum dynamic flow problem consists in finding a feasible dynamic flow that maximizes this objective function.

To solve the considered problem we propose an approach, which is based on the reduction of the maximum dynamic flow problem to a well studied maximum static flow problem. We show that our problem on network N can be reduced to a classical problem on the time-expanded network N^{T}, which is defined in the following way:

1. $V^{\mathrm{T}}: = \{v(t)\,|\,v\in V,\ t\in\mathbb{T}\}$;
2. $E^{\mathrm{T}}: = \{(v(t),z(t+\tau_e))\,|\,e=(v,z)\in E,\ 0\leq t\leq T-\tau_e\}$;
3. ${u_{e(t)}}^{\mathrm{T}}: = u_e(t)$ for $e(t)\in E^{\mathrm{T}}$;
4. ${y_{v(t)}}^{\mathrm{T}}: = y_v(t)$ for $v(t)\in V^{\mathrm{T}}$.

The correspondence between feasible flows in the dynamic network N and feasible flows in the time-expanded network N^{T} is defined by (13.13).

Lemma 13.3. *The correspondence (13.13) is a bijection from the set of feasible flows in the dynamic network N onto the set of feasible flows in the time-expanded network N^{T}.*

Proof. It is easy to see that the correspondence (13.13) is a bijection from the set of T-horizon functions in the dynamic network N onto the set of functions in the time-expanded network N^{T}. In the following, we have to verify that each flow in the dynamic network N is put into the correspondence with a flow in the time-expanded network N^{T} and vice-versa. Moreover, we have to show that a feasible flow in the dynamic network N is a feasible flow in the time-expanded network N^{T} and vice-versa.

Henceforward, we define

$$d_v(t) = \begin{cases} y_v(t), \ v \in V_+, \\ 0, \ v \in V_*, \\ -y_v(t), \ v \in V_-, \end{cases} \qquad \forall t \in \mathbb{T}, \ \forall v \in V,$$

and continue the proof of the lemma in the way similar to the one for Lemma 13.1.

The lemma is proved. □

Remark 13.1. The following condition is true:

$$\sum_{t \in \mathbb{T}} \sum_{v \in V_-} y_v(t) = \sum_{t \in \mathbb{T}} \sum_{v \in V_+} y_v(t).$$

The total value of the static flow in the time-expanded network N^{T} is defined as follows:

$$|x^{\mathrm{T}}| = \sum_{t \in \mathbb{T}} \sum_{v(t) \in V_+^{\mathrm{T}}} y_{v(t)}{}^{\mathrm{T}}.$$

Theorem 13.3. *If x is a flow in the dynamic network N and x^{T} is a corresponding flow in the time-expanded network N^{T}, then*

$$|x| = |x^{\mathrm{T}}|.$$

Moreover, for each maximum flow x^* in the dynamic network N there is a corresponding maximum flow $x^{*\mathrm{T}}$ in the static network N^{T} such that

$$|x^*| = |x^{*\mathrm{T}}|$$

and vice-versa.

Proof. Let $x : E \times T \to R_+$ be an arbitrary dynamic flow in N. Then according to Lemma 13.3 the unique flow x^{T} in N^{T} corresponds to the flow x in N, and we obtain:

$$|x| = \sum_{t \in \mathbb{T}} \sum_{v \in V_+} y_v(t) = \sum_{t \in \mathbb{T}} \sum_{v(t) \in V_+^{\mathrm{T}}} y_{v(t)}{}^{\mathrm{T}} = |x^{\mathrm{T}}|.$$

So, the first part of the theorem is proved. The second part of the theorem is proved in the way similar to the one for Theorem 13.1. The theorem is proved. □

In such a way, the maximum flow problem on dynamic networks can be solved by applying network flow optimization methods and algorithms for static flows directly to the time-expanded network. To solve the maximum flow problem on dynamic network we have to construct the time-expanded network, after what to solve the classical maximum flow problem on the static network, using one of the known algorithms [1, 11, 13, 14, 22, 27, 31, 34, 43, 46, 48], and then to reconstruct the solution of the static problem to the dynamic problem.

13.4 Dynamic Multicommodity Flow Problems and Algorithms for their Solving

In this section we study dynamic versions of the minimum cost multicommodity flow problem and the maximum multicommodity flow problem on networks. These problems are considered on dynamic networks with time-varying capacities of arcs and transit times on arcs that depend on sort of commodity entering them. For the minimum cost multicommodity dynamic flow problem we assume that cost functions, defined on arcs, are nonlinear and depend on time and flow, and demand-supply functions depend on time. We also consider optimal multicommodity flow problems on dynamic networks with transit time functions that depend on flow and time. For solving the considered dynamic problems we propose methods and algorithms based on reduction of dynamic problems to static ones on an auxiliary time-expanded network. The algorithm for construction a special reduced time-expanded network for an acyclic network is also proposed.

13.4.1 The Minimum Cost Multicommodity Flow Problem on Dynamic Network

We consider a dynamic network $N = (V, E, K, \tau, d, u, w, \varphi)$, determined by directed graph $G = (V, E)$, where V is a set of vertices and E is a set of arcs, set of commodities $K = \{1, 2, \ldots, q\}$ that must be routed through the same network, transit time function τ: $E \times K \to R_+$, demand-supply function d: $V \times K \times T \to R$, mutual capacity function u: $E \times T \to R_+$, individual capacity function w: $E \times K \times T \to R_+$ and cost function φ: $E \times R_+ \times T \to R_+$. So, $\tau_e = (\tau_e^1, \tau_e^2, \ldots, \tau_e^q)$ is a vector, each component of which reflects the transit time on arc $e \in E$ for commodity $k \in K$. We consider the discrete time model, where all times are integral and bounded by horizon T, which defines the set $T = \{0, 1, \ldots, T\}$ of time moments.

In order for the flow to exist it is required that $\sum_{t\in T}\sum_{v\in V} d_v^k(t) = 0, \forall k \in K$. As above without losing generality we consider that for every commodity $k \in K$ the set of vertices V is divided into three disjoint subsets V_+^k, V_-^k, V_*^k, such that:

V_+^k consists of nodes $v \in V$, for which $d_v^k(t) \geq 0$ for $t \in T$ and there exists at least one moment of time $t_0 \in T$ such that $d_v^k(t_0) > 0$;

V_-^k consists of nodes $v \in V$, for which $d_v^k(t) \leq 0$ for $t \in T$ and there exists at least one moment of time $t_0 \in T$ such that $d_v^k(t_0) < 0$;

V_*^k consists of nodes $v \in V$, for which $d_v^k(t) = 0$ for every $t \in T$.

So, V_+^k is a set of sources, V_-^k is a set of sinks and V_*^k is a set of intermediate nodes for the commodity $k \in K$ in the network N.

A feasible dynamic multicommodity flow in the network N is determined by a function x: $E \times K \times T \to R_+$ that satisfies the following conditions:

$$\sum_{e\in E^-(v)} x_e^k(t) - \sum_{\substack{e\in E^+(v)\\ t-\tau_e^k\geq 0}} x_e^k(t-\tau_e^k) = d_v^k(t),\ \forall t \in T,\ \forall v \in V,\ \forall k \in K; \quad (13.19)$$

$$\sum_{k\in K} x_e^k(t) \leq u_e(t),\ \forall t \in T,\ \forall e \in E; \quad (13.20)$$

$$0 \leq x_e^k(t) \leq w_e^k(t),\ \ \forall t \in T,\ \forall e \in E,\ \forall k \in K; \quad (13.21)$$

$$x_e^k(t) = 0,\ \forall e \in E,\ t = \overline{T - \tau_e^k + 1, T},\ \forall k \in K. \quad (13.22)$$

Here, the function x defines the value $x_e^k(t)$ of flow of commodity k entering arc e at moment of time t. Condition (13.22) ensures that the flow of commodity k does not enter arc e at time t if it has to leave the arc after time horizon T. Individual and mutual capacity constraints (13.21) and (13.20) are called weak and strong forcing constraints, respectively. Conditions (13.19) represent flow conservation constraints.

The total cost of the dynamic multicommodity flow x in the network N is defined as follows:

$$F(x) = \sum_{t\in T}\sum_{e\in E} \varphi_e(x_e^1(t), x_e^2(t), \ldots, x_e^q(t), t). \quad (13.23)$$

The minimum cost dynamic multicommodity flow problem consists in finding a feasible dynamic multicommodity flow that minimizes the objective function (13.23).

13.4.2 The Algorithm for Solving the Minimum Cost Dynamic Multicommodity Flow Problem

To solve the formulated problem we propose an approach based on the reduction of the dynamic problem to a static problem. We show that the minimum cost multicommodity flow problem on network N can be reduced to a static problem on a special auxiliary network N^{T}. In the case of the minimum cost multicommodity flow problem on dynamic network with different transit times on an arc for different commodities the auxiliary time-expanded network $N^{\mathrm{T}} = (V^{\mathrm{T}}, E^{\mathrm{T}}, K, d^{\mathrm{T}}, u^{\mathrm{T}}, w^{\mathrm{T}}, \varphi^{\mathrm{T}})$ is defined in the following way:

1. $\overline{V}^{\mathrm{T}} := \{v(t) \,|\, v \in V,\ t \in T\}$;
2. $\widetilde{V}^{\mathrm{T}} := \{e(v(t)) \,|\, v(t) \in \overline{V}^{\mathrm{T}},\ e \in E^-(v),\ t \in T \setminus \{T\}\}$;
3. $V^{\mathrm{T}} := \overline{V}^{\mathrm{T}} \cup \widetilde{V}^{\mathrm{T}}$;
4. $\widetilde{E}^{\mathrm{T}} := \{\widetilde{e}(t) = (v(t), e(v(t))) \,|\, v(t) \in \overline{V}^{\mathrm{T}}$ and corresponding $e(v(t)) \in \widetilde{V}^{\mathrm{T}},\ t \in T \setminus \{T\}\}$;
5. $\overline{E}^{\mathrm{T}} := \{e^k(t) = (e(v(t)), z(t+\tau_e^k)) \,|\, e(v(t)) \in \widetilde{V}^{\mathrm{T}},\ z(t+\tau_e^k) \in \overline{V}^{\mathrm{T}},\ e = (v,z) \in E,\ 0 \le t \le T - \tau_e^k,\ k \in K\}$;
6. $E^{\mathrm{T}} := \overline{E}^{\mathrm{T}} \cup \widetilde{E}^{\mathrm{T}}$;
7. ${d^k_{v(t)}}^{\mathrm{T}} := d_v^k(t)$ for $v(t) \in \overline{V}^{\mathrm{T}},\ k \in K$;

   ${d^k_{e(v(t))}}^{\mathrm{T}} := 0$ for $e(v(t)) \in \widetilde{V}^{\mathrm{T}},\ k \in K$;
8. ${u_{\widetilde{e}(t)}}^{\mathrm{T}} := u_e(t)$ for $\widetilde{e}(t) \in \widetilde{E}^{\mathrm{T}}$;

   ${u_{e^k(t)}}^{\mathrm{T}} := \infty$ for $e^k(t) \in \overline{E}^{\mathrm{T}}$;
9. ${w^l_{e^k(t)}}^{\mathrm{T}} := \begin{cases} w_e^k(t), & \text{if } l = k \text{ for } e^k(t) \in \overline{E}^{\mathrm{T}},\ l \in K; \\ 0, & \text{if } l \neq k \text{ for } e^k(t) \in \overline{E}^{\mathrm{T}},\ l \in K; \end{cases}$

   ${w^l_{\widetilde{e}(t)}}^{\mathrm{T}} = \infty$ for $\widetilde{e}(t) \in \widetilde{E}^{\mathrm{T}},\ l \in K$;
10. ${\varphi_{\widetilde{e}(t)}}^{\mathrm{T}}({x^1_{\widetilde{e}(t)}}^{\mathrm{T}}, {x^2_{\widetilde{e}(t)}}^{\mathrm{T}}, \ldots, {x^q_{\widetilde{e}(t)}}^{\mathrm{T}}) := \varphi_e(x_e^1(t), x_e^2(t), \ldots, x_e^q(t), t)$ for $\widetilde{e}(t) \in \widetilde{E}^{\mathrm{T}}$;

    ${\varphi_{e^k(t)}}^{\mathrm{T}}({x^1_{e^k(t)}}^{\mathrm{T}}, {x^2_{e^k(t)}}^{\mathrm{T}}, \ldots, {x^q_{e^k(t)}}^{\mathrm{T}}) := 0$ for $e^k(t) \in \overline{E}^{\mathrm{T}}$.

In the following, we construct the time-expanded network N^{T} for the dynamic network N given in Fig. 13.1 with set of two commodities $K = \{1,2\}$, set of time moments $T = \{0,1,2,3\}$ and transit times $\tau_{e_1}^1 = 2$, $\tau_{e_1}^2 = 1$, $\tau_{e_2}^1 = 1$, $\tau_{e_2}^2 = 3$, $\tau_{e_3}^1 = 1$, $\tau_{e_3}^2 = 2$. The mutual capacity, individual capacity, demand–supply, and cost functions are considered to be known. The constructed time-expanded network N^{T} is presented in Fig. 13.5.

Lemma 13.4. *Let $x^{\mathrm{T}}\colon E^{\mathrm{T}} \times K \to R_+$ be a multicommodity flow in the static network N^{T}. Then the function $x\colon E \times K \times T \to R_+$ defined in the following way:*

$$x_e^k(t) = {x^k_{e^k(t)}}^{\mathrm{T}} = {x^k_{\widetilde{e}(t)}}^{\mathrm{T}}$$

$$\textit{for } e = (v,z) \in E,\ e^k(t) = (e(v(t)), z(t+\tau_e^k)) \in \overline{E}^{\mathrm{T}},$$

$$\widetilde{e}(t) = (v(t), e(v(t))) \in \widetilde{E}^{\mathrm{T}},\ k \in K,\ t \in T,$$

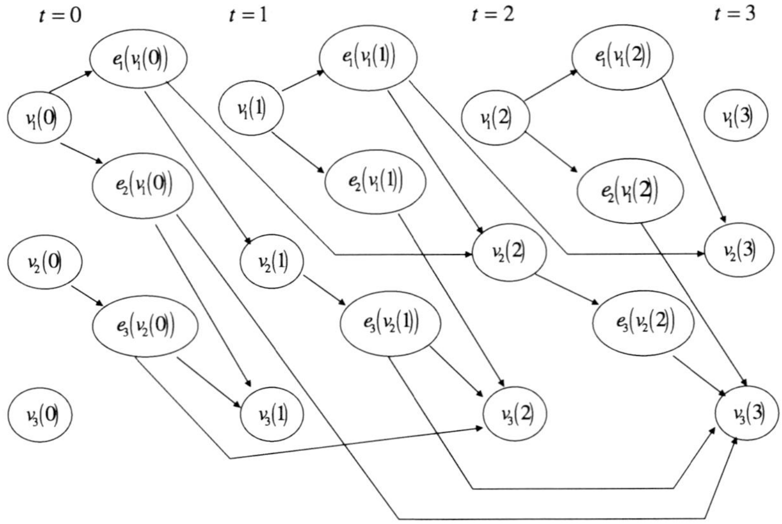

Fig. 13.5 The time-expanded network (case of different transit times on an arc for different commodities)

represents a multicommodity flow in the dynamic network N.

Let x: $E \times K \times T \rightarrow R_+$ be a multicommodity flow in the dynamic network N. Then the function x^{T}: $E^{\mathrm{T}} \times K \rightarrow R_+$ defined in the following way:

$$x^{k}_{\tilde{e}(t)}{}^{\mathrm{T}} = x^{k}_{e}(t) \; for \; \tilde{e}(t) = (v(t), e(v(t))) \in \widetilde{E}^{\mathrm{T}}, \; e = (v,z) \in E, \; k \in K, \; t \in T;$$

$$x^{k}_{e^k(t)}{}^{\mathrm{T}} = x^{k}_{e}(t); \; x^{l}_{e^k(t)}{}^{\mathrm{T}} = 0, \; l \neq k$$

$$for \; e^k(t) = (e(v(t)), \; z(t + \tau^k_e)) \in \overline{E}^{\mathrm{T}}, \; e = (v,z) \in E, \; l,k \in K, \; t \in T,$$

represents a multicommodity flow in the static network N^{T}.

Proof. To prove the first part of the lemma we have to show that conditions (13.19)–(13.22) for the defined above x in the dynamic network N are true. These conditions evidently result from the following definition of multicommodity flows in the static network N^{T}:

$$\sum_{e(t) \in E^{-}(v(t))} x^{k}_{e(t)}{}^{\mathrm{T}} - \sum_{e(t-\tau^k_e) \in E^{+}(v(t))} x^{k}_{e(t-\tau^k_e)}{}^{\mathrm{T}} = d^{k}_{v(t)}{}^{\mathrm{T}},$$

$$\forall v(t) \in V^{\mathrm{T}}, \; \forall k \in K; \qquad (13.24)$$

$$\sum_{k\in K} x^{k}_{e(t)}{}^{\mathrm{T}} \leq u_{e(t)}{}^{\mathrm{T}},\ \forall e(t)\in E^{\mathrm{T}}; \tag{13.25}$$

$$0 \leq x^{k}_{e(t)}{}^{\mathrm{T}} \leq w^{k}_{e(t)}{}^{\mathrm{T}},\ \forall e(t)\in E^{\mathrm{T}},\ \forall k\in K; \tag{13.26}$$

$$x^{k}_{e(t)}{}^{\mathrm{T}} = 0,\ \forall e(t)\in E^{\mathrm{T}},\ t=\overline{T-\tau^{k}_{e}+1\ T},\ \forall k\in K, \tag{13.27}$$

where by $v(t)$ and $e(t)$ we denote $\overline{v}(t)$ or $\widetilde{v}(t)$ and $\overline{e}(t)$ or $\widetilde{e}(t)$, respectively, against context.

In order to prove the second part of the lemma it is sufficiently to show that conditions (13.24)–(13.27) hold for the defined above x^{T}. Correctness of these conditions results from the procedure of constructing the time-expanded network, the correspondence between flows in static and dynamic networks and the satisfied conditions (13.19)–(13.22).

The lemma is proved. □

Theorem 13.4. *If $x^{*\mathrm{T}}$ is a minimum cost multicommodity flow in the static network N^{T}, then the corresponding according to Lemma 13.4 multicommodity flow x^{*} in the dynamic network N is also a minimum cost one and vice-versa.*

Proof. Taking into account the correspondence between static and dynamic multicommodity flows on the basis of Lemma 13.4, we obtain that costs of the static multicommodity flow in the time-expanded network N^{T} and the corresponding dynamic multicommodity flow in the dynamic network N are equal. To solve the minimum cost multicommodity flow problem on the static time-expanded network N^{T}, we have to solve the following problem:

$$F^{\mathrm{T}}(x^{\mathrm{T}}) = \sum_{t\in T}\sum_{e(t)\in E^{\mathrm{T}}} \varphi_{e(t)}{}^{\mathrm{T}}\left(x^{1}_{e(t)}{}^{\mathrm{T}}, x^{2}_{e(t)}{}^{\mathrm{T}}, \ldots, x^{q}_{e(t)}{}^{\mathrm{T}}\right) \to \min$$

subject to (13.24)–(13.27). □

In the case of the minimum cost multicommodity flow problem on dynamic network with common transit times on an arc for different commodities the time-expanded network N^{T} can be constructed more simply:

1. $V^{\mathrm{T}}:=\{v(t)\,|\,v\in V,\ t\in T\}$;
2. $E^{\mathrm{T}}:=\{e(t)=(v(t),z(t+\tau_e))\,|\,v(t)\in V^{\mathrm{T}},\ z(t+\tau_e)\in V^{\mathrm{T}},\ e=(v,z)\in E,\ 0\leq t\leq T-\tau_e\}$;
3. $d^{k}_{v(t)}{}^{\mathrm{T}}:=d^{k}_{v}(t)$ for $v(t)\in V^{\mathrm{T}},\ k\in K$;
4. $u_{e(t)}{}^{\mathrm{T}}:=u_e(t)$ for $e(t)\in E^{\mathrm{T}}$;
5. $w^{k}_{e(t)}{}^{\mathrm{T}}:=w^{k}_{e}(t)$ for $e(t)\in E^{\mathrm{T}},\ k\in K$;
6. $\varphi_{e(t)}{}^{\mathrm{T}}(x^{1}_{e(t)}{}^{\mathrm{T}}, x^{2}_{e(t)}{}^{\mathrm{T}},\ldots,x^{q}_{e(t)}{}^{\mathrm{T}}):=\varphi_e(x^{1}_{e}(t),x^{2}_{e}(t),\ldots,x^{q}_{e}(t),t)$ for $e(t)\in E^{\mathrm{T}}$.

The following lemma and theorem can be considered as particular cases of Lemma 13.4 and Theorem 13.4.

Lemma 13.5. *Let* x^{T}*:* $E^{\mathrm{T}} \times K \rightarrow R_+$ *be a multicommodity flow in the static network* N^{T}*. Then the function* x*:* $E \times K \times T \rightarrow R_+$ *defined as follows:*

$$x_e^k(t) = x_{e(t)}^{k\ \mathrm{T}} \text{ for } e \in E,\ e(t) \in E^{\mathrm{T}},\ k \in K,\ t \in T,$$

represents the multicommodity flow in the dynamic network N*.*

Let x*:* $E \times K \times T \rightarrow R_+$ *be a multicommodity flow in the dynamic network* N*. Then the function* x^{T}*:* $E^{\mathrm{T}} \times K \rightarrow R_+$ *defined as follows:*

$$x_{e(t)}^{k\ \mathrm{T}} = x_e^k(t) \text{ for } e(t) \in E^{\mathrm{T}},\ e \in E,\ k \in K,\ t \in T,$$

represents the multicommodity flow in the static network N^{T}*.*

Theorem 13.5. *If* $x^{*\mathrm{T}}$ *is a minimum cost multicommodity flow in the static network* N^{T}*, then the corresponding according to Lemma 13.5 multicommodity flow* x^* *in the dynamic network* N *is also a minimum cost one and vice-versa.*

In such a way, to solve the minimum cost multicommodity flow problem on dynamic networks we have:

1. To build the time-expanded network N^{T} for the given dynamic network N.
2. To solve the classical minimum cost multicommodity flow problem on the static network N^{T} [4, 8–10, 14, 15, 44].
3. To reconstruct the solution of the static problem on N^{T} to the dynamic problem on N.

The complexity of this algorithm depends on the complexity of the algorithm used for the minimum cost multicommodity flow problem on the static network. If such an algorithm has complexity $O(f(n',m'))$, where n' is the number of vertices and m' is the number of arcs in the network, then the complexity of solving the minimum cost multicommodity flow problem with different transit times on arcs for different commodities on the time-expanded network employing the same algorithm is $O(f((n+m)T, m(k+1)T))$, where n is the number of vertices in the dynamic network, m is the number of arcs in the dynamic network and k is the number of commodities.

13.4.3 The Construction of the Time-Expanded Network for Acyclic Graphs

In this subsection, we consider the minimum cost multicommodity flow problem on the acyclic dynamic network $N = (V, E, K, \tau, d, u, w, \varphi)$ with time horizon $T = +\infty$ and common transit times on an arc for different commodities. Without losing generality, we assume that no arcs enter sources or exit sinks. Let $T^* = \max\{|L|\} =$

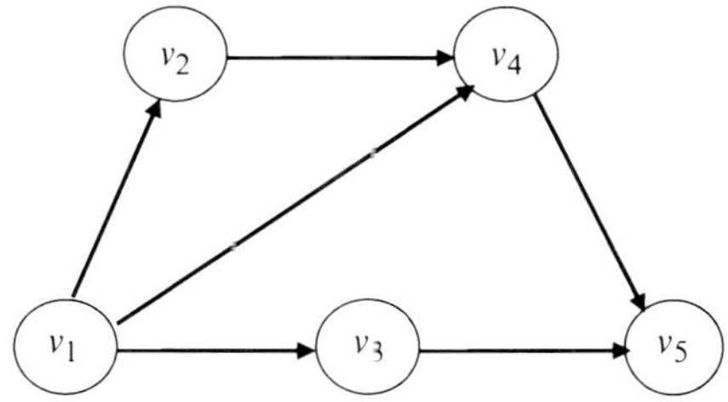

Fig. 13.6 The dynamic network N with five nodes, and six arcs

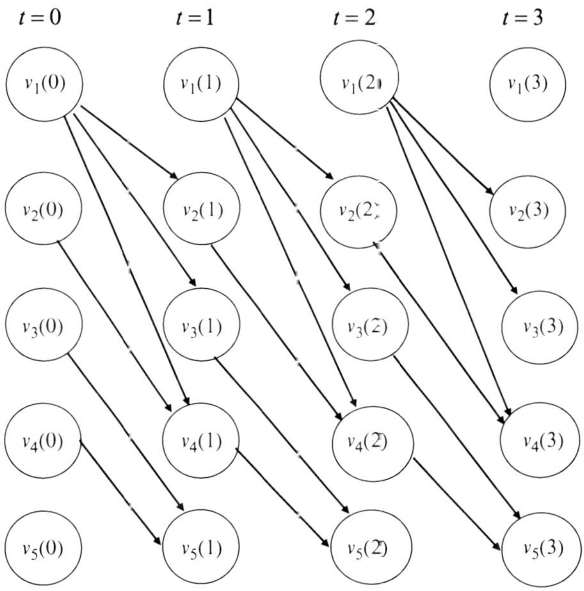

Fig. 13.7 The time-expanded network N^{T^*} for the dynamic network N

$\max\{\sum_{e\in L}\tau_e\}$, where L is a directed path in the graph $G=(V,E)$. It is not difficult to show that $x_e^k(t)=0$ for $\forall e\in E$, $\forall k\in K$, $\forall t\geq T^*$. This fact allows us to replace the infinite time horizon with the finite one, by substituting T^* for the positive infinity.

In many cases, a big number of nodes is not connected with a directed path both to a sink and a source. Removing such nodes from the considered network does not influence the set of flows in this network. These nodes are called irrelevant to the flow problem. Nodes that are not irrelevant are relevant. The static network obtained by eliminating the irrelevant nodes and all arcs adjacent to them from the time-expanded network is called the reduced time-expanded network.

The network in Fig. 13.6 is a dynamic network that conforms to the definition of the acyclic dynamic network, with $V_+=\{v_1\}$ and $V_-=\{v_5\}$. Let us consider that all transit times are equal to 1, and accordingly $T^*=3$.

The time-expanded network built according to the definition is presented in Fig. 13.7. This network has 20 nodes and 18 arcs.

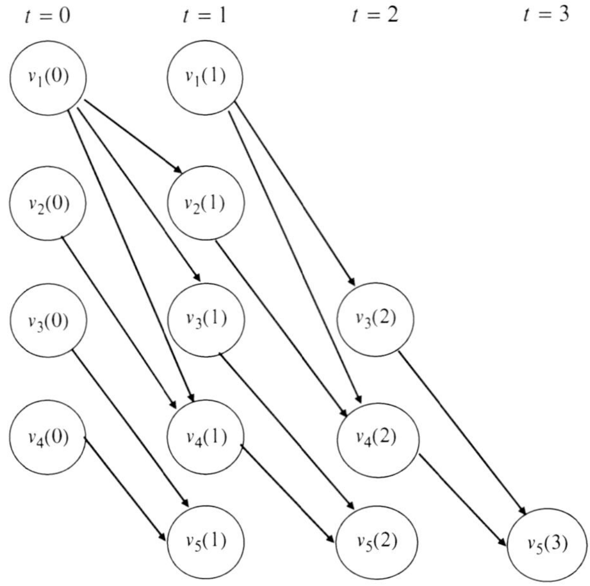

Fig. 13.8 The reduced time-expanded network N^{rT^*}

If we exclude the irrelevant nodes we obtain a smaller static network depicted in Fig. 13.8 with 13 nodes and 13 arcs.

It is proposed the following algorithm for constructing the reduced network $N^{\mathrm{rT}^*} = (V^{\mathrm{rT}^*}, E^{\mathrm{rT}^*}, d^{\mathrm{rT}^*}, u^{\mathrm{rT}^*}, w^{\mathrm{rT}^*}, \varphi^{\mathrm{rT}^*})$ which is based on the process of elimination of irrelevant nodes from the time-expanded network:

1. To build the time-expanded network N^{T^*} for the given dynamic network N.
2. To perform a breadth-first parse of the nodes for each source from the time expanded-network. The result of this step is the set $V_-(V_-^{\mathrm{T}^*})$ of the nodes that can be reached from at least a source in V^{T^*}.
3. To perform a breadth-first parse of the nodes beginning with the sink for each sink and parsing the arcs in the direction opposite to their normal orientation. The result of this step is the set $V_+(V_+^{\mathrm{T}^*})$ of nodes from which at least a sink in V^{T^*} can be reached.
4. The reduced network will consist of a subset of nodes V^{T^*} and arcs from E^{T^*} determined in the following way:

$$V^{\mathrm{rT}^*} = V^{\mathrm{T}^*} \cap V_-(V_-^{\mathrm{T}^*}) \cap V_+(V_+^{\mathrm{T}^*}),$$

$$E^{\mathrm{rT}^*} = E^{\mathrm{T}^*} \cap (V^{\mathrm{rT}^*} \times V^{\mathrm{rT}^*}).$$

5. ${d^{\mathrm{r}k}_{v(t)}}^{\mathrm{T}^*} := d_v^k(t)$ for $v(t) \in V^{\mathrm{rT}^*}$, $k \in K$.
6. ${u^{\mathrm{r}}_{e(t)}}^{\mathrm{T}^*} := u_e(t)$ for $e(t) \in E^{\mathrm{rT}^*}$.

7. ${w^{\text{r}k}_{e(t)}}^{\text{T}^*}: = w_e^k(t)$ for $e(t) \in E^{\text{rT}^*}$, $k \in K$.
8. ${\varphi^{\text{r}}_{e(t)}}^{\text{T}^*}({x^1_{e(t)}}^{\text{T}*}, {x^2_{e(t)}}^{\text{T}*}, \ldots, {x^q_{e(t)}}^{\text{T}*}): = \varphi_e(x_e^1(t), x_e^2(t), \ldots, x_e^q(t), t)$ for $e(t) \in E^{\text{rT}^*}$.

The complexity of this algorithm can be estimated to be the same as the complexity of constructing the time-expanded network It can be proven by using the similar approach as in [39] that the reduced network can be used in place of the time-expanded network.

We'd like to mention that the proposed above approach with some modifications can be used for constructing the reduced time-expanded network for the optimal single-commodity dynamic flow problems and the optimal multicommodity dynamic flow problems with different transit times on an arc for different commodities.

13.4.4 Multicommodity Dynamic Networks with Transit Time Functions that Depend on Flow and Time

We propose an approach for solving the minimum cost multicommodity dynamic flow problem with transit time functions that depend on flow and time. This problem is considered on dynamic networks with time-varying lower and upper capacity functions, time-varying mutual capacity function and time-varying demand–supply function. It is assumed that cost functions, defined on arcs, are nonlinear and depend on flow and time. The transit time function $\tau_e^k(x_e^k(t),t)$ is considered to be a non-negative non-decreasing left-continuous step function for each commodity $k \in K$.

The method for solving the minimum cost multicommodity dynamic flow problem with transit time functions that depend on flows and time is based on the reduction of the dynamic problem to a static problem on an auxiliary time-expanded network $N^{\text{T}} = (V^{\text{T}}, E^{\text{T}}, d^{\text{T}}, u^{\text{T}}, w'^{\text{T}}, w''^{\text{T}}, \varphi^{\text{T}})$ which is defined as follows:

1. $\overline{V}^{\text{T}}: = \{v(t) \,|\, v \in V,\ t \in T\}$;
2. $\widetilde{V}^{\text{T}}: = \{e(v(t)) \,|\, v(t) \in \overline{V}^{\text{T}},\ e \in E^-(v),\ t \in T \setminus \{T\}\}$;
3. $V^{\text{T}}: = \overline{V}^{\text{T}} \cup \widetilde{V}^{\text{T}}$;
4. $\widetilde{E}^{\text{T}}: = \{\widetilde{e}(t) = (v(t), e(v(t))) \,|\, v(t) \in \overline{V}^{\text{T}}$ and corresponding $e(v(t)) \in \widetilde{V}^{\text{T}},\ t \in T \setminus \{T\}\}$;
5. $\overline{E}^{\text{T}}: = \{e^{k,p}(t) = (e(v(t)), z(t + \tau_e^{k,p}(x_e^k(t),t))) \,|\, e(v(t)) \in \widetilde{V}^{\text{T}},\ z(t + \tau_e^{k,p}(x_e^k(t),t)) \in \overline{V}^{\text{T}}, e = (v,z) \in E,\ 0 \le t \le T - \tau_e^{k,p}(x_e^k(t),t),\ p \in P_{e,x}^k -$ set of numbers of steps of the transit time function $\tau_e^k(x_e^k(t),t),\ k \in K\}$;
6. $E^{\text{T}}: = \overline{E}^{\text{T}} \cup \widetilde{E}^{\text{T}}$;

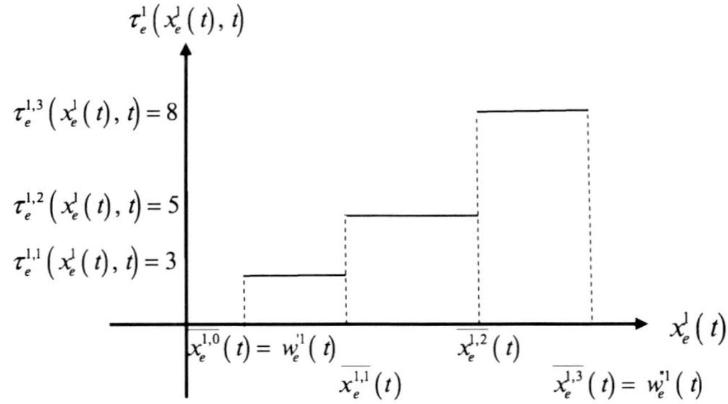

Fig. 13.9 The transit time function for commodity 1 for the fixed moment of time t and the given arc $e = (v, z)$

7. ${d^k_{v(t)}}^{\mathrm{T}} := d^k_v(t)$ for $v(t) \in \overline{V}^{\mathrm{T}}$, $k \in K$;
   ${d^k_{e(v(t))}}^{\mathrm{T}} := 0$ for $e(v(t)) \in \widetilde{V}^{\mathrm{T}}$, $k \in K$;
8. ${u_{\widetilde{e}(t)}}^{\mathrm{T}} := u_e(t)$ for $\widetilde{e}(t) \in \widetilde{E}^{\mathrm{T}}$;
   ${u_{e^{k,p}(t)}}^{\mathrm{T}} := \infty$ for $e^{k,p}(t) \in \overline{E}^{\mathrm{T}}$;
9. $${w'^{l}_{e^{k,p}(t)}}^{\mathrm{T}} := \begin{cases} \overline{x_e^{k,p-1}}(t), \text{ if } l = k \text{ for } e^{k,p}(t) \in \overline{E}^{\mathrm{T}},\ l \in K, \\ \qquad \text{where } \overline{x_e^{k,0}}(t) = w'^{k}_e(t); \\ 0, \qquad \text{if } l \neq k \text{ for } e^{k,p}(t) \in \overline{E}^{\mathrm{T}},\ l \in K; \end{cases}$$
   $${w''^{l}_{e^{k,p}(t)}}^{\mathrm{T}} := \begin{cases} \overline{x_e^{k,p}}(t), \text{ if } l = k \text{ for } e^{k,p}(t) \in \overline{E}^{\mathrm{T}},\ l \in K; \\ 0, \qquad \text{if } l \neq k \text{ for } e^{k,p}(t) \in \overline{E}^{\mathrm{T}},\ l \in K; \end{cases}$$
   ${w'^{l}_{\widetilde{e}(t)}}^{\mathrm{T}} = -\infty$; ${w''^{l}_{\widetilde{e}(t)}}^{\mathrm{T}} = +\infty$ for $\widetilde{e}(t) \in \widetilde{E}^{\mathrm{T}}$, $l \in K$;
10. ${\varphi_{\widetilde{e}(t)}}^{\mathrm{T}}({x^1_{\widetilde{e}(t)}}^{\mathrm{T}}, {x^2_{\widetilde{e}(t)}}^{\mathrm{T}}, \ldots, {x^q_{\widetilde{e}(t)}}^{\mathrm{T}}) := \varphi_e(x^1_e(t), x^2_e(t), \ldots, x^q_e(t), t)$ for $\widetilde{e}(t) \in \widetilde{E}^{\mathrm{T}}$;
    ${\varphi_{e^{k,p}(t)}}^{\mathrm{T}}({x^1_{e^{k,p}(t)}}^{\mathrm{T}}, {x^2_{e^{k,p}(t)}}^{\mathrm{T}}, \ldots, {x^q_{e^{k,p}(t)}}^{\mathrm{T}}) := \varepsilon^{k,p}$ for $e^{k,p}(t) \in \overline{E}^{\mathrm{T}}$, where $\varepsilon^{k,1} < \varepsilon^{k,2} < \cdots < \varepsilon^{k,|P^k_{e,t}|}$ are small numbers.

For example, let us consider the transit time functions for an arc $e = (v, z)$ at the moment of time t presented in Figs. 13.9 and 13.10, which correspond to commodities 1 and 2, respectively.

The constructed part of the time-expanded network for the fixed moment of time t for the arc $e = (v, z)$ is presented in Fig. 13.11.

The following lemma and theorem give us the relationship between flows in network N and flows in network N^{T}.

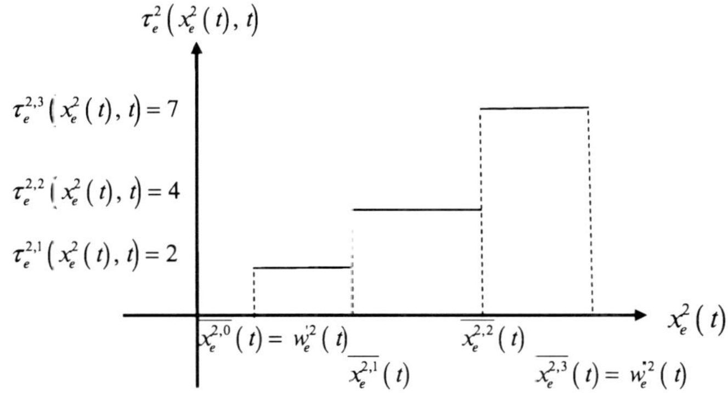

Fig. 13.10 The transit time function for commodity 2 for the fixed moment of time t and the given arc $e = (v, z)$

Fig. 13.11 The part of the constructed time-expanded network N^{T} for the fixed moment of time t for the arc $e = (v, z)$

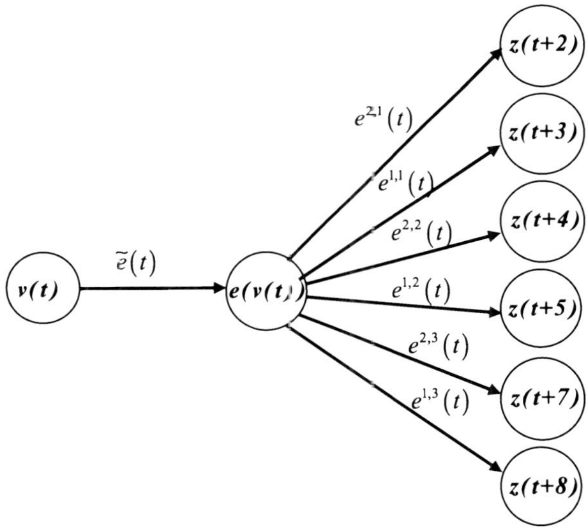

Lemma 13.6. *Let* $x^{\mathrm{T}}: E^{\mathrm{T}} \times K \to R_+$ *be a multicommodity flow in the static network* N^{T}*. Then the function* $x: E \times K \times T \to R_+$ *defined in the following way:*

$$x_e^k(t) = x_{\widetilde{e}(t)}^{k\ \ \mathrm{T}} = x_{e^{k,p}(t)}^{k\ \ \mathrm{T}}$$

$$\text{for } e = (v,z) \in E,\ \widetilde{e}(t) = (v(t), e(v(t))) \in \widetilde{E}^{\mathrm{T}},$$

$$e^{k,p}(t) = (e(v(t)), z(t + \tau_e^{k,p}(x_e^k(t),t))) \in \overline{E}^{\mathrm{T}},$$

$$p \in P_{e,t}^k \text{ is such that } x_{\widetilde{e}(t)}^{k\ \ \mathrm{T}} \in \left(\overline{x_e^{k,p-1}}(t), \overline{x_e^{k,p}}(t)\right],\ t \in T,\ k \in K,$$

represents a multicommodity flow in the dynamic network N.

Let x: $E \times K \times T \rightarrow R_+$ be a multicommodity flow in the dynamic network N. Then the function x^{T}: $E^{\mathrm{T}} \times K \rightarrow R_+$ defined in the following way:

$$x^{k\ \mathrm{T}}_{\widetilde{e}(t)} = x^k_e(t)$$

$$\text{for } \widetilde{e}(t) = (v(t), e(v(t))) \in \widetilde{E}^{\mathrm{T}},\ e = (v,z) \in E,\ k \in K,\ t \in T;$$

$$x^{l\ \mathrm{T}}_{e^{k,p}(t)} = 0,\ l \neq k;$$

$$x^{k\ \mathrm{T}}_{e^{k,p}(t)} = x^k_e(t) \text{ for such } p \in P^k_{e,t} \text{ that } x^k_e(t) \in \left(\overline{x^{k,p-1}_e}(t), \overline{x^{k,p}_e}(t)\right],$$

$$x^{k\ \mathrm{T}}_{e^{k,p}(t)} = 0 \text{ for all other } p \in P^k_{e,t}$$

$$\text{for } e^{k,p}(t) = (e(v(t)), z(t + \tau^{k,p}_e(x^k_e(t),t))) \in \overline{E}^{\mathrm{T}},$$
$$e = (v,z) \in E,\ l,k \in K,\ t \in T,$$

represents a multicommodity flow in the static network N^{T}.

Theorem 13.6. *If $x^{*\mathrm{T}}$ is a minimum cost multicommodity flow in the static network N^{T}, then the corresponding according to Lemma 13.6 multicommodity flow x^* in the dynamic network N is also a minimum cost one and vice-versa.*

13.4.5 The Maximum Multicommodity Dynamic Flow Problem

We show that the proposed time-expanded network method can be used for the maximum multicommodity dynamic flow problem. The scope of this problem is to find the maximum flow of a set of commodities within a given time bound through a network without violating capacity constraints of arcs. As above, time is measured in discrete steps, the set of time moments is $\mathbb{T} = \{0, 1, \ldots, T\}$.

We consider a network N determined by a directed graph $G = (V, E)$ and a set of commodities K that must be routed through the same network. Each arc $e \in E$ has a nonnegative time-varying capacity $w^k_e(t)$ which bounds the amount of flow of each commodity $k \in K$ allowed on arc $e \in E$ at every moment of time $t \in \mathbb{T}$. We also consider that every arc $e \in E$ has a nonnegative time-varying capacity for all commodities, which is known as the mutual capacity $u_e(t)$. Moreover, each arc $e \in E$ has an associated nonnegative transit time τ^k_e which determines the amount of time it takes for flow of commodity $k \in K$ to travel through the arc.

A feasible multicommodity dynamic flow in N is determined by a function x: $E \times K \times \mathbb{T} \rightarrow R_+$ that satisfies conditions (13.20)–(13.22) and the following conditions:

$$\sum_{e \in E^-(v)} x^k_e(t) - \sum_{\substack{e \in E^+(v) \\ t - \tau^k_e \geq 0}} x^k_e(t - \tau^k_e) = \begin{cases} y^k_v(t),\ v \in V^k_+, \\ 0,\ v \in V^k_*, \\ -y^k_v(t),\ v \in V^k_-, \end{cases}$$

$$\forall t \in \mathbb{T},\ \forall v \in V,\ \forall k \in K;$$

$$y_v^k(t) \geq 0,\ \forall v \in V, \forall t \in \mathbb{T},\ \forall k \in K.$$

The total value of the multicommodity dynamic flow x in the network N is defined as follows:

$$|x| = \sum_{k \in K} \sum_{t \in \mathbb{T}} \sum_{v \in V_+^k} y_v^k(t).$$

The object of the formulated problem is to find a feasible multicommodity dynamic flow that maximizes this objective function.

To solve the maximum multicommodity dynamic flow problem by its reduction to a static one we define the time-expanded network N^{T} in a similar way as for the minimum cost multicommodity dynamic flow problem. The correspondence between flows in the dynamic network N and the static network N^{T} is also determined as above.

Using the same reasoning we obtain that if $x^{*\mathrm{T}}$ is a maximum multicommodity flow in the static network N^{T}, then the corresponding multicommodity flow x^* in the dynamic network N is also a maximum one and vice-versa.

In such a way, the maximum multicommodity flow problem on dynamic networks can be solved by applying network flow optimization techniques for static flows directly to the expanded network. To solve the maximum multicommodity flow problem on N we have to build the time-expanded network N^{T} for the given dynamic network N, after what to solve the classical maximum multicommodity flow problem on the static network N^{T}, using one of the known algorithms [8–10, 15, 44] and then to reconstruct the solution of the static problem on N^{T} to the dynamic problem on N.

13.5 Game-Theoretic Approach for Solving Multiobjective Multicommodity Flow Problems on Networks

In this section, we consider the game-theoretic formulation of the multiobjective multicommodity flow problem. If we associate to each commodity a player, we can regard this problem as a game version of the problem, where players interact between them and the choices of one player influence the choices of the others. Each player seeks to optimize his own vector utility function in response to the actions of the other players and at the same time players are interested to preserve Nash optimality principle when they interact between them. The game theory fits perfectly in the realm of such a problem, and an equilibrium or stable operating point of the system has to be found.

13.5.1 Pareto–Nash Equilibria for Multiobjective Games

In order to investigate the multiobjective multicommodity flow problem we will use the game-theoretic concept from [2, 38].

The multiobjective game with q players is denoted by $\overline{\Gamma} = (X_1, X_2, \ldots, X_q, \overline{F}_1, \overline{F}_2, \ldots, \overline{F}_q)$, where X_k is a set of strategies of player k, $k \in \{1,2,\ldots,q\}$, and $\overline{F}_k = (F_k^1, F_k^2, \ldots, F_k^{r_k})$ is a vector payoff function of player k, $k \in \{1,2,\ldots,q\}$, defined on the set of situations $X = X_1 \times X_2 \times \cdots \times X_q$:

$$\overline{F}_k \,:\, X_1 \times X_2 \times \cdots \times X_q \to R^{r_k},\; k \in \{1,2,\ldots,q\}.$$

Each component F_k^i of $\overline{F}_k$ corresponds to a partial criterion of player k and represents a real function defined on the set of situations $X = X_1 \times X_2 \times \cdots \times X_q$:

$$F_k^i \,:\, X_1 \times X_2 \times \cdots \times X_q \to R^1,\; i = \overline{1, r_k},\; k \in \{1,2,\ldots,q\}.$$

In [38], the solution of the multiobjective game $\overline{\Gamma} = (X_1, X_2, \ldots, X_q, \overline{F}_1, \overline{F}_2, \ldots, \overline{F}_q)$ is called the Pareto–Nash equilibrium and is defined in the following way.

Definition 13.1. The situation $x^* = (x_1^*, x_2^*, \ldots, x_q^*) \in X$ is called the Pareto–Nash equilibrium for the multiobjective game $\overline{\Gamma} = (X_1, X_2, \ldots, X_q, \overline{F}_1, \overline{F}_2, \ldots, \overline{F}_q)$ if for every $k \in \{1,2,\ldots,q\}$ the strategy x_k^* represents the Pareto solution for the following multicriterion problem:

$$opt_{x_k \in X_k} \to \overline{f}_{x^*}^k(x_k) = (f_{x^*}^{k1}(x_k), f_{x^*}^{k2}(x_k), \ldots, f_{x^*}^{kr_k}(x_k)),$$

where

$$f_{x^*}^{ki}(x_k) = F_k^i(x_1^*, x_2^*, \ldots, x_{k-1}^*, x_k, x_{k+1}^*, \ldots, x_q^*),\; i = \overline{1, r_k},\; k \in \{1,2,\ldots,q\}.$$

This definition generalizes the well-known Nash equilibrium for classical noncooperative games (single-objective games) and the Pareto optimum for multicriterion problems. If $r_k = 1$, $k \in \{1,2,\ldots,q\}$, then $\overline{\Gamma}$ becomes the classical noncooperative game, where x^* represents a Nash equilibrium solution; in the case $q = 1$ the game $\overline{\Gamma}$ becomes the Pareto multicriterion problem, where x^* is a Pareto solution.

In the following, we present the theorem from [38] which represents an extension of the Nash theorem for the multiobjective version of the game.

Theorem 13.7. *Let $\overline{\Gamma} = (X_1, X_2, \ldots, X_q, \overline{F}_1, \overline{F}_2, \ldots, \overline{F}_q)$ be a multiobjective game, where $X_1, X_2, \ldots, X_q$ are convex compact sets and $\overline{F}_1, \overline{F}_2, \ldots, \overline{F}_q$ represent continuous vector payoff functions. Moreover, let us assume that for every $k \in \{1,2,\ldots,q\}$ each component $F_k^i(x_1, x_2, \ldots, x_{k-1}, x_k, x_{k+1}, \ldots, x_q)$, $i = \overline{1, r_k}$, of the vector function $\overline{F}_k(x_1, x_2, \ldots, x_{k-1}, x_k, x_{k+1}, \ldots, x_q)$ represents a concave function with respect to x_k*

on X_k *for fixed* $x_1, x_2, \dots, x_{k-1}, x_{k+1}, \dots, x_q$. *Then for the multiobjective game* $\overline{\Gamma} = (X_1, X_2, \dots, X_q, \overline{F}_1, \overline{F}_2, \dots, \overline{F}_q)$ *there exists the Pareto–Nash equilibrium situation* $x^* = (x_1^*, x_2^*, \dots, x_q^*) \in X_1 \times X_2 \times \cdots \times X_q$.

The proof of Theorem 13.7 in [38] is based on the reduction of the multiobjective game $\overline{\Gamma} = (X_1, X_2, \dots, X_q, \overline{F_1}, \overline{F_2}, \dots, \overline{F_q})$ to an auxiliary game $\Gamma = (X_1, X_2, \dots, X_q, f_1, f_2, \dots, f_q)$ for which Nash theorem from [45] can be applied. In order to reduce the multiobjective game $\overline{\Gamma}$ to an auxiliary game Γ linear convolution criteria for vector payoff functions are used.

In such a way, if conditions of Theorem 13.7 are satisfied then Pareto–Nash equilibrium solution for the multiobjective game can be found by using the following algorithm:

1. Fix an arbitrary set of real numbers $\alpha_{11}, \alpha_{12}, \dots, \alpha_{1r_1}, \alpha_{21}, \alpha_{22}, \dots, \alpha_{2r_2}, \dots, \alpha_{q1}, \alpha_{q2}, \dots, \alpha_{qr_q}$, which satisfy conditions:

$$\begin{cases} \sum_{i=1}^{r_k} \alpha_{ki} = 1, & k \in \{1, 2, \dots, q\}; \\ \alpha_{ki} > 0, & i = \overline{1, r_k},\ k \in \{1, 2, \dots, q\}. \end{cases} \tag{13.28}$$

2. Form the single objective game $\Gamma = (X_1, X_2, \dots, X_q, f_1, f_2, \dots, f_q)$, where

$$f_k(x_1, x_2, \dots, x_q) = \sum_{i=1}^{r_k} \alpha_{ki} F_k^i(x_1, x_2, \dots, x_q),\ k \in \{1, 2, \dots, q\}.$$

3. Find the Nash equilibrium $x^* = (x_1^*, x_2^*, \dots, x_q^*)$ for the noncooperative game $\Gamma = (X_1, X_2, \dots, X_q, f_1, f_2, \dots, f_q)$ and fix x^* as the Pareto–Nash equilibrium solution for the multiobjective game $\overline{\Gamma} = (X_1, X_2, \dots, X_q, \overline{F}_1, \overline{F}_2, \dots, \overline{F}_q)$.

This algorithm finds only one of the solutions of the multiobjective game $\overline{\Gamma} = (X_1, X_2, \dots, X_q, \overline{F}_1, \overline{F}_2, \dots, \overline{F}_q)$. In order to find all solutions in the Pareto–Nash sense it is necessary to apply the algorithm for every $\alpha_{11}, \alpha_{12}, \dots, \alpha_{1r_1}, \alpha_{21}, \alpha_{22}, \dots, \alpha_{2r_2}, \dots, \alpha_{q1}, \alpha_{q2}, \dots, \alpha_{qr_q}$, which satisfy condition (13.28), and then to form the union of all obtained solutions.

13.5.2 The Multiobjective Multicommodity Flow Models

In the following, we formulate the multiobjective multicommodity flow problem on static and dynamic networks. Such problem consists of shipping a given set of commodities from their respective sources to their sinks through a network in order to optimize different criteria so that the total flow going through arcs does not exceed their capacities.

13.5.2.1 The Static Model

The aim is to send the flow through the network in order to optimize the vector utility function $\overline{F}_k = (F_k^1, F_k^2, \ldots, F_k^{r_k})$ for every commodity $k \in \{1, 2, \ldots, q\}$:

$$\overline{F}_k \; : \; X_1 \times X_2 \times \cdots \times X_q \to R^{r_k},$$
$$F_k^i \; : \; X_1 \times X_2 \times \cdots \times X_q \to R^1, \; i = \overline{1, r_k},$$

where X_k is a set of flows of commodity k, r_k is a number of criteria for commodity k.

13.5.2.2 The Dynamic Model

The purpose is to transport the flow through the network in order to optimize the vector utility function $\overline{F}_k = (F_k^1, F_k^2, \ldots, F_k^{r_k})$ for every commodity $k \in \{1, 2, \ldots, q\}$:

$$\overline{F}_k \; : \; (X_1 \times \mathbb{T}) \times (X_2 \times \mathbb{T}) \times \cdots \times (X_q \times \mathbb{T}) \to R^{r_k},$$
$$F_k^i \; : \; (X_1 \times \mathbb{T}) \times (X_2 \times \mathbb{T}) \times \cdots \times (X_q \times \mathbb{T}) \to R^1, \; i = \overline{1, r_k}.$$

where X_k is a set of flows of commodity k, $\mathbb{T}$ is a set of considered time moments, r_k is a number of criteria for commodity k.

In the framework of the game theory each commodity is associated with a player. We consider a general model with q players, each of which wishes to optimize his own vector utility function $\overline{F}_k$, $k \in \{1, 2, \ldots, q\}$, defined on the set of strategies of all players. Every component F_k^i, $i = \overline{1, r_k}$, $k \in \{1, 2, \ldots, q\}$, of the vector utility function F_k of player k corresponds to a partial criterion of player k. The cost of transportation of a given resource, the time necessary to transport it to its destination as well as the quality of the transportation play the role of the components of the vector utility function of a player in the game-theoretic formulation of the problem.

Each player competes in a Nash equilibrium manner so as to optimize his own criteria in the task of transporting flow from its origins to its destinations. In our problem each player has several objectives, so we use the Pareto–Nash equilibrium concept extended to networks. In such a way, players intend to optimize their utility functions in the sense of Pareto and at the same time players are interested to preserve Nash optimality principle when they interact between them. So, control decisions are made by each player according to its own individual performance objectives and depending on the choices of the other players.

13.5.3 Comments

In real-life problems, users have to make decision concerning routing as well as type and amount of resources that they wish to transport. Different sets of parameters may suit the service requirements of a user. However, the performance measures depend not only on the user's choices, but also on the decisions of other connected users, where this dependence is often described as a function of some network "state." In this setting the game paradigm and the Pareto–Nash equilibrium concept become the natural choice at the user level.

Game-theoretic models are widely employed in the context of flow control, routing, virtual path bandwidth allocation and pricing in modem networking. Flow problems in multimedia applications (teleconferencing, digital libraries) over high-speed broadband networks can serve a good example of this. In a multimedia network telecommunication companies carrying different traffic types (voice, data, and video) may share the limited common network resources such as buffers or transmission lines. These companies may have different objectives of maximizing packet throughput or minimizing packet blocking probability. A Pareto–Nash equilibrium may be reached when companies achieve their objectives in such a way that no company can improve its own performance by unilaterally changing its traffic load admission and routing strategies.

The problem of providing bandwidth which will be shared by many users [35,42] is one of the most important problems. As it is typical for games in such a problem the interaction among the users on their individual strategies has to be imposed. This can be done using a utility function that depends on the availability of bandwidth and other factors in the network.

13.6 Conclusions

In this chapter, the minimum cost flow problem and the maximum flow problem on dynamic networks, that generalize classical optimal flow problems on static networks, were investigated. The minimum cost flow problem was considered in the case when demand-supply and capacity functions depend on time and cost functions on arcs are nonlinear and depend both on time and on flow. The maximum flow problem was considered on dynamic networks with time-dependent capacities of arcs. The dynamic model with transit time functions that depend on the amount of flow and the entering time-moment of flow in the arc was formulated and studied. The properties of the optimal flows were stated and on their basis the methods and algorithms for solving the considered optimal dynamic flow problems were proposed. The time-expanded network method was generalized for the dynamic versions of the minimum cost multicommodity flow problem and the maximum multicommodity flow problem on networks. The dynamic multicommodity problems were studied on networks with time-varying capacities of arcs and transit times on arcs that depend on sort of commodity entering them. For the minimum cost multicommodity dynamic flow problem it was assumed that cost functions, defined on arcs, are nonlinear and depend on time and flow, and demand-supply functions depend on time. The case when transit time functions depend on flow and time was also analyzed. The methods and algorithms for solving the considered optimal dynamic multicommodity flow problems were developed.

The multiobjective version of the optimal multicommodity flow problem was considered. Investigation of this problem was effectuated on the basis of the concept of multiobjective games using the notion of the Pareto–Nash equilibrium.

Acknowledgements I express my gratitude to Dmitrii Lozovanu for close collaboration.

References

1. Ahuja, R., Magnati, T., Orlin, J.: Network flows. Prentice-Hall, Englewood Cliffs (1993)
2. Altman, E., Boulogne, T., El-Azouzi, R., Jimenez, T., Wynter, L.: A survey on networking games in telecommunications. Computers and Operations Research. **33(2)**, 286-311 (2006).
3. Aronson, J.: A survey of dynamic network flows. Ann. Oper. Res. **20**, 1–66 (1989)
4. Assad, A.: Multicommodity network flows: A survey. Networks. **8**, 37–92 (1978)
5. Bland, R.G., Jensen, D.L.: On the computational behavior of a polynomial-time network flow algorithm. Technical Report 661, School of Operations Research and Industrial Engineering, Cornell University (1985)
6. Cai, X., Sha, D., Wong, C.K.: Time-varying minimum cost flow problems. European Journal of Operational Research. **131**, 352–374 (2001)
7. Carey, M., Subrahmanian, E.: An approach to modelling time-varying flows on congested networks. Transportation Research Part B. **34**, 157–183 (2000)
8. Castro J.: A specialized interior-point algorithm for multicommodity network flows. Siam J. Optim. **10(3)**, 852–877 (2000)
9. Castro, J.: Solving difficult multicommodity problems with a specialized interior-point algorithm. Annals of Operations Research. **124**, 35–48 (2003)
10. Castro, J., Nabona, N.: An implementation of linear and nonlinear multicommodity network flows. European Journal of Operational Research Theory and Methodology. **92**, 37–53 (1996)
11. Cherkasskij, B.V.: Algorithm for construction of maximal flow in networks with complexity of $0(V^2\sqrt{E})$ operations. Math. Methods Sol. Econ. Probl. **7**, 112–125 (1977)
12. Christofides, N.: Graph theory. An algorithmic approach. Academic Press, New York, London, San Francisco (1975)
13. Edmonds, J., Karp, R.M.: Theoretical improvements in algorithmic efficiency for network flow problems. J. Assoc. Comput. Mach. **19**, 248–264 (1972)
14. Ermoliev, Iu., Melnic, I.: Extremal problems on graphs. Naukova Dumka, Kiev (1968)
15. Fleisher, L.: Approximating multicommodity flow independent of the number of commodities. Siam J. Discrete Math. **13(4)**, 505–520 (2000)
16. Fleischer, L.: Universally maximum flow with piecewise-constant capacities. Networks. **38(3)**, 115–125 (2001)
17. Fleischer, L.: Faster algorithms for the quickest transshipment problem. Siam J. Optim. **12(1)** 18–35 (2001)
18. Fleisher, L., Skutella, M.: The quickest multicommodity flow problem. Integer programming and combinatorial optimization. Springer, Berlin, 36–53 (2002)
19. Fonoberova, M., Lozovanu, D.: Minimum cost multicommodity flows in dynamic networks and algorithms for their finding. The Bulletin of Academy of Sciences of Moldova, Mathe-matics. **1(53)**, 107–119 (2007)
20. Ford, L., Fulkerson, D.: Constructing maximal dynamic flows from static flows. Operation Res. **6**, 419–433 (1958)
21. Ford, L., Fulkerson, D.: Flows in Networks. Princeton University Press, Princeton, NJ (1962)
22. Gabow, H.N.: Scaling algorithms for network problems. J. Comput. Syst. Sci. **31**, 148–168 (1985)
23. Glockner, G., Nemhauser, G.: A dynamic network flow problem with uncertain arc capacities: formulation and problem structure. Operations Research. **48(2)**, 233–242 (2000)

24. Goldberg, A.: An efficient implementation of a scaling minimum cost flow algorithm. J. Algorithms. **22(1)**, 1–29 (1997)
25. Goldberg, A.V., Tarjan, R.E.: Solving minimum-cost flow problems by successive approximation. Proc. 19th ACM STOC. 7–18 (1987)
26. Goldberg, A.V., Tarjan, R.E.: Finding minimum-cost circulations by canceling negative cycles. Technical Report CS-TR 107-87, Department of Computer Science, Princeton University (1987)
27. Goldberg, A., Tarjan, R.: A new approach to the maximum-flow problem. Journal of the Association for Computing Machinery. **35(4)**, 921–940 (1988)
28. Goljshtein, E., Iudin, D.: Linear programming problems of transport type. Nauka, Moscow (1969)
29. Hoffman, A.: Some recent applictions of the theory of linear inequalities to extremal combinatorial analysis. Proc. Symposia on Applied Math. **10** (1960)
30. Hoppe, B., Tardos, E.: The quickest transshipment problem. Mathematics of Operations Research. **25**, 36–62. (2000)
31. Hu, T.: Integer programming and network flows. Addison-Wesley Publishing Company, Reading, Massachusetts, Menlo Park, California London Don Mills, Ontario (1970)
32. Klinz, B., Woeginger, C.: Minimum cost dynamic flows: the series parallel case. Integer programming and combinatorial optimization (Copenhagen, 1995). Springer, Berlin, 329–343 (1995)
33. Klinz, B., Woeginger, C.: One, two, three, many, or: complexity aspects of dynamic network flows with dedicated arcs. Operations Research Letters. **22**, 119–127 (1998)
34. Kumar, S., Gupta, P.: An incremental algorithm for the maximum flow problem. Journal of Mathematical Modelling and Algorithms. **2**, 1–16 (2003)
35. Lazar, A., Orda, A., Dimitrios, E.: Virtual path bandwidth allocation in multiuser networks. IEEE/ACM transaction on networking. **5(6)**, 861–871 (1997)
36. Lozovanu, D.: Extremal-combinatorial problems and algorithms for their solving. Stiinta, Chisinau (1991)
37. Lozovanu, D., Fonoberova, M.: Optimal Flows in Dynamic Networks. Chisinau, CEP USM (2006)
38. Lozovanu, D., Pickl, S.: Pareto-Nash equilibrium for multiobjective games. European Journal of Operational Research. **181**, 1214–1232 (2007)
39. Lozovanu, D., Stratila, D.: The minimum cost flow problem on dynamic networks and algorithm for its solving. Bul. Acad. Ştiinţe Repub. Mold., Mat. **3**, 38–56 (2001)
40. Lozovanu, D., Stratila, D.: Optimal flow in dynamic networks with nonlinear cost functions on edges. Analysis and optimization of differential systems. Kluwer Academic Publishers, 247–258 (2003)
41. Ma, Z., Cui, D., Cheng, P.: Dynamic network flow model for short-term air traffic flow management. IEEE Transactions on Systems, Man, and Cybernetics – Part A: Systems and Humans. **34(3)**, 351–358 (2004)
42. MacKie-Mason, J., Varian, H.: Pricing congestible network resources. IEEE Journal of selected areas in communications. **13(7)**, 1141–1149 (1995)
43. Mazzoni, G., Pallottino, S., Scutella, M.: The maximum flow problem: A max-preflow approach. European Journal of Operational Research. **53**, 257–278 (1991)
44. McBride, R.: Progress made in solving the multicommodity flow problems. Siam J. Optim. **8(4)**, 947–955 (1998)
45. Nash, J.: Non cooperative games. Annals of Mathematics. **2**, 236–295 (1951)
46. Papadimitrou, C., Steiglitz, K.: Combinatorial optimization: algorithms and complexity. Prentice-Hall Inc., Englewood Cliffs, N.J. (1982)
47. Powell, W., Jaillet, P., Odoni, A.: Stochastic and dynamic networks and routing. In: M. O. Ball, T. L. Magnanti, C. L. Monma, and G. L. Nemhauser, (eds.) Network Routing, volume 8 of Handbooks in Operations Research and Management Science, chapter 3, pp. 141–295. North Holland, Amsterdam, The Netherlands (1995)
48. Tarjan, R.E.: A simple version of Karzanov's blocking flow algorithm. Oper. Res. Lett. **2**, 265–268 (1984)

Chapter 14
Some Distributed Approaches to the Service Facility Location Problem in Dynamic and Complex Networks

Ioannis Stavrakakis

Abstract The need to efficiently accommodate over the Internet the ever exploding (user-generated) content and services, calls for the development of service placement schemes that are distributed and of low complexity. As the derivation of the optimal placement in such environments is prohibitive due to the global topology and demand requirement and the large scale and dynamicity of the environment, feasible and efficient solutions of low complexity are necessary even at the expense of non-guaranteed optimality. This chapter presents three such approaches that migrate the service along cost-reducing paths by utilizing topology and demand information that is strictly local or confined to a small neighborhood: the neighbor hopping migration requires strictly local information and guarantees optimality for topologies of unique shortest path tree; the r-hop neighborhood migration appears to be more effective for general topologies and can also address jointly the derivation of both the number and locations of services to be deployed; the generalized neighborhood migration approach opens up new possibilities in defining localities, other than topological ones, that contain the most relevant candidates for the optimal placement, by exploiting emerging metrics and structures associated with complex and social networks. The underlying assumptions, strengths, efficiency and applicability of each of these approaches are discussed and some indicative results are shown.

14.1 Introduction

The problem of determining the location of facilities (factories, merchandise distribution centers, etc.) so that their operational cost is minimized is an old one. It has appeared since the early years of the industrial revolution and has been

I. Stavrakakis (✉)
Department of Informatics and Telecommunications, National & Kapodistrian University of Athens, Ilissia, 157 84 Athens, Greece
e-mail: ioannis@di.uoa.gr

M.T. Thai and P.M. Pardalos (eds.), *Handbook of Optimization in Complex Networks: Theory and Applications*, Springer Optimization and Its Applications 57,
DOI 10.1007/978-1-4614-0754-6_14,

extensively pursued by the operations research scientific community. Depending on the driving application, this problem has yielded various variants and given rise to numerous formulations. The most widely considered and studied have been those of the uncapacitated k-median and the uncapacitated Facility Location (FL), [1]. The uncapacitated k-median problem prescribes the locations for instantiating a fixed number of facilities so as to minimize the distance between users and the closest facility that can serve them. In the uncapacitated facility location problem, the number of facilities is not fixed, but jointly derived along with the locations, as part of a solution that minimizes the combined facility opening and accessing costs.

The vast majority of work on the general facility location problem until recently has considered a centralized environment. All the information needed as an input to the problem, such as the demand for service and the access cost, is considered to be centrally available. The challenge then has been to devise computational efficient approaches to solving or approximating the high complexity (or NP hard) optimization problems, [1–3]. The centralized approaches focus on greedy heuristics [4–7], on linear programming [8–10], on primal-dual [11, 12], local search [13, 14], and other techniques [15–20] that have been proposed in the past to deal with the increased complexity of the facility location problem.

Recently emerged distributed environments (such as those of networked users demanding services from network service providing facilities) have motivated the consideration of distributed approaches to solving the facility location problem. One recently initiated research thread relates to the approximability of distributed approaches to the facility location problem. The work in [21] draws on a primal-dual approach earlier devised in [12], to derive a distributed algorithm that trades-off the approximation ratio with the communication overhead under the assumption of $O(logn)$ bits message size, where n is the number of clients. More recently, an alternative distributed algorithm was derived in [22] that compares favorably with the one in [21] in resolving the same trade-off.

Several specific application-oriented approaches to the distributed facility location (service placement) problem have appeared in the literature, such as: [23] (deployment of multicast reflectors), [24] (deployment of mirrored web content), [25] (on-line multi-player network games), [26] (constrained mirror placement), and [27] (cache placement). Relevant is also the work in [28] on systems aspects of a distributed shared platform for service deployment, and [29] on the overheads of updating replica placements under non-stationary demand.

The aforementioned works on distributed approaches to solving the service placement problem either are applicable to a specific application scenario or aim at providing provable bounds for the run time and the quality of the solutions. In this chapter we present some distributed approaches to the service placement problem that are of broad applicability to general and, complex networking environments, rely on local or other limited network topology and demand information and employ heuristics to yield the optimal or near-optimal solutions. More specifically, this chapter presents three approaches to addressing the service placement problem in a networking environment, based on the kind of limited (topology and demand) information that is available. Although some of the approaches are directly applicable to the case of multiple facilities or that of unknown number of facilities, the presentation here is limited to the k-median formulation (with $k = 1$ in some cases) to be kept simple. Pointers to extensions are provided occasionally. The terms *service, facility, service facility, host* and occasionally *content* will be used interchangeably in this chapter.

Distributed solutions to the service placement problem are needed today in the design and operation of several applications in modern networks, such as for the large-scale timely distribution of customized software due to a software update (e.g., Microsoft Windows update). Such an update operation not only delivers immense amounts of data to millions of users, but it also has to incorporate complex decision processes for customizing the delivered updates to the peculiarities of different users [30] with respect to localization, previously-installed updates, compatibilities, and optional components, among others. This complex update process goes beyond the dissemination of a single large file, which could also be carried out through a peer-to-peer approach [31]. As it is unlikely that software providers will be willing to trust intermediaries to undertake such a responsibility, the software update (and other tasks) are likely to be undertaken by dedicated or virtual hosts, such as servers offered for lease through third-party overlay networks (Akamai or Planet Lab), or the newest breed of Cloud Computing platforms (e.g., Amazon EC2). To that end, distributed solutions to the service placement problem would be necessary to optimize the operational cost and improve end user experience.

The general environment considered here is that of a network of nodes and links over which some service located at a specific node (referred to also as the *host*) is provided. The path employed is assumed to be the shortest path between the node requesting the service and the node hosting it. The information required to solve the global optimization problem that will yield the optimal location for the service is the per node demand for the service and the costs of the links of the network; that is, full topology and demand information. Even if such information were available to a central entity, it would have required the solution of a very large (for realistic networking environments) and complex optimization problem, which is cumbersome if not impossible. Furthermore, the dynamicity of typical networking environments today would render any such centralized and global information-based approach useless as it would soon loose its optimality.

The approaches presented in this chapter have the following common characteristics. The service placement process is initiated by assuming an initial solution (location of the service). Given this location, some limited service demand and topology information is considered (which may be collected or be assumed to be a priori available) and a critical calculation (that captures the cost for providing the service from the particular host) is carried out based on it. The outcome of this calculation dictates whether the service should be placed at another node (and which) or not, anticipating a reduced service provisioning from the new location. The procedure repeats from the new location until the cost reduction achieved by moving to the new location falls below a small value or it starts increasing. This approach is not only a reasonable one for potentially converging to an optimal solution by starting from a random one, but is also one that matches well the intricacies of modern networking environments, where the service entity is generated at an almost random networked node and can be equipped with certain autonomicity features to allow it to migrate to and be hosted by pretty much any other network node. The starting location in the solution approach then corresponds to the physical location launching initially the service.

In Sect. 14.2, a distributed solution that utilizes strictly local information (i.e., own knowledge only) to solve the 1-median problem is presented. This approach yields the optimal location for specific networking environments, while yielding good performing solutions, although sub-optimal, for more general ones. This work was originally introduced in [32] where more details may also be found. In Sect. 14.3, a different distributed approach is presented that utilizes information from a broader locality and is shown to yield a solution that although not provably the optimal is shown to approximate it well. This work was originally introduced in [33] where more details may also be found. Finally, Sect. 14.4 presents a different approach that exploits the "social" standing of the nodes in terms of their significance in relaying the service between the nodes and the host of the service, to define "social" localities to engage in the limited complexity approach. This work was originally introduced in [34] where more details may also be found.

14.2 Neighbor-Hopping Service Migration

The key assumption here is that the node that executes the (distributed) algorithm for solving the service placement problem (i.e., the host) requires no topology or demand information to be communicated to it, besides that locally available at the specific node. As it will be shown, this very limited complexity approach reaches provably the optimal solution for certain network topologies. For more general topologies, an extension to this approach is presented that is of higher complexity and requires knowledge from the neighboring nodes.

14.2.1 Problem Formulation and Algorithm

The network topology throughout the chapter is represented by a connected undirected graph $G(V,E)$, where V and E denote the set of nodes and links between them, respectively; let $|V| = N$ denote the number of nodes in the network. The definitions illustrated in Table 14.1 will be adopted. To facilitate the discussion, some key quantities are also depicted in Fig. 14.1.

It is assumed that a shortest path is established by an underlying routing protocol between the host and any network node. Eventually, a generally non-unique shortest path tree is created, rooted at the host and including all network nodes. After a service movement, some of the (parts of the) shortest paths appearing between the nodes and the new host would change, while some new alternative ones of equal

Table 14.1 Notation used

Symbol	Meaning
S_v	The set of neighbor nodes of node v (i.e., nodes having a link with node v)
(u,v)	The link between two neighbor nodes u and v
$w(u,v) \geq 0$	The *weight* associated with link (u,v)
$d(x,y)$	The *distance* between node x and node y, derived as the summation of the weights of the links along a *shortest path* between these nodes, with $d(x,x) = w(x,x) = 0$
λ_v	The traffic load generated by node v as a result of requesting and receiving the service
K_t	The host at time t, where t represents the discrete times at which the service placement algorithm is executed and a (next) location decision is taken
$T_t(x)$	The set of all shortest path trees rooted at node x at time t
T_t^x	The shortest path tree in $T_t(x)$ that is actually utilized for data exchanges in connection with the service between the nodes and the host x at time t
$v \in T_t^x$	Indicates that node v is served by host x
$C_t(x)$	The cost incurred at time t for providing service by host x to all nodes $v \in T_t^x$ (i.e., by utilizing the specific shortest path tree T_t^x); it is given by $C_t(x) = \sum_{\forall v \in T_t^x} \lambda_v d(v,x)$
C	The minimal service provisioning cost possible, incurred when the hosts location is the optimal; the latter is can be determined by solving the classical 1-median problem by using full topology and demand information
$a_t = \frac{C_t}{C}$	The *approximation ratio* of the cost induced at time t by the host K_t, over the minimum (optimal) one; the closer the value of a_t to 1, the closer the induced cost at time t to the optimal one. This ratio will be used as a benchmark to establish the efficiency of the neighbor-hopping service migration solution
$I^y(T_t^x)$	The subtree – which is also a tree rooted at neighbor node y – that carries all the demand that reaches host x through neighbor y
$\Lambda\,(I^y(T_t^x))$	The *aggregate service demands* that are forwarded to host x through link (x,y) (for some neighbor node $y \in S_x$, and $y \in T_t^x$) over subtree $I^y(T_t^x)$; $\Lambda\,(I^y(T_t^x))$ is equal to the summation of the service demands of the individual nodes of the corresponding subtree, i.e., $\Lambda\,(I^y(T_t^x)) = \sum_{\forall v \in I^y(T_t^x)} \lambda_v$

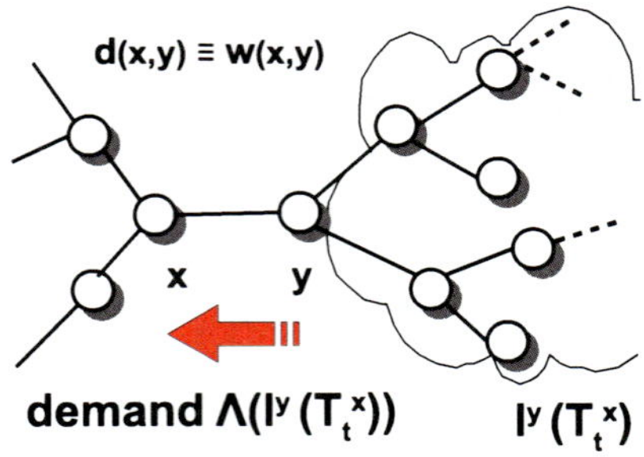

Fig. 14.1 Some key quantities assuming the host at node x at time t

distance to the ones used before the movement may appear. It is reasonable to assume that the underlying routing protocol would try to minimize the overhead introduced by a service movement by not modifying previously utilized shortest paths provided that they are not worse than any of the new shortest paths that emerged as a result of the facility movement; this assumption will be adopted here and will be referred to as the *migration rule*. Notice that $\Lambda\left(I^y(T_t^x)\right)$ can be available to host x using a monitoring mechanism that captures the incoming and outgoing packets or, in case λ_v is known to node v, by communicating these values to x (e.g., through piggybacking). It will be assumed that host x has knowledge of $\Lambda\left(I^y(T_t^x)\right)$ associated with all neighbor nodes $y \in S_x$. This locally available information is utilized by the neighbor-hopping service migration policy.

The key idea behind neighbor-hopping service migration is to establish conditions, based on information *locally available at the host*, under which a cost reduction would be achieved or not by moving the facility to a neighbor node y. Such local information which is sufficient for this is shown below to be $\Lambda\left(I^y(T_t^x)\right)$.

Let $x \to y(t)$ denote a *facility movement* – initiated at time t – from host x at t to its neighbor node y that becomes the host at time $t+1$. Let $C_{t+1}^{T_t^x}(y)$ denote a *hypothetical cost* assuming that (a) the facility moves to node y at time $t+1$ and (b) the corresponding shortest path tree over which data are forwarded towards host y (which should have been T_{t+1}^y, if facility movement $x \to y(t)$ had actually taken place) remains the current one (i.e., T_t^x). For this hypothetical cost, let the distance between any node v that is served by facility y over the shortest path tree T_t^x be denoted by $d^{T_t^x}(v,y)$ instead of $d(v,y)$ and consequently $C_{t+1}^{T_t^x}(y) = \sum_{\forall v \in T_t^x} \lambda_v d^{T_t^x}(v,y)$. Note that in general shortest path trees are different for different roots (i.e., $T_{t+1}^y \neq T_t^x$), except for the special case of topologies with unique shortest path trees [35]. *Unique shortest path tree topologies* are those for which $T_t(x) = T_t(y)$, for all pairs of nodes $x, y \in V$, at any time t. The following lemmas (see [32]) are the basis for the migration policy presented later.

Lemma 14.1. *Assuming node x is the host at t and $y \in S_x$, then $C_{t+1}^{T_t^x}(y) \geq C_{t+1}(y)$, with the equality holding for unique shortest path tree topologies; in addition, the difference between cost $C_{t+1}^{T_t^x}(y)$ and cost $C_t(x)$ is given by:*

$$C_{t+1}^{T_t^x}(y) - C_t(x) = \left(\Lambda\left(T_t^x \setminus I^y(T_t^x)\right) - \Lambda\left(I^y(T_t^x)\right)\right) w(x,y). \tag{14.1}$$

Sketch of Proof. The distance between node y and any node v over any shortest path of T^y_{t+1} is smaller than or equal to the distance over a shortest path of any other shortest path tree of different root (e.g., T^x_t). $d(v,y) \leq d^{T^x_t}(v,y)$, $\forall v \in V$, and $C_{t+1}(y) \leq C^{T^x_t}_{t-1}(y)$. The equality holds for the particular case that $T^y_{t+1} = T^x_t$, as it is the case for unique shortest path tree topologies. For any node $v \in I^y(T^x_t)$, $d^{T^x_t}(v,y) = d(v,x) - w(x,y)$, while for any node $v \in T^x_t \setminus I^y(T^x_t)$, $d^{T^x_t}(v,y) = d(v,x) + w(x,y)$. From the above it is derived that, $C^{T^x_t}_{t+1}(y) - C_t(x) = -\sum_{\forall v \in I^y(T^x_t)} \lambda_v w(x,y) + \sum_{\forall v \in T^x_t \setminus I^y(T^x_t)} \lambda_v w(x,y) = (\Lambda(T^x_t \setminus I^y(T^x_t)) - \Lambda(I^y(T^x_t)))\, w(x,y)$.

The right part of (14.1) depends on the link weight $w(x,y)$, the aggregate service demands that are forwarded to node x through node y (i.e., $\Lambda\left(I^y(T^x_t)\right)$) and the rest of the aggregate service demands that arrive through the other neighbor nodes of x (i.e., set $S_x \setminus \{y\}$) plus the service demands of node x itself (i.e., $\sum_{\forall i \in S_x \setminus \{y\}} \Lambda\left(I^v(T^x_t)\right) + \lambda_x = \Lambda\left(T^x_t \setminus I^y(T^x_t)\right)$, since $\cup_{\forall v \in S_x \setminus \{y\}} I^v(T^x_t) \cup \{x\} = T^x_t \setminus I^y(T^x_t)$). As mentioned before, both $\Lambda\left(T^x_t \setminus I^y(T^x_t)\right)$ and $\Lambda\left(I^y(T^x_t)\right)$ are locally available at node x (i.e., strictly local information).

In view of Lemma 14.1, two interesting observations can be made regarding the cost difference shown there. First, this difference does not depend on the weights of the links of the network, apart from the weight of the link between the two involved neighboring nodes, i.e., $w(x,y)$. Second, it depends on the difference between the aggregate service demands. Consequently, global knowledge of the network (i.e., knowledge of the weights of each link and the service demands of each node in the network) is not necessary to determine the differences in costs associated with neighboring hosting nodes and, eventually, determine the host that induces the lowest cost among all neighboring nodes. Even knowledge of $w(x,y)$ is not necessary, as it is shown later in Theorem 14.1. What is actually required is information regarding the aggregate service demands, which can be available at the host. The following theorem (proved in [32]) provides the conditions that need to be checked by the host, in order to decide or not to move the service to some neighbor node.

Theorem 14.1. *Assuming node x is the host at t and $y \in S_x$, then a cost reduction is achieved by moving the service to node y, i.e., $C_{t+1}(y) < C_t(x)$, provided that $\Lambda\left(T^x_t \setminus I^y(T^x_t)\right) < \Lambda\left(I^y(T^x_t)\right)$.*

Sketch of Proof. In view of Lemma 14.1 and since $w(x,y) > 0$, if $\Lambda(T^x_t \setminus I^y(T^x_t)) < \Lambda(I^y(T^x_t))$ then $C^{T^x_t}_{t+1}(y) < C_t(x)$. Since $C^{T^x_t}_{t+1}(y) \geq C_{t+1}(y)$, $C_{t+1}(y) < C_t(x)$ is also satisfied.

In view of Theorem 14.1, the following neighbor-hopping migration policy can be employed to solve the 1-median problem in a large-scale, distributed networking environment using strictly local information.

The neighbor-hopping migration strategy: Assuming node x is the host at t, then the service is moved from node x to some neighbor node $y \in S_x$ iff $\Lambda\left(T^x_t \setminus I^y(T^x_t)\right) < \Lambda\left(I^y(T^x_t)\right)$; this move results in cost reduction per Theorem 14.1. Moving a service under the conditions stated in Theorem 14.1 and achieving overall cost reduction does not necessarily mean that the service will eventually reach the optimal position in the general case. The latter is shown to be guaranteed for networks with a unique shortest path tree as stated next and shown in [32, 36].

Theorem 14.2. *In a network consisted of a unique shortest path tree, a single service facility always arrives at the optimal location under the neighbor-hopping migration strategy.*

As shown in [32], the neighbor-hopping service migration strategy can also be applied in the case of two or more service facilities, as a distributed and of low-complexity approximate approach to solving the k-median problem, $k > 1$. The following theorem, proved in [32], shows this strategy moves the service facilities along cost decreasing paths.

Theorem 14.3. *In a network of more than one facilities, if a facility located at some node x at time t moves under the neighbor-hopping service strategy to some neighbor node y, then $C_{t+1}(y) < C_t(x)$.*

14.2.2 Assessment and Extensions of the Neighbor-Hopping Migration Strategy

While the neighbor-hopping migration strategy provably moves the service along a monotonically cost decreasing path, it does not guarantee that the service will move all the way till the optimal location, except from the case of one service and a network topology consisted of a unique shortest path tree (Theorem 14.2). Notice that unique shortest path tree topologies (e.g., trees) are not uncommon; in fact, trees are formed frequently as a result of routing protocols in dynamic environments (e.g., mobile ad-hoc networks [37]).

In the case of network topologies with non-unique shortest path trees, the neighbor-hopping migration strategy is not guaranteed to reach the optimal location. For such topologies the efficiency of the neighbor-hopping migration strategy is measured in terms of the divergence of the *approximation ratio* $a_t = \frac{C_t}{C}$ from 1; C denotes the minimum cost induced when the service is in the optimal location and C_t denotes the cost induced by the neighbor-hopping migration strategy as a consequence of the location the strategy has moved the service to at time step t.

Simulation results under various network topologies (trees, grids, geometric random graphs [38], Erdős–Rényi random graphs [39], and Albert–Barabási graphs [40]) are derived to illustrate the behavior of the neighbor-hopping service migration strategy and their accordance with the analytical study. The initial service location is randomly selected and the demand of the nodes for the service is uniformly distributed.

Figure 14.2 presents simulation results under the neighbor-hopping migration strategy for cases with 1 and more than one (2 and 3) service facilities and for tree and grid network topologies. In Fig. 14.2, a results are shown under a tree topology of 100 nodes with equal link weights. Notice that all three curves for a_t

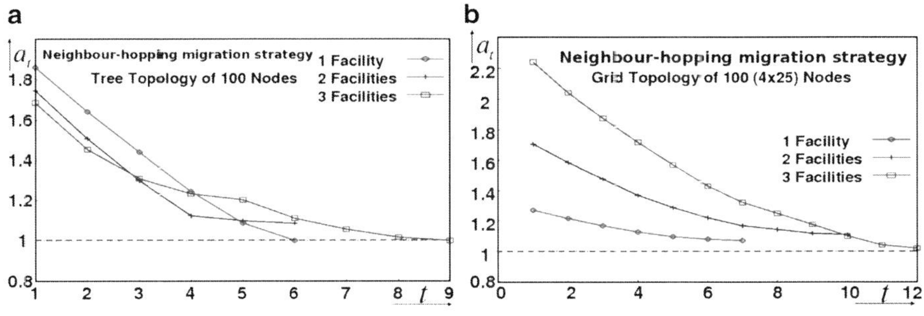

Fig. 14.2 Neighbor-hopping migration strategy in tree and grid topologies of 100 nodes

are monotonically decreasing with time, which is consistent with Theorems 14.1 and 14.3; that is, facilities are moving to neighbor nodes of smaller overall cost. For a single service facility, the approximation ratio eventually (here after six movements) becomes 1 (i.e., the optimal location is reached), as anticipated in view of Theorem 14.2. For the case of two service facilities, facility movements stop at time 6 when $a_6 > 1$, implying that they did not arrive at the optimal location; notice that the analysis did not provide any guarantee for that. For the case of three facilities, facility movements stop at time 9 when $a_9 = 1$, implying that the facilities arrived (and remained) at their optimal locations. Note that according to the analysis (Theorem 14.3), if facilities do move under the neighbor-hopping migration strategy, overall cost reduction is always achieved; nevertheless, they may or may not finally arrive at the optimal locations. In Fig. 14.2, b results are shown under a grid topology of 100 nodes with equal link weights. Notice that although all facilities move along a monotonically cost decreasing path, they fail in this particular case to arrive at the optimal positions ($a_t > 1$ in all cases).

Besides the tree and grid topologies results are also shown under other popular and relevant networking topologies, such as the geometric random graphs (suitable for studying mobile ad hoc networks [38]), Erdős–Rényi random graphs (suitable for comparison reasons [39]), and Albert–Barabási graphs (power-law graphs that model many modern networks including the Internet [40]). More specifically, geometric random graphs are created considering a connectivity radius $r_c = 0.21$ around each node in the square plane $[0,1] \times [0,1]$, Erdős–Rényi random graphs considering connectivity probability $p_c = 0.1$ and Albert–Barabási based on preferential attachment [40].

Figure 14.3.a presents simulation results under the neighbor-hopping migration strategy for various network sizes. Link weights are equal, which typically leads to a large number of shortest path trees and, consequently shortcuts, that can make the neighbor-hopping migration strategy stop prematurely and miss a better sub-optimal or even the optimal location (see later). Let T denote the *termination time* and let a_T be the value of a_t at termination time. Based on Fig. 14.3.a, it is observed that the approximation ratio is not affected as the network size increases. It is also interesting to observe that the approximation ratio remains below 1.5, which is

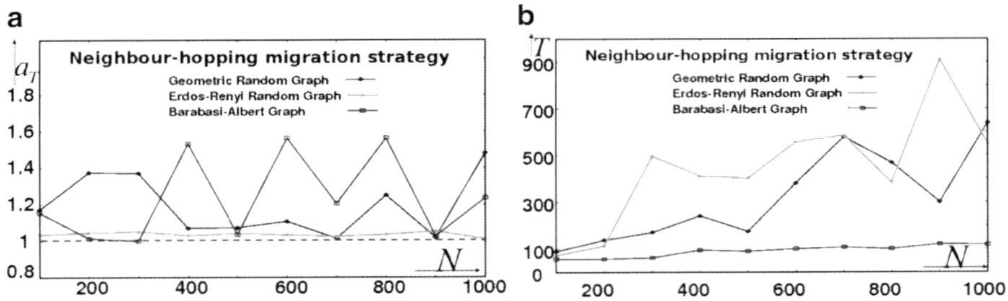

Fig. 14.3 Approximation ratio and termination time as a function of N

small considering the fact that equal link topologies do not allow for significant improvements under the neighbor-hopping migration strategy.

The performance of the neighbor-hopping migration strategy can be improved using extensions and heuristics that may be found in [32]. The basic idea behind them is to try to address to some extent the premature termination of the course along a cost decreasing path of the neighbor-hopping migration strategy due to shortcuts. This issue is briefly elaborated next.

As mentioned earlier, in the case of network topologies with non-unique shortest path trees, the neighbor-hopping migration strategy is not guaranteed to reach the optimal location, as it may stop prematurely its course along the cost decreasing path and fail to further reduce the cost. One reason that this may happen is because while a cost reducing neighboring node to the host exists, this is wrongly not detected by the neighbor-hopping migration strategy, as it fails to account for routing shortcuts that could reduce the cost by moving the facility to the neighbor. Shortcuts appear in topologies with a non-unique shortest path tree. Suppose that node x is the host at time t when the facility moves to neighbor node y (at $t+1$). If the shortest path tree of root node y is different from that of root node x (i.e., $T_{t+1}^{y} \neq T_t^{x}$), then this indicates that some nodes have preferred a *shortcut*, i.e., a shortest path towards the new host y that is shorter than that towards node x plus the weight $w(x,y)$.

One extension (referred to as Migration Policy E in [32]) can help alleviate the problem of missing shortcuts by the neighbor-hopping migration strategy, or eliminate it entirely in the case of topologies with equal weights. This is achieved through tentative movements to a neighbor node and utilization of neighbors' local information. In general topologies, this extension can be invoked when the neighbor-hopping migration strategy comes to a stop and potentially move it further along the cost decreasing path.

14.3 The r-hop Neighborhood Service Migration

14.3.1 Introduction and Overview

In this subsection, a different distributed approach to service migration is presented that utilizes information from a broader locality compared to the neighborhood-hopping one. Through the broader information considered and the formulation developed, this approach is expected to be applicable to and deal more effectively with more general networking environments. On the positive side, the neighbor-hopping approach requires strictly local information and is provably reaching the optimal location for single shortest path tree topologies and a single service facility. Nevertheless, reaching the final location may take some time due to the neighbor by neighbor-hopping constraint, it may fail to move further to a decreased cost neighbor due to the shortcuts, it does not address the problem of jointly determining the location and *number* of facilities to be deployed, etc. By paying a slightly higher overhead in information gathering, the r-hop neighborhood service migration approach presented here can reach potentially faster a more efficient location as its local optimization scope is broader than the 1-hop neighborhood constrained one. Furthermore, it has a common framework for handling the pursue for the optimal or near-optimal location of k facilities, as well as determining an optimal or near optimal number of required facilities.

The r-hop neighborhood service migration scheme was presented originally and in detail in [33]. According to this strategy, an initial set of service facilities are allowed to migrate adaptively to (ideally) the best network locations, and optionally to increase/decrease in number so as to best service the current demand while also minimizing the service deployment (facility opening) expenses. The basic idea is to develop distributed versions of the (uncapacitated) k-median (for the case in which the total number of facilities is fixed) and the (uncapacitated) Facility Location (FL) problem when additional facilities can be opened at a price or some of them be closed, to yield a lesser global cost for providing the service.

Both problems are combined under a common framework with the following characteristics. An existing facility gathers the topology information of its immediate surrounding area, which is defined by nodes that are within a *radius* of r hops from the facility. The facility also monitors the demand that it receives from the nodes that have it as the closest facility. It keeps an exact account of the demand from within its r-hop neighborhood, and an approximate and aggregate account of the demand of all the nodes outside the r-hop neighborhood that receive service from it (nodes on the *ring* of its r-hop neighborhood). The latter is accomplished by increasing the demand of the nodes on the *"surface"* of the r-hop neighborhood to account for the aggregate demand that flows through those nodes within the r-hop neighborhood from outside it. When multiple r-hop neighborhoods (when multiple facilities are considered) intersect, they merge to form more complex sub-graphs referred to as *r-shapes*. The observed topology and demand information is then used to re-optimize the current facility locations (and their number if the FL problem is pursued) by solving the uncapacitated k-median UKM (or the uncapacitated FL) problem within the r-shape.

It should be pointed out that reducing the radius r also decreases the amount of topological / demand information that needs to be gathered and processed centrally at any stage of the process, which makes the approach more scalable. On the other hand, reducing the radius r impacts negatively on the prospects for finding the optimal solution or the overall effectiveness of the solution, as compared to centralized solutions that consider the entire topological information. This trade-off is investigated experimentally using synthetic (Erdös–Rényi [39] and Barabási–Albert [40]) and real (AS-level [41]) topologies. It is shown that even for very small radius, e.g., $r = 1$, or $r = 2$, the performance of the distributed approach tracks closely that of the centralized one. Thus, increasing r substantially is not that necessary for performance, while it might result in large complexities since r-shapes typically increase fast with the radius due to the small, typically $O(\log n)$, diameter of most networks, including the aforementioned ones.

14.3.2 Problem Formulation, Projection of the World Outside and Algorithm

The r-hop neighborhood service migration approach to the facility location problem is described in some detail here. The focus of the discussion here will be on the case of a fixed number of facilities, that is the k-median problem. Nevertheless, the formulation is directly applicable to the case of the uncapacitated facility location problem that determines the optimal number of facilities in addition to their optimal location [33].

As already said the basic idea is to provide a distributed and of low-complexity, yet efficient if not optimal, solution to the k-median problem, by requiring the tentative hosts (of the service facilities) to have exact knowledge of the topology and demand of nodes in their r-hop neighborhood and approximate knowledge of the aggregate demand from nodes on the ring surrounding their r-hop neighborhood (see below). The approach described will be based on an iterative method in which the locations of the hosts may change between iterations.

The following definitions are employed below, where a superscript m denotes the step of the iteration. Let $F^{(m)} \subseteq V$ denote the set of hosts at the mth iteration, containing the locations of the k service facilities at this iteration, or of the currently available service facilities in the case of the uncapacitated facility location problem. Let $V_i^{(m)}$ denote the *r-hop neighborhood* of host v_i and $U_i^{(m)}$ denote its *ring* v_i, i.e., the set of nodes not contained in $V_i^{(m)}$, which are being served by host v_i, or equivalently, the nodes that have v_i as their closest facility; the *domain* $W_i^{(m)} = V_i^{(m)} \bigcup U_i^{(m)}$ of host v_i consists of its r-hop neighborhood and the surrounding ring.

From the previous definitions it is easy to see that $V = V^{(m)} \bigcup U^{(m)}$, where $V^{(m)} = \bigcup_{v_i \in F^{(m)}} V_i^{(m)}$, $U^{(m)} = \bigcup_{v_i \in F^{(m)}} U_i^{(m)}$.

The r-hop neighborhood service migration algorithm in the case of the k-median considers k service facilities located initially in randomly selected locations (hosts); these locations are refined iteratively through relocation until a (locally) optimal solution is reached. It includes the following steps:

An initial set $F^{(0)} \subseteq V$ of $k_0 = |F^{(0)}|$ nodes are randomly picked to act as hosts. Let $\mathscr{F} = F^{(0)}$ denote a temporary variable containing the "unprocessed" hosts during the current iteration; that is, the hosts for which the small scale k-median problem associated with their r-hop neighborhood, or r-shape, is not yet pursued. Also, let $\mathscr{F}^- = F^{(0)}$ denote a variable containing the current hosts.

At each iteration m the following steps are executed for each host $v_i \in \mathscr{F}$:

1. The r-hop neighborhood is formed by employing some neighborhood discovery protocol (e.g., [42]).
2. It is examined whether its r-hop neighborhood can be merged with that of other nearby hosts. Two or more hosts can be merged (i.e., their r-hop neighborhoods can be merged), if their r-hop neighborhoods intersect, that is when there exists at least one node that is part of these two or more neighborhoods. Let $J \subseteq F^{(m)}$ denote a set composed of v_i and the hosts that can be merged with it. J induces an r-shape $G_J = (V_J, E_J)$, defined as the sub-graph of G composed of the service facilities in J, their neighbors up to distance r, and the edges between them. Constraints on the maximal size of r-shapes could be placed to guarantee that it is always much smaller than $O(n)$.
3. The r-shape G_J is re-optimized by solving for the $|J|$-median within the r-shape, which can produce a new set of hosts (i.e., locations for the $|J|$ facilities). The re-optimization is carried out by using a centralized algorithm, such as the Integer Linear Programming (ILP) formulations [1] or local-search heuristics [14], for solving the k-median within r-shapes.
4. Processed hosts (both the original v_i and the ones merged with it) are removed from the set of unprocessed hosts of the current iteration, i.e., set $\mathscr{F} = \mathscr{F} \setminus (J \bigcap \mathscr{F}^-)$. Also $F^{(m)}$ is updated with the new hosts (service facility locations) after the re-optimization.
5. After all the hosts have been processed and provided that there has been a change in the hosts in the current iteration, another iteration is carried out. Otherwise (i.e., if no host change has been observed), the search is terminated yielding the (local) optimal solution to the k-median problem.

The input to a k-median problem is defined completely by a tuple $\langle V, s, k \rangle$, containing the topology, the demand, and the number of allowed medians, respectively. For optimizing the r-shapes or r-hop neighborhoods that are formed during the execution of the algorithm, the topology and number of medians (service facilities) are set as $V = V_J$, and $k = |J|$. Determining the demand input in these smaller scale optimization problems is a less straightforward issue and is discussed next.

The most straightforward approach would be to retain in the re-optimization of an r-shape the original demand of the nodes contained in that r-shape, i.e., set $s = \{s(v_j) : \forall v_j \in V_J\}$. Such an approach would, nonetheless, be inaccurate (as was also confirmed through results) since the hosts within an r-shape serve the demand of the nodes contained in the r-shape, *as well as those in the corresponding ring of the r-shape*. Since the number of hosts k is expected to typically be small, each one of them would serve a potentially large number of nodes (*e.g.*, of order $O(n)$), and thus the rings would typically be much larger than the corresponding r-shapes. Re-optimizing the locations of the hosts within an r-shape without considering the demand that flows-in from the ring would, therefore, amount to disregarding too much information, as compared to the information considered by a centralized solution yielding the optimal locations. Including the nodes of the ring into the optimization is, of course, not an option, as the ring could then become arbitrarily large ($O(n)$) and this would contradict our prime objective to solve the facility location problem in a scalable, distributed manner.

In order to account for the impact of the nodes of the ring on the solution, the demand of the ring is implicitly mapped into the local demand of the nodes that constitute the *surface* of the r-shape. The surface consists of nodes on the border (or edge) of the r-shape, i.e., nodes of the r-shape that have direct links to nodes of the ring. This mapping bridges the gap between absolute disregard for the ring, and full consideration of its exact topology. More details of the mapping may be found in [33].

14.3.3 Assessment of the r-hop Neighborhood Service Migration Strategy

In this part, some discussions are presented on the performance of the algorithm: the load mapping error and the closeness of the delivered solution to the optimal one.

It is not hard to show that the iterative algorithm presented earlier converges in a finite number of iterations. Since the solution space is finite, it suffices to show that there cannot be loops, i.e., repeated visits to the same configuration of hosts. A sufficient condition for this is that the cost be monotonically decreasing between successive iterations, i.e., $c^{(m)} \geq c^{(m+1)}$. It is not hard to show (see [33]) that this is the case for the 1-median applied to r-shapes; the case of k-median ($k > 1$) applied to r-shapes follows from straightforward generalizations.

In the sequel, it is shown how to control the convergence speed so as to adapt it to the requirements of practical systems. Specifically, the condition that the cost is reduced at least by a factor of α could be imposed, in order for the iteration to be accepted and continue the optimizing process; i.e., accept the outcome from the re-optimization of an r-shape at the mth iteration, only if $c^{(m)} \geq (1+\alpha)c^{(m+1)}$. In this case it can be shown [33] that the r-hop neighborhood service migration algorithm converges in $O(\log_{1+\alpha} n)$ steps.

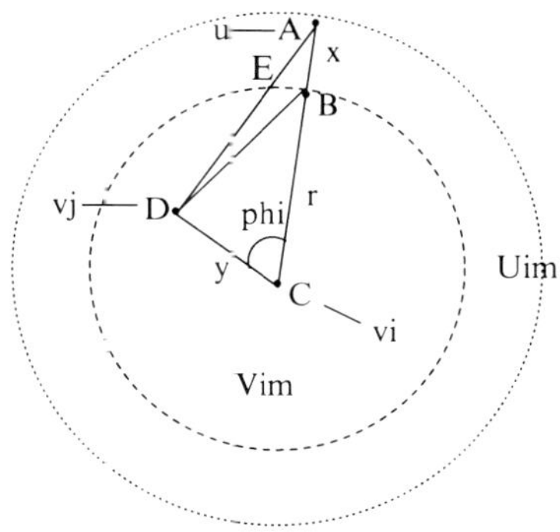

Fig. 14.4 Example of a possible facility movement from node v_i to node v_j with respect to a particular node $u \in U_i$

The largest price paid for gaining in complexity by sequentially optimizing the solution within an r-shape, is the potential for not arriving at the optimal global solution due to the approximate consideration of the demand of the ring through its mapping on the surface of the r-hop neighborhood. Under the centralized approach, the amount of demand generated by or attributed to a node is not affected by the particular configuration of the hosts within the graph, since all nodes in the network are included and considered with their original demand. Under the r-hop neighborhood approach, however, the amount of demand attributed to a surface node can be affected by the particular configuration of hosts within the r-shape. Figure 14.4 illustrates why this is the case. Node u on the ring has a shortest path to facility node v_i that intersects the surface of v_i's r-hop neighborhood (drawn as a circle with radius r in Fig. 14.4) at point B, thereby increasing the demand of a local node at B by $s(u)$. As the locations of the facilities may change during the various steps of the local optimizing process (e.g., the facility moves from C to D, Fig. 14.4), the node on the surface along the shortest path between u and the new location of the facility may change (node/point E in Fig. 14.4). Consequently, a demand *mapping error* is introduced by keeping the mapping fixed (as initially determined) throughout the location optimization process.

The mapping error could be eliminated by re-computing the surface mapping at each stage of the optimizing process (i.e., for each new intermediate facility configuration). Such an approach not only would add to the computational cost but – most important – would be practically extremely difficult to implement; it would require the collection of demand statistics under each new facility placement, delaying the optimization process and inducing substantial overhead.

A more detailed discussion on the mapping error may be found in [33], where it is shown that this mapping error is upper bounded by $\Delta_i(r) \leq 2\pi^2 r^3 (R^2 - r^2)$, where R is the radius of the particular domain W_i (assumed for simplicity to be also a circle), under the assumption that nodes are scattered in a uniform and continuous manner over this domain. This upper bound for $\Delta_i(r)$ is close to 0, when $r \to 0$ or $r \to R$. Since small values of r are to be used, a small mapping error and performance penalty is expected.

In the sequel, some results on the performance of the r-hop neighborhood service migration are presented by employing synthetic Erdös–Rényi (ER) [39] and Barabási–Albert (BA) [40] graphs. In the particular set of results, these graphs were generated by employing the BRITE generator [43], under which the ER graph is constructed by assuming that the probability of existence of a direct link between two nodes is given by $P(u,v) = \alpha \cdot e^{-d/(\beta L)}$, where d is the Euclidean distance between u and v, and L is the maximum distance between any two nodes [44]. The default values of BRITE $\alpha = 0.15$, $\beta = 0.2$ combined with an incremental model in which each node connects to $m = 2$ other nodes is used; the same incremental growth with $m = 2$ is also used for the BA graphs. This parametrization creates graphs in which the number of (undirected) links is almost double the number of vertices. The latter is also observed in real traces associated with Autonomous Systems (AS) of the Internet.

For network sizes $n = 400, 600, 800, 1{,}000$ – which are typical sizes of Internet ASs – it turns out that a substantial fraction of the total node population lays within a relatively small number of hops r from any node. For instance, for ER graphs, $r = 2$ covers $2 - 10\%$ of the nodes, whereas $r = 3$ increases the coverage to $10 - 32\%$, depending on network size. The coverage is even higher in BA graphs, where $r = 2$ covers $4 - 15\%$, whereas $r = 3$ covers $20 - 50\%$, depending again on network size. These observations are explained by the fact that larger networks exhibit longer shortest paths and diameters and also because BA graphs possess shorter shortest paths and diameters than corresponding ER graphs of the same link density, due to their highly skewed (power-law) degree distribution.

In the sequel the performance of the r-hop neighborhood service migration is compared to the centralized k-median solution utilizing full knowledge. Consider a network size of $n = 400$ nodes and assume that all nodes generate the same amount of service demand $s(v) = 1, \forall v \in V$. For scalability reasons, the radius values are limited to $r = 1$ and $r = 2$, to avoid running into r-shapes involving more than 10% of the total nodes. In the cases considered, the number of hosts (k) take values $k/n = 0.1\%, 0.5\%, 1\%, 2\%$, and 5%. The cost induced under the r-hop neighborhood service migration approach (denoted by $c(\mathrm{d}UKM(r))$ normalized with respect to that under the centralized k-median approach (denoted by $c(UKM(r))$ is depicted on the left-hand-side of Fig. 14.5, with the plot on top for ER graphs and the plot on the bottom for BA graphs. For both ER and BA graphs, the performance of the distributed approach tracks closely that of the centralized one, with the difference diminishing fast as r and k increase. The normalized performance for BA graphs converges faster (i.e., at smaller k for a given r) to ratios that approach 1, which is attributed to the existence of highly-connected nodes ("hubs") in BA graphs. Creating service facilities in few of the hubs is sufficient for approximating closely the performance of the centralized k-median. The two plots on the right-hand-side of Fig. 14.5 depict the number of iterations needed for the distributed approach to converge. A smaller value of r requires more iterations as it leads to the creation of a large number of small sub-problems (re-optimizations of many small r-shapes). BA graphs converge in fewer iterations, since for the same value of r BA graphs induce larger r-shapes.

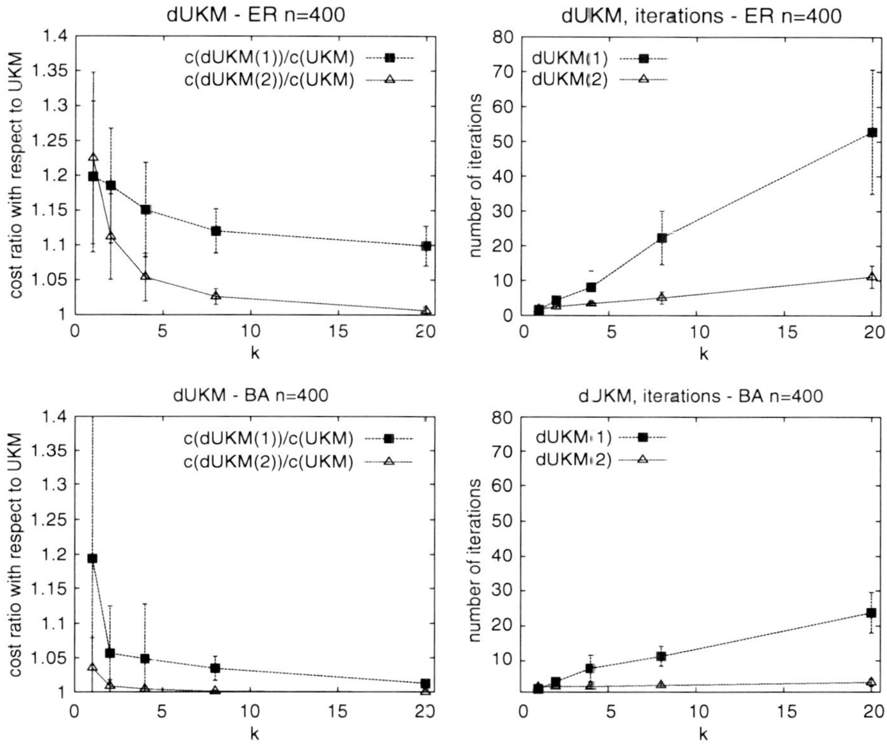

Fig. 14.5 The relative cost performance between the r-hop neighborhood service migration approach and the centralized k-median approach, and the number of iterations needed for the convergence of the former, for $r = 1$ and $r = 2$, and different facility densities $k/n = 0.1\%, 0.5\%, 1\%, 2\%$, and 5% under ER and BA graphs

More results on the performance of the distributed approach may be found in [33] for real topologies of ASs in the Internet and employing real traces, as well as for the case of non-stationary demands and under imperfect re-direction. Although perfect redirection is feasible using route triangulation and DNS [45], it may be costly to implement or perform sub-optimally due to faults or excessive load. Imperfect re-direction amounts to allowing that the demand is not always served by the closest host; it may easily emerge under host migration, introducing performance penalties. The effect of imperfect re-direction may be investigated by assuming that there exists a certain amount of *lag* between the time a host migrates to a new node and the time that the migration is communicated to the affected clients. During this time interval, a node might be receiving service from its previously closest host which, however, may have ceased to be optimal due to one or several migrations. Notice that under the existence of lag, even under stationary demand, the optimization is no longer guaranteed to be loop-free, as indicated earlier. Further discussion on this topic and some results showing a smooth performance degradation due to imperfect redirections may be found in [33].

14.4 Generalized Neighborhood Service Migration

14.4.1 Introduction and Motivation

The r-hop neighborhood service migration approach presented earlier aims at solving the large scale facility location optimization problem by solving sequentially smaller ones defined over r-hop topological localities around the current host (service location). The main rational for working with *r-hop topological* localities is that information about such locality and associated computations are expected to incur low overhead. No other node selection criterion, besides its topological location, is applied in forming these localities.

In this chapter, the locality over which the smaller facility location problems are solved is generalized in the sense that the criteria for including a node in this generalized locality are broadened. This generalization is motivated and largely enabled by recent trends in networking that create new (overlay) structures to be exploited in defining generalized localities and collecting relevant information. Complex Network Analysis (CNA) insights are employed to naturally define these generalized neighborhoods and construct the sub-graph that is typically now not spread almost symmetrically around the current host (as under the r-hop neighborhood approach) but asymmetrically as needed to include the most important nodes according to the selection criterion applied. The contrast between the r-hop neighborhood and the generalized neighborhood is depicted in Fig. 14.6.

In complex networking environments (including online and mobile social networks), nodes may exhibit fairly diverse characteristics with respect to their (statistical) connectivity properties (e.g., degree distribution) that eventually determine their links with other nodes. From a communication standpoint, this means they can have different roles as intermediaries, as, e.g., nodes having a higher number of links might be key in helping establish links between other nodes. Such nodes appear to hold critical positions throughout the network topology that helps them exhibit relatively high service demand concentration power and should probably be major players in the solution of the facility location problem. The CNA methods applied to identify such nodes and subsequently construct this subgraph, introduce a new thread of heuristic solutions to effectively address global location optimization problems.

Let G^i denote the subgraph of nodes that constitute the generalized neighborhood of some host i. The generalized neighborhood service migration strategy amounts to sequentially solving the small scale 1-median problems over G^i, determining the optimal location for the host in G^i, moving the host there, forming the new generalized neighborhood associated with the new host and repeating the process until the cost reduction achieved is below a threshold. The cost that is minimized in each step is given in terms of the minimum cost of the path linking the host i and a node n, $d(i,n)$, and some weight $w_{\text{eff}}(n)$ (to be discussed later) associated with node n, for $n \in G^i$.

$$Cost(i) = \sum_{n \in G^i} w_{\text{eff}}(n) \cdot d(i,n). \tag{14.2}$$

Notice that the global optimal location of the service facility is obtained by minimizing the following cost function over the entire network, V, and weights w_n that are equal to the service demand associated with each network node:

$$Cost(k) = \sum_{n \in V} w(n) \cdot d(k,n). \tag{14.3}$$

14.4.2 The Generalized Neighborhood

The generalized neighborhood around the current host is defined as the sub-graph that includes the host and all network nodes that meet a certain criterion. This criterion involves an innovative Centrality metric that:

1. Identifies the nodes that are seen to contribute the most to the aggregate service access cost by holding a central position within the network topology and/or route large amounts of the demand for the service; such nodes are expected to pull the service strongly in their direction in order to reduce the service provisioning cost and eventually optimize the location of the host;

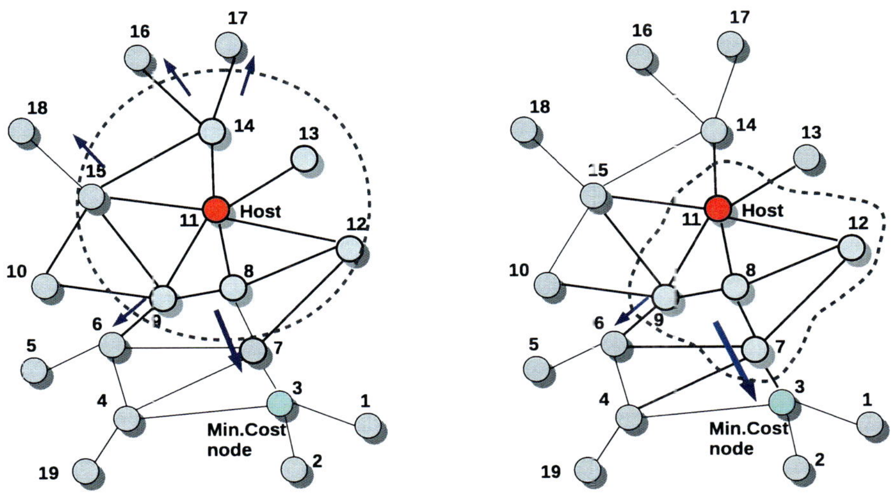

Fig. 14.6 A pictorial comparison of the *r*-hop (*left*) and generalized (*right*) neighborhoods, depicting also the pulling forces from outside them

2. Properly projects the intensity of the attraction forces these nodes exert to the service under the current location of the host and, thus, expected to facilitate the migration steps towards the optimal location.

Centrality indices are widely used in social network analysis, since Freeman's late 1970s influential articles [46, 47]; they are usually used as graph-theoretic tools in order to explain social phenomena. They are defined either on the nodes or edges of a graph and are usually based on geodesic paths that link members of a network, aiming at providing a measure of their importance. Different measures have been introduced to capture a variety of a node's attributes, such as its ability to reach numerous nodes via relative short paths or its popularity [47].

Betweenness centrality is a centrality index that measures the extent to which a node lies on the (shortest) paths linking other nodes and is defined as follows. Let σ_{st} denote the number of shortest paths between any two nodes s and t in a connected graph $G = (V, E)$. If $\sigma_{\mathrm{st}}(u)$ is the number of shortest paths passing through node $u \in V$, then the betweenness centrality index is given by:

$$BC(u) = \sum_{s=1}^{|V|} \sum_{t=1}^{s-1} \frac{\sigma_{\mathrm{st}}(u)}{\sigma_{\mathrm{st}}} \tag{14.4}$$

$BC(u)$ captures a node's ability to control or help establish paths between pairs of nodes; this is an average over all network pairs. When the host node is t, the traffic flow of relevance – that shapes the resulting cost of service provisioning from that host – is the one between all node pairs (x, t), $\forall\, x \in V$, for the fixed node t, and not all possible pairs, as it is the case with the betweeness centrality index. Consequently, it would make sense to include in the subgraph of nodes over which the small-scale optimization problem would be solved, the nodes that stand between the most paths linking the network nodes to the specific host. Under a uniform demand pattern, the presence of such nodes would reflect the fact that relatively high demand (that shapes the resulting cost) is coming through such nodes. The conditional betweenness centrality (*CBC*) index defined below, can be employed in the selection of the nodes to be included in the subgraph:

$$\mathrm{CBC}(u;t) = \sum_{s \in V, u \neq t} \frac{\sigma_{\mathrm{st}}(u)}{\sigma_{\mathrm{st}}} \tag{14.5}$$

Effectively, *CBC* assesses the extent to which a node u acts as a shortest path aggregator towards the current host t, by enumerating the shortest paths from all other network nodes to host t that pass through node u. This metric has been firstly introduced to facilitate the subgraph extraction [48] in synthetic graph topologies characterized by nodes producing equal demand load, and furthermore proved to yield (generalized) localities that lead the optimization sequence towards efficient locations.

In general, a high number of shortest paths through some node u does not necessarily mean that equally high demand load stems from the sources of those paths. Naturally, the pure topology-aware metric shown in (14.5) should be enhanced in order to also account for the service demand that will eventually be served by the shortest paths routes towards the host. To this end, a weighted conditional betweenness centrality (*wCBC*) has been introduced in [34], where the shortest path ratio of $\sigma_{st}(u)/\sigma_{st}$ in (14.5), is modulated by the demand load generated by each node, as given by:

$$wCBC(u;t) = \sum_{s \in V, u \neq t} w(s) \cdot \frac{\sigma_{st}(u)}{\sigma_{st}}. \tag{14.6}$$

Therefore, *wCBC* assesses the extent to which a node can serve as demand load concentrator towards a given service location. Clearly, when a service is requested equally by all nodes in the network (uniform demand) the *wCBC* metric degenerates to the *CBC* one, within a constant factor.

By employing the *CBC* or *wCBC* metric, the generalized neighborhood can be defined to include only nodes with high such values, as opposed to *all* the nodes that are topologically close (e.g., up to r-hops away) to the current host. More specifically the generalized neighborhood around the host over which the smaller scale 1-median problems are sequentially solved is determined by requiring that only a small percentage of the top network nodes be included. Some discussion on the nature and the construction of the generalized neighborhood is presented later; more details on these and related discussion may be found in [34].

After having selected the generalized neighborhood, a critical issue that comes up – as it was the case with the r-hop neighborhood service migration strategy – is how to represent the neglected nodes outside the generalized neighborhood in the small scale set up confined to the nodes and associated links included in the generalized neighborhood only. That is, how to set the coefficients $w_{\mathrm{eff}}(n)$ for $n \in G^i$ that are the keys to the cost formulation (see (14.2)). This is discussed next.

14.4.3 Projection of the World Outside the Generalized Neighborhood

As it was also stated in the presentation of the r-hop neighborhood service migration approach, by restricting the solution domain to a subgraph of nodes and their corresponding demand, the contribution to the service provisioning cost of the nodes outside the subgraph is totally neglected. This would most likely result in trapping the solution to a local optimum with little or no chance for moving out of it, since potentially strong (due to high service demand) cost-reducing forces exercised by such outside nodes (see Fig. 14.6) that would pull the host under the global problem formulation towards them are neglected. To allow for the inclusion of the outside forces, which would potentially pull the location of the host away from a local and towards the global optimal, the demand for service from the nodes outside the subgraph were mapped on the surface of the r-hop neighborhood; this way, the relative impact on the solution under the small scale formulation of the different outside forces was reasonably well represented.

A mapping of the outside demand as above should also be considered in the generalized neighborhood service migration approach. Besides using the *CBC* metric to determine which nodes to include in the generalized neighborhood (or subgraph G^i for host i) and be considered as candidate hosts, this metric also allows for mapping directly the demand of the rest of the network nodes on the ones forming the subgraph over which the facility location problem is solved. To this end, the weights involved in the description of the service cost to be minimized under the small scale set up (i.e., $W_{\text{eff}}(n)$ in (14.2)) consists of two terms. One term that is equal to the service demand load generated locally by node $n \in G^i$ and another term that brings in a properly identified part of the influence of the outside world. The second one corresponds to the contribution of the remaining outer network nodes which is captured by a new quantity defined for each node in the subgraph based on a modification of the original *CBC* metric, as detailed in [34] and discussed briefly here. To properly capture the impact on the small scale solution of some node z that is not part of the generalized neighborhood, its host-attraction power (analogous to its service demand) is "delegated" or "credited" to the first node contained in the subgraph that is encountered on each shortest path from z towards the service host.

14.4.4 Assessment of the Generalized Neighborhood Service Migration Strategy

The CNA-driven metric adopted for the identification of the generalized neighborhood and, furthermore, the quality of the derived solutions (i.e., degree of convergence to the optimal) are expected to be heavily dependent on two factors: the network topology and service demand distribution. The joint impact of the latter may enforce or suppress strong service demand attractors and assist or impede the progress of the service migration process towards the optimal location in the network. Evaluation of the generalized neighborhood service migration strategy has been carried out based on simulation on a set of physical topologies (snapshots of ISP networks); extensive results are reported in [34], along with a proof of the convergence of the strategy in a finite number of iterations.

Graphs of real-world networks provide testbeds for highly realistic and practice-oriented experiments. From a structural point of view, those networks do not have the predictable properties of the commonly used synthetic graphs and may differ substantially one from another. The ISP topology dataset [49] employed includes, among others, a number of Tier-1 ISP network topology files; high-degree nodes and considerable variance in the connectivity properties of nodes are typically present across such network snapshots.

To assess the generalized neighborhood service migration strategy under the simultaneous influence of network topology and service demand dynamics, asymmetry in the service demand distribution within the network is introduced. A Zipf distribution of the service demand is employed to model the diverse preference

Table 14.2 Results derived by the generalized neighborhood service migration strategy

	Min. Subgraph size for solutions within 2.5% of the optimal	
Size of physical topology	$s = 0$	$s = 1$
76	4	4
100	5	5
180	5	4
184	4	4
216	4	4
339	7	6
378	5	5

of nodes to a given service. Practically, the distribution could correspond to the normalized request rate for a given service by each network node. By increasing the value of the Zipf parameter s from 0 to 1, the distribution asymmetry grows from zero (uniform demand) towards higher values.

As already stated earlier in this chapter, there is a clear trade-off between the size of the neighborhood or subgraph considered (in number of nodes) and the goodness of the approximation to the optimal solution by the one derived by solving sequentially the smaller scale optimization problems. To put it in another way, it is important to determine the minimum number of nodes required to participate in the subgraph so as to yield an (almost) optimal solution. The results shown in Table 14.2 present the minimum number of subgraph nodes required to achieve a solution that induces a cost that lies within 2.5% of that induced by the optimal solution, for different levels of asymmetry in the service demand distribution. The main observation is that this number shows a remarkable insensitivity to both topological structure and service demand dynamics. Although the considered ISP topologies differ significantly in size and diameter [34], the number of nodes that need to be included in the generalized neighborhood over which the 1-median problem is solved, does not change. On the contrary, about half a dozen nodes suffice to yield very good accuracy even under uniform demand distribution, which is the least favorable scenario, as no high contrast demands – to intensify the pulling of the location towards the optimal – are then present. Likewise, the required minimum number remains practically invariable with the demand distribution skewness. Although for larger values of s, more nodes exhibit larger asymmetries in the demand and, thus, become stronger attractors towards the optimal solution (location), the added value for the algorithms accuracy is negligible.

These results suggest that there is already adequate topological structure of these real-world topologies. The high-degree nodes that are present, can easily be "identified" by the generalized neighborhood service migration strategy as low-cost hosts for the migrating service; even for small 1-median subgraph sizes, their attraction forces appear to be strong enough to pave a cost-effective service migration path. Moreover, the performance of this migration strategy is robust to possibly inaccurate estimates of the service demand generated by each node. Regarding implementation considerations and overheads associated with the generalized neighborhood service migration strategy, it is clear that the strategy is highly decentralized; all nodes that are candidates for hosting the service share the decision-making process for optimally placing the service in the network. It is also scalable, as it formulates and solves small scale optimization problems and avoids the computational burden related to the (global) solution of the 1-median problem; the latter may become a prohibitive task for large-scale networks, especially when changes in the service demand characteristics call for its repeated execution.

Nevertheless, some topological and demand information still needs to be shared in the network when implementing the generalized neighborhood service migration strategy, to enable the computation of the CNA metrics that serve as the criterion for forming the neighborhoods. For small-size networks, topological information is available through the operation of link-state routing protocols that distribute and use global topology information. For larger-scale networks, one way to acquire topology information would be through the deployment of some kind of source-routing or path-switching protocol that carries information about the path it traverses on its headers. Likewise, information about the interest of end-users in contents and services has become more abundant recently through social infrastructures such as online social networks. User-profiling mechanisms could interact with the generalized neighborhood process as built in components of some peer-to-peer protocol, so that information that is already available could be reused at no additional cost.

14.5 Concluding Remarks

Motivated by the explosion of content and services generated, accommodated and provided over the Internet, this chapter presents some distributed approaches to the optimal/efficient service placement within a large-scale networking structure, by relying only on local topology and demand information. The approaches assume an initial host (i.e., node hosting the service) that progressively migrates along cost-reducing paths and stops at some location which in the general case is not provably the optimal. The lack of provable optimality is compensated for by (a) showing that the achieved solution induces a cost that is close to that under the optimal solution, and (b) the tremendous reduction in the complexity compared to a traditional approach yielding the optimal solution. The latter statement is substantiated by the fact that while the latter traditional approach requires full network topology (weights of all links) and service demand information, the presented approaches require such information (possibly somewhat modified based on information that is locally available through standard networking operations) confined to a small locality.

The neighbor-hopping service migration strategy requires no information outside the host; based on locally available neighbor information and locally observable demand request flows, the host can determine if migrating the service to a neighbor node would reduce the service provisioning cost. For single shortest path topologies (e.g., trees) this strategy provably reaches the optimal location. For general topologies the migration may stop prematurely due to unidentifiable routing short-cuts; results on various topologies have shown that the cost reduction gain can be substantial and close to that of under the optimal location. Some heuristics to help unlock a premature termination of the migration are also pointed to in [32].

The r-hop and generalized neighborhood service migration strategies [33, 34], share a common framework in that a neighborhood around the host is identified and topology and demand information associated with that neighborhood is utilized in solving a small-scale k-median problem confined to that neighborhood. This low complexity solution identifies the optimal location within the neighborhood and migrates the host accordingly. At the end, the host migrates along a cost reducing path in jumps that are not restricted to (one-hop) neighbors. A great challenge associated with this approach is the representation of the world (key information associated with nodes) outside the neighborhood.

Under the r-hop neighborhood service migration strategy, the demand associated with a node outside the neighborhood is mapped on the node that acts as the entry point of the shortest path connecting the outside node and the host; the topology information outside the neighbor is indirectly employed in identifying the afore-mentioned entry node. In practice, the operation of the network and the routing protocol employed determine those entry nodes, which also can record the demand load passing through them. Thus, the mapping is easily implemented.

The generalized neighborhood service migration approach opens up new possibilities in defining localities that contain the most relevant candidates for the placement solution. In the particular approach presented here, emerging metrics and structures associated with complex and social networks are exploited to identify an appropriate subgraph (neighborhood) over which the k-median problem could be solved. Again, the operation of the network and information available for establishing overlay structures (and in particular, social networking structures, here), are exploited in order to identify the generalized neighborhoods as well as implement effectively the mapping of the outside world inside.

Finally, it should be noted that the migration approach to the distributed and low complexity approximation to the optimal placement solution matches well a natural implementation strategy in real networking environments. As the service/content is expected to be generated at pretty much any network node (e.g., user-generated content), the demand for this content would not be in the general case such that the location of the generation of the service would be the one minimizing the service provisioning cost. The original host would then run one of the migration strategies to determine a cost reducing location within the locality considered. Following the initial migration, the new host will eventually define its own broad locality, run one of the migration strategies and the migration procedure will continue until no further migration is detected. Occasionally, the host can redefine its locality and collect demand statistics and run the migration strategy again, in case the dynamicity of the topology or the demand have created new conditions which may lead to the identification of a better host. Thus, the presented migration strategy can naturally follow topological and demand changes in a dynamic environment and adapt to them by restarting the service migration strategy. Considering the fact that current networking designs seek to equip the various networking elements and entities with autonomicity features or, since autonomicity emerges naturally on its own in modern networking structures services are expected to be equipped with autonomicity features that would allow for the running of the migration strategy by service-residing functions and only utilize computing/communication resources of the host nodes and locally available input. Thus, the migration strategy could easily become an autonomic feature loaded to the service itself.

Acknowledgements This work has been supported in part by the IST-FET project SOCIALNETS (FP7-IST-217141) and the IST-FET project RECOGNITION (FP7-IST-257756).

References

1. P.B. Mirchandani and R.L. Francis, "Discrete Location Theory," John Wiley and Sons, (1990).
2. O. Kariv, and S.L. Hakimi, "An algorithmic approach to network location problems, II: The p-medians," SIAM Journal on Applied Mathematics, 37, 3, 539–560, (1979).
3. A. Tamir, "An O(pn) algorithm for p-median and related problems on tree graphs," Operations Research Letters, 19:59–64, (1996).
4. D. S. Hochbaum, "Heuristics for the fixed cost median problem," Mathematica Programming, 22:148–162, 1982.
5. S. Guha, and S. Khuller, "Greedy strikes back: improved facility location algorithms," J. Algorithms 31 (1999) 228–248.
6. K. Jain, M. Mahdian, and A. Saberi, "A new greedy approach for facility location problems," In Proc. ACM Symposium on the Theory of Computing (STOC'02), 2002.
7. K. Jain, M. Mahdian, E. Markakis, A. Saberi, and V. V. Vazirani, "Greedy facility location algorithms analyzed using dual fitting with factor-revealing LP," Journal of the ACM, 2003.
8. D. B. Shmoys, E. Tardos,, and K. Aardal, "Approximation algorithms for facility location problems," In Proceedings of the 29th ACM Symposium on Theory of Computing, pages 265–274, 1997.
9. F. Chudak and D. Shmoys. Improved approximation algorithms for capacitated facility location problem. In Proceedings of the 10th Annual ACM-SIAM Symposium on Discrete Algorithms, pages 875–876, 1999.
10. M. Sviridenko, "An improved approximation algorithm for the metric uncapacitated facility location problem," In W.J. Cook and A.S. Schulz, editors, Integer Programming and Combinatorial Optimization, Volume 2337 of Lecture Notes in Computer Science, pages 240–257, Springer, Berlin, 2002.
11. K. Jain and V. V. Vazirani, "Primal-dual approximation algorithms for metric facility location and k-median problems," in Proc of IEEE FOCS 99, New York City, NY, USA, 1999.
12. K. Jain and V. Vazirani, "Approximation algorithms for metric facility location and kmedian problems using the primal-dual schema and Lagrangian relaxation," Journal of ACM, 48(2):274–296, 2001.
13. M. R. Korupolu, C. G. Plaxton, and R. Rajaraman, "Analysis of a local search heuristic for facility location problems," Proc. 9th Annual ACM-SIAM Symposium on Discrete Algorithms, 1–10, 1998.
14. V. Arya, N. Garg, R. Khandekar, K. Munagala, A. Meyerson, and V. Pandit, "Local search heuristic for k-median and facility location problems," In Proceedings of the 33rd Annual Symposium on Theory of Computing (ACM STOC), pages 21–29. ACM Press, 2001.
15. J.-H. Lin, and J.S. Vitter, "Approximation Algorithms for Geometric Median Problems," Information Processing Letters, 44:245–249, 1992.

16. M. Charikar, S. Guha, D. B. Shmoys, and E. Tardos, "A constant factor approximation algorithm for the k-median problem," in Proc. of ACM STOC 99, Atlanta, GA, USA, 1999, pp. 110.
17. I.D. Baev, and R. Rajaraman, "Approximation algorithms for data placement in arbitrary networks," In Proceedings of the 12th Annual Symposium on Discrete Algorithms (ACM-SIAM SODA), pages 661–670, January 2001.
18. M. Mahdian, Y. Ye, and J. Zhang, "Improved Approximation Algorithms for Metric Facility Location Problems," 5th International Workshop on Approximation Algorithms for Combinatorial Optimization, 2002.
19. M. Mahdian and M. Pal, "Universal facility location," in Proc. of ESA 03, Budapest, Hungary, 2003, pp. 409–421.
20. N. Garg, R. Khandekar, and V. Pandit, "Improved approximation for universal facility location," in Proc of ACM-SIAM SODA 05, 2005, pp. 959–960.
21. Thomas Moscibroda and Roger Wattenhofer, "Facility Location: Distributed Approximation," 24th ACM Symposium on the Principles of Distributed Computing (PODC), Las Vegas, Nevada, USA, 2005.
22. S. Pandit and S. Pemmaraju, "Return of the primal-dual: distributed metric facility location ," 28th ACM Symposium on the Principles of Distributed Computing (PODC), Calgary, Alberta, Canada, 2009.
23. Lidia Yamamoto and Guy Leduc, "Autonomous Reflectors Over Active Networks: Towards Seamless Group Communication," AISB, vol. 1, no. 1, pp. 125–146, 2001.
24. Michael Rabinovich and Amit Aggarwal, "RaDaR: A Scalable Architecture for a Global Web Hosting Service," in Proc. of WWW '99, Toronto, Canada, 1999.
25. Chris Chambers and Wu-chi Feng and Wu-chang Feng and Debanjan Saha, "A Geographic Redirection Service for On-line Games," in Proc. of ACM MULTIMEDIA '03, Berkeley, CA, USA, 2003.
26. Cronin, Eric and Jamin, Sugih and Jin, Cheng and Kurc, Anthony R. and Raz, Danny and Shavitt, Yuval, "Constraint Mirror Placement on the Internet," IEEE Journal on Selected Areas in Communications, vol. 20, no. 7, 2002.
27. Krishnan, P. and Raz, Danny and Shavit, Yuval, "The Cache Location Problem," IEEE/ACM Transactions on Networking, vol. 8, no. 5, pp. 568–581, 2002.
28. David Oppenheimer and Brent Chun and David Patterson and Alex C. Snoeren and Amin Vahdat, "Service Placement in a Shared Wide-area Platform," in Proc. of USENIX '06, Boston, MA, 2006.
29. Loukopoulos, Thanasis and Lampsas, Petros and Ahmad, Ishfaq, "Continuous Replica Placement Schemes in Distributed Systems," in Proc. of ACM ICS '05, Boston, MA, 2005.
30. C. Gkantsidis, T. Karagiannis, P. Rodriguez, and M. Vojnovic, "Planet Scale Software Updates," in Proc. of ACM SIGCOMM 06, Pisa, Italy, 2006.
31. D. Kostić, A. Rodriguez, J. Albrecht, and A. Vahdat, "Bullet: High Bandwidth Data Dissemination Using an Overlay Mesh," in Proc. of SOSP 03, Bolton Landing, NY, USA, 2003.
32. K. Oikonomou and I. Stavrakakis, "Scalable service migration in autonomic network environments,"IEEE J.Sel. A. Commun., vol. 28, no. 1, pp. 84–94, 2010.
33. G. Smaragdakis, N. Laoutaris, K. Oikonomou, I.Stavrakakis and A. Bestavros, "Distributed Server Migration for Scalable Internet Service Deployment," IEEE/ACM Transactions on Networking. (to appear), 2011.
34. P. Pantazopoulos, M. Karaliopoulos, and I. Stavrakakis, in *Scalable distributed service migration via complex networks analysis*, Technical Report NKUA. (Available online), http://cgi.di.uoa.gr/~istavrak/publications.html.Cited10Mar2011.
35. D. Bertsekas, and R. Gallager, "Data networks," 2nd edition, Prentice-Hall, Inc., 1992.
36. K. Oikonomou and I. Stavrakakis, "Scalable Service Migration: The Tree Topology Case," The Fifth IFIP Annual Mediterranean Ad Hoc Networking Workshop (Med-Hoc-Net 2006), Sicily, Italy, June 14–17, 2006.
37. G. Wittenburg, and J. Schiller, "A Survey of Current Directions in Service Placement in Mobile Ad-hoc Networks," Proceedings of the Sixth Annual IEEE International Conference on Pervasive Computing and Communications (PerCom '08), Hong Kong, 17–21 March 2008.

38. M. Penrose, "Random geometric graphs," Oxford Univarsity Press Inc., 2003.
39. Paul Erdös and Alfred Rényi, On random graphs I, Publ. Math. Debrecen, vol. 6, pp. 290–297, 1959.
40. R. Albert and A. Barabási, "Statistical mechanics of complex networks," Rev. Mod. Phys., 2002.
41. L. Subramanian, S. Agarwal, J. Rexford, and R. H. Katz, Characterizing the Internet Hierarchy from Multiple Vantage Points, in Proc. of IEEE INFOCOM 02, New York City, NY, 2002.
42. M. Naor and U. Wieder, Know Thy Neighbors Neighbor: Better Routing for Skip-Graphs and Small Worlds, in Proc. of IPTPS, 2004.
43. A. Medina, A. Lakhina, I. Matta, and J. Byers, BRITE: An Approach to Universal Topology Generation, in Proc. of MASCOTS 01, Cincinnati, OH, 2001.
44. B. M. Waxman, Routing of multipoint connections, IEEE Journal on Selected Areas in Communications, vol. 6, no. 9, pp. 16171622, 1988.
45. N. Faber and R. Sundaram, MOVARTO: Server Migration across Networks using Route Triangulation and DNS, in Proc. of VMworld07, San Francisco, CA, 2007.
46. L. C. Freeman, "A set of measures of centrality based on betweenness," Sociometry, vol. 40, no. 1, pp. 3541, 1977.
47. L. C. Freeman, "Centrality in social networks: Conceptual clarification," Social Networks, vol. 1, no. 3, pp. 215239.
48. P. Pantazopoulos, I. Stavrakakis, A. Passarella, and M. Conti, "Efficient social-aware content placement for opportunistic networks," in IFIP/IEEE WONS, Kranjska Gora, Slovenia, February, 3–5 2010.
49. J.-J. Pansiot, P. Mrindol, B. Donnet, and O. Bonaventure, "Extracting intra-domain topology from mrinfo probing," in Proc. Passive and Active Measurement Conference (PAM), April 2010.

Part IV
Applications

Chapter 15
Modeling Epidemic Spreading in Complex Networks: Concurrency and Traffic

Sandro Meloni, Alex Arenas, Sergio Gómez, Javier Borge-Holthoefer, and Yamir Moreno

Abstract The study of complex networks sheds light on the relation between the structure and function of complex systems. One remarkable result is the absence of an epidemic threshold in infinite-size scale-free networks, which implies that any infection will perpetually propagate regardless of the spreading rate. However, real-world networks are finite and experience indicates that infections do have a finite lifetime. In this chapter, we will provide with two new approaches to cope with the problem of concurrency and traffic in the spread of epidemics. We show that the epidemic incidence is shaped by contact flow or traffic conditions. Contrary to the classical assumption that infections are transmitted as a diffusive process from nodes to all neighbors, we instead consider the scenario in which epidemic pathways are defined and driven by flows. Extensive numerical simulations and theoretical predictions show that whether a threshold exists or not depends directly on contact flow conditions. Two extreme cases are identified. In the case of low traffic, an epidemic threshold shows up, while for very intense flow, no epidemic threshold

S. Meloni (✉)
Department of Informatics and Automation, University of Rome "Roma Tre", Rome 00146, Italy

A. Arenas • J. Borge-Holthoefer
Departament d'Enginyeria Informàtica i Matemàtiques, Universitat Rovira i Virgili, 43007 Tarragona, Spain

Instituto de Biocomputación y Física de Sistemas Complejos (BIFI), Universidad de Zaragoza, 50009 Zaragoza, Spain

S. Gómez
Departament d'Enginyeria Informàtica i Matemàtiques, Universitat Rovira i Virgili, 43007 Tarragona, Spain

Y. Moreno
Instituto de Biocomputación y Física de Sistemas Complejos (BIFI), Universidad de Zaragoza, 50009 Zaragoza, Spain

Departamento de Física Teórica, Universidad de Zaragoza, 50009 Zaragoza, Spain
e-mail: yamir.moreno@gmail.com

M.T. Thai and P.M. Pardalos (eds.), *Handbook of Optimization in Complex Networks: Theory and Applications*, Springer Optimization and Its Applications 57,
DOI 10.1007/978-1-4614-0754-6_15,

appears. In this way, the classical mean-field theory for epidemic spreading in scale free networks is recovered as a particular case of the proposed approach. Our results explain why some infections persist with low prevalence in scale-free networks, and provide a novel conceptual framework to understand dynamical processes on complex networks.

15.1 Introduction

The problem of modeling how diseases spread among individuals has been intensively studied for many years [2,20,31,39]. The development of mathematical models to guide our understanding of the disease dynamics has allowed to address important issues such as immunization and vaccination policies [2,22,32]. Physicist's approaches to problems in epidemiology involve statistical physics, the theory of phase transitions and critical phenomena [53], which have been extremely helpful to grasp the macroscopic behavior of epidemic outbreaks [4,15,24,34,37,38,41,45,46]. The main artifice of this success has been the Mean-Field (MF) approximation, where local homogeneities of the ensemble are used to average the system, reducing degrees of freedom. It consists of coarse-grained vertices within degree classes and considers that all nodes in a degree class have the same dynamical properties; the approach also assumes that fluctuations can be neglected.

The study of complex networks [6, 21, 42] has provided new grounds to the understanding of contagion dynamics. Particularly important in nature are scale-free (SF) networks, whose degree distribution follows a power law $P(k) \sim k^{-\gamma}$ for the number of connections, k, an individual has. SF networks include patterns of sexual contacts [33], the Internet [47], as well as other social, technological and biological networks [10]. SF networks [3, 6, 21] are characterized by the presence of hubs, which are responsible for several striking properties for the propagation of information, rumors or infections [4, 24, 34, 38, 41, 45]. The HMF approach analytically predicts the critical rate β_c at which the disease spreads, i.e. the epidemic threshold.

Theoretical modeling of how diseases spread in complex networks is largely based on the assumption that the propagation is driven by reaction processes, in the sense that the transmission occurs from every infected through all its neighbors at each time step, producing a diffusion of the epidemics on the network. However, this approach overlooks the notion that the network substrate is a fixed snapshot of all the possible connections between nodes, which does not imply that all nodes are concurrently active [26]. Many networks observed in nature [6, 21], including those in society, biology and technology, have nodes that temporally interact only with a subset of its neighbors [1, 43]. For instance, hub proteins do not always interact with all their neighbor proteins at the same time [30], just as individuals in a social network [33] do not interact simultaneously with all of their acquaintances. Likewise, Internet connections being utilized at a given time depends on the specific traffic and routing protocols. Given that transport is one of the most common functions of networked systems, a proper consideration of this issue will irreparably affect how a given dynamical process evolves.

In this chapter, we present a theoretical framework for contact-based spreading of diseases in complex networks. This formulation, Microscopic Markov-Chain Approach (MMCA), is based on probabilistic discrete-time Markov chains, generalizes existing HMF approaches and applies to weighted and unweighted complex networks [26]. Within this context, in addition to capturing the global dynamics of the different contact models and its associated critical behavior, it is now possible to quantify the *microscopic dynamics* at the individual level by computing the probability that any node is infected in the asymptotic regime. MC simulations corroborate that the formalism here introduced reproduces correctly the *whole* phase diagram for model and real-world networks. Moreover, we capitalize on this approach to address how the spreading dynamics depends on the number of contacts actually used by a node to propagate the disease.

After that, we introduce a theoretical approach to investigate the outcome of an epidemic spreading process driven by transport instead of reaction events [37]. To this end, we analyze a paradigmatic abstraction of epidemic contagion, the so-called Susceptible–Infected–Susceptible (SIS) model [40], which assumes that contagion occurs through the eventual contact or transmission between connected partners that are using their connections at the time of propagation. This is achieved by considering a quantized interaction at each time step. Mathematically, we set up the model in a flow scenario where contagion is carried by interaction packets traveling across the network. We consider two possible scenarios that encompass most of real traffic situations: (1) unbounded delivery rate and (2) bounded delivery rate, of packets per unit time. We derive the equation governing the critical threshold for epidemic spreading in SF networks, which embeds, as a particular case, previous theoretical findings. For unbounded delivery rate, it is shown that the epidemic threshold decreases in *finite* SF networks when traffic flow increases. In the bounded case, nodes accumulate packets at their queues when traffic flow overcomes the maximal delivery rate, i.e. when congestion arises. From this moment on, the results show that both the epidemic threshold and the infection prevalence are bounded due to congestion.

15.2 Microscopic Markov-Chain Approach to Disease Spreading

The critical properties of an epidemic outbreak in SF networks can be addressed using the heterogeneous MF (HMF) prescription [4, 24, 34, 37, 38, 41, 45, 46]. It consists of coarse-grained vertices within degree classes and considers that all nodes in a degree class have the same dynamical properties; the approach also assumes that fluctuations can be neglected. Specifically, if β is the rate (probability per unit time) at which the disease spreads, it follows that the epidemic threshold in uncorrelated SF networks is given [45] by $\beta_c = \langle k \rangle / \langle k^2 \rangle$, leading to $\beta_c \to 0$ as $N \to \infty$ when $2 < \gamma \leq 3$.

MF approaches are extremely useful to assess the critical properties of epidemic models however, they are not designed to give information about the probability of individual nodes but about classes of nodes. Then, questions concerning the probability that a given node be infected are not well posed in this framework. To obtain more details at the individual level of description, one has to rely on Monte Carlo (MC) simulations, which have also been used to validate the results obtained using MF methods. Restricting the scope of epidemiological models to those based in two states [20, 31, 39] –susceptible (S) and infected (I)–, the current theory concentrates on two specific situations, the contact process [7, 11–13, 29, 35] (CP) and the reactive process [14, 18, 19, 23] (RP). A CP stands for a dynamical process that involves an individual stochastic contagion per infected node per unit time, while in the RP there are as many stochastic contagions per unit time as neighbors a node has. This latter process underlies the abstraction of the susceptible-infected-susceptible (SIS) model [20, 31, 39]. However, in real situations, the number of stochastic contacts per unit time is surely a variable of the problem itself [26]. In this first part of the chapter, we develop a microscopic model, based on Markov-Chains, to cope with the concurrency problem in the spreading of epidemics.

15.2.1 Contact-Based Epidemic Spreading Models

Let us suppose we have a complex network, undirected or directed, made up of N nodes, whose connections are represented by the entries $\{a_{ij}\}$ of an N-by-N adjacency matrix **A**, where $a_{ij} \in \{0, 1\}$. Unlike standard HMF approaches, our formalism allows the analysis of weighted networks, thus we denote by $\{w_{ij}\}$ the non-negative weights ($w_{ij} \geq 0$) of the connections between nodes, being $w_i = \sum_j w_{ij}$ the total output strength [5] of node i. The above quantities completely define the structure of the underlying graph. The dynamics we consider is a discrete two-state contact-based process, where every node is either in a susceptible (S) or infected (I) state. Each node of the network represents an individual (or a place, a city, an airport, etc.) and each edge is a connection along which the infection spreads. At each time step, an infected node makes a number λ of trials to transmit the disease to its neighbors with probability β per unit time, and then has a probability μ of recovering to the susceptible state. This forms a Markov chain where the probability of a node being infected depends only on the last time step, hence the name Microscopic Markov-Chain Approach (MMCA). After some transient time, the previous dynamics sets the system into a stationary state in which the average density of infected individuals, ρ, defines the prevalence of the disease.

We are interested in the probability $p_i(t)$ that any given node i is infected at time step t. We denote by r_{ij} the probability that a node i is in contact with a node j, defining a matrix **R**. These entries represent the probabilities that existing links in the network are used to transmit the infection. If i and j are not connected, then $r_{ij} = 0$. With these definitions, the discrete-time version of the evolution of the probability of infection of any node i reads

$$p_i(t+1) = (1 - q_i(t))(1 - p_i(t)) + (1 - \mu)p_i(t) + \mu(1 - q_i(t))p_i(t), \tag{15.1}$$

where $q_i(t)$ is the probability of node i not being infected by any neighbor at time t,

$$q_i(t) = \prod_{j=1}^{N}(1 - \beta r_{ji} p_j(t)). \tag{15.2}$$

The first term on the right hand side of (15.1) is the probability that node i is susceptible $(1 - p_i(t))$ and is infected $(1 - q_i(t))$ by at least a neighbor. The second term stands for the probability that node i is infected at time t and does not recover, and finally the last term takes into account the probability that an infected node recovers $(\mu p_t(t))$ but is re-infected by at least a neighbor $(1 - q_i(t))$. Within this formulation, we are assuming the most general situation in which recovery and infection occur on the same time scales, allowing then reinfection of individuals during a discrete time window (for instance, one MC step). This formulation generalizes previous approximations where one time step reinfections can not occur.

The formulation so far relies on the assumption that the probabilities of being infected p_i are independent random variables. This hypothesis turns out to be valid in the vast majority of complex networks because the inherent topological disorder makes dynamical correlations not persistent. The dynamical system ((15.1) and (15.2)) corresponds to a family of possible models, parameterized by the explicit form of the contact probabilities r_{ij}. Without loss of generality, it is instructive to think of these probabilities as the transition probabilities of random walkers on the network. The general case is represented by λ_i random walkers leaving node i at each time step:

$$r_{ij} = 1 - \left(1 - \frac{w_{ij}}{w_i}\right)^{\lambda_i}. \tag{15.3}$$

The Contact Process (CP) corresponds to a model dynamics of one contact per unit time, $\lambda_i = 1$, $\forall i$ in (15.3) thus $r_{ij} = w_{ij}/w_i$.[1] In the Reactive Process (RP), all neighbors are contacted, which corresponds, in this description, to set the limit $\lambda_i \to \infty$, $\forall i$ resulting on $r_{ij} = a_{ij}$ regardless of whether the network is weighted or not. Other prescriptions for λ_i conform the spectrum of models that can be obtained using this unified framework. The phase diagram of every model is simply obtained solving the system formed by (15.1) for $i = 1, \ldots, N$ at the stationary state,

[1]Strictly speaking, when $\lambda = 1$, our model is not exactly the standard CP, since in that case reinfections are not considered. However, we will refer to it as a CP since only one neighbor is contacted at each time step and the critical points of both variants are the same.

$$p_i = (1 - q_i) + (1 - \mu) p_i q_i , \tag{15.4}$$

$$q_i = \prod_{j=1}^{N} (1 - \beta r_{ji} p_j) . \tag{15.5}$$

This equation has always the trivial solution $p_i = 0, \forall i = 1, \ldots, N$. Other non-trivial solutions are reflected as non zero fixed points of (15.4) and (15.5), and can be easily computed numerically by iteration. The macroscopic order parameter is given by the expected fraction of infected nodes ρ, computed as

$$\rho = \frac{1}{N} \sum_{i=1}^{N} p_i . \tag{15.6}$$

15.2.2 Numerical Results

To show the validity of the MMCA model here discussed, we have performed MC simulations on different SF networks for RP. In Fig. 15.1, the phase diagram of the system obtained by MC simulations is compared with the numerical solution of (15.4) and (15.5). To model the epidemic dynamics on the described topologies we incorporate a SIS model in which, at each time step, each node can be susceptible or infected. Each simulation starts with a fraction ρ_0 of randomly chosen infected individuals ($\rho_0 = 0.05$ in our simulations), and time is discretized in time-steps. At each time step an infected node i infects with the same probability β all its neighbors and recovers at a rate μ. The simulation runs until a stationary state for the density of susceptible individuals, $\rho(t)$ is reached. The agreement between both curves is matchless. Moreover, the formalism also captures the microscopic dynamics as given by the p_i's, see the inset of Fig. 15.1. While the computational cost of the MC simulations is considerably large, the numerical solution of the fix point (15.4) and (15.5), by iteration, is fast and accurate.

In Fig. 15.2, we analyze our formalism on top of the airports network data set, composed of passenger flights operating in the time period November 1, 2000, to October 31, 2001 compiled by OAG Worldwide (Downers Grove, IL) and analyzed previously by Prof. Amaral's group [28]. It consists of 3,618 nodes (airports) and 14,142 links, we used the weighted network in our analysis. Airports corresponding to a metropolitan area have been collapsed into one node in the original database. We show the density of infected individuals ρ as a function of β for different values of λ. Both the critical points and the shape of the $\rho - \beta$ phase diagrams greatly change at varying the number of stochastic contacts (λ). We observe a moderate disease prevalence in the case of small values of λ, even for large values of the spreading rate β. In contrast, when the number of trials is of order 10^3 the situation is akin to a RP.

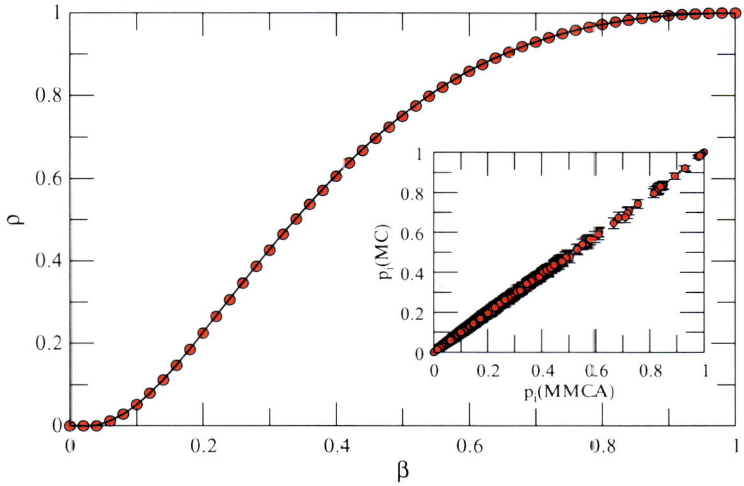

Fig. 15.1 Average fraction of infected nodes ρ as a function of the infection rate β for $N = 10^4$. *Lines* stand for the MMCA solutions (with $\lambda = \infty$) and *symbols* correspond to MC simulations of the SIS model on top of random scale-free networks with $\gamma = 2.7$ (error bars are smaller than the size of the symbol). In the inset, scatter plot for the probability that a node is infected using results of MC simulations (the y-axis) and the solutions (x-axis) of (15.4) and (15.5). Both results have been obtained for $\mu = 1$, the inset is for $\beta = 0.1$. After [26]

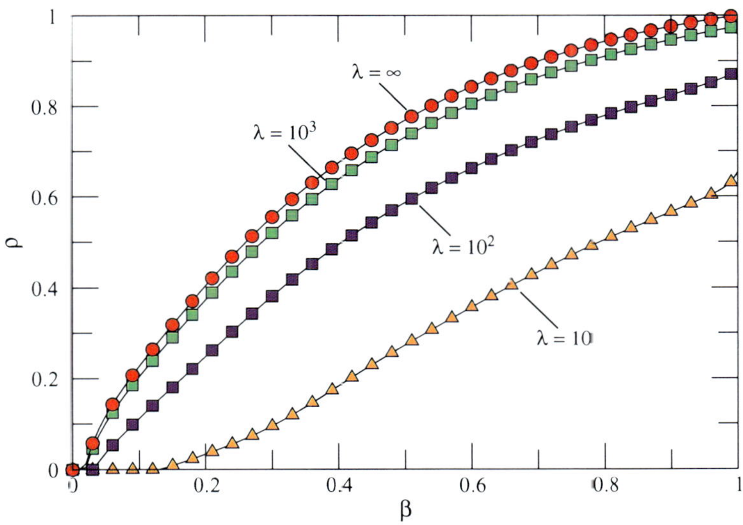

Fig. 15.2 Density of infected individuals ρ as a function of β for different values of λ in the air transportation network [28]. The smallest epidemic threshold and largest incidence is obtained for the RP, in which the matrix **R** corresponds to the adjacency matrix. This implies that the SIS on unweighted networks is a worst case scenario for the epidemic spreading in real weighted networks. ρ is calculated according to (15.6) once the p_i's are obtained, μ is set to 1. After [26]

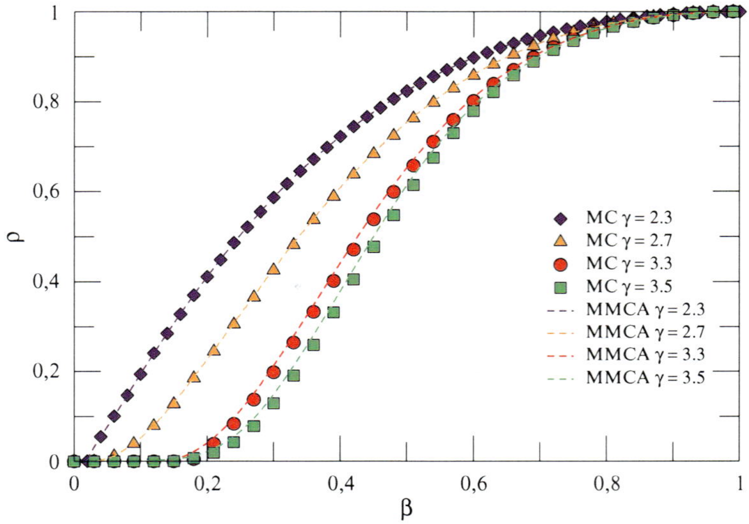

Fig. 15.3 Phase diagram for the SIS model ($\lambda = \infty$) in a random scale free network for different γ's. The networks size is $N = 10^4$ nodes and $\mu = 1$. MC results are averages over 10^2 realizations. *Dashed lines* corresponds to the theoretical prediction and symbols to MC results. After [26]

Finally, we compare the results of the formalism for different random scale-free networks satisfying $P(k) \sim k^{-\gamma}$, which have been generated using the configuration model [6, 21] with a fixed size of $N = 10^4$ nodes. Figure 15.3 shows the phase diagram for $\mu = 1$ and several values of the exponent γ, both below and above $\gamma = 3$. Symbols correspond to MC simulations, whereas dotted lines represent the results obtained using the analytical approximation. As it can be seen, the agreement between both methods is remarkable, even for values of $\gamma < 2.5$ where structural changes are extremely relevant [51]. On the other hand, one may explore the dependency with the system size while fixing the degree distribution exponent γ. This is what is shown in Fig. 15.4, where we have depicted the phase diagram for networks with $\gamma = 2.7$ for several system sizes ranging from $N = 500$ to $N = 10^5$. Except for $N = 500$, where MC results have a large standard deviation close to the critical point, the agreement is again excellent in the whole range of β values.

15.2.3 Epidemic Threshold

Let us now assume the existence of a critical point β_c for fixed values of μ and λ_i such that $\rho = 0$ if $\beta < \beta_c$ and $\rho > 0$ when $\beta > \beta_c$. The calculation of this critical

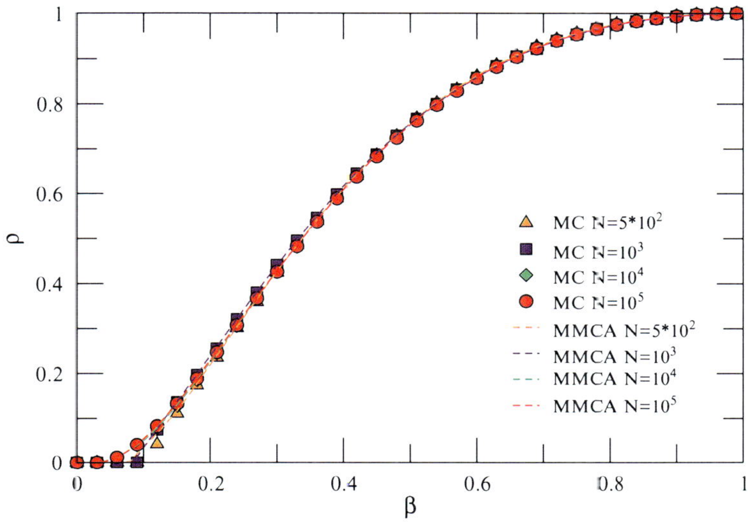

Fig. 15.4 Phase diagram for the SIS model ($\lambda = \infty$) in a random scale free network for different system sizes as indicated. The networks have a power-law degree distribution with an exponent $\gamma = 2.7$ and $\mu = 1$. MC results are averages over 10^2 realizations. After [26]

point is performed by considering that when $\beta \to \beta_c$, the probabilities $p_i \approx \varepsilon_i$, where $0 \le \varepsilon_i \ll 1$, and then after substitution in (15.2) one gets

$$q_i \approx 1 - \beta \sum_{j=1}^{N} r_{ji} \varepsilon_j \, . \tag{15.7}$$

Inserting (15.7) in (15.4), and neglecting second order terms in ε we get

$$\sum_{j=1}^{N} \left(r_{ji} - \frac{\mu}{\beta} \delta_{ji} \right) \varepsilon_j = 0 \, , \qquad \forall i = 1, \ldots, N \, , \tag{15.8}$$

where δ_{ij} stands for the Kronecker delta. The system (15.8) has non trivial solutions if and only if μ/β is an eigenvalue of the matrix $\mathbf{R}$. Since we are looking for the onset of the epidemic, the lowest value of β satisfying (15.8) is

$$\beta_c = \frac{\mu}{\Lambda_{\max}} \, , \tag{15.9}$$

where $\Lambda_{\max}$ is the largest eigenvalue of the matrix $\mathbf{R}$. Equation (15.9) defines the epidemic threshold of the disease spreading process.

With the previous development it is worth analyzing the two limiting cases of CP and RP above. In the first case, one must take into account that the matrix

R is a transition matrix whose maximum eigenvalue is always $\Lambda_{\max} = 1$. Thus, the trivial result that the only non-zero solution corresponds to $\beta_c = \mu$. For the RP corresponding to the SIS spreading process usually adopted [45], the classical result for uncorrelated SF networks is recovered because, in this case, the largest eigenvalue [16, 48] is $\Lambda_{\max} = \langle k^2 \rangle / \langle k \rangle$.

15.2.4 Mesoscopic Equations at the Critical Point

Once the general framework given by the dynamical system ((15.1) and (15.2)) has been proposed, it is instructive to approximate it using the hypotheses underlying HMF. These hypotheses consist of: (1) coarse-graining the system in classes of nodes by degree, assuming that the dynamical properties within each class are the same and (2) neglecting fluctuations. To obtain the mesoscopic description we consider the second order approximation of (15.4) and (15.5), and proceed as in the previous section. Therefore,

$$q_i \approx 1 - \beta \sum_j r_{ji} \varepsilon_j + \beta^2 \sum_{j<l} r_{ji} r_{li} \varepsilon_j \varepsilon_l \,. \tag{15.10}$$

After substitution in (15.4) and reordering terms one gets

$$0 = -\mu \varepsilon_i + \beta (1 - \varepsilon_i) \sum_j r_{ji} \varepsilon_j + \mu \beta \varepsilon_i \sum_j r_{ji} \varepsilon_j - \beta^2 \sum_{j<l} r_{ji} r_{li} \varepsilon_j \varepsilon_l \,, \tag{15.11}$$

which are the equations governing the dynamics of the contact-based epidemic spreading process at the microscopic level. It is possible to write (15.11) at the commonly used mesoscopic (degree class) level for unweighted, undirected heterogeneous networks. The interactions then take place between classes of nodes. Defining the average density of infected nodes with degree k as $\rho_k = \frac{1}{N_k} \sum_{k_i = k} p_i$, where N_k is the number of nodes with degree k and the sum runs over the set of nodes of degree k, we obtain the generalized HMF equation near criticality.

To simplify the notation, we define the function

$$R_\lambda(x) = 1 - (1 - x)^\lambda \,. \tag{15.12}$$

Thus, the values of r_{ji} may be expressed as

- Weighted networks:

$$r_{ji} = R_\lambda \left(\frac{w_{ji}}{w_j} \right) \,. \tag{15.13}$$

- Unweighted networks:

$$r_{ji} = R_\lambda\left(\frac{a_{ji}}{k_j}\right) = a_{ji}R_\lambda\left(\frac{1}{k_j}\right) = a_{ji}R_\lambda(k_j^{-1}). \qquad (15.14)$$

15.2.4.1 Homogeneous Networks

For homogeneous unweighted undirected networks, $\varepsilon_i = \varepsilon$ and $k_i \approx \langle k\rangle$ for all nodes. Thus, $\rho = \frac{1}{N}\sum_j \varepsilon_j = \varepsilon$ and

$$0 = -\mu\rho + \beta\rho(1-\rho)\sum_j r_{ji} + \mu\beta\rho^2\sum_j r_{ji} - \beta^2\rho^2\sum_{j<l} r_{ji}r_{li}. \qquad (15.15)$$

The terms involving values of r_{ji} are

$$r_{ji} \approx a_{ji}R_\lambda(\langle k\rangle^{-1}), \qquad (15.16)$$

$$\sum_j r_{ji} \approx \langle k\rangle R_\lambda(\langle k\rangle^{-1}), \qquad (15.17)$$

$$\sum_{j<l} r_{ji}r_{li} \approx \frac{1}{2}\langle k\rangle(\langle k\rangle - 1)R_\lambda(\langle k\rangle^{-1})^2. \qquad (15.18)$$

Now, (15.15) becomes

$$\begin{aligned} 0 = & -\mu\rho + \beta\rho(1-\rho)\langle k\rangle R_\lambda(\langle k\rangle^{-1}) \\ & + \mu\beta\rho^2\langle k\rangle R_\lambda(\langle k\rangle^{-1}) - \beta^2\rho^2\frac{1}{2}\langle k\rangle(\langle k\rangle - 1)R_\lambda(\langle k\rangle^{-1})^2, \end{aligned} \qquad (15.19)$$

which may be considered as the MF approximation of our model for homogeneous networks.

If $\lambda = 1$ then $R_1(\langle k\rangle^{-1}) = \frac{1}{\langle k\rangle}$ and (15.19) becomes

$$0 = -\mu\rho + \beta\rho(1-\rho) + \mu\beta\rho^2 - \frac{\langle k\rangle - 1}{2\langle k\rangle}\beta^2\rho^2. \qquad (15.20)$$

If $\lambda \to \infty$ then $R_\infty(\langle k\rangle^{-1}) = 1$ and (15.19) reads

$$0 = -\mu\rho + \beta\rho(1-\rho)\langle k\rangle + \mu\beta\rho^2\langle k\rangle - \frac{1}{2}\beta^2\rho^2\langle k\rangle(\langle k\rangle - 1). \qquad (15.21)$$

In both cases, the first two terms correspond to the standard CP and RP models (previously reported in the literature), respectively, and the additional terms are second order contributions corresponding to reinfections and multiple infections.

15.2.4.2 Heterogeneous Networks

Now we will concentrate on the class of heterogeneous unweighted undirected networks completely specified by their degree distribution $P(k)$ and by the conditional probability $P(k'|k)$ that a node of degree k is connected to a node of degree k'. Of course, the normalization conditions $\sum_k P(k) = 1$ and $\sum_{k'} P(k'|k) = 1$ must be fulfilled. In this case, the average number of links that goes from a node of degree k to nodes of degree k' is $kP(k'|k)$.

In these heterogeneous networks, it is supposed that all nodes of the same degree behave equally, thus $\varepsilon_i = \varepsilon_j$ if $k_i = k_j$, and the density ρ_k of infected nodes of degree k is given by $\rho_k = \frac{1}{N_k}\sum_{i\in K}\varepsilon_i = \varepsilon_j\,, \quad \forall j \in K$, where $N_k = P(k)N$ is the expected number of nodes with degree k. Here, we have made use of K to denote the set of nodes with degree k. This notation allows to group the sums by the degrees of the nodes. For instance, if the degree of node i is $k_i = k$ then

$$\sum_j a_{ji}\varepsilon_j = \sum_{k'}\sum_{j\in K'} a_{ji}\rho_{k'} = \sum_{k'}\rho_{k'}\sum_{j\in K'} a_{ij} = \sum_{k'}\rho_{k'}kP(k'|k) = k\sum_{k'}P(k'|k)\rho_{k'}\,. \qquad (15.22)$$

Now, let us find the mean field equation for heterogeneous networks. First we substitute (15.14) in (15.11)

$$\begin{aligned} 0 = {} & -\mu\varepsilon_i + \beta(1-\varepsilon_i)\sum_j a_{ji}R_\lambda(k_j^{-1})\varepsilon_j + \mu\beta\varepsilon_i\sum_j a_{ji}R_\lambda(k_j^{-1})\varepsilon_j \\ & -\beta^2\sum_{j<l} a_{ji}a_{li}R_\lambda(k_j^{-1})R_\lambda(k_l^{-1})\varepsilon_j\varepsilon_l\,. \end{aligned} \qquad (15.23)$$

It is convenient to analize separately the summatory terms in (15.23), supposing node i has degree k:

$$\begin{aligned} \sum_j a_{ji}R_\lambda(k_j^{-1})\varepsilon_j &= \sum_{k'}\sum_{j\in K'} a_{ji}R_\lambda(k'^{-1})\rho_{k'} \\ &= \sum_{k'}R_\lambda(k'^{-1})\rho_{k'}\sum_{j\in K'} a_{ij} \\ &= k\sum_{k'}P(k'|k)R_\lambda(k'^{-1})\rho_{k'}\,, \end{aligned} \qquad (15.24)$$

$$\begin{aligned} & \sum_{j<l} a_{ji}a_{li}R_\lambda(k_j^{-1})R_\lambda(k_l^{-1})\varepsilon_j\varepsilon_l \\ & = \frac{1}{2}\sum_j\sum_l a_{ji}a_{li}R_\lambda(k_j^{-1})R_\lambda(k_l^{-1})\varepsilon_j\varepsilon_l - \frac{1}{2}\sum_j a_{ji}^2R_\lambda(k_j^{-1})^2\varepsilon_j^2 \\ & = \frac{1}{2}\sum_{k'}\sum_{k''}\sum_{j\in K'}\sum_{l\in K''} a_{ji}a_{li}R_\lambda(k'^{-1})R_\lambda(k''^{-1})\rho_{k'}\rho_{k''} - \frac{1}{2}\sum_{k'}\sum_{j\in K'} a_{ji}^2R_\lambda(k'^{-1})^2\rho_{k'}^2 \end{aligned}$$

$$= \frac{1}{2}\sum_{k'}\sum_{k''} R_\lambda(k'^{-1})R_\lambda(k''^{-1})\rho_{k'}\rho_{k''}\sum_{j\in K'} a_{ij}\sum_{l\in K''} a_{il} - \frac{1}{2}\sum_{k'} R_\lambda(k'^{-1})^2\rho_{k'}^2\sum_{j\in K'} a_{ij}^2$$

$$= \frac{1}{2}k^2\sum_{k'}\sum_{k''} R_\lambda(k'^{-1})R_\lambda(k''^{-1})P(k'|k)P(k''|k)\rho_{k'}\rho_{k''}$$

$$- \frac{1}{2}k\sum_{k'} R_\lambda(k'^{-1})^2 P(k'|k)\rho_{k'}^2 . \tag{15.25}$$

Substitution in (15.23) leads to the generalized HMF equation

$$0 = -\mu\rho_k + \beta k(1-\rho_k)\sum_{k'} P(k'|k)R_\lambda(k'^{-1})\rho_{k'}$$

$$+ \mu\beta k\rho_k\sum_{k'} P(k'|k)R_\lambda(k'^{-1})\rho_{k'}$$

$$- \frac{1}{2}\beta^2k^2\sum_{k'}\sum_{k''} R_\lambda(k'^{-1})R_\lambda(k''^{-1})P(k'|k)P(k''|k)\rho_{k'}\rho_{k''}$$

$$+ \frac{1}{2}\beta^2 k\sum_{k'} R_\lambda(k'^{-1})^2 P(k'|k)\rho_{k'}^2 . \tag{15.26}$$

If $\lambda = 1$, then $R_1(k^{-1}) = \frac{1}{k}$ and (15.26) becomes

$$0 = -\mu\rho_k + \beta k(1-\rho_k)\sum_{k'}\frac{1}{k'}P(k'|k)\rho_{k'}$$

$$+\mu\beta k\rho_k\sum_{k'}\frac{1}{k'}P(k'|k)\rho_{k'} + \frac{1}{2}\beta^2 k\sum_{k'}\frac{1}{k'^2}P(k'|k)\rho_{k'}^2$$

$$-\frac{1}{2}\beta^2k^2\left(\sum_{k'}\frac{1}{k'}P(k'|k)\rho_{k'}\right)^2 . \tag{15.27}$$

If $\lambda \to \infty$, then $R_\infty(k^{-1}) = 1$ and (15.26) reads

$$0 = -\mu\rho_k + \beta k(1-\rho_k)\sum_{k'} P(k'|k)\rho_{k'}$$

$$+\mu\beta k\rho_k\sum_{k'} P(k'|k)\rho_{k'} + \frac{1}{2}\beta^2 k\sum_{k'} P(k'|k)\rho_{k'}^2$$

$$-\frac{1}{2}\beta^2k^2\left(\sum_{k'} P(k'|k)\rho_{k'}\right)^2 . \tag{15.28}$$

Again, the first two terms in both cases correspond to the standard CP and RP HMF equations, respectively, and the additional terms are second order contributions corresponding to reinfections and multiple infections.

15.3 Traffic-Driven Epidemic Spreading in Complex Networks

In the second part of this chapter, we investigate the outcome of an epidemic spreading process driven by transport instead of diffusion. To this end, we analyzed a paradigmatic abstraction of epidemic contagion, the so-called Susceptible–Infected–Susceptible (SIS) model, which assumes that contagion occurs through the eventual contact or transmission between connected partners that are using their connections at the time of propagation. This is achieved by considering a quantized interaction at each time step. Mathematically, we set up the model in a flow scenario where contagion is carried by interaction packets traveling across the network.

15.3.1 The Model

In the first place, two different types of SF networks are generated. On one hand, we build random uncorrelated SF networks using the configuration model [6, 21]. On the other hand, small-world, SF and highly clustered networks – all properties found in many real-world networks [6, 21] such as the Internet – are also generated using a class of recently developed network models [9, 50], in which nearby nodes in a hidden metric space are connected. This metric space can represent social, geographical or any other relevant distance between the nodes of the simulated networks. Specifically, in the model currently at study, nodes are uniformly distributed in a one-dimensional circle by assigning them a random polar angle θ distributed uniformly in the interval $[0, 2\pi)$ and assigned an expected degree k. The expected degrees of the nodes are then drawn from some distribution $x(k)$ and the network is completed by connecting two nodes with hidden coordinates (θ, k) and (θ', k') with probability $r(\theta, k, \theta', k') = \left(1 + \frac{d(\theta,\theta')}{\eta' k k'}\right)^{-\alpha}$, where $\eta' = (\alpha - 1)/2\langle k\rangle$, $d(\theta, \theta')$ is the geodesic distance between the two nodes on the circle, and $\langle k\rangle$ is the average degree. Finally, choosing $x(k) = (\gamma - 1)k_0^{\gamma-1}k^{-\gamma}$, $k > k_0 \equiv (\gamma - 2)\langle k\rangle/(\gamma - 1)$ generates random networks with a power law distribution with exponent $\gamma > 2$. In most of the simulations, $\gamma = 2.7$ $\langle k\rangle = 3$ and $\alpha = 2$ are fixed.

Once the networks are built up, the traffic process is implemented in the following way. At each time step, $p = \Lambda N$ new packets are created with randomly chosen origins and destinations. For the sake of simplicity, packets are considered non-interacting so that no queues are used. The routing of information is modeled through even a shortest path delivery strategy or a greedy algorithm [8, 9]. In the latter, the second class of SF networks is used and a node i forwards a packet to node j in its neighborhood, which is the closest node (in the hidden metric space) to the final packet destination. Results are insensitive to the two routing protocols implemented.

To model the spreading dynamics we have implemented the aforementioned Susceptible-Infected-Susceptible model, in which each node can be in two possible states: healthy (S) or infected (I). Starting from an initial fraction of infected individuals $\rho_0 = I_0/N$, the infection spreads in the system as the nodes interact. A susceptible node has a probability β of becoming infected every time it interact with an infected neighbors. We also assume that infected nodes are recovered at a rate μ, which we fix to 1 for most of the simulations. After a transient time, we compute the average density of infected individuals, ρ, which is the prevalence of disease in the system. To account for link concurrency, we consider that two nodes do not interact at all times t, but only when they exchange at least a packet. This situation is reminiscent of disease transmission on air transportation networks; if an infected individual did not travel between two cities, then regardless of whether or not those cities are connected by a direct flight, the epidemic will not spread from one place to the other. In this way, although a node can potentially interact with as many contacts as it has and as many times as packets it exchanges with its neighbors, the effective interactions are driven by a second dynamics (traffic). The more packets travel through a link, the more likely the disease will spread through it. On the other hand, once an interaction is at work, the epidemics spreads from infected to susceptible nodes with probability β. For example, if at time t node i is infected and a packet is traveling from node i to one of its neighbors node j, then at the next time step, node j will be infected with probability β. Therefore, susceptible and infected states are associated with the nodes, whereas the transport of packets is the mechanism responsible for the propagation of the disease at each time step.

15.3.2 Unbounded Delivery Rate

We firstly concentrate on an unbounded delivery rate scenario, in which every node can handle as much packets it receives. In this situation, congestion can not arise in the system. Figure 15.5 shows the results for the stationary density of infected nodes ρ as a function of β and the traffic generation rate Λ for SF networks.

In this case, the traffic level determines the value of both the epidemic incidence and the critical thresholds and it's important to notice the emergence of an epidemic threshold under low traffic conditions. This implies that for a fixed value of Λ, the epidemic dies out if the spreading rate is below a certain critical value $\beta_c(\Lambda)$. More intense packet flows yield lower epidemic thresholds. The reason for the dependence of the critical spreading rates on Λ is rooted in the effective topological paths induced by the flow of packets through the network. At low values of Λ, there are only a few packets traveling throughout the system, so the epidemic simply dies out because many nodes do not participate in the interaction via packets exchanges. As Λ grows, more paths appear between communicating nodes, thus spreading the

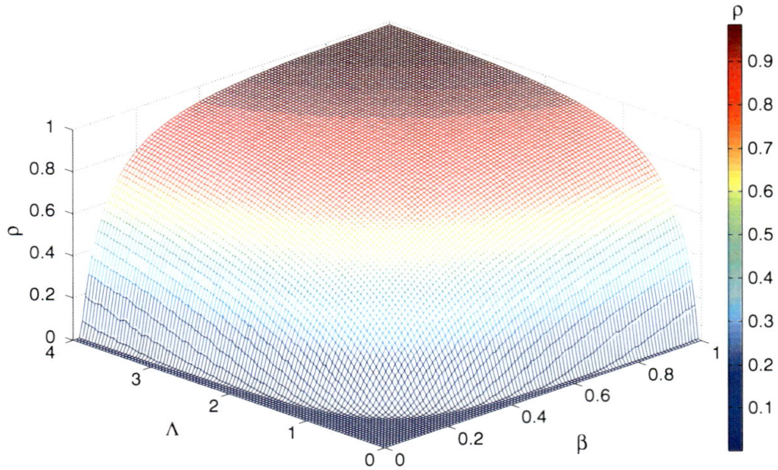

Fig. 15.5 Dependence of epidemic incidence on traffic conditions for unbounded delivery rate. The density of infected nodes, ρ, is shown as a function of the spreading rate β and the intensity of flow Λ in SF networks. Flow conditions (controlled by Λ) determine both the prevalence level and the values of the epidemic thresholds. Increasing the number of packets traveling through the system has a malicious effect: the epidemic threshold decreases as the flow increases. Each curve is an average of 10^2 simulations starting from an initial density of infected nodes $\rho_0 = 0.05$. The network is made up of 10^3 nodes using the model in [9], results correspond to the greedy routing scheme. The remaining parameters are $\alpha = 2$, $\gamma = 2.6$ and $\langle k \rangle = 3$. After [37]

infection to a larger portion of the network. Therefore, in traffic-driven epidemic processes the infection is constrained to propagate only through links that transmit a packet, and thus the number of attempts to transmit the infection depends on the flow conditions at a local level, namely, on the number of active communication channels at each time step. As a consequence, the effective network that spreads the infection is no longer equivalent to the complete underlying topology. Instead, it is a map of the dynamical process associated with packet traffic flow. The conclusion is that the disease propagation process has two dynamical components: one intrinsic to the disease itself (β) and the other to the underlying traffic dynamics (the flow). To theorize about these effects we next formulate the analytical expression for the dependence of the epidemic threshold on the amount of traffic injected into the system, following a mean-field approach akin to the conventional analysis of the reaction driven case. Mathematically, the fraction of paths traversing a node given a certain routing protocol [27], the so-called algorithmic betweenness, b^k_{alg}, defines the flow pathways. Let us consider the evolution of the relative density, $\rho_k(t)$, of infected nodes with degree k. Following the heterogeneous mean-field approximation [45], the dynamical rate equations for the SIS model are

$$\partial_t \rho_k(t) = -\mu \rho_k(t) + \beta \Lambda b^k_{\text{alg}} N \left[1 - \rho_k(t)\right] \Theta(t). \tag{15.29}$$

The first term in (15.29) is the recovery rate of infected individuals (we set henceforth $\mu = 1$). The second term takes into account the probability that a node with k links belongs to the susceptible class, $[1 - \rho_k(t)]$, and gets the infection via packets traveling from infected nodes. The latter process is proportional to the spreading probability β, the probability $\Theta(t)$ that a packet travels through a link pointing to an infected node and the number of *packets* received by a node of degree k. This, in turns, is proportional to the total number of packets in the system, $\sim \Lambda N$, and the algorithmic betweenness of the node, b^k_{alg}. Note that the difference with the standard epidemic spreading model is given by these factors, as now the number of contacts per unit time of a node is not proportional to its connectivity but to the number of packets that travel through it. Finally, $\Theta(t)$ takes the form

$$\Theta(t) = \frac{\sum_k b^k_{\text{alg}} P(k) \rho_k(t)}{\sum_k b^k_{\text{alg}} P(k)}. \tag{15.30}$$

Equation (15.29) has been obtained assuming: (1) that the network is uncorrelated $P(k'|k) = k'P(k')/\langle k \rangle$ and (2) that the algorithmic flow between the classes of nodes of degree k and k' factorizes $b^{kk'}_{\text{alg}} \sim b^k_{\text{alg}} b^{k'}_{\text{alg}}$. Although no uncorrelated networks exist, this approximation allows us to identify the governing parameters of the proposed dynamics. The second approximation is an upper bound to the actual value of the $b^{kk'}_{\text{alg}}$, whose mathematical expression is, in general, unknown. The validity of the theory even with these approximations is notable as confirmed by the numerical simulations.

By imposing stationarity $[\partial_t \rho_k(t) = 0]$, (15.29) yields

$$\rho_k = \frac{\beta \Lambda b^k_{\text{alg}} N \Theta}{1 + \beta \Lambda b^k_{\text{alg}} N \Theta}, \tag{15.31}$$

from which a self-consistent equation for Θ is obtained as

$$\Theta = \frac{1}{\sum_k b^k_{\text{alg}} P(k)} \sum_k \frac{(b^k_{\text{alg}})^2 P(k) \beta \Lambda N \Theta}{1 + \beta \Lambda b^k_{\text{alg}} N \Theta}. \tag{15.32}$$

The value $\Theta = 0$ is always a solution. In order to have a non-zero solution, the condition

$$\frac{1}{\sum_k b^k_{\text{alg}} P(k)} \frac{\mathrm{d}}{\mathrm{d}\Theta} \left(\sum_k \frac{(b^k_{\text{alg}})^2 P(k) \beta \Lambda N \Theta}{1 + \beta \Lambda b^k_{\text{alg}} N \Theta} \right) \Bigg|_{\Theta = 0} > 1 \tag{15.33}$$

must be fulfilled, from which the epidemic threshold is obtained as

$$\beta_{\text{c}} = \frac{\langle b_{\text{alg}} \rangle}{\langle b^2_{\text{alg}} \rangle} \frac{1}{\Lambda N}, \tag{15.34}$$

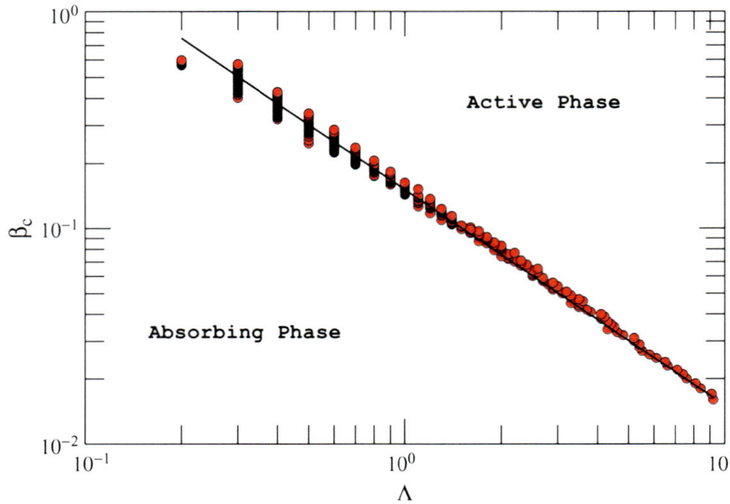

Fig. 15.6 Comparison between numerical and theoretical critical points. Log–log plot of the critical thresholds, β_c, as a function of the rate at which packets are injected into the system, Λ. Two regions are differentiated: an active and an absorbing phase as indicated. The *solid line* corresponds to (15.34) with $\frac{\langle b_{\text{alg}} \rangle}{\langle b^2_{\text{alg}} \rangle} \frac{1}{N} = 0.154$. The agreement is remarkable even though (15.34) is derived using a MF approach. The underlying network, infection spreading mechanism and routing protocol are the same as in Fig. 15.5. Each curve is an average of 10^2 simulations. Remaining parameters are the same as in Fig. 15.5. After [37]

below which the epidemic dies out, and above which there is an endemic state. In Fig. 15.6a comparison between the theoretical prediction and numerical observations is presented. Here, we have explicitly calculated the algorithmic betweenness for the greedy routing as it only coincides with the topological betweenness for shortest paths routing. The obtained curve separates two regions: an absorbing phase in which the epidemic disappears, and an active phase where the infection is endemic.

Equation (15.34) is notably simple but has profound implications: the epidemic threshold decreases with traffic and eventually vanishes in the limit of very large traffic flow in finite systems, in contrast to the expected result of a finite-size reminiscent threshold in the classical reactive–diffusive framework. Admittedly, this is a new feature with respect to previous results on epidemic spreading in SF networks. It is rooted in the increase of the effective epidemic spreading rate due to the flow of packets. This is a genuine effect of traffic-driven epidemic processes and generalizes the hypothesis put forward in the framework of a reaction-diffusion process [18] on SF networks. It implies that an epidemic will pervade the (finite) network whatever the spreading rate is if the load on it is high enough. Moreover, (15.34) reveals a new dependence. The critical threshold depends on the topological features of the graph, but at variance with the standard case, through the first two moments of the algorithmic betweenness distribution. As noted above, the algorithmic betweenness of a node is given by the number of packets traversing that node given a routing protocol. In other words, it has two components: a topological one which is given by the degree of the node and a dynamical component defined by the routing protocol.

Within our formulation, the classical result [45]

$$\beta_c = \frac{\langle k \rangle}{\langle k^2 \rangle}, \tag{15.35}$$

can be obtained for a particular protocol and traffic conditions, although we note that the microscopic dynamics of our model is different from the classical SIS. To see this, assume a random protocol. If packets of information are represented as w random walkers traveling in a network with average degree $\langle k \rangle$, then under the assumption that the packets are not interacting, it follows that the average number of walkers at a node i in the stationary regime (the algorithmic betweenness) is given by [36, 44] $b^i_{\text{alg}} = \frac{k_i}{N\langle k \rangle} w$. The effective critical value is then $(\beta\Lambda)_c = \langle k \rangle^2/(\langle k^2 \rangle w)$, that recovers, when $w = \langle k \rangle$, the result in (15.35).

Results are robust for other network models and different routing algorithms. We have also made numerical simulations of the traffic-driven epidemic process on top of Barabási–Albert and random SF networks implementing a shortest paths delivery scheme. In this case, packets are diverted following the shortest path (in the actual topological space) from the packets' origins to their destinations. The rest of model parameters and rules for epidemic spreading remain the same. Figures 15.7 and 15.8 show the results obtained for random SF networks generated via the configuration model and the Barabási–Albert model, respectively. As can be seen, the phenomenology is the same for both types of networks: the epidemic threshold depends on the amount of traffic in the network such that the higher the flow is, the smaller the epidemic threshold separating the absorbing and active phases. On the other hand, for processes in which the delivery of packets follows a shortest path algorithm, (15.34) looks like

$$\beta_c = \frac{\langle b_{\text{top}} \rangle}{\langle b^2_{\text{top}} \rangle} \frac{1}{\Lambda N}, \tag{15.36}$$

where b_{top} is the topological betweenness. To further confirm our findings on a realistic topology we run the model on top of the Air Transportation Network (ATN) [28]. The network composed by the direct flies between more the 3,000 airports in the world, in which each node represents an airport and the links represents the direct connection between them. Although in the ATN links have weights accounting for the annual number of passengers voyaging on each connection, we considered the network as un-weighted and the shortest-path routing protocol. Also in this case the results are confirmed as shown in Fig. 15.9. Figure 15.10 also shows the agreement between the analytical prediction and the numerical simulations.

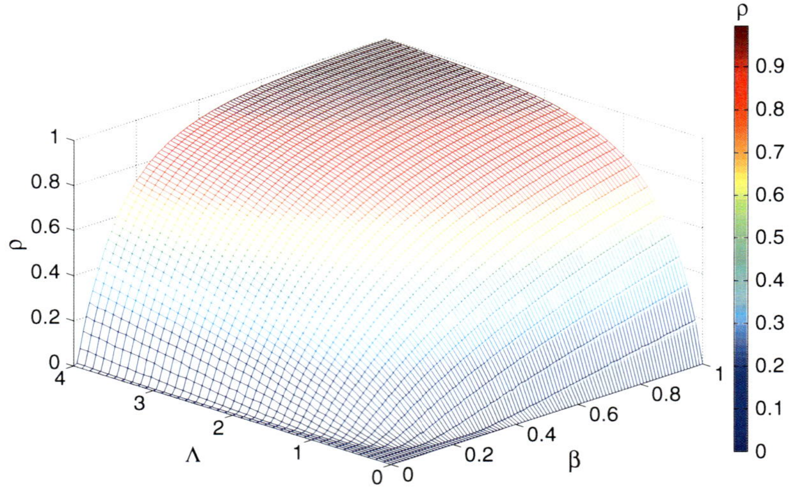

Fig. 15.7 Density of infected nodes, ρ, as a function of traffic flow (determined by Λ) and the epidemic spreading rate β for random scale-free networks and a shortest paths routing scheme for packets delivery. Each point is the result of 10^2 averages over different networks and initial conditions. The exponent of the degree distribution of the network is $\gamma = 2.7$. After [37]

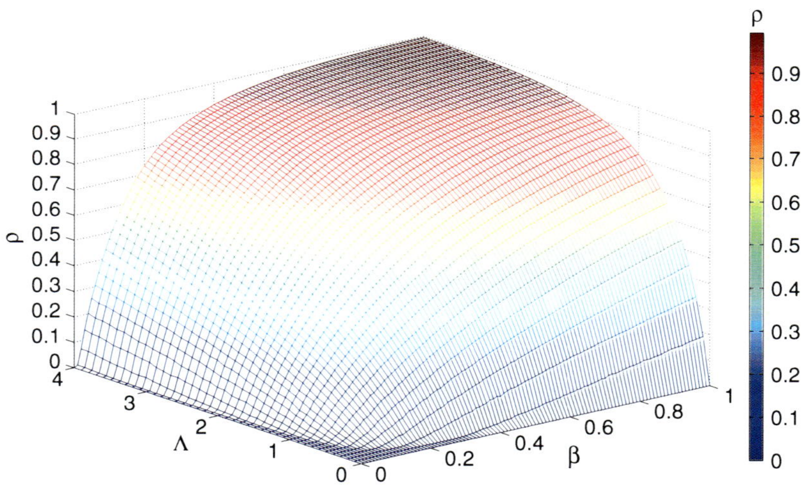

Fig. 15.8 Density of infected nodes, ρ, as a function of traffic flow (determined by Λ) and the epidemic spreading rate β for BA scale-free networks and a shortest paths routing scheme for packets delivery. Each point is the result of 10^2 averages over different networks and initial conditions. After [37]

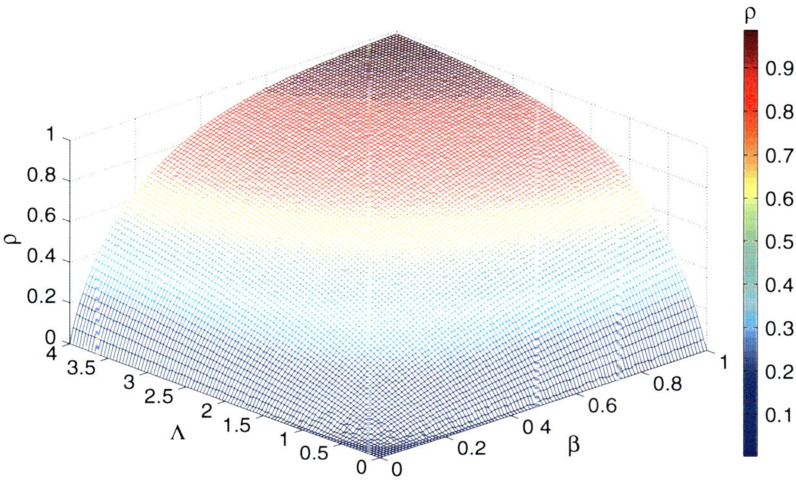

Fig. 15.9 Density of infected nodes, ρ, as a function of traffic flow (determined by Λ) and the epidemic spreading rate β for the ATN (considered as unweighted) and a shortest paths routing scheme for packets delivery. Each point is the result of 10^2 averages over different networks and initial conditions. After [37]

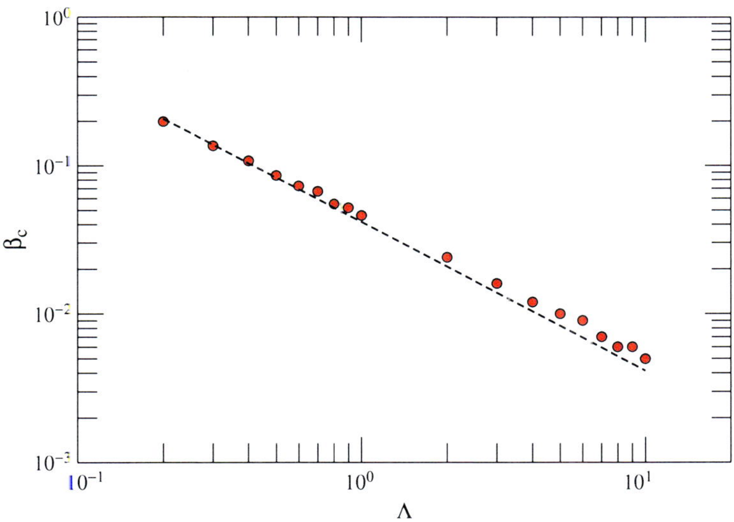

Fig. 15.10 Comparison between numerical and theoretical critical points in the ATN. Log–log plot of the critical thresholds, β_c, as a function of the rate at which packets are injected into the system, Λ for the ATN. The *dashed line* corresponds to (15.34) with $\frac{\langle \wp_{alg} \rangle}{\langle \wp_{alg}^2 \rangle}\frac{1}{N} = 0.041$. Despite existing degree correlations in the network, the agreement is remarkable. Each point is an average of 10^2 simulations. After [37]

15.3.3 Bounded Delivery Rate

Equation (15.36) allows us to investigate also the equivalent scenario in the presence of congestion. Let us consider the same traffic process above but with nodes having queues that can store as many packets as needed but can deliver, on average, only a finite number of them at each time step. It is known that there is a critical value of Λ above which the system starts to congest [27]

$$\Lambda_{\mathrm{c}} = \frac{(N-1)}{b^*_{\mathrm{alg}}}. \tag{15.37}$$

Equation (15.37) gives the traffic threshold that defines the onset of congestion, which is governed by the node with maximum algorithmic betweenness b^*_{alg}. Substituting (15.37) in (15.34) we obtain a critical threshold for an epidemic spreading process bounded by congestion. Increasing the traffic above Λ_{c} will gradually congest all the nodes in the network up to a limit in which the traffic is stationary and the lengths of queues grow without limit.

To illustrate this point, let us assume that the capacities for processing and delivering information are heterogeneously distributed [49, 52, 54] so that the larger the number of paths traversing a node, the larger its capability to deliver the packets. Specifically, each node i of the network delivers at each time step a maximum of $\lceil c_i = 1 + k_i^{\eta} \rceil$ packets, where η is a parameter of the model. In this case, the critical value of Λ in (15.37) is multiplied by the maximum delivery capacity [54]. Moreover, without loss of generality, we will explore the behavior of the model in random SF networks where the routing is implemented by shortest paths $b_{\mathrm{alg}} = b_{\mathrm{top}} \sim k^{\nu}$, being ν usually between 1.1 and 1.3 [47]. The previous assumption for the delivery capability thus allows to explore as a function of η the situations in which the delivery rate is smaller or larger than the arrival rate (defined by the algorithmic betweenness). Phenomenologically, these two scenarios correspond to the cases in which the traffic is in a free flow regime (if $\eta > \nu$) or when the network will congest (if $\eta < \nu$). We also note that the adopted approach is equivalent to assume a finite length for the queues at the nodes.

Figure 15.11 shows the fraction of active packets on the network, as a function of the spreading rate β and the rate at which packets are generated Λ for two different values of η using a shortest path delivery scheme on top of random SF networks. For $\eta = 0.8$, the epidemic incidence is significantly small for all values of the parameters Λ and β as compared with the results obtained when the rate of packets delivery is unbounded. On the contrary, when $\eta = 1.7$ the phase diagram is qualitatively the same as for the unbounded case, including the result that the epidemic incidence vanishes when Λ is large enough. A closer look at the dynamical evolution unveils an interesting, previously unreported, feature – when the rate at which packets are delivered is smaller than the rate at which they arrive, the average value of infected nodes saturates beyond a certain value of the traffic flow rate Λ. This effect is due to the emergence of traffic congestion. When the flow of packets into the system is

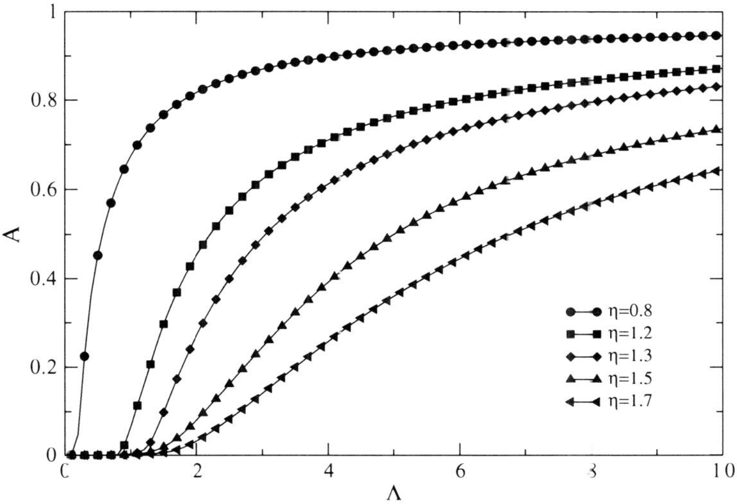

Fig. 15.11 Fraction of active packets as a function of the traffic flow with bounded delivery rate. It represents the fraction of active packets A: packets still traveling in the network over the total amount of generated packets in a time period τ, as function of the traffic injected in the system Λ for different values of the delivery capacity η. The underlying network and the routing protocol are the same as in Fig. 15.7

such that nodes are not able to deliver at least as many packets as they receive, their queues start growing and packets pile up. This in turns implies that the spreading of the disease becomes less efficient, or in other words, the spreading process slows down. The consequence is that no matter whether more packets are injected into the system, the average level of packets able to move from nodes to nodes throughout the network is roughly constant and so is the average level of infected individuals.

Figure 15.12 illustrates the phenomenological picture described above. It shows the epidemic incidence ρ for a fixed value of $\beta = 0.15$ as a function of Λ for different values of η. The figure clearly evidences that congestion is the ultimate reason of the behavior described above. Therefore, the conclusion is that in systems where a traffic process with finite delivery capacity is coupled to the spreading of the disease the *epidemic incidence is bounded*. This is good news as most of the spreading processes in real-world networks involves different traffic flow conditions. Further evidence of this phenomenology is given in Fig. 15.13, where we have depicted the epidemic threshold as a function of Λ for two different values of η, less and greater than v. When $\eta < v$ congestion arises, and the contrary holds for $\eta > v$ where the diagram is equivalent to that of unbounded traffic. The onset of congestion determines the value of β above which congestion starts. It is clearly visualized as the point beyond which the power law dependence in (15.34) breaks down. The plateau of β_c corresponds to the stationary situation of global congestion. A comparison for different values of η in the bounded delivery rate model is presented in Fig. 15.14.

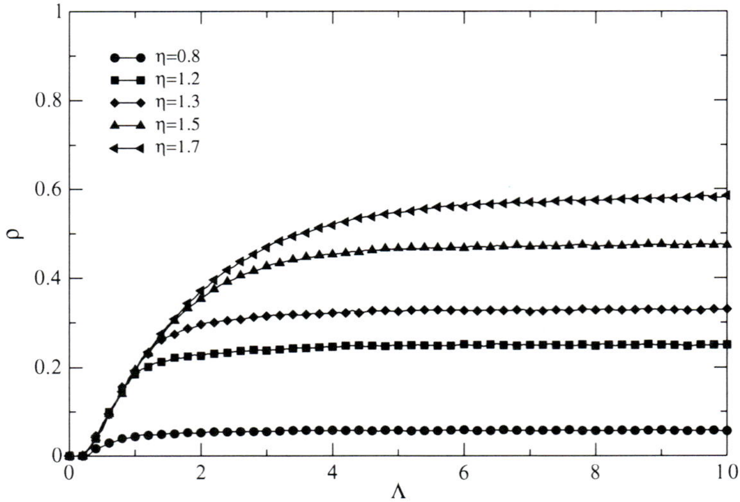

Fig. 15.12 Epidemic incidence in traffic-driven epidemic processes with bounded delivery rate. The figure represents the average fraction of infected nodes ρ as a function of Λ for different delivery rates at fixed $\beta = 0.15$. When congestion arises, the curves depart from each other and the epidemic incidence saturates soon afterwards. After [37]

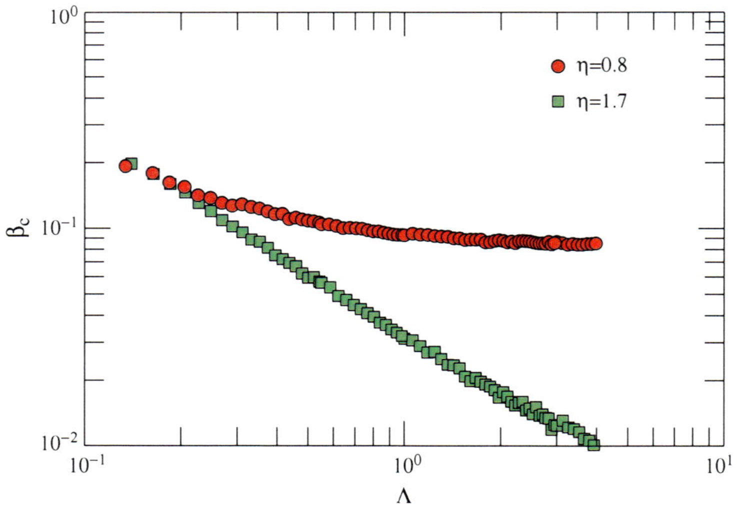

Fig. 15.13 Epidemic thresholds as a function of Λ for two values of η. The onset of congestion marks the point, $\Lambda_c \approx 0.150$, at which the curve for $\eta = 0.8$ departs from (15.34), i.e., when the power law dependence breaks down. Soon afterwards congestion extends to the whole network leading to a bounded (from below) epidemic threshold. On the contrary, when the delivery rate is large enough (as in the case of $\eta = 1.7$), (15.34) holds for all values of Λ, thus resembling the unbounded delivery rate case. After [37]

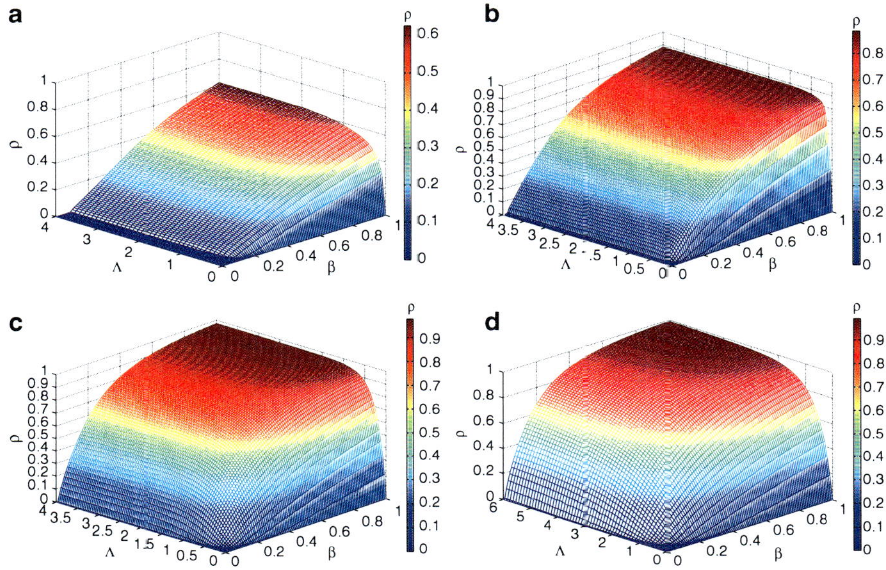

Fig. 15.14 Comparison between different delivery capacity in the bounded delivery rate model. The plot represents density of infected nodes, ρ, as a function of traffic flow Λ and the epidemic spreading rate β for random scale-free networks and a shortest paths routing scheme with different values of the delivery capacity η: panel (**a**) $\eta = 0.8$, (**b**) $\eta = 1.0$, (**c**) $\eta = 1.5$ and (**d**) $\eta = 1.7$, for the random SF network and the shortest path delivery scheme. After [37]

15.4 Conclusions

In the first part of this chapter, we have presented a novel framework, the Microscopic Markov-Chain Approach, to study disease spreading in networks. By defining a set of discrete-time equations for the probability of individual nodes to be infected, we construct a dynamical system that generalizes from an individual contact process to the classical case in which all connections are concurrently used, for any complex topology. The whole phase diagram of the system can be found solving the equations at the stationary state. The numerical solution of the analytic equations overcomes the computational cost of MC simulations. Moreover, the formalism allows to gain insight on the behavior of the critical epidemic threshold for different values of the probability of contacting a fraction λ of neighbors per time step.

The MMCA model deals with infections driven by direct contacts between nodes, but not with traffic situations where nodes transmit the epidemics by flow communication with others [37]. In this latter case, the routing protocol of traffic between nodes is absolutely relevant and can change the critical point of the epidemic spreading.

In the second part of this chapter, we have developed a framework in the scope of MF theories to cope with the problem of assessing the impact of epidemics when the routing of traffic is considered. We have argued both analytically and numerically the conditions for the emergence of an epidemic outbreak in scale-free networks when disease contagion is driven by traffic or interaction flow. The study provides a more general theory of spreading processes in complex heterogeneous networks that includes the previous results as a particular case of diffusive spreading. Moreover, we have shown that the situation in which the epidemic threshold vanishes in finite scale-free networks is also plausible, thus, providing an explanation to the long-standing question of why some viruses prevail in the system with a low incidence.

The new approach presented here provides a novel framework to address related problems. For instance, in the context of air-transportation networks [17], a similar mechanism to the one reported here could explain the observed differences in the impact of a disease during a year [25]. One might even expect that, due to seasonal fluctuations in flows, the same disease could not provoke a system-wide outbreak if the flow were not high enough during the initial states of the disease contagion. Incorporating the non-diffusive character of the spreading process into current models has profound consequences for the way the system functions. Also the theory could help designing new immunization algorithms or robust protocols; one in particular being quarantining highly sensitive traffic nodes. On more general grounds, our conclusions point to the need of properly dealing with link concurrency. Further exploring this challenge will have far-reaching consequences for the study of dynamical processes on networks, and especially the relationship between structure and dynamics for networked systems. Ultimately, this paves the way towards a more complete theoretical framework of complex networks.

Acknowledgements We acknowledge the group of Prof. L. A. N. Amaral for sharing the airports data set. This work was supported by Spanish MICINN FIS2009-13730-C02-02, FIS2008-01240 and FIS2009-13364-C02-01, and the Generalitat de Catalunya 2009-SGR-838. A. A. acknowledges partial support by the Director, Office of Science, Computational and Technology Research, U.S. Department of Energy under Contract DE-AC02-05CH11231. Y. M. acknowledges support from the DGA through Project PI038/08 and a grant to FENOL.

References

1. Amaral, L. A. N. , Scala, A., Barthélemy, M., Stanley, H. E. Classes of small-world networks. *Proc. Nat. Acad. Sci. USA* 97:11149-11152. (2000)
2. Anderson, R. M., May, R. M. *Infectious diseases of humans: Dynamics and Control*. (Oxford University Press, Oxford). (1992)
3. Barabási, A. L., Albert, R. Emergence of scaling in random networks. *Science* 286:509-512. (1999)
4. Barthélemy, M., Barrat, A., Pastor-Satorras, R., Vespignani A. Velocity and hierarchical spread of epidemic outbreaks in scale-free networks. *Phys. Rev. Lett.* 92:178701. (2004)

5. Barrat, A., Barthélemy, M., Pastor-Satorras, R., Vespignani, A. The architecture of complex weighted networks. *Proc. Natl. Acad. Sci. USA* 101:3747-3752. (2004)
6. Boccaletti, S., Latora, V., Moreno, Y., Chávez, M., Hwang, D. U. Complex Networks: Structure and Dynamics. *Phys. Rep.* 424:175-308. (2006)
7. Boguña, M., Castellano, C., Pastor-Satorras, R. Langevin approach for the dynamics of the contact process on annealed scale-free networks. *Phys. Rev. E* 79:036110. (2009)
8. Boguñá, M., Krioukov, D. Navigating Ultrasmall Worlds in Ultrashort Time. *Phys. Rev. Lett.* 102:058701. (2009)
9. Boguñá, M., Krioukov, D., Claffy, K. C. Navigability of Complex Networks. *Nature Physics* 5:74-80. (2009)
10. Caldarelli, G. *Scale-Free Networks*. (Oxford University Press, Oxford). (2007)
11. Castellano, C., Pastor-Satorras, R. Non-mean-field behavior of the contact process on scale-free networks. *Phys. Rev. Lett.* 96:038701. (2006)
12. Castellano, C., Pastor-Satorras, R. Reply: Non-mean-field behavior of the contact process on scale-free networks. *Phys. Rev. Lett.* 98:029802. (2007)
13. Castellano, C., Pastor-Satorras, R. Routes to thermodynamic limit on scale-free networks. *Phys. Rev. Lett.* 100:148701. (2008)
14. Catanzaro, M., Boguña, M., Pastor-Satorras, R. Diffusion-annihilation processes in complex networks. *Phys. Rev. E* 71:056104. (2005)
15. Chakrabarti, D., Wang, Y., Wang, C., Leskovec, J., Faloutsos, C. Epidemic thresholds in real networks. *ACM Trans. Inf. Syst. Secur.* 10(4):13. (2008)
16. Chung, F., Lu, L., Vu, V. Spectra of random graphs with given expected degrees. *Proc. Natl. Acad. Sci. USA* 100:6313-6318. (2003)
17. Colizza, V., Barrat, A., Barthélemy, M., Vespignani, A. Predictability and epidemic pathways in global outbreaks of infectious diseases: the SARS case study. *BMC Medicine* 5:34. (2007)
18. Colizza, V., Pastor-Satorras, R., Vespignani A. Reaction-diffusion processes and metapopulation models in heterogeneous networks. *Nature Physics* 3:276-282. (2007)
19. Colizza, V., Vespignani, A. Epidemic modeling in metapopulation systems with heterogeneous coupling pattern: Theory and simulations. *Journal of Theoretical Biology* 251:450-467. (2008)
20. Daley, D. J., Gani, J. *Epidemic Modelling*. (Cambridge University Press, Cambridge). (1999)
21. Dorogovtsev, S. N, Goltsev, A. V., Mendes, J. F. F. Critical phenomena in complex networks. *Rev. Mod. Phys.* 80:1275-1336. (2008)
22. Eubank, S., Guclu, H., Anil-Kumar, V. S., Marathe, M. V., Srinivasan, A., Toroczkai, Z., Wang N. Modelling disease outbreaks in realistic urban social networks. *Nature* 429:180-184. (2004)
23. Gallos, L. K., Argyrakis, P. Absence of Kinetic Effects in Reaction-diffusion processes in scale-free Networks. *Phys. Rev. Lett.* 92:138301. (2004)
24. Gardeñes, J.G., Latora, V., Moreno, Y., Profumo, E. Spreading of sexually transmitted diseases in heterosexual populations. *Proc. Nat. Acad. Sci. USA* 105:1399-1404. (2008)
25. Grais, R. F., Ellis, J. H., Kress, A., Glass, G. E. Modeling the spread of annual influenza epidemics in the U.S.: The potential role of air travel. *Health Care Management Science* 7:127-134. (2004)
26. Gómez, S., Arenas, A., Borge-Holthoefer, J., Meloni, S., Moreno, Y. Discrete-time Markov chain approach to contact-based disease spreading in complex networks. *Europhys. Lett.* 89:38009. (2010)
27. Guimerà, R., Díaz-Guilera, A., Vega-Redondo, F., Cabrales, A., Arenas, A. Optimal network topologies for local search with congestion. *Phys. Rev. Lett.* 89:248701. (2002)
28. Guimerà, R., Mossa, S., Turtschi, A., Amaral, L. A. N. The worldwide air transportation network: Anomalous centrality, community structure, and cities' global roles. *Proc. Natl. Acad. Sci. USA* 102:7794-7799. (2005)
29. Ha, M., Hong, H., Park, H. Comment: Non-mean-field behavior of the contact process on scale-free networks. *Phys. Rev. Lett.* 98:029801. (2007)
30. Han, J.-D. J., Bertin, N., Hao, T., Goldberg, D. S. et al Evidence for dynamically organized modularity in the yeast protein-protein interaction network. *Nature* 430: 88-93. (2004)
31. Hethcote, H. W. The mathematics of infectious diseases. *SIAM Review* 42:599-653. (2000)

32. Hufnagel, L., Brockmann, D., Geisel T. Forecast and control of epidemics in a globalized world. *Proc. Natl. Acad. Sci. USA* 101:1512415129. (2004)
33. Liljeros F., Edling C. R., Amaral L. A. N., Stanley H. E., Aberg Y. The Web of Human Sexual Contacts. *Nature* 411:907-908. (2001)
34. LLoyd, A. L., May, R. M. How viruses spread among computers and people. *Science* 292:1316-1317. (2001)
35. Marro, J., Dickman, R. *Nonequilibrium phase transitions in lattice models*. (Cambridge University Press, Cambridge). (1999)
36. Meloni, S., Gardeñes, J.G., Latora, V., Moreno, Y. Scaling Breakdown in Flow Fluctuations on Complex Networks. *Phys. Rev. Lett.* 100:208701. (2008)
37. Meloni, S., Arenas, A., Moreno Y. Traffic-Driven Epidemic Spreading in Finite-Size Scale-Free Networks. *Proc. Natl. Acad. Sci. USA* 106:16897-16902. (2009)
38. Moreno, Y., Pastor-Satorras, R. Vespignani, A. Epidemic outbreaks in complex heterogeneous networks. *Eur. Phys. J. B* 26:521-529. (2002)
39. Murray J. D. *Mathematical Biology*. (Springer-Verlag, Germany, Berlin). (2002)
40. Murray, J. D. *Mathematical Biology*. (Springer-Verlag, 3rd Edition). (2007)
41. Newman, M. E. J. The spread of epidemic disease on networks. *Phys. Rev. E* 66:016128. (2002)
42. Newman, M. E. J. The structure and function of complex networks. *SIAM Review* 45:167-256. (2003)
43. Newman, M. E. J., Forrest, S., Balthrop, J. Email networks and the spread of computer viruses. *Phys. Rev. E* 66:035101. (2002)
44. Noh, J. D., Rieger, H. Random Walks on Complex Networks. *Phys. Rev. Lett.* 92:118701. (2004)
45. Pastor-Satorras, R., Vespignani, A. Epidemic spreading in scale-free networks. *Phys. Rev. Lett.* 86:3200-3203. (2001)
46. Pastor-Satorras, R., Vespignani A. Epidemic dynamics and endemic states in complex networks. *Phys. Rev. E* 63:066117. (2001)
47. Pastor-Satorras, R., Vespignani A. *Evolution and Structure of the Internet: a statistical physics approach*. (Cambridge University Press, Cambridge). (2004)
48. Restrepo, J. G., Ott, E., Hunt, B. R. Approximating the largest eigenvalue of network adjacency matrices. *Phys. Rev. E* 76:056119. (2007)
49. Rosato, V., Meloni, S., Issacharoff, L., Tiriticco, F. Is the topology of the internet network really fit to its function? *Physica A* 387:1689-1704. (2008)
50. Serrano, M. A. , Krioukov, D., Boguñá, M. Self-Similarity of Complex Networks and Hidden Metric Spaces. *Phys. Rev. Lett.* 100:078701. (2008)
51. Shao, J., Buldyrev, S. V., Braunstein, L. A., Havlin, S., Stanley, E. Structure of shells in complex networks. *Phys. Rev. E* 80:036105. (2009)
52. Sreenivasan, S., Cohen, R., Lopez, E., Toroczkai, Z., Stanley, H. E. Structural Bottlenecks for Communication in Networks. *Phys. Rev. E* 75:036105. (2007)
53. Stanley, H. E. *Introduction to Phase Transitions and Critical Phenomena*. (Oxford University Press, Oxford). (1987)
54. Zhao, L., Lai, Y.-C., Park, K., Ye, N. Onset of traffic congestion in complex networks. *Phys. Rev. E* 71:026125. (2005)

Chapter 16
Theory of Citing

M.V. Simkin and V.P. Roychowdhury

Abstract We present empirical data on misprints in citations to 12 high-profile papers. The great majority of misprints are identical to misprints in articles that earlier cited the same paper. The distribution of the numbers of misprint repetitions follows a power law. We develop a stochastic model of the citation process, which explains these findings and shows that about 70–90% of scientific citations are copied from the lists of references used in other papers. Citation copying can explain not only why some misprints become popular, but also why some papers become highly cited. We show that a model where a scientist picks few random papers, cites them, and copies a fraction of their references accounts quantitatively for empirically observed distribution of citations.

16.1 Statistics of Misprints in Citations

> Now let us come to those references to authors, which other books have, and you want for yours. The remedy for this is very simple: You have only to look out for some book that quotes them all, from A to Z ..., and then insert the very same alphabet in your book, and though the imposition may be plain to see, because you have so little need to borrow from them, that is no matter; there will probably be some simple enough to believe that you have made use of them all in this plain, artless story of yours. At any rate, if it answers no other purpose, this long catalogue of authors will serve to give a surprising look of authority to your book. Besides, no one will trouble himself to verify whether you have followed them or whether you have not, being no way concerned in it...
>
> Miguel de Cervantes, Don Quixote

M.V. Simkin (✉) • V.P. Roychowdhury
Department of Electrical Engineering, University of California, Los Angeles, CA 90095-1594, USA
e-mail: simkin@ee.ucla.edu; vwani@ee.ucla.edu

M.T. Thai and P.M. Pardalos (eds.), *Handbook of Optimization in Complex Networks: Theory and Applications*, Springer Optimization and Its Applications 57,
DOI 10.1007/978-1-4614-0754-6_16,

Table 16.1 Papers, misprints in citing which we studied

Number	Reference
1	K.G. Wilson, Phys. Rev. 179, 1499 (1969)
2	K.G. Wilson, Phys. Rev. B 4, 3174 (1971)
3	K.G. Wilson, Phys. Rev. B 4, 3184 (1971)
4	K.G. Wilson, Phys. Rev. D 10, 2445 (1974)
5	J.M. Kosterlitz and D.J. Thouless, J. Phys. C 6, 1181 (1973)
6	J.M. Kosterlitz, J. Phys. C 7, 1046 (1974)
7	M.J. Feigenbaum, J. Stat. Phys. 19, 25 (1978)
8	M.J. Feigenbaum, J. Stat. Phys. 21, 669 (1979)
9	P. Bak, J. von Boehm, Phys. Rev. B 21, 5297 (1980)
10	P. Bak, C. Tang, and K. Wiesenfeld, Phys. Rev. Lett. 59, 381 (1987)
11	P. Bak, C. Tang, and K. Wiesenfeld, Phys. Rev. A 38, 364 (1988)
12	P. Bak and C. Tang, J. Geophys. Res. B 94, 15635 (1989)

When scientists are writing their scientific articles, they often use the method described in the above quote. They can do this and get away with it until one day they copy a citation, which carries in it a DNA of someone else's misprint. In such case, they can be identified and brought to justice, similar to how biological DNA evidence helps to convict criminals, who committed more serious offences than that.

Our initial report [1] led to a lively discussion[1] on whether copying a citation is a proof of not reading the original paper. Alternative explanations are worth exploring; however, such hypotheses should be supported by data and not by anecdotal claims. It is indeed most natural to assume that a copying citer also failed to read the paper in question (albeit this cannot be rigorously proved). *Entities must not be multiplied beyond necessity.* Having thus shaved the critique with Occam's razor, we will proceed to use the term non-reader to describe a citer who copies.

As misprints in citations are not too frequent, only celebrated papers provide enough statistics to work with. Let us have a look at the distribution of misprints in citations to one renowned paper (number 5 in Table 16.1), which at the time of our initial inquiry [1], that is in late 2002, had accumulated 4,301 citations. Out of these citations 196 contained misprints, out of which only 45 were distinct. The most popular misprint in a page number appeared 78 times.

As a preliminary attempt, one can estimate the ratio of the number of readers to the number of citers, R, as the ratio of the number of *distinct* misprints, D, to the *total number* of misprints, T. Clearly, among T citers, $T - D$ copied, because they repeated someone else's misprint. For the D others, with the information at hand, we have no evidence that they did not read, so according to the presumed innocent

[1] See, for example, the discussion "Scientists Don't Read the Papers They Cite" on Slashdot: http://science.slashdot.org/article.pl?sid=02/12/14/0115243&mode=thread&tid=134.

Table 16.2 Citation and misprint statistics together with estimates of R for 12 studied papers

Number	Citations	Misprints Total	Misprints Distinct	M(%)	R (16.1)	R (16.8)	R (16.22)	R MC	Percentile rank for R = 0.2(%)
1	1,291	61	29	2.2	0.48	0.46	0.44	0.37	15
2	861	33	13	1.5	0.39	0.38	0.35	0.28	44
3	818	38	11	1.3	0.29	0.28	0.22	0.22	68
4	2,578	263	32	1.2	0.12	0.11	–	0.10	95
5	4,301	196	45	1.0	0.23	0.22	0.17	0.15	76
6	1,673	40	12	0.7	0.30	0.29	0.25	0.22	65
7	1,639	36	21	1.3	0.58	0.58	0.57	0.49	6
8	837	55	18	2.2	0.33	0.31	0.26	0.22	57
9	419	20	8	1.9	0.40	0.39	0.34	0.29	50
10	1,717	33	14	0.8	0.42	0.42	0.40	0.31	36
11	1,348	78	27	2.0	0.35	0.33	0.29	0.23	47
12	397	61	18	4.5	0.30	0.26	0.17	0.19	69

The citation data were retrieved from the ISI database in late 2002 and early 2003. The way we count misprints is look at the whole sequence of volume, page number and the year, which amounts to between 8 and 11 digits for different studied papers. That is, two misprints are distinct if they are different in any of the places, and they are repeats if they agree on all of the digits

principle, we assume that they did. Then in our sample, we have D readers and T citers, which lead to:

$$R \approx D/T. \tag{16.1}$$

Substituting $D = 45$ and $T = 196$ in (16.1), we obtain that $R \approx 0.23$. The values of R for the rest of the dozen studied papers are given in Table 16.2.

As we pointed out in [2] the above reasoning would be convincing if the people who introduced original misprints had always read the original paper. It is more reasonable to assume that the probability of introducing a new misprint in a citation does not depend on whether the author had read the original paper. Then, if the fraction of read citations is R, the number of readers in our sample is RD, and the ratio of the number of readers to the number of citers in the sample is RD/T. What happens to our estimate, (16.1)? It is correct, just the sample is not representative: the fraction of read citations among the misprinted citations is less than in general citation population.

Can we still determine R from our data? Yes. From the misprint statistics we can determine the average number of times, n_p, a typical misprint propagates:

$$n_p = \frac{T - D}{D}. \tag{16.2}$$

The number of times a misprint had propagated is the number of times the citation was copied from either the paper which introduced the original misprint, or from one of subsequent papers, which copied (or copied from copied, etc.) from it.

A misprinted citation should not differ from a correct citation as far as copying is concerned. This means that a selected at random citation, on average, is copied (including copied from copied, etc.) n_p times. The read citations are no different from unread citations as far as copying goes. Therefore, every read citation, on average, was copied n_p times. The fraction of read citations is thus

$$R = \frac{1}{1+n_p}. \tag{16.3}$$

After substituting (16.2) into (16.3), we recover (16.1).

Note, however, that the average number of times a misprint propagates is not equal to the number of times the citation was copied, but to the number of times it was copied *correctly*. Let us denote the average number of citations copied (including copied from copied etc) from a particular citation as $n_{c.}$ It can be determined from n_p the following way. The n_c consists of two parts: n_p (the correctly copied citations) and misprinted citations. If the probability of making a misprint is M and the number of correctly copied citations is n_p then the total number of copied citations is $n_p/(1-M)$ and the number of misprinted citations is $(n_pM)/(1-M)$. As each misprinted citation was itself copied n_c times, we have the following self-consistency equation for n_c:

$$n_c = n_p + n_p \times \frac{M}{1-M} \times (1+n_c) \tag{16.4}$$

It has the solution

$$n_c = \frac{n_p}{1-M-n_p \times M} \tag{16.5}$$

After substituting (16.2) into (16.5) we get:

$$n_c = \frac{T-D}{D-MT}. \tag{16.6}$$

From this, we get:

$$R = \frac{1}{1+n_c} = \frac{D}{T} \times \frac{1-(MT/D)}{1-M} \tag{16.7}$$

The probability of making a misprint we can estimate as $M = D/N$, where N is the total number of citations. After substituting this into (16.7) we get:

$$R = \frac{D}{T} \times \frac{N-T}{N-D}. \tag{16.8}$$

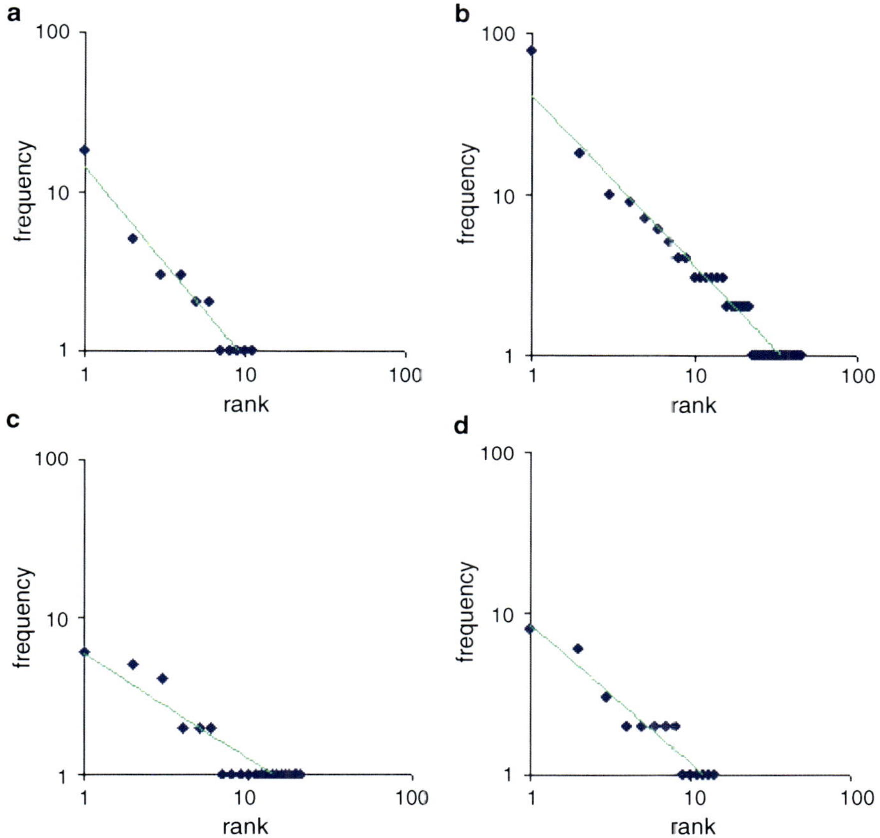

Fig. 16.1 Rank-frequency distributions of misprints in referencing four high-profile papers (here the rank is determined by the frequency so that the most popular misprint has rank 1, second most frequent misprint has rank 2 and so on). Figures (**a**–**d**) are for papers 2, 5, 7, and 10 of Table 16.1. *Solid lines* are fits to Zipf Law with exponents **a** 1.20; **b** 1.05; **c** 0.66; **d** 0.85

Substituting $D = 45$, $T = 196$, and $N = 4301$ in (16.8), we get $R \approx 0.22$, which is very close to the initial estimate, obtained using (16.1). The values of R for the rest of the papers are given in Table 16.2. They range between 11% and 58%.

In the next section we introduce and solve the stochastic model of misprint propagation. The model explains the power law of misprint repetitions (see Fig. 16.1). If you do not have time to read the whole chapter, you can proceed after Sect. 16.2.1 right to Sect. 16.3.1. There we formulate and solve the model of random-citing scientists (RCS). The model is as follows: when scientist is writing a manuscript he picks up several random papers, cites them, and copies a fraction of their references. The model can explain why some papers are far more cited than others. After that, you can directly proceed to discussion in Sect. 16.5. If you have questions, you can find answers to some of them in other sections. The results of Sect. 16.1 are exact in the limit of infinite number of citations. Since this number is obviously finite, we need to study finite size effects, which affect our estimate of R. This is done in Section 16.2.2 using complicated mathematical methods and in Sect. 16.2.3 using Monte Carlo simulations. The limitations of the simple model arising from the instances like, for example, the same author repeats the same misprint, are discussed in Sect. 16.2.4. In Sect. 16.2.5, we review the previous work on identical misprints. In short: some people did notice repeat misprints and attributed them to citation copying, but nobody derived (16.1) before us. The RCS model of Sect. 16.3 can explain a power law in overall citation distribution, but cannot explain a power-law distribution in citations to the papers of the same age. Section 16.4.1 introduces the modified model of random-citing scientist (MMRCS), which solves the problem. The model is as follows: when a scientist writes a manuscript, he picks up several random *recent* papers, cites them, and also copies some of their references. The difference with the original model is the word *recent*. In Sect. 16.4.2 the MMRCS is solved using theory of branching processes and the power-law distribution of citations to the papers of the same age is derived. Section 16.4.3 considers the model where papers are not created equal but have Darwinian fitness that affects their citability. Section 16.4.4 studies effects of literature growth (yearly increase of the number of published papers) on citation distribution. Section 16.4.5 describes numerical simulations of MMRCS, which perfectly match real citation data. Section 16.4.6 shows that MMRCS can explain the phenomenon of literature aging that is why papers become less cited as they get older. Section 16.4.7 shows that MMRCS can explain the mysterious phenomenon of sleeping beauties in science (papers that are at first hardly noticed suddenly awake and get a lot of citations). Section 16.4.8 describes the connection of MMRCS to the Science of Self-Organized Criticality (SOC).

16.2 Stochastic Modeling of Misprints in Citations

16.2.1 Misprint Propagation Model

Our misprint propagation model (MPM) [1, 3] which was stimulated by Simon's [4] explanation of Zipf Law and Krapivsky–Redner [5] idea of link redirection, is as follows. Each new citer finds the reference to the original in any of the papers that already cite it (or it can be the original paper itself). With probability R he gets the citation information from the original. With probability $1 - R$ he copies the citation to the original from the paper he found the citation in. In either case, the citer introduces a new misprint with probability M.

Let us derive the evolution equations for the misprint distribution. The only way to increase the number of misprints that appeared only once, N_1, is to introduce a new misprint. So, with each new citation N_1 increases by 1 with probability M.

The only way to decrease N_1, is to copy correctly one of misprints that appeared only once, this happens with probability $\alpha \times (N_1/N)$, where

$$\alpha = (1-R) \times (1-M) \tag{16.9}$$

is the probability that a new citer copies the citation without introducing a new error, and N is the total number of citations. For the expectation value, we thus have:

$$\frac{dN_1}{dN} = M - \alpha \times \frac{N_1}{N}. \tag{16.10}$$

The number of misprints that appeared K times, N_K, (where $K > 1$) can be increased only by copying correctly a misprint which appeared $K-1$ times. It can only be decreased by copying (again correctly) a misprint which appeared K times. For the expectation values, we thus have:

$$\frac{dN_K}{dK} = \alpha \times \frac{(K-1) \times N_{K-1} - K \times N_K}{N} \quad (K > 1). \tag{16.11}$$

Assuming that the distribution of misprints has reached its stationary state, we can replace the derivatives (dN_K/dN) by ratios (N_K/N) to get:

$$\frac{N_1}{N} = \frac{M}{1+\alpha}; \quad \frac{N_{K+1}}{N_K} = \frac{K}{1+1/\alpha+K} \quad (K > 1). \tag{16.12}$$

Note that for large K: $N_{K+1} \approx N_K + dN_K/dK$, therefore (16.12) can be rewritten as:

$$\frac{dN_K}{dK} \approx -\frac{1+1/\alpha}{1+1/\alpha+K} N_k \approx \frac{1+1/\alpha}{K} N_k.$$

From this follows that the misprints frequencies are distributed according to a power law:

$$N_K \sim 1/K^{\gamma}, \tag{16.13}$$

where

$$\gamma = 1 + \frac{1}{\alpha} = 1 - \frac{1}{(1-R)} \times (1-M). \tag{16.14}$$

Relationship between γ and α in (16.14) is the same as the one between exponents of number-frequency and rank-frequency distributions.[2] Therefore the parameter

[2]Suppose that the number of occurrences of a misprint (K), as a function of the rank (r), when the rank is determined by the above frequency of occurrence (so that the most popular misprint has rank 1, second most frequent misprint has rank 2 and so on), follows a Zipf law: $K(r) = C/r^{\alpha}$. We want to find the number-frequency distribution, i.e. how many misprints appeared n times. The number of misprints that appeared between K_1 and K_2 times is obviously $r_2 - r_1$ where $K_1 = C/r_1^{\alpha}$ and

α, which was defined in (16.9), turned out to be the Zipf law exponent. An exact formula for N_k can also be obtained by iteration of (16.12) to get:

$$\frac{N_K}{N} = \frac{\Gamma(K)\Gamma(\gamma)}{\Gamma(K+\gamma)} \times \frac{M}{\alpha} = \mathrm{B}(K,\gamma) \times \frac{M}{\alpha} \tag{16.15}$$

Here Γ and B are Euler's Gamma and Beta functions. Using the asymptotic for constant γ and large K

$$\frac{\Gamma(\gamma)}{\Gamma(K+\gamma)} \sim K^{-\gamma} \tag{16.16}$$

we recover (16.13).

The rate equation for the total number of misprints is:

$$\frac{dT}{dN} = M + \alpha \times \frac{T}{N}. \tag{16.17}$$

The stationary solution of (16.17) is:

$$\frac{T}{N} = \frac{M}{1-\alpha} = \frac{M}{R+M-RM}. \tag{16.18}$$

The expectation value for the number of distinct misprints is obviously

$$D = N \times M. \tag{16.19}$$

From (16.18) and (16.19) we obtain:

$$R = \frac{D}{T} \times \frac{N-T}{N-D}, \tag{16.20}$$

which is identical to (16.8).

One can ask why we did not choose to extract R using (16.9) or (16.14). This is because α and γ are not very sensitive to R when it is small (in fact (16.9) gives negative values of R for some of the fittings in Fig. 16.1). In contrast, T scales as $1/R$.

We can slightly modify our model and assume that original misprints are only introduced when the reference is derived from the original paper, while those who copy references do not introduce new misprints (e.g., they do cut and paste). In this case one can show that $T = N \times M$ and $D = N \times M \times R$. As a consequence (16.1) becomes exact (in terms of expectation values, of course).

$K_2 = C/r_2^{\alpha}$. Therefore, the number of misprints that appeared K times, N_k, satisfies $N_K dK = -dr$ and hence, $N_K = -dr/dK \sim K^{-1/\alpha-1}$.

16.2.2 Finite-Size Corrections

Preceding analysis assumes that the misprint distribution had reached its stationary state. Is this reasonable? Equation (16.17) can be rewritten as:

$$\frac{d(T/N)}{M-(T/N)\times(1-\alpha)}=d\ln N. \tag{16.21}$$

Naturally the first citation is correct (it is the paper itself). Then the initial condition is $N=1; T=0$. Equation (16.21) can be solved to get:

$$\frac{T}{N}=\frac{M}{1-\alpha}\times\left(1-\frac{1}{N^{1-\alpha}}\right)=\frac{M}{R+M-M\times R}\times\left(1-\frac{1}{N^{R+M-M\times R}}\right) \tag{16.22}$$

This should be solved numerically for R. The values obtained using (16.22) are given in Table 16.2. They range between 17% and 57%. Note that for one paper (No.4) no solution to (16.22) was found.[3] As N is not a continuous variable, integration of (16.17) is not perfectly justified, particularly when N is small. Therefore, we reexamine the problem using a rigorous discrete approach due to Krapivsky and Redner [6]. The total number of misprints, T, is a random variable that changes according to

$$T(N+1)=\begin{cases}T(N) & \text{with probability} \quad 1-M-\frac{T(N)}{N}\alpha\\ T(N)+1 & \text{with probability} \quad M+\frac{T(N)}{N}\alpha\end{cases} \tag{16.23}$$

after each new citation. Therefore, the expectation values of T obey the following recursion relations:

$$\langle T(N+1)\rangle=\langle T(N)\rangle+\frac{\langle T(N)\rangle}{N}\alpha+M \tag{16.24}$$

To solve (16.24) we define a generating function:

$$\chi(\omega)=\sum_{n=1}^{\infty}\langle T(N)\rangle\omega^{N-1} \tag{16.25}$$

[3] Why did this happen? Obviously, T reaches maximum when R equals zero. Substituting $R=0$ in (16.22) we get: $T_{\mathrm{MAX}}=N(1-1/N^{M})$. For paper No.4 we have $N=2,578$, $M=D/N=32/2,578$. Substituting this into the preceding equation, we get $T_{\mathrm{MAX}}=239$. The observed value $T=263$ is therefore higher than an expectation value of T for any R. This does not immediately suggest discrepancy between the model and experiment but a strong fluctuation. In fact out of 1,000,000 runs of Monte Carlo simulation of MPM with the parameters of the mentioned paper and $R=0.2$ exactly 49,712 runs (almost 5%) produced $T\geq 263$.

After multiplying (16.24) by $N\omega^{N-1}$ and summing over $N \geq 1$ the recursion relation is converted into the differential equation for the generating function

$$(1-\omega)\frac{d\chi}{d\omega} = (1+\alpha)\chi + \frac{M}{(1-\omega)^2} \tag{16.26}$$

Solving (16.26) subject to the initial condition $\chi(0) = \langle T(1) \rangle = 0$ gives

$$\chi(\omega) = \frac{M}{1-\alpha}\left(\frac{1}{(1-\omega)^2} - \frac{1}{(1-\omega)^{1-\alpha}}\right) \tag{16.27}$$

Finally we expand the right-hand side of (16.27) in Taylor series in ω and equating coefficients of ω^{N-1} obtain:

$$\frac{\langle T(N) \rangle}{N} = \frac{M}{1-\alpha}\left(1 - \frac{\Gamma(N+\alpha)}{\Gamma(1+\alpha)\Gamma(N+1)}\right) \tag{16.28}$$

Using (16.16) we obtain that for large N

$$\frac{\langle T(N) \rangle}{N} = \frac{M}{1-\alpha}\left(1 - \frac{1}{\Gamma(1+\alpha)} \times \frac{1}{N^{1-\alpha}}\right) \tag{16.29}$$

This is identical to (16.22) except for the pre-factor $1/\Gamma(1+\alpha)$. Parameter α (it is defined in (16.9)) ranges between 0 and 1. Therefore, the argument of Gamma function ranges between 1 and 2. Because $\Gamma(1) = \Gamma(2) = 1$ and between 1 and 2 Gamma function has just one extremum $\Gamma(1.4616\ldots) = 0.8856\ldots$, the continuum approximation (16.22) is reasonably accurate.

16.2.3 Monte Carlo Simulations

In the preceding section, we calculated the expectation value of T. However, it does not always coincide with the most likely value when the probability distribution is not Gaussian. To get a better idea of the model's behavior for small N and a better estimate of R we did numerical simulations. To simplify comparison with actual data the simulations were performed in a "micro-canonical ensemble," i.e., with a fixed number of distinct misprints. Each paper is characterized by the total number of citations, N, and the number of distinct misprints, D. At the beginning of a simulation, D misprints are randomly distributed between N citations and chronological numbers of the citations with misprints are recorded in a list. In the next stage of the simulation for each new citation, instead of introducing a misprint with probability M, we introduce a misprint only if its chronological number is included in the list created at the outset. This way one can ensure that the number

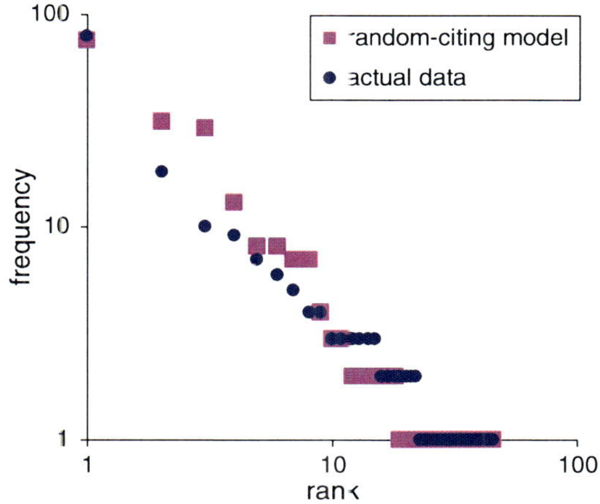

Fig. 16.2 A typical outcome of a single simulation of the MPM (with $R = 0.2$) compared to the actual data for paper 5 in Table 16.1

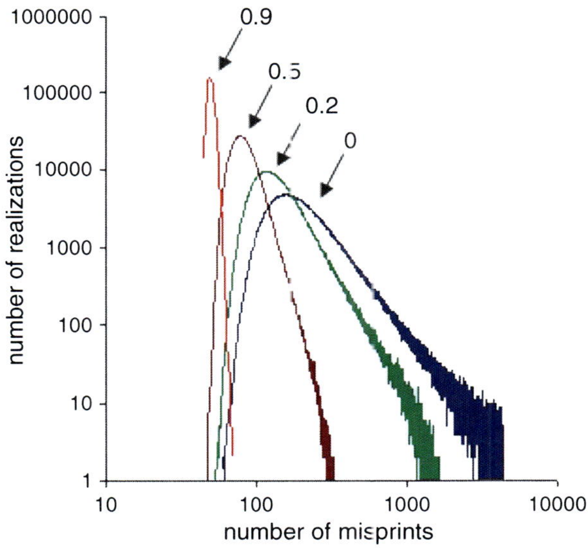

Fig. 16.3 The outcome of 1,000,000 runs of the MPM with $N = 4301$, $D = 45$ (parameters of paper 5 from Table 16.1) for four different values of R (0.9, 0.5, 0.2, 0 from left to right)

of distinct misprints in every run of a simulation is equal to the actual number of distinct misprints for the paper in question. A typical outcome of such simulation for paper 5 is shown in Fig. 16.2.

To estimate the value of R, 1,000,000 runs of the random-citing model with $R = 0$, $0.1, 0.2 \ldots, 0.9$ were done. An outcome of such simulations for one paper is shown in Fig. 16.3. The number of times, N_R, when the simulation produced a total number of

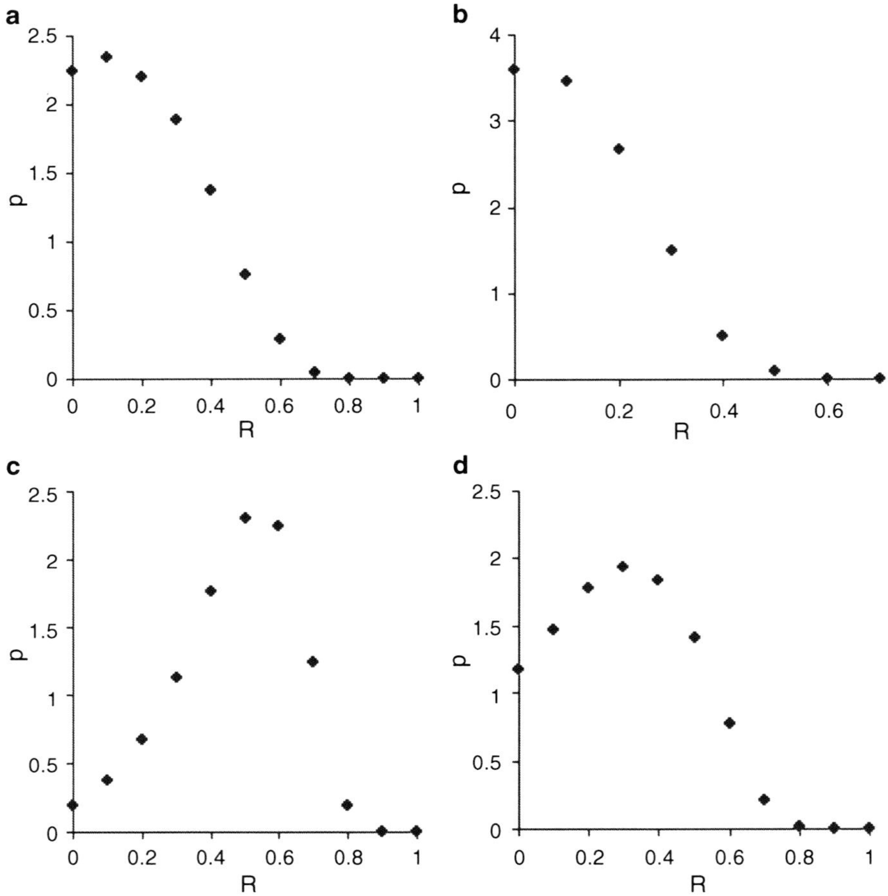

Fig. 16.4 Bayesian inference for the probability density of the readers/citers ratio, R, computed using (16.30). Figures. (**a**–**d**) are for papers 2, 5, 7, and 10 (Table 16.1)

misprints equal to the one actually observed for the paper in question was recorded for each R. Bayesian inference was used to estimate the probability of R:

$$P(R) = \frac{N_R}{\sum_R N_R} \tag{16.30}$$

Estimated probability distributions of R, computed using (16.30) for four sample papers are shown in Fig. 16.4. The median values are given in Table 16.2 (see the MC column). They range between 10% and 49%.

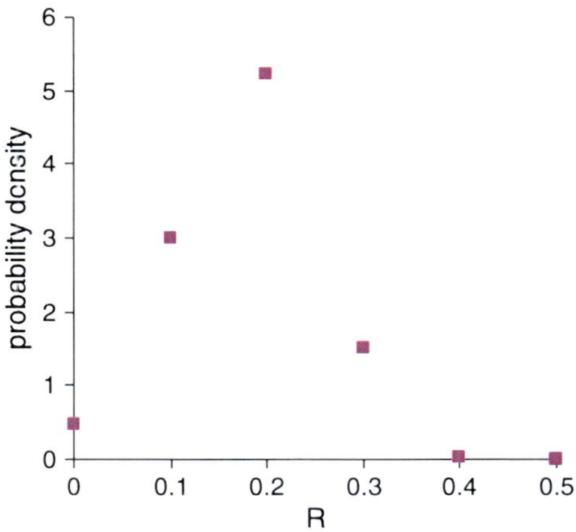

Fig. 16.5 Bayesian inference for the readers/citers ratio, R, based on 12 studied papers computed using (16.31)

Now let us assume R to be the same for all 12 papers and compute Bayesian inference:

$$P(R) = \frac{\prod_{i=1}^{12} N_R^i}{\sum_R \prod_{i=1}^{12} N_R^i} \tag{16.31}$$

The result is shown in Fig. 16.5. $P(R)$ is sharply peaked around $R = 0.2$. The median value of R is 18% and with 95% probability R is less than 34%.

But is the assumption that R is the same for all 12 papers reasonable? The estimates for separate papers vary between 10% and 50 %! To answer this question we did the following analysis. Let us define for each paper a "percentile rank." This is the fraction of the simulations of the MPM (with $R = 0.2$) that produced T, which was less than actually observed T. Actual values of these percentile ranks for each paper are given in Table 16.2 and their cumulative distribution is shown in Fig. 16.6. Now if we claim that MPM with same $R = 0.2$ for all papers indeed describes the reality – then the distribution of these percentile ranks must be uniform. Whether or not the data is consistent with this, we can check using Kolmogorov–Smirnov test [7]. The maximum value of the absolute difference between two cumulative distribution functions (D-statistics) in our case is $D = 0.15$. The probability for D to be more than that is 91%. This means that the data is perfectly consistent with the assumption of $R = 0.2$ for all papers.

One can notice that the estimates of M (computed as $M = D/N$) for different papers (see Table 16.2) are also different. One may ask if it is possible that M is the same for all papers and different values of D/N are results of fluctuations. The answer is that the data is totally inconsistent with single M for all papers. This is not unexpected, because some references can be more error-prone, for example, because they are longer. Indeed, the most-misprinted paper (No.12) has two-digit volume number and five-digit page number.

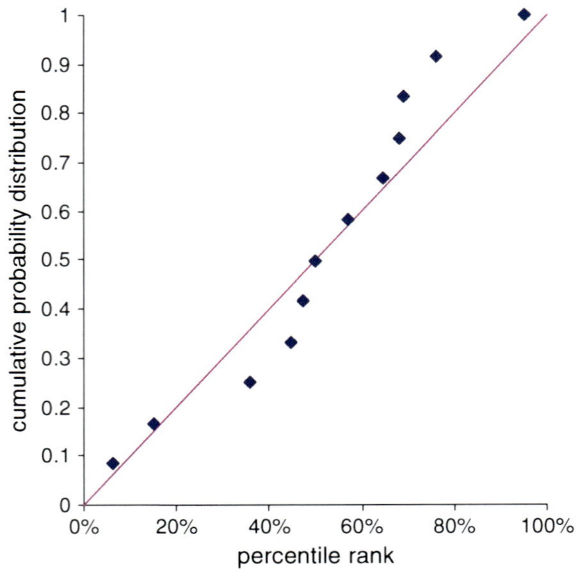

Fig. 16.6 Cumulative distribution of the percentile ranks of the observed values of T with regard to the outcomes of the simulations of the MPM with $R = 0.2$ (*diamonds*). For comparison the cumulative function of the uniform distribution is given (a line)

16.2.4 Operational Limitations of the Model

Scientists copy citations because they are not perfect. Our analysis is imperfect as well. There are occasional repeat identical misprints in papers, which share individuals in their author lists. To estimate the magnitude of this effect we took a close look at all 196 misprinted citations to paper 5 of Table 16.1. It turned out that such events constitute a minority of repeat misprints. It is not obvious what to do with such cases when the author lists are not identical: should the set of citations be counted as a single occurrence (under the premise that the common co-author is the only source of the misprint) or as multiple repetitions. Counting all such repetitions as only a single misprint occurrence results in elimination of 39 repeat misprints. The number of total misprints, T, drops from 196 to 157, bringing the upper bound for R (16.1) from $45/196 \cong 23\%$ up to $45/157 \cong 29\%$. An alternative approach is to subtract all the repetitions of each misprint by the originators of that misprint from non-readers and add it to the number of readers. There were 11 such repetitions, which increases D from 45 up to 56 and the upper bound for R (16.1) rises to $56/196 \cong 29\%$, which is the same value as the preceding estimate. It would be desirable to redo the estimate using (16.20) and (16.22), but the MPM would have to be modified to account for repeat citations by same author and multiple authorships of a paper. This may be a subject of future investigations.

Another issue brought up by the critics [8] is that because some misprints are more likely than others, it is possible to repeat someone else's misprint purely by chance. By examining the actual data, one finds that about two-third of distinct misprints fall in to the following categories:

(a) One misprinted digit in volume, page number, or in the year.
(b) One missing or added digit in volume or page number.
(c) Two adjacent digits in a page number are interchanged.

The majority of the remaining misprints are combinations of (a–c), for example, one digit in page number omitted and one digit in year misprinted.[4] For a typical reference, there are over 50 aforementioned likely misprints. However, even if probability of certain misprint is not negligibly small but 1 in 50, our analysis still applies. For example, for paper 5 (Table 16.1) the most popular error appeared 78 times, while there were 196 misprints in total. Therefore, if probability of certain misprint is 1/50, there should be about $196/50 \approx 4$ such misprints, not 78. In order to explain repeat misprints distribution by higher probability of certain misprint this probability should be as big as $78/196 \approx 0.4$. This is extremely unlikely. However, finding relative propensities of different misprints deserves further investigation.

Smith noticed [9] that some misprints are in fact introduced by the ISI. To estimate the importance of this effect we explicitly verified 88 misprinted (according to ISI) citations in the original articles. Seventy-two of them were exactly as in the ISI database, but 16 were in fact correct citations. To be precise some of them had minor inaccuracies, like second initial of the author was missing, while page number, volume, and year were correct. Apparently, they were victims of an "erroneous error correction" [9]. It is not clear how to consistently take into account these effects, specifically because there is no way to estimate how many wrong citations have been correctly corrected by ISI [10]. But given the relatively small percentage of the discrepancy between ISI database and actual articles ($16/88 \cong 18\%$) this can be taken as a noise with which we can live.

It is important to note that within the framework of the MPM R is not the ratio of readers to citers, but the probability that a citer consults the original paper, provided that he encountered it through another paper's reference list. However, he could encounter the paper directly. This has negligible effect for highly-cited papers, but is important for low-cited papers. Within the MPM framework the probability of such an event for each new citation is obviously $1/n$, where n is the current total number of citations. The expectation value of the true ratio of readers to citers is therefore:

$$R^*(N) = R + (1-R) \times \frac{\sum_{n=1}^{n} \frac{1}{n}}{N} \approx R + (1-R) \times \frac{\ln(2N-1)}{N}. \qquad (16.32)$$

The values of R^* for papers with different total numbers of citations, computed using (16.32), are shown in Fig. 16.7. For example, on average, about four people have read a paper which was cited ten times. One can use (16.32) and empirical

[4] There are also misprints where author, journal, volume, and year are perfectly correct, but the page number is totally different. Probably, in such case the citer mistakenly took the page number from a neighboring paper in the reference list he was lifting the citation from.

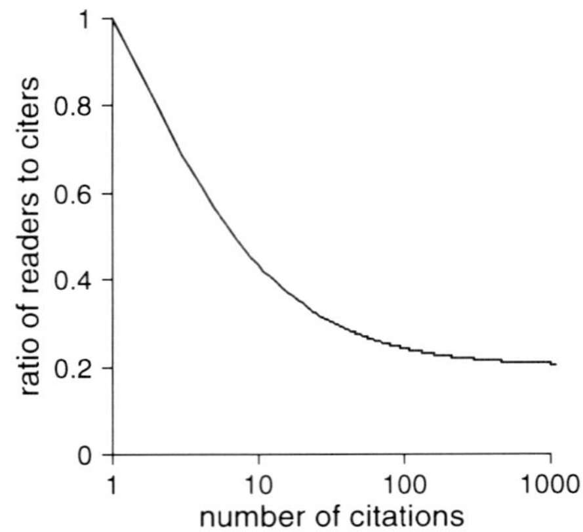

Fig. 16.7 Ratio of readers to citers as a function of total amount of citations for $R = 0.2$, computed using (16.32)

citation distribution to estimate an average value of R^* for the scientific literature in general. The formula is:

$$\langle R^* \rangle = \frac{\sum R^*(N_i) \times N_i}{\sum N_i} \tag{16.33}$$

Here the summation is over all of the papers in the sample and N_i is the number of citations that ith paper had received. The estimate, computed using citation data for *Physical Review D* [11] and (16.32) and (16.33) (assuming $R = 0.2$), is $\langle R^* \rangle \approx 0.33$.

16.2.5 Comparison with the Previous Work

The bulk of previous literature on citations was concerned with their counting. After extensive literature search we found only a handful of papers which analyzed misprints in citations (the paper by Steel [12], titled identically to our first misprint paper, i.e., "Read before you cite," turned out to use the analysis of the content of the papers, not of the propagation of misprints in references). Broadus [13] looked through 148 papers, which cited both the renowned book, which misquoted the title of one of its references, and that paper, the title of which was misquoted in the book. He found that 34 or 23% of citing papers made the same error as was in the book. Moed and Vries [14] (apparently independent of Broadus, as they do not refer to his work), found identical misprints in scientific citations and attributed them to citation copying. Hoerman and Nowicke [15] looked through a number of papers, which deal with the so-called Ortega Hypothesis of Cole and Cole. When Cole and Cole quoted a passage from the book by Ortega they introduced three distortions. Hoerman and Nowicke found seven papers which cite Cole and Cole and also quote that passage from Ortega. In six out of these seven papers all of the distortions made by Cole and Cole were repeated. According to [15] in this process even the original meaning of the quotation was altered. In fact, information is sometimes *defined* by its property to deteriorate in chains [16].

While the fraction of copied citations found by Hoerman and Nowicke [15], $6/7 \cong 86\%$ agrees with our estimate, Boadus' number, 23%, seems to disagree with it. Note, however, that Broadus [13] assumes that citation, if copied – was copied from the book (because the book was renowned). Our analysis indicates that majority of citations to renowned papers are copied. Similarly, we surmise, in the Broadus' case citations to both the book and the paper were often copied from a third source.

16.3 Copied Citations Create Renowned Papers?

16.3.1 The Model of Random-Citing Scientists

During the "Manhattan project" (the making of nuclear bomb), Fermi asked Gen. Groves, the head of the project, what is the definition of a "great" general [17]. Groves replied that any general who had won five battles in a row might safely be called great. Fermi then asked how many generals are great. Groves said about three out of every hundred. Fermi conjectured that considering that opposing forces for most battles are roughly equal in strength, the chance of winning one battle is $1/2$ and the chance of winning five battles in a row is $1/2^5 = 1/32$. "So you are right, General, about three out of every hundred. Mathematical probability, not genius." The existence of military genius also questioned Lev Tolstoy in his book "War and Peace."

A commonly accepted measure of "greatness" for scientists is the number of citations to their papers [18]. For example, SPIRES the High-Energy Physics literature database, divides papers into six categories according to the number of citations they receive. The top category, "Renowned papers" are those with 500 or more citations. Let us have a look at the citations to roughly eighteen and a half thousand papers,[5] published in *Physical Review D* in 1975–1994 [11]. As of 1997 there were about 330 thousands of such citations: 18 per published paper on average. However, 44 papers were cited 500 times or more. Could this happen if *all papers are created equal*? If they indeed are then the chance to win a citation is one in 18,500. What is the chance to win 500 cites out of 330,000? The calculation is

[5]In our initial report [22] we mentioned "over 24 thousand papers." This number is incorrect and the reader surely understands the reason: misprints. In fact, out of 24,295 "papers" in that dataset only 18,560 turned out to be real papers and 5,735 "papers" turned out to be misprinted citations. These "papers" got 17,382 out of 351,868 citations. That is every distinct misprint on average appeared three times. As one could expect, cleaning out misprints lead to much better agreement between experiment and theory: compare Fig.16.8 and Fig. 1 of [22].

slightly more complex than in the militaristic case,[6] but the answer is 1 in 10^{500}, or, in other words, it is zero. One is tempted to conclude that those 44 papers, which achieved the impossible, are great.

In the preceding sections, we demonstrated that copying from the lists of references used in other papers is a major component of the citation dynamics in scientific publication. This way a paper that already was cited is likely to be cited again, and after it is cited again it is even more likely to be cited in the future. In other words, "unto every one which hath shall be given" [Luke 19:26]. This phenomenon is known as "Matthew effect",[7] "cumulative advantage" [20], or "preferential attachment" [21].

The effect of citation copying on the probability distribution of citations can be quantitatively understood within the framework of *the model of random-citing scientists (RCS)* [22],[8] which is as follows. When a scientist is writing a manuscript he picks up m random articles,[9] cites them, and also copies some of their references, each with probability p.

The evolution of the citation distribution (here N_k denotes the number of papers that were cited K times, and N is the total number of papers) is described by the following rate equations:

$$\begin{aligned} \frac{dN_0}{dN} &= 1 - m \times \frac{N_0}{N}, \\ \frac{dN_k}{dN} &= m \times \frac{(1+p(K-1))N_{k-1} - (1+pK)N_k}{N}, \end{aligned} \tag{16.34}$$

which have the following stationary solution:

$$N_0 = \frac{N}{m+1}; N_k = \frac{1+p(K-1)}{1+1/m+pK} N_{k-1}. \tag{16.35}$$

---

[6]If one assumes that all papers are created equal then the probability to win m out of n possible citations when the total number of cited papers is N is given by the Poisson distribution: $P = ((n/N)^m/m!) \times e^{-n/N}$. Using Stirling's formula one can rewrite this as: $\ln(P) \approx m\ln(ne/Nm) - (n/N)$. After substituting $n = 330,000$, $m = 500$ and $N = 18500$ into the above equation we get: $\ln(P) \approx -1,180$, or $P \approx 10^{-512}$.

[7]Sociologist of science Robert Merton observed [19] that when a scientist gets recognition early in his career he is likely to get more and more recognition. He called it "Matthew Effect" because in Gospel according to Mathew (25:29) appear the words: "unto every one that hath shall be given". The attribution of a special role to St. Matthew is unfair. The quoted words belong to Jesus and also appear in Luke and Mark's gospels. Nevertheless, thousands of people who did not read The Bible copied the name "Matthew Effect."

[8]From the mathematical perspective, almost identical to RCS model (the only difference was that they considered an undirected graph, while citation graph is directed) was earlier proposed in [23].

[9]The analysis presented here also applies to a more general case when m is not a constant, but a random variable. In that case m in all of the equations that follow should be interpreted as the mean value of this variable.

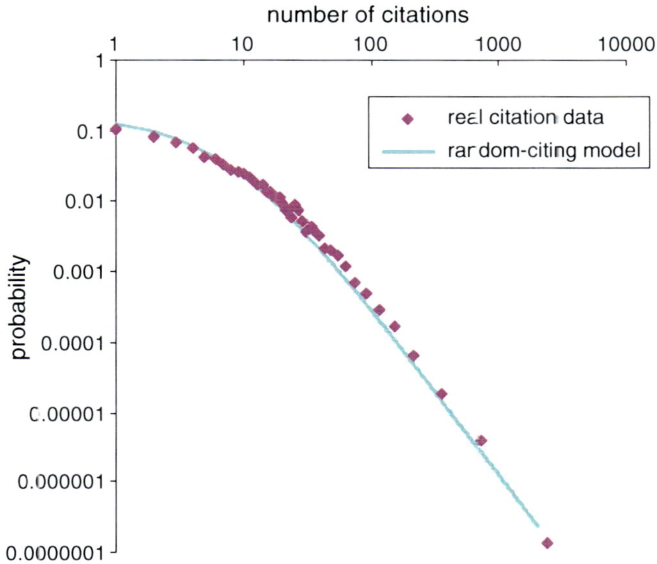

Fig. 16.8 Outcome of the model of random-citing scientists (with $m = 5$ and $p = 0.14$) compared to actual citation data. Mathematical probability rather than genius can explain why some papers are cited a lot more than the others

For large K it follows from (16.35) that:

$$N_k \sim 1/K^{\gamma}; \gamma = 1 + \frac{1}{m \times p}. \tag{16.36}$$

Citation distribution follows a power law, empirically observed in [24–26].

A good agreement between the RCS model and actual citation data [11] is achieved with input parameters $m = 5$ and $p = 0.14$ (see Fig. 16.8). Now what is the probability for an arbitrary paper to become "renowned," i.e., receive more than 500 citations? Iteration of (16.35) (with $m = 5$ and $p = 0.14$) shows that this probability is 1 in 420. This means that about 44 out of 18,500 papers should be renowned. Mathematical probability, not genius.

On one incident [27] Napoleon (incidentally, he was the military commander, whose genius was questioned by Tolstoy) said to Laplace "They tell me you have written this large book on the system of the universe, and have never even mentioned its Creator." The reply was "I have no need for this hypothesis." It is worthwhile to note that Laplace was not against God. He simply did not need to postulate. His existence in order to explain existing astronomical data. Similarly, the present work is not blasphemy. Of course, in some spiritual sense, great scientists do exist. It is just that even if they would not exist, citation data would look the same.

16.3.2 Relation to Previous Work

Our original paper on the subject [22] was stimulated by the model introduced by Vazquez [27]. It is as follows. When scientist is writing a manuscript, he picks up a paper, cites it, follows its references, and cites a fraction p of them. Afterward he repeats this procedure with each of the papers that he cited. And so on. In two limiting cases ($p = 1$ and $p = 0$) the Vazquez model is exactly solvable [27]. Also in these cases it is identical to the RCS model ($m = 1$ case), which in contrast can be solved for any p. Although theoretically interesting, the Vazquez model cannot be a realistic description of the citation process. In fact, the results presented in two preceding sections indicate that there is essentially just one "recursion," that is, references are copied from the paper at hand, but hardly followed. To be precise, results of two preceding sections could support a generalized Vazquez model, in which the references of the paper at hand are copied with probability p, and afterward the copied references are followed with probability R. However, given the low value of this probability ($R \approx 0.2$), it is clear that the effect of secondary recursions on the citation distribution is small.

The book of Ecclesiastes says: "Is there any thing whereof it may be said, See, this is new? It hath been already of old time, which was before us." The discovery reported in this section is no exception. Long ago Price [20], by postulating that the probability of paper being cited is somehow proportional to the amount of citations it had already received, explained the power law in citation frequencies, which he had earlier observed [22]. However, Price did not propose any mechanism for that. Vasquez did propose a mechanism, but it was only a hypothesis. In contrast, our paper is rooted in facts.

16.4 Mathematical Theory of Citing

16.4.1 Modified Model of Random-Citing Scientists

> … citations not only vouch for the authority and relevance of the statements they are called upon to support; they embed the whole work in context of previous achievements and current aspirations. It is very rare to find a reputable paper that contains no reference to other research. Indeed, one relies on the citations to show its place in the whole scientific structure just as one relies on a man's kinship affiliations to show his place in his tribe.
>
> John M. Ziman, FRS [28]

In spite of its simplicity, the model of RCS appeared to account for the major properties of empirically observed distributions of citations. A more detailed analysis, however, reveals that some features of the citation distribution are not accounted for by the model. The cumulative advantage process would lead to oldest papers being most highly cited [5, 21, 29].[10] In reality, the average citation rate decreases as the paper in question gets older [24, 30–32]. The cumulative advantage process would also lead to an exponential distribution of citations to papers of the same age [5, 29]. In reality citations to papers published during the same year are distributed according to a power-law (see the ISI dataset in Fig.16.1a in [26]).

In this section, we study the *modified* model of random-citing scientists (*MM-RCS*) [33]: when a scientist writes a manuscript, he picks up several random *recent* papers, cites them, and also copies some of their references. The difference with the original model is the word *recent*. We solve this model using methods of the theory of branching processes [34] (we review its relevant elements in Appendix A), and show that it explains both the power-law distribution of citations to papers published during the same year and literature aging. A similar model was earlier proposed by Bentley, Hahn, and Shennan [35] in the context of patents citations. However they just used it to explain a power law in citation distribution (for what the usual cumulative advantage model will do) and did not address the topics we just mentioned.

While working on a paper, a scientist reads current issues of scientific journals and selects from them the references to be cited in it. These references are of two sorts:

- *Fresh papers he had just read* – to embed his work in the context of current aspirations.
- *Older papers that are cited in the fresh papers* he had just read – to place his work in the context of previous achievements.

It is not a necessary condition for the validity of our model that the citations to old papers are copied, but the paper itself remains unread (although such opinion is supported by the studies of misprint propagation). The necessary conditions are as follows:

- Older papers are considered for possible citing only if they were recently cited.
- If a citation to an old paper is followed and the paper is formally read – the scientific qualities of that paper do not influence its chance of being cited.

A reasonable estimate for the length of time a scientist works on a particular paper is one year. We will thus assume that "recent" in the MMRCS means the preceding year. To make the model mathematically tractable we enforce time-discretization with a unit of one year. The precise model to be studied is as follows. Every year N papers are published. There is, on average, N_{ref} references in a published paper (the actual value is somewhere between 20 and 40). Each year, a fraction α of references

[10]Some of these references do not deal with citing, but with other social processes, which are modeled using the same mathematical tools. Here we rephrase the results of such papers in terms of citations for simplicity.

goes to randomly selected preceding year papers (the estimate[11] from actual citation data is $\alpha \approx 0.1$(see Fig. 4 in [24]) or $\alpha \approx 0.15$ (see Fig. 6 in [36])). The remaining citations are randomly copied from the lists of references used in the preceding year papers.

16.4.2 Branching Citations

When N is large, this model leads to the first-year citations being Poisson-distributed. The probability to get n citations is

$$p(n) = \frac{{\lambda_0}^n}{n!} e^{-\lambda_0}, \tag{16.37}$$

where λ_0 is the average expected number of citations

$$\lambda_0 = \alpha N_{\text{ref}}. \tag{16.38}$$

The number of the second-year citations, generated by each first year citation (as well as, third-year citations generated by each second year citation and so on), again follows a Poisson distribution, this time with the mean

$$\lambda = (1 - \alpha). \tag{16.39}$$

Within this model, citation process is a branching process (see Appendix A) with the first-year citations equivalent to children, the second-year citations to grand children, and so on.

As $\lambda < 1$, this branching process is subcritical. Figure 16.9 shows a graphical illustration of the branching citation process.

Substituting (16.37) into (16.78) we obtain the generating function for the first year citations:

$$f_0(z) = e^{(z-1)\lambda_0}. \tag{16.40}$$

Similarly, the generating function for the later-years citations is:

$$f(z) = e^{(z-1)\lambda}. \tag{16.41}$$

[11] The uncertainty in the value of α depends not only on the accuracy of the estimate of the fraction of citations which goes to previous year papers. We also arbitrarily defined recent paper (in the sense of our model), as the one published within a year. Of course, this is by order of magnitude correct, but the true value can be anywhere between half a year and 2 years.

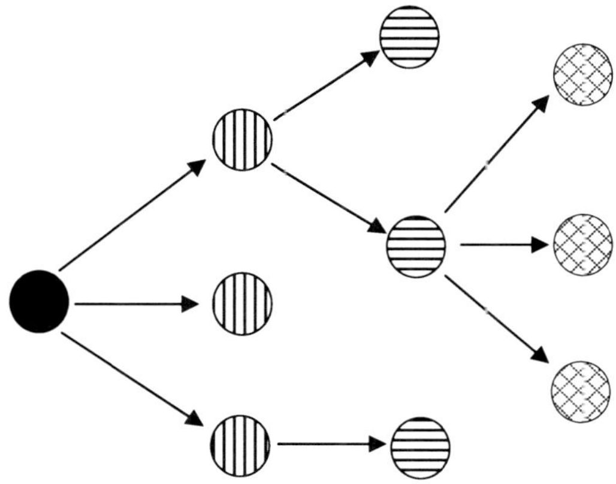

Fig. 16.9 An illustration of the branching citation process, generated by the modified model of random-citing scientists. During the first year after publication, the paper was cited in three other papers written by the scientists who have read it. During the second year one of those citations was copied in two papers, one in a single paper and one was never copied. This resulted in three second year citations. During the third year, two of these citations were never copied, and one was copied in three papers

The process is easier to analyze when $\lambda = \lambda_0$, or $\lambda_0/\lambda = (\alpha/(1-\alpha))N_{\text{ref}} = 1$, as then we have a simple branching process, where all generations are governed by the same offspring probabilities. The case when $\lambda \neq \lambda_0$ we study in Appendix B.

16.4.2.1 Distribution of Citations to Papers Published During the Same Year

Theory of branching processes allows us to analytically compute the probability distribution, $P(n)$, of the total number of citations the paper receives before it is forgotten. This should approximate the distribution of citations to old papers. Substituting (16.41) into (16.85) we get:

$$P(n) = \frac{1}{n!}\left[\frac{d^{n-1}}{d\omega^{n-1}}e^{n(\omega-1)\lambda}\right]_{\omega=0} = \frac{(n\lambda)^{n-1}}{n!}e^{-\lambda n}. \tag{16.42}$$

Applying Stirling's formula to (16.42), we obtain the large n asymptotic of the distribution of citations:

$$P(n) \propto \frac{1}{\sqrt{2\pi}\lambda n^{3/2}}e^{-(\lambda-1-\ln\lambda)n}. \tag{16.43}$$

When $1-\lambda \ll 1$ we can approximate the factor in the exponent as:

$$\lambda - 1 - \ln\lambda \approx (1-\lambda)^2/2 \tag{16.44}$$

As $1-\lambda \ll 1$, the above number is small. This means that for $n \ll 2/(1-\lambda)^2$ the exponent in (16.43) is approximately equal to 1 and the behavior of $P(n)$ is dominated by the $1/n^{3/2}$ factor. In contrast, when $n \ll 2/(1-\lambda)^2$ the behavior of $P(n)$ is dominated by the exponential factor. Thus citation distribution changes from a power law to an exponential (suffers an exponential cut-off) at about

$$n_c = \frac{1}{\lambda - 1 - \ln \lambda} \approx \frac{2}{(1-\lambda)^2} \tag{16.45}$$

citations. For example, when $\alpha = 0.1$ (16.39) gives $\lambda = 0.9$ and from (16.45) we get that the exponential cut-off happens at about 200 citations. We see that, unlike the cumulative advantage model, our model is capable of qualitative explanation of the power-law distribution of citations to papers of the same age. The exponential cut-off at 200, however, happens too soon, as the actual citation distribution obeys a power law well into thousands of citations. In the following sections we show that taking into account the effects of literature growth and of variations in papers' Darwinian fitness can fix this.

In the cumulative advantage (AKA preferential attachment) model, a power-law distribution of citations is only achieved because papers have different ages. This is not immediately obvious from the early treatments of the problem [4, 20], but is explicit in later studies [5, 21, 29]. In that model, the oldest papers are the most cited ones. The number of citations is mainly determined by paper's age. At the same time, distribution of citations to papers of the same age is exponential [5, 29]. The key difference between that model and ours is as follows. In the cumulative advantage model, the rate of citation is proportional to the number of citations the paper had accumulated since its publication. In our model, the rate of citation is proportional to the number of citations the paper received during preceding year. This means that if an unlucky paper was not cited during previous year – it will never be cited in the future. This means that its rate of citation will be less than that in the cumulative advantage model. On the other hand, the lucky papers, which were cited during the previous year, will get all the citation share of the unlucky papers. Their citation rates will be higher than in the cumulative advantage model. There is thus more stratification in our model than in the cumulative advantage model. Consequently, the resulting citation distribution is far more skewed.

16.4.2.2 Distribution of Citations to Papers Cited During the Same Year

We denote as $N(n)$ the number of papers cited n times during given year. The equilibrium distribution of $N(n)$ should satisfy the following self-consistency equation:

$$N(n) = \sum_{m=1}^{\infty} N(m) \frac{(\lambda m)^n}{n!} e^{-\lambda m} + N \frac{(\lambda_0)^n}{n!} e^{-\lambda_0} \tag{16.46}$$

Here the first term comes from citation copying and the second from citing previous year papers. In the limit of large n the second term can be neglected and the sum can be replaced with integral to get:

$$N(n) = \frac{1}{n!}\int_0^\infty dmN(m)(\lambda m)^n e^{-\lambda m} \tag{16.47}$$

In the case $\lambda = 1$ one solution of (16.47) is $N(m) = C$, where C is an arbitrary constant. Clearly, the integral becomes a gamma function and the factorial in the denominator cancels out. However, this solution is, meaningless since the total number of citations per year, which is given by

$$N_{\text{cite}} = \sum_{m=1}^{\infty} mN(m) \tag{16.48}$$

diverges. In the case $\lambda < 1$, $N(m) = C$ is no longer a solution since the integral gives C/λ. However $N(m) = C/m$ is a solution. This solution is again meaningless because the total number of yearly citations given by (16.48) again diverges. One can look for a solution of the form

$$N(m) = \frac{C}{m}\exp(-\mu m) \tag{16.49}$$

After substituting (16.49) into (16.47) we get that $N(n)$ is given by the same function but instead of μ with

$$\mu' = \ln(1 + \mu/\lambda) \tag{16.50}$$

The self-consistency equation for μ is thus

$$\mu = \ln(1 + \mu/\lambda). \tag{16.51}$$

The obvious solution is $\mu = 0$ which gives us the previously rejected solution $N(m) = C/m$. It is also easy to see that this stationary solution is unstable. If μ slightly deviates from zero (16.50) gives us $\mu' = \mu/\lambda$. Since $\lambda < 1$ the deviation from stationary shape will increase the next year. Another solution of (16.51) can be found by expansion of the logarithm up to the second order in μ. It is $\mu \approx 2(1-\lambda)$. One can show that this solution is stable. Thus, we get:

$$N(m) \approx \frac{C}{m}\exp(-2(1-\lambda)m) \tag{16.52}$$

After substituting this into (16.48) we get

$$C \approx 2(1-\lambda)N_{\text{cite}} \tag{16.53}$$

The solution which we just presented was stimulated by that obtained by Wright [37], who studied the distribution of alleles (alternative forms of a gene) in a population. In Wright's model, the gene pool at any generation has constant size N_g. To form the next generation we N_g times select a random gene from current generation pool and copy it to next generation pool. With some probability, a gene can mutate during the process of copying. The problem is identical to ours with an allele replaced with a paper and mutation with a new paper. Our solution follows that of Wright but is a lot simpler. Wright considered finite N_g. and as a consequence got Binomial distribution and a Beta function in his analog of (16.47). The simplification was possible because in the limit of large N_g Binomial distribution becomes Poissonian. Alternative derivations of (16.52) can be found in [33] and [38].

16.4.3 Scientific Darwinism

Now we proceed to investigate the model, where papers are not created equal, but each has a specific *Darwinian fitness*, which is a bibliometric measure of scientific fangs and claws that help a paper to fight for citations with its competitors. While this parameter can depend on factors other than the intrinsic quality of the paper, the fitness is the only channel through which the quality can enter our model. The fitness may have the following interpretation. When a scientist writes a manuscript he needs to include in it a certain number of references (typically between 20 and 40, depending on implicit rules adopted by a journal where the paper is to be submitted). He considers random scientific papers one by one for citation, and when he has collected the required number of citations, he stops. Every paper has specific probability to be selected for citing, once it was considered. We will call this probability a Darwinian fitness of the paper. Defined in such way, fitness is bounded between 0 and 1.

In this model a paper with fitness ϕ will on average have

$$\lambda_0(\phi) = \alpha N_{\text{ref}} \phi / \langle \phi \rangle_p \tag{16.54}$$

first-year citations. Here we have normalized the citing rate by the *average fitness of published papers*, $\langle \varphi \rangle_p$, to insure that the fraction of citations going to previous year papers remained α. The fitness distribution of references is different from the fitness distribution of published papers, as papers with higher fitness are cited more often. This distribution assumes an asymptotic form $p_r(\varphi)$, which depends on the distribution of the fitness of published papers, $p_p(\varphi)$, and other parameters of the model.

During later years there will be on average

$$\lambda(\varphi) = (1 - \alpha)\varphi / \langle \varphi \rangle_r \tag{16.55}$$

next-year citations per one current year citation for a paper with fitness ϕ. Here, $\langle \varphi \rangle_r$ is the *average fitness of a reference*.

16.4.3.1 Distribution of Citations to Papers Published During the Same Year

Let us start with the self-consistency equation for $p_r(\varphi)$, the equilibrium fitness distribution of references:

$$p_r(\varphi) = \alpha \frac{\varphi \times p_p(\varphi)}{\langle \varphi \rangle_p} + (1-\alpha) \frac{\varphi \times p_r(\varphi)}{\langle \varphi \rangle_r} \tag{16.56}$$

solution of which is:

$$p_r(\varphi) = \alpha \frac{\alpha \times \varphi \times p_p(\varphi)/\langle \varphi \rangle_p}{1-(1-\alpha)\varphi/\langle \varphi \rangle_r} \tag{16.57}$$

One obvious self-consistency condition is that

$$\int p_r(\varphi) d\varphi = 1. \tag{16.58}$$

Another is:

$$\int \phi \times p_r(\phi) d\phi = \langle \phi \rangle_r.$$

However, when the condition of (16.58) is satisfied the above equation follows from (16.56).

Let us consider the simplest case when the fitness distribution, $p_p(\varphi)$, is uniform between 0 and 1. This choice is arbitrary, but we will see that the resulting distribution of citations is close to the empirically observed one. In this case, the average fitness of a published paper is $\langle \varphi \rangle_p = 0.5$. After substituting this into (16.56), the result into (16.58), and performing integration we get:

$$\alpha = -\frac{((1-\alpha)/\langle \varphi \rangle_r)^2/2}{\ln(1-(1-\alpha)/\langle \varphi \rangle_r) + (1-\alpha)/\langle \varphi \rangle_r} \tag{16.59}$$

Since α is close to 0, $\langle \varphi \rangle_r$ must be very close to $1-\alpha$, and we can replace it with the latter everywhere but in the logarithm to get:

$$\frac{1-\alpha}{\langle \varphi \rangle_r} = 1 - e^{\frac{1}{2\alpha}-1} \tag{16.60}$$

For papers of fitness φ, citation distribution is given by (16.42) or (16.43) with λ replaced with $\lambda(\varphi)$, given by (16.55):

$$P(n,\phi) \propto \frac{1}{\sqrt{2\pi}\lambda(\phi) n^{3/2}} e^{-(\lambda(\phi)-1-\ln\lambda(\phi))n}. \tag{16.61}$$

When $\alpha = 0.1$, (16.60) gives $(1-\alpha)/\langle\varphi\rangle_r \approx 1 - 2\times 10^{-3}$. From (16.55) it follows that $\lambda(1) = (1-\alpha)/\langle\varphi\rangle_r$. Substituting this into (16.45) we get that the exponential cut-off for the fittest papers ($\varphi = 1$) starts at about 300,000 citations. In contrast, for the unfit papers the cut-off is even stronger than in the model without fitness. For example, for papers with fitness $\varphi = 0.1$ we get $\lambda(0.1) = 0.1(1-\alpha)/\langle\varphi\rangle_r \approx 0.1$ and the decay factor in the exponent becomes $\lambda(0.1) - 1 - \ln\lambda(0.1) \approx 2.4$. This cut-off is so strong than not even a trace of a power-law distribution remains for such papers.

To compute the overall probability distribution of citations we need to average (16.61) over fitness:

$$P(n) \propto \frac{1}{\sqrt{2\pi}n^{3/2}} \int_0^1 \frac{d\phi}{\lambda(\phi)} e^{-(\lambda(\phi)-1-\ln\lambda(\phi))n}. \tag{16.62}$$

We will concentrate on the large n asymptotic. Then only highest-fitness papers, which have $\lambda(\varphi)$ close to 1, are important and we can approximate the integral in (16.62), using (16.44), as:

$$\int_0^1 d\varphi \exp\left(-\left[1-\varphi\frac{1-\alpha}{\langle\varphi\rangle_r}\right]^2\frac{n}{2}\right) = \frac{\langle\varphi\rangle_r}{1-\alpha}\sqrt{\frac{2}{n}} \int_{\left(1-\frac{1-\alpha}{\langle\varphi\rangle_r}\right)\sqrt{\frac{n}{2}}}^{\sqrt{\frac{n}{2}}} dz e^{-z^2}$$

We can replace the upper limit in the above integral with infinity when n is large. The lower limit can be replaced with zero when $n \ll n_c$, where

$$n_c = 2\left(1 - \frac{1-\alpha}{\langle\varphi\rangle_r}\right)^{-2}. \tag{16.63}$$

In that case the integral is equal to $\sqrt{\pi}/2$, and (16.62) gives:

$$P(n) \propto \frac{\langle\varphi\rangle_r}{2(1-\alpha)}\frac{1}{n^2}. \tag{16.64}$$

In the opposite case,$n >> n_c$, we get:

$$P(n) \propto \frac{\langle\varphi\rangle_r}{4(1-\alpha)}\frac{\sqrt{n_c}}{n^{2.5}}e^{-\frac{n}{n_c}}. \tag{16.65}$$

When $\alpha = 0.1$, $n_c = 3\times 10^5$.

Compared to the model without fitness, we have a modified power-law exponent (2 instead of 3/2) and a much relaxed cut-off of this power law. This is consistent with the actual citation data shown in the Fig. 16.10.

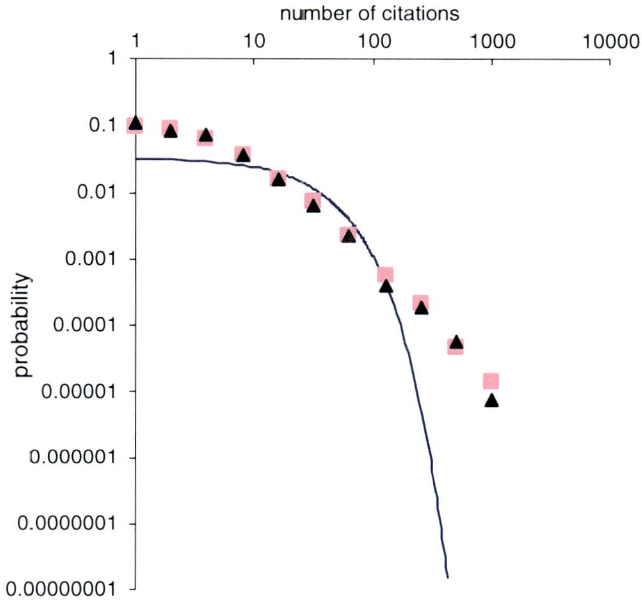

Fig. 16.10 Numerical simulations of the *modified* model of random-citing scientists (*triangles*) compared to actual citation data for papers *published during a single year* (*squares*). The *solid line* is the prediction of the cumulative advantage (AKA preferential attachment) model

Table 16.3 The onset of exponential cut-off in the distribution of citations, n_c, as a function of α, computed using (16.63) and (16.60)

α	0.3	0.25	0.2	0.15	0.1	0.05
n_c	167	409	1,405	9,286	3.1E + 05	7.2E + 09

As was already mentioned, because of the uncertainty of the definition of "recent" papers, the exact value of α is not known. Therefore, we give n_c for a range of values of α in Table 16.3. As long as $\alpha \leq 0.5$ the value of n_c does not contradict existing citation data.

The major results, obtained for the uniform distribution of fitness, also hold for a non-uniform distribution, which approaches some finite value at its upper extreme $p_p(\varphi = 1) = a > 0$. In [33] we show that in this case $(1-\alpha)/\langle\varphi\rangle_r$ is very close to unity when α is small. Thus we can treat (16.62) the same way as in the case of the uniform distribution of fitness. The only change is that (16.64) and (16.65) acquire a pre-factor of a. Things turn out a bit different when $p_p(1) = 0$. In Appendix C we consider the fitness distribution, which vanishes at $\varphi = 1$ as a power law: $p_p(\varphi) = (\theta+1)(1-\varphi)^\theta$. When θ is small ($\theta < \frac{2\times\alpha}{1-\alpha}$) the behavior of the model is similar to what was in the case of a uniform fitness distribution. The distribution of the fitness of cited papers $p_r(\varphi)$ approaches some limiting form with $(1-\alpha)/\langle\varphi\rangle_r$ being very close to unity when α is small. The exponent of the power law is, however, no longer 2 as it was in the case of a uniform fitness distribution (16.64), but becomes $2+\theta$. However, when $\theta > \frac{2\times\alpha}{1-\alpha}$ the model behaves differently: $(1-\alpha)/\langle\varphi\rangle_r$ strictly equals 1. This means that the power law does not have an exponential cut-off. Thus, a wide class of fitness distributions produces citation distributions very similar to the experimentally observed one. More research is needed to infer the actual distribution of the Darwinian fitness of scientific papers.

The fitness distribution of references $p_r(\varphi)$ adjusts itself in a way that the fittest papers become critical. This is similar to what happens in the SOC model [39] where the distribution of sand grains adjusts itself that the avalanches become critical. Recently we proposed a similar SOC-type model to describe the distribution of links in blogosphere [40].

16.4.3.2 Distribution of Citations to Papers Cited During the Same Year

This distribution in the case without fitness is given in (16.52). To account for fitness we need to replace λ with $\lambda(\varphi)$ in (16.52) and integrate it over φ. The result is:

$$p(n) \sim \frac{1}{n^2} e^{-n/n_c^*}, \tag{16.66}$$

where

$$n_c^* = \frac{1}{2}\left(1 - \frac{1-\alpha}{\langle\varphi\rangle_r}\right)^{-1}. \tag{16.67}$$

Note that $n_c^* \sim \sqrt{n_c}$. This means that the exponential cut-off starts much sooner for the distribution of citation to papers *cited* during the same year, then for citation distribution for papers *published* during the same year.

The above results qualitatively agree with the empirical data for papers cited in 1961 (see Fig. 2 in [24]). The exponent of the power law of citation distribution reported in that work is, however, between 2.5 and 3. Quantitative agreement thus may be lacking.

16.4.4 Effects of Literature Growth

Up to now we implicitly assumed that the yearly volume of published scientific literature does not change with time. In reality, however, it grows, and does so exponentially. To account for this, we introduce a Malthusian parameter, β, which is yearly percentage increase in the yearly number of published papers. From the data on the number of items in the Mathematical Reviews Database [41], we obtain that the literature growth between 1970 and 2000 is consistent with $\beta \approx 0.045$. From the data on the number of source publications in the ISI database (see Table 1 in [30]) we get that the literature growth between 1973 and 1984 is characterized by $\beta \approx 0.03$. One can argue that the growth of the databases reflected not only growth of the volume of scientific literature, but also increase in activities of Mathematical Reviews and ISI and true β must be less. One can counter-argue that may be ISI and Mathematical Reviews could not cope with literature growth and β must be more. Another issue is that the average number of references in papers also grows. What is important for our modeling is the yearly increase not in number of papers, but in the number of citations these papers contain. Using the ISI data we get that this increase is characterized by $\beta \approx 0.05$. As we are not sure of the precise value of β, we will be giving quantitative results for a range of its values.

16.4.4.1 Model Without Fitness

At first, we will study the effect of β in the model without fitness. Obviously, (16.38) and (16.39) will change into:

$$\lambda_0 = \alpha(1+\beta)N_{\text{ref}}, \tag{16.68}$$

$$\lambda = (1-\alpha)(1+\beta). \tag{16.69}$$

The estimate of the actual value of λ is: $\lambda \approx (1-0.1)(1+0.05) \approx 0.945$. Substituting this into (16.45) we get that the exponential cut-off in citation distribution now happens after about 660 citations.

A curious observation is that when the volume of literature grows in time the average amount of citations a paper receives, N_{cit}, is bigger than the average amount of references in a paper, N_{ref}. Elementary calculation gives:

$$N_{\text{cit}} = \sum_{m=0}^{\infty} \lambda_0 \lambda^m = \frac{\lambda_0}{1-\lambda} = \frac{\alpha(1+\beta)N_{\text{ref}}}{1-(1-\alpha)(1+\beta)}. \tag{16.70}$$

As we see $N_{\text{cit}} = N_{\text{ref}}$ only when $\beta = 0$ and $N_{\text{cit}} > N_{\text{ref}}$ when $\beta > 0$. There is no contradiction here if we consider an infinite network of scientific papers, as one can show using methods of the set theory that there are one-to-many mappings of an infinite set on itself. When we consider real, i.e., finite, network where the number of citations is obviously equal to the number of references we recall that N_{cit}, as computed in (16.70), is the number of citations accumulated by a paper during its cited lifetime. So recent papers did not yet receive their share of citations and there is no contradiction again.

Table 16.4 Critical value of the Malthusian parameter β_c as a function of α computed using (16.73). When $\beta > \beta_c$ the fittest papers become supercritical

α	0.3	0.25	0.2	0.15	0.1	0.05
β_c	0.12	0.075	0.039	0.015	2.6E-03	1.7E-05

16.4.4.2 Model with Darwinian Fitness

Taking into account literature growth leads to transformation of (16.54) and (16.55) into:

$$\lambda_0(\phi) = \alpha(1+\beta)N_{\text{ref}}\phi/\langle\phi\rangle_p, \tag{16.71}$$

$$\lambda(\phi) = (1-\alpha)(1+\beta)\phi/\langle\phi\rangle_r. \tag{16.72}$$

As far as the average fitness of a reference, $\langle\varphi\rangle_r$, goes, β has no effect. Clearly, its only result is to increase the number of citations to all papers (independent of their fitness) by a factor $1+\beta$. Therefore $\langle\varphi\rangle_r$ is still given by (16.59). While, $\lambda(\varphi)$ is always less than unity in the case with no literature growth, it is no longer so when we take this growth into account. *When β is large enough, some papers can become supercritical.* The critical value of β, i.e., the value which makes papers with $\varphi = 1$ critical, can be obtained from (16.72):

$$\beta_c = \langle\phi\rangle_r/(1-\alpha) - 1 \tag{16.73}$$

When $\beta > \beta_c$, a finite fraction of papers becomes supercritical. The rate of citing them will increase with time. Note, however, that it will increase always slower than the amount of published literature. Therefore, the relative fraction of citations to those papers to the total number of citations will decrease with time.

Critical values of β for several values of α are given in Table 16.4. For realistic values of parameters ($\alpha \leq 0.15$ and $\beta \geq 0.03$) we have $\beta > \beta_c$ and thus our model predicts the existence of supercritical papers. Note, however, that this conclusion also depends on the assumed distribution of fitness.

It is not clear whether supercritical papers exist in reality or are merely a pathological feature of the model. Supercritical papers probably do exist if one generalizes "citation" to include references to a concept, which originated from the paper in question. For instance, these days a negligible fraction of scientific papers which use Euler's Gamma function contain a reference to Euler's original paper. It is very likely that the number of papers mentioning Gamma function is increasing year after year.

Let us now estimate the fraction of supercritical papers predicted by the model. As $(1-\alpha)/\langle\varphi\rangle_r$ is very close to unity, it follows from (16.72) that papers with fitness $\varphi > \varphi_c \approx 1/(1+\beta) \approx 1-\beta$ are in the supercritical regime. As $\beta \approx 0.05$, about 5% of papers are in such regime. This does not mean that 5% of papers will be cited forever, because being in supercritical regime only means having extinction probability less than one. To compute this probability we substitute (16.72) and (16.41) into (16.80) and get:

$$p_{\text{ext}}(\phi) = \exp((1+\beta) \times \phi \times (p_{\text{ext}}(\varphi) - 1)).$$

It is convenient to rewrite the above equation in terms of survival probability:

$$1 - p_{\text{surv}}(\phi) = \exp(-(1+\beta) \times \phi \times p_{\text{surv}}(\phi)).$$

As $\beta \ll 1$ the survival probability is small and we can expand the RHS of the above equation in powers of p_{surv}. We limit this expansion to terms up to $(p_{\text{surv}})^2$ and after solving the resulting equation get:

$$p_{\text{surv}}(\phi) \approx 2\frac{\phi - \frac{1}{1+\beta}}{(1+\beta)\phi} \approx 2(\phi - 1 + \beta).$$

The fraction of forever-cited papers is thus: $\int\limits_{1-\beta}^{1} 2(\varphi - 1 + \beta)d\varphi = \beta^2$. For $\beta \approx 0.05$ this will be one in four hundred. By changing the fitness distribution $p_p(\varphi)$ from a uniform this fraction can be made much smaller.

16.4.5 Numerical Simulations

The analytical results are of limited use, as they are exact only for infinitely old papers. To see what happens with finitely old papers, one has to do numerical simulations. Figure 16.10 shows the results from such simulations (with $\alpha = 0.1$, $\beta = 0.05$, and uniform between 0 and 1 fitness distribution), i.e., distributions of citations to papers published within a single year, 22 years after publication. Results are compared with actual citation data for *Physical Review D* papers published in 1975 (as of 1997) [11]. Prediction of the cumulative advantage [20] (AKA preferential attachment [21]) model is also shown. As we mentioned earlier, that model leads to exponential distribution of citations to papers of same age, and thus cannot account for highly-skewed distribution empirically observed.

16.4.6 Aging of Scientific Literature

Scientific papers tend to get less frequently cited as time passes since their publication. There are two ways to look at the age distribution of citations. One can take all papers *cited* during a particular year, and study the distribution of their ages. In Bibliometrics this is called *synchronous* distribution [30]. One can take all the papers *published* during a particular distant year, and study the distribution of the citations to these papers with regard to time difference between citation and publication. Synchronous distribution is steeper than the distribution of citation to papers published during the same year (see Figs. 16.2 and 16.3 in [30]). For example, if one looks at a synchronous distribution, then 10-year-old papers appear to be cited three times less than 2-year-old papers. However, when one looks at the distribution of citations to papers published during the same year the number of citations 10 years after publication is only 1.3 times less than 2 years after publication. The apparent discrepancy is resolved by noting that the number of published scientific papers had grown 2.3 times during 8 years. When one plots not total number of citations to papers published in a given year, but the ratio of this number to the annual total of citations than resulting distribution (it is called *diachronous* distribution [30]) is symmetrical to the synchronous distribution.

Recently, Redner [36] who analyzed a century worth of citation data from *Physical Review* had found that the synchronous distribution (he calls it citations *from*) is exponential, and the distribution of citations to papers published during the same year (he calls it citations *to*) is a power law with an exponent close to 1. If one were to construct a diachronous distribution using Redner's data, – it would be a product of a power law and an exponential function. Such distribution is difficult to tell from an exponential one. Thus, Redner's data may be consistent with synchronous and diachronous distributions being symmetric.

The predictions of the mathematical theory of citing are as follows. First, we consider the model without fitness. The average number of citations a paper receives during the kth year since its publication, C_k, is:

$$C_k = \lambda_0 \lambda^{k-1}, \tag{16.74}$$

and thus, decreases exponentially with time. This is in qualitative agreement with Nakamoto's [30] empirical finding. Note, however, that the exponential decay is empirically observed after the second year, with average number of the second-year citations being higher than the first year. This can be understood as a mere consequence of the fact that it takes about a year for a submitted paper to get published.

Let us now investigate the effect of fitness on literature aging. Obviously, (16.74) will be replaced with:

$$C_k = \int_0^1 d\phi \lambda_0(\phi) \lambda^{k-1}(\phi). \tag{16.75}$$

Table 16.5 The number of years, after which the decrease in average citing rate will change from a power law to an exponential, k_c, computed using (16.77), as a function of α

α	0.3	0.25	0.2	0.15	0.1	0.05
k_c	9	14	26	68	392	59,861

Substituting (16.54) and (16.55) into (16.75) and performing integration we get:

$$C_k = \frac{\alpha N_{\text{ref}}}{\langle\phi\rangle_p}\left(\frac{1-\alpha}{\langle\phi\rangle_r}\right)^{k-1}\frac{1}{k+1}. \tag{16.76}$$

The average rate of citing decays with paper's age as a power law with an exponential cut-off. This is in agreement with Redner's data (See Fig. 7 of [36]), though it contradicts the older work [30], which found exponential decay of citing with time.

In our model, the transition from hyperbolic to exponential distribution occurs after about

$$k_c = -1/\ln((1-\alpha)/\langle\varphi\rangle_r) \tag{16.77}$$

years. The values of k_c for different values of α are given in Table 16.5. The values of k_c for $\alpha \leq 0.2$ do not contradict the data reported by Redner [36].

We have derived literature aging from a realistic model of scientist's referencing behavior. Stochastic models had been used previously to study literature aging, but they were of artificial type. Glänzel and Schoepflin [31] used a modified cumulative advantage model, where the rate of citing is proportional to the product of the number of accumulated citations and some factor, which decays with age. Burrell [42], who modeled citation process as a non-homogeneous Poisson process had to postulate some obsolescence distribution function. In both these cases, aging was inserted by hand. In contrast, in our model, literature ages naturally.

16.4.7 Sleeping Beauties in Science

Figure 16.11 shows two distinct citation histories. The paper, whose citation history is shown by the squares, is an ordinary paper. It merely followed some trend. When 10 years later that trend got out of fashion the paper got forgotten. The paper, whose citation history is depicted by the triangles, reported an important but premature [43] discovery, significance of which was not immediately realized by scientific peers. Only 10 years after its publication did the paper get recognition, and got cited widely and increasingly. Such papers are called "Sleeping Beauties" [44]. Surely, the reader has realized that both citation histories are merely the outcomes of numerical simulations of the MMRCS.

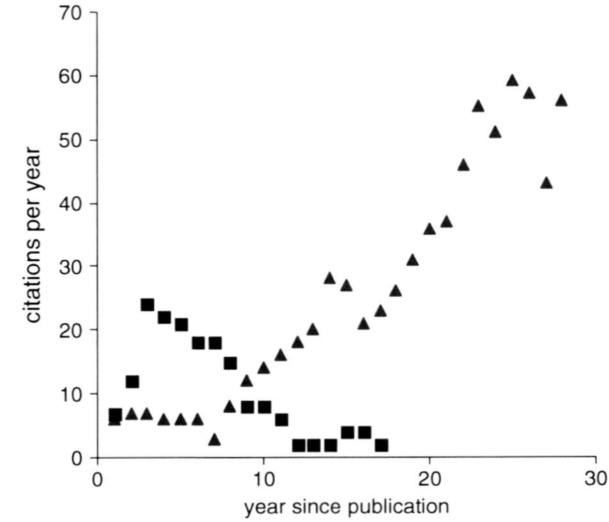

Fig. 16.11 Two distinct citation histories: an ordinary paper (*squares*) and a "Sleeping Beauty" (*triangles*)

16.4.8 Relation to Self-Organized Criticality

Three out of twelve high-profile papers misprints in citing which we studied in Sect. 16.1 (see papers 10, 11, and 12 in Tables 16.1 and 16.2) advance the science of SOC [39]. Interestingly this science itself is directly related to the theory of citing. We model scientific citing as a random branching process. In its mean-field version, SOC can also be described as a branching process [45]. Here the sand grains, which are moved during the original toppling, are equivalent to sons. These displaced grains can cause further toppling, resulting in the motion of more grains, which are equivalent to grandsons, and so on. The total number of displaced grains is the size of the avalanche and is equivalent to the total offspring in the case of a branching process. The distribution of offspring sizes is equivalent to the distribution of avalanches in SOC.

Bak [46] himself had emphasized the major role of chance in works of Nature: one sand grain falls, – nothing happens; another one (*identical*) falls, – and causes an avalanche. Applying these ideas to biological evolution, Bak and Sneppen [47] argued that no cataclysmic external event was necessary to cause a mass extinction of dinosaurs. It could have been caused by one of many minor external events. Similarly, in the model of random-citing scientists: one paper goes unnoticed, but another one (*identical in merit*), causes an avalanche of citations. Therefore apart from explanations of $1/f$ noise, avalanches in sandpiles, and extinction of dinosaurs, the highly cited Science of self-organized criticality can also account for its own success.

16.5 Discussion

The conclusion of this study that a scientific paper can become famous due to ordinary law of chances independently of its content may seem shocking to some people. Here we present more facts to convince them.

Look at the following example. The writings of J. Lacan (10,000 citations) and G. Deleuze (8,000 citations) were exposed by Sokal and Bricmont [48] as nonsense. At the same time, the work of the true scientists is far less cited: A. Sokal – 2,700 citations, J. Bricmont – 1,000 citations.

Additional support for the plausibility of this conclusion gives us the statistics of the very misprints in citations the present study grew from. Few citation slips repeat dozens of times, while most appear just once (see Fig 16.1). Can one misprint be more seminal than the other?

More support comes from the studies of popularity of other elements of culture. A noteworthy case where prominence is reached by pure chance is the statistics of baby-names. Hahn and Bentley [49] observed that their frequency distribution follows a power law, and proposed a copying mechanism that can explain this observation. For example, during the year 2000 34,448 new-born American babies were named Jacob, while only 174 were named Samson [50]. This means that the name "Jacob" is 200 times more popular than the name "Samson." Is it intrinsically better?

A blind test was administered offering unlabeled paintings, some of which were famous masterpieces of Modern art while others were produced by the author of the test [51]. Results indicate that people cannot tell great art from chaff when the name of a great artist is detached from it. One may wonder if a similar test with famous scientific articles would lead to similar results. In fact there is one forgotten experiment though not with scientific articles, but with a scientific lecture. Naftulin, Ware, and Donnelly [52] programmed an actor to teach on a subject he knew nothing. They presented him to a scientific audience as Dr. Myron Fox, an authority on application of mathematics to human behavior (we would like to note that in practice the degree of authority of a scientist is determined by the number of citations to his papers). He read a lecture and answered questions and nobody suspected anything wrong. Afterward the attendees were asked to rate the lecturer and he got high grades. They indicated that they learned a lot from the lecture and one of respondents even indicated that he had read Dr. Fox's articles.

To conclude let us emphasize that the Random-citing model is used not to ridicule the scientists, but because it can be exactly solved using available mathematical methods, while yielding a better match with data than any existing model. This is similar to the random-phase approximation in the theory of an electron gas. Of course, the latter did not arouse as much protest, as the model of random-citing scientists, – but this is only because electrons do not have a voice. What is an electron? – Just a green trace on the screen of an oscilloscope. Meanwhile, within itself, electron is very complex and is as inexhaustible as the universe. When an electron is annihilated in a lepton collider, the whole universe dies with it. And as for the random-phase approximation: Of course, it accounts for the experimental facts – but so does the model of random-citing scientists.

Appendix A: Theory of Branching Processes

Let us consider a model where in each generation, $p(0)$ percent of the adult males have no sons, $p(1)$ have one son and so on. The problem is best tackled using the method of generating functions [34], which are defined as:

$$f(z) = \sum_{n=0}^{\infty} p(n)z^n. \tag{16.78}$$

These functions have many useful properties, including that the generating function for the number of grandsons is $f_2(z) = f(f(z))$. To prove this, notice that if we start with two individuals instead of one, and both of them have offspring probabilities described by $f(z)$, their combined offspring has generating function $(f(z))^2$. This can be verified by observing that the nth term in the expansion of $(f(z))^2$ is equal to $\sum_{m=0}^{n} p(n-m)p(m)$, which is indeed the probability that the combined offspring of two people is n. Similarly one can show that the generating function of combined offspring of n people is $(f(z))^n$. The generating function for the number of grandsons is thus:

$$f_2(z) = \sum_{n=0}^{\infty} p(n)(f(z))^n = f(f(z)).$$

In a similar way one can show that the generating function for the number of grand-grandsons is $f_3(z) = f(f_2(z))$ and in general:

$$f_k(z) = f(f_{k-1}(z)). \tag{16.79}$$

The probability of extinction, p_{ext}, can be computed using the self-consistency equation:

$$p_{\text{ext}} = \sum_{n=0}^{\infty} p(n)p_{\text{ext}}^n = f(p_{\text{ext}}). \tag{16.80}$$

The fate of families depends on the average number of sons $\lambda = \sum np(n) = [f\prime(z)]_{z=1}$. When $\lambda < 1$, (16.80) has only one solution, $p_{\text{ext}} = 1$, that is all families get extinct (this is called subcritical branching process). When $\lambda > 1$, there is a solution where $p_{\text{ext}} < 1$, and only some of the families get extinct, while others continue to exist forever (this is called supercritical branching process). The intermediate case, $\lambda = 1$, is critical branching process, where all families get extinct, like in a subcritical process, though some of them only after very long time.

For a subcritical branching process we will also be interested in the probability distribution, $P(n)$, of total offspring, which is the sum of the numbers of sons, grandsons, grand-grandsons, and so on (to be precise we include the original individual in this sum just for mathematical convenience). We define the corresponding generating function [54]:

$$g(z) = \sum_{n=1}^{\infty} p(n) z^n. \tag{16.81}$$

Using an obvious self-consistency condition (similar to the one in (16.80)) we get:

$$zf(g) = g. \tag{16.82}$$

We can solve this equation using Lagrange expansion (see [53]), which is as follows. Let $z = F(g)$ and $F(0) = 0$ where $F'(0) \neq 0$, then:

$$\Phi(g(z)) = \sum_{n=0}^{\infty} \frac{1}{n!} \frac{d^{n-1}}{dg^{n-1}} \left(\Phi'(g) \left(\frac{g}{F(g)} \right)^n \right) |_{g=0} \, z^n. \tag{16.83}$$

Substituting $F(g) = g/F(g)$ (see (16.82)) and $\Phi(g) = g$ into (16.83) we get:

$$g = \sum_{n=1}^{\infty} \frac{z^n}{n!} \left[\frac{d^{n-1}}{d\omega^{n-1}} (f(\omega))^n \right]_{\omega=0}. \tag{16.84}$$

Using (16.81) we get:

$$P(n) = \frac{1}{n!} \left[\frac{d^{n-1}}{d\omega^{n-1}} (f(\omega))^n \right]_{\omega=0}. \tag{16.85}$$

Theory of branching processes can help to understand scientific citation process. The first-year citations correspond to sons. Second year citations, which are copies of the first year citations, correspond to grandsons, and so on.

Appendix B

Let us consider the case when $\lambda \neq \lambda_0$, i.e., a branching process were the generating function for the first generation is different from the one for subsequent generations. One can show that the generating function for the total offspring is:

$$\tilde{g}(z) = z f_0(f(z)). \tag{16.86}$$

In the case $\lambda = \lambda_0$ we have $f(z) = f_0(z)$ and because of (16.82) $\tilde{g}(z) = g(z)$. We can compute $f_0(g(z))$ by substituting f_0 for Φ in (16.83)

$$f_0(g(z)) = \sum_{n=0}^{\infty} \frac{1}{n!} \frac{d^{n-1}}{dg^{n-1}} (f_0'(g)(f(g))^n) |_{g=0} \, z^n. \tag{16.87}$$

After substituting (16.40) and (16.41) into (16.87) and the result into (16.86) we get

$$\tilde{P}(n) = \frac{\lambda_0((n-1)\lambda + \lambda_0)^{n-2}}{(n-1)!} \tag{16.88}$$

The large n asymptotic of (16.88) is

$$\tilde{P}(n) \propto \frac{\lambda_0}{\lambda}\exp\left(\frac{\lambda_0}{\lambda} - 1 + \lambda - \lambda_0\right) P(n), \tag{16.89}$$

where $P(n)$ is given by (16.43). We see that having different first generation offspring probabilities does not change the functional form of the large-n asymptotic, but merely modifies the numerical pre-factor. After substituting $\alpha \approx 0.1$ and $N_{\mathrm{ref}} \approx 20$ into (16.38) and (16.39) and the result into (16.89) we get $\tilde{P}(n) \approx 2.3P(n)$.

Appendix C

Let us investigate the fitness distribution

$$p_p(\phi) = (\theta + 1)(1 - \phi)^{\theta}. \tag{16.90}$$

After substituting (16.90) into (16.57) we get:

$$p_r(\phi) = \frac{\alpha(\theta - 1)(\theta + 2)\phi(1 - \phi)^{\theta}}{1 - ((1 - \alpha)/\langle\phi\rangle_r)\phi}. \tag{16.91}$$

After substituting this into (16.58) we get:

$$1 = \alpha(\theta + 1)(\theta + 2)\left(\frac{\langle\phi\rangle_r}{1 - \alpha}\right)^2 \int_0^1 \frac{(1 - \phi)^{\theta} dx}{\frac{\langle\phi\rangle_r}{1-\alpha} + \phi} - \alpha(\theta + 2)\frac{\langle\phi\rangle_r}{1 - \alpha}. \tag{16.92}$$

As acceptable values of $\langle\varphi\rangle_r$ are limited to the interval between $1 - \alpha$ and 1, it is clear that when α is small the equality in (16.92) can only be attained when the integral is large. This requires $\langle\phi\rangle_r/(1 - \alpha)$ being close to 1. And this will only help if θ is small. In such case the integral in (16.92) can be approximated as

$$\int_{\frac{\langle\phi\rangle_r}{1-\alpha}}^{1} \frac{\phi^{\theta} dx}{\phi} = \frac{1}{\theta}\left(1 - \left(\frac{\langle\phi\rangle_r}{1 - \alpha} - 1\right)^{\theta}\right).$$

Substituting this into (16.92) and replacing in the rest of it $\langle\phi\rangle_r/(1-\alpha)$ with unity we can solve the resulting equation to get:

$$\frac{\langle\phi\rangle_r}{1-\alpha}-1\approx\left(\frac{\alpha-\frac{\theta}{\theta+2}}{\alpha(\theta+1)}\right)^{\frac{1}{\theta}}. \tag{16.93}$$

For example, when $\alpha=0.1$ and $\theta=0.1$ we get from (16.93) that $\langle\phi\rangle_r/(1-\alpha)-1\approx 6\times10^{-4}$. However (16.93) gives a real solution only when

$$\alpha\geq\frac{\theta}{\theta+2}. \tag{16.94}$$

The R.H.S. of (16.91) has a maximum for all values of φ when $\langle\varphi\rangle_r=1-\alpha$. After substituting this into (16.91) and integrating we get that the maximum possible value of $\int_0^1 p_r(\varphi)d\varphi$ is $\alpha((\theta+2)/\theta)$. We again get a problem when the condition of (16.94) is violated. Remember, however, that when we derived (16.57) from (16.56) we divided by $1-(1-\alpha)\varphi/\langle\varphi\rangle_r$, which, in the case $\langle\varphi\rangle_r=1-\alpha$, is zero for $\varphi=1$. Thus, (16.57) is correct for all values of φ, except for 1. The solution of (16.56) in the case when the condition of (16.94) is violated is:

$$p_r(\phi)=\alpha(\theta+1)(\theta+2)\phi(1-\phi)^{\theta-1}+\left(1-\alpha\frac{\theta+2}{\theta}\right)\delta(\phi-1). \tag{16.95}$$

References

1. Simkin MV, Roychowdhury VP (2003) Read before you cite! Complex Systems **14**: 269–274. Alternatively available at http://arxiv.org/abs/cond-mat/0212043
2. Simkin MV, Roychowdhury VP (2006) An introduction to the theory of citing. Significance 3: 179–181. Alternatively available at http://arxiv.org/abs/math/0701086
3. Simkin MV, Roychowdhury VP (2005) Stochastic modeling of citation slips. Scientometrics 62: 367–384. Alternatively available at http://arxiv.org/abs/cond-mat/0401529
4. Simon HA (1957) Models of Man. New York: Wiley.
5. Krapivsky PL, Redner S (2001) Organization of growing random networks. *Phys. Rev.* E 63, 066123; Alternatively available at http://arxiv.org/abs/cond-mat/0011094
6. Krapivsky PL, Redner S (2002) Finiteness and Fluctuations in Growing Networks. J. Phys. A 35: 9517; Alternatively available at http://arxiv.org/abs/cond-mat/0207107
7. Press WH, Flannery BP, Teukolsky SA, Vetterling WT (1992) Numerical Recipes in FORTRAN: The Art of Scientific Computing. Cambridge: University Press (see Chapt. 14.3, p.617–620).
8. Simboli B (2003) http://listserv.nd.edu/cgi-bin/wa?A2=ind0305&L=pamnet&P=R2083. Accessed on 7 Sep 2011
9. Smith A (1983) Erroneous error correction. New Library World 84: 198.
10. Garfield E (1990) Journal editors awaken to the impact of citation errors. How we control them at ISI. Essays of Information Scientist 13:367.
11. SPIRES (http://www.slac.stanford.edu/spires/) data, compiled by H. Galic, and made available by S. Redner: http://physics.bu.edu/∼redner/projects/citation. Accessed on 7 Sep 2011

12. Steel CM (1996) Read before you cite. The Lancet 348: 144.
13. Broadus RN (1983) An investigation of the validity of bibliographic citations. Journal of the American Society for Information Science 34: 132.
14. Moed HF, Vriens M (1989) Possible inaccuracies occurring in citation analysis. Journal of Information Science 15:95.
15. Hoerman HL, Nowicke CE (1995) Secondary and tertiary citing: A study of referencing behaviour in the literature of citation analyses deriving from the Ortega Hypothesis of Cole and Cole. Library Quarterly 65: 415.
16. Kåhre J (2002) The Mathematical Theory of Information. Boston: Kluwer.
17. Deming WE (1986) Out of the crisis. Cambridge: MIT Press.
18. Garfield E (1979) Citation Indexing. New York: John Wiley.
19. Merton RK (1968) The Matthew Effect in Science. Science 159: 56.
20. Price D de S (1976) A general theory of bibliometric and other cumulative advantage process. Journal of American Society for Information Science 27: 292.
21. Barabasi A-L, Albert R (1999) Emergence of scaling in random networks. Science 286: 509.
22. Simkin MV, Roychowdhury VP (2005) Copied citations create renowned papers? Annals of Improbable Research 11:24–27. Alternatively available at http://arxiv.org/abs/cond-mat/0305150
23. Dorogovtsev SN, Mendes JFF (2004) Accelerated growth of networks. http://arxiv.org/abs/cond-mat/0204102 (see Chap. 0.6.3)
24. Price D de S (1965) Networks of Scientific Papers. Science 149: 510.
25. Silagadze ZK (1997) Citations and Zipf-Mandelbrot law. Complex Systems 11: 487
26. Redner S (1998) How popular is your paper? An empirical study of citation distribution. Eur. Phys. J. B 4: 131.
27. Vazquez A (2001) Disordered networks generated by recursive searches. Europhys. Lett. 54: 430.
28. Ziman JM (1969) Information, communication, knowledge. Nature, 324: 318.
29. Günter R, Levitin L, Schapiro B, Wagner P (1996) Zipf's law and the effect of ranking on probability distributions. International Journal of Theoretical Physics 35: 395
30. Nakamoto H (1988) Synchronous and diachronous citation distributions. In. Egghe L and Rousseau R (eds) Informetrics 87/88. Amsterdam: Elsevier.
31. Glänzel W., Schoepflin U. (1994). A stochastic model for the ageing of scientific literature. Scientometrics 30: 49–64.
32. Pollmann T. (2000). Forgetting and the aging of scientific publication. Scientometrics 47: 43.
33. Simkin M. V., Roychowdhury V. P. (2007) A mathematical theory of citing. Journal of the American Society for Information Science and Technology 58:1661–1673.
34. Harris T.E. (1963). The theory of branching processes. Berlin: Springer.
35. Bentley R. A., Hahn, M.W., Shennan S.J. (2004). Random drift and culture change. Proceedings of the Royal Society B: Biological Sciences 271: 1443 – 1450.
36. Redner S. (2004). Citation Statistics From More Than a Century of Physical Review. http://arxiv.org/abs/physics/0407137
37. Wright S (1931) Evolution in Mendelian populations. Genetics 16: 97–159.
38. Simkin M. V., Roychowdhury V. P. (2010) An explanation of the distribution of inter-seizure intervals. EPL 91: 58005
39. Bak P, Tang C, Wiesenfeld K (1988) Self-organized criticality. Phys. Rev. A 38: 364–374.
40. Simkin M. V., Roychowdhury V. P. (2008) A theory of web traffic. EPL 62: 28006. Accessed on 7 Sep 2011
41. Some Statistics about the MR Database http://www.ams.org/publications/60ann/FactsandFigures.html
42. Burrell Q L (2003) Predicting future citation behavior. Journal of the American Society for Information Science and Technology 54: 372–378.
43. Garfield E (1980) Premature discovery or delayed recognition -Why? Current Contents 21: 5–10.
44. Raan AFJ van (2004) Sleeping Beauties in science. Scientometrics 59: 467–472

45. Alstrøm P (1988). Mean-field exponents for self-organized critical phenomena. Phys. Rev. A 38: 4905–4906.
46. Bak P (1999). How Nature Works the Science of Self-Organized Criticality. New York: Copernicus.
47. Bak P, Sneppen, K (1993). Punctuated equilibrium and criticality in a simple model of evolution. Physical Review Letters 71: 4083–4086
48. Sokal A, Bricmont J (1998) Fashionable Nonsense. New York: Picador.
49. Hahn MW, Bentley RA (2003) Drift as a mechanism for cultural change: an example from baby names. Proc. R. Soc. Lond. B (Suppl.), Biology Letters DOI 10.1098/rsbl.2003.0045.
50. Social Security Administration: Popular Baby Names http://www.ssa.gov/OACT/babynames/. Accessed on 7 Sep 2011
51. Simkin MV (2007) My Statistician Could Have Painted That! A Statistical Inquiry into Modern Art. Significance 14:93–96. Also available at http://arxiv.org/abs/physics/0703091
52. Naftulin DH, Ware JE, Donnelly FA (1973) The Doctor Fox Lecture: A Paradigm of Educational Seduction. Journal of Medical Education 48: 630–635.
53. Encyclopaedia of Mathematics (Ed. M. Hazewinkel). See: Bürmann–Lagrange series: http://eom.springer.de/b/b017790.htm. Accessed on 7 Sep 2011
54. Otter R (1949) The multiplicative process. The Annals of Mathematical Statistics 20: 206

Chapter 17
TTLed Random Walks for Collaborative Monitoring in Mobile and Social Networks

Yaniv Altshuler, Shlomi Dolev, and Yuval Elovici

Abstract Complex network and complex systems research has been proven to have great implications in practice in many scopes including Social Networks, Biology, Disease Propagation, and Information Security. One can use complex network theory to optimize resource locations and optimize actions. Randomly constructed graphs and probabilistic arguments lead to important conclusions with a possible great social and financial influence. Security in online social networks has recently become a major issue for network designers and operators. Being "open" in their nature and offering users the ability to compose and share information, such networks may involuntarily be used as an infection platform by viruses and other kinds of malicious software. This is specifically true for mobile social networks, that allow their users to download millions of applications created by various individual programers, some of which may be malicious or flawed. In order to detect that an application is malicious, monitoring its operation in a real environment for a significant period of time is often required. As the computation and power resources of mobile devices are very limited, a single device can monitor only

Y. Altshuler (✉)
Department of Information Systems Engineering and Deutsche Telekom Laboratories, Ben Gurion University of the Negev, POB 653, Beer Sheva 84105, Israel

MIT Media Laboratory, 77 Mass. Ave., E15 Cambridge, MA 02139, USA
e-mail: yanival@mit.edu

S. Dolev
Computer Science Department, Ben Gurion University of the Negev, POB 653, Beer Sheva 84105, Israel
e-mail: dolev@bgu.ac.il

Y. Elovici
Department of Information Systems Engineering and Deutsche Telekom Laboratories, Ben Gurion University of the Negev, POB 653, Beer Sheva 84105, Israel
e-mail: elovici@bgu.ac.il

M.T. Thai and P.M. Pardalos (eds.), *Handbook of Optimization in Complex Networks: Theory and Applications*, Springer Optimization and Its Applications 57, DOI 10.1007/978-1-4614-0754-6_17, a limited number of potentially malicious applications locally. In this work, we propose an efficient collaborative monitoring scheme that harnesses the collective resources of many mobile devices, generating a "vaccination"-like effect in the network. We suggest a new local information flooding algorithm called *Time-to-Live Probabilistic Propagation* (TPP). The algorithm is implemented in any mobile device, periodically monitors one or more applications and reports its conclusions to a small number of other mobile devices, who then propagate this information onward, whereas each message has a predefined "Time-to-Live" (TTL) counter. The algorithm is analyzed, and is shown to outperform the existing state of the art information propagation algorithms, in terms of convergence time as well as network overhead. We then show both analytically and experimentally that implementing the proposed algorithm significantly reduces the number of infected mobile devices. Finally, we analytically prove that the algorithm is tolerant to the presence of adversarial agents that inject false information into the system.

17.1 Introduction

The market share of Smart-phones is rapidly increasing and is expected to increase even faster with the introduction of 4th generation mobile networks, reaching from 350 million in 2009 to one billion by 2012 [13]. Companies that are distributing new mobile device operating systems had created a variety of marketplaces that motivate individuals and other companies to introduce new applications (such as Apple's *App Store* Google's *Android Market*, Nokia's *Ovi Store*, and others). The main assumption behind these marketplaces is that users will prefer a mobile device based on an operating system with larger marketplace offerings. It is expected that in the future, various communities will develop additional alternative marketplaces that will not be regulated. These marketplaces will allow mobile users to download from a variety of millions of new applications. An example for such a marketplace is *GetJar*, offering 60,000 downloadable applications for over 2,000 mobile devices, counting a total of over 900 million downloads by August 2010 [27]. The content of most marketplaces is currently not verified by their operators, and thus some applications may be malicious or contain faulty code segments. Downloading a malicious application from the marketplace is not the only way that a mobile device may be infected by a malicious code. This may also happen as a result of a malicious code that manages to exploit a vulnerability in the operating systems and applications or through one of the mobile phone communication channels such as Bluetooth, Wi-Fi, etc. [19,30,32,37,58]. The number of currently active malware variants is estimated to be above 370 [31], most of which are introduced as a result of an infected application installation. McAfee's Mobile Security Report for 2008 states that nearly 14% of global mobile users were directly infected or have known someone who had been infected by a mobile virus (this number had increased in the followed year) [1, 2]. In many cases, in order to detect that an application is malicious, monitoring its operation in a real environment for a significant period of time is required. The data that results from this monitoring is being processed using advanced machine learning algorithms in order to assess the maliciousness of the application. For a variety of local monitoring techniques for mobile phone applications, see [33, 36, 45–47, 49].

In recent years, most of the prominent security and privacy threats for communication networks have relied on the use of collaborative methods (e.g., Botnets). The danger stemming from such threats is expected to significantly increase in the near future, as argued in [37, 65] and others. The amount of resources a single unit may allocate in order to defend from such threats without interfering with its routine work is very limited. Therefore, the development of efficient "collaborative defense infrastructures" is strongly required.

In this work, we propose such a collaborative application monitoring infrastructure, providing high efficiency, scalability and fault tolerance for known adversarial and Byzantine attacks.

The efficiency of the proposed algorithm stems from the generation of an implicit collaboration between a group of random walking agents who are released from different sources in the network (and at different times). This technique is new, as most related works discussing random walkers did not take into consideration that agents may be released collaboratively from different sources (e.g., [20, 21]). Hence, the analysis of such systems was limited to the probabilistic analysis of the movements of the agents.

We present a simple collaborative monitoring algorithm, called *TPP – Time-to-Live Probability Propagation*. The algorithm uses a "Time-to-Live" counter for each message, defining the number of time a message may be forwarded before it is deleted, that is logarithmic in the number of the network's devices, n. A successful completion of the mission using these short living messages is then proved (Theorem 17.2). In fact, the upper bounds for the algorithm's completion time and overall number of messages are $O(\log n)$ and $O(n \log n)$, respectively (Corollaries 17.1 and 17.2). The benefit factor of participating in the proposed collaborative scheme is shown to monotonically increase with the size of the network, n. Specifically, we show that by sending $O(\ln n)$ messages, the number of applications a device has to monitor locally is reduced by a factor of $O(\ln n)$ (see Corollary 17.2).

In addition, the algorithm is shown to be partially fault tolerant to the presence of Byzantine devices, that are capable of distributing messages concerning benign applications.[1] (see Theorem 17.5 in Sect. 17.6 discussing devices who distribute false messages).

Furthermore, we show that in addition to providing a mechanism for defending against adversarial abuse, the efforts the algorithm requires for overcoming such attacks grow asymptotically slower than the efforts required in order to increase the strength of such attacks, making the algorithm highly scalable (see more details in Sect. 17.6 and specifically an illustration that appears in Fig. 17.8).

---

[1] We assume that interference in the messages' content, or generation of messages using false identity are impossible, due to, say, the use of cryptographic means.

The rest of this work is organized as follows: A related work section and a comparison to state-of-the-art techniques is presented in Sect. 17.2, in which it is shown that in many cases the *TPP* algorithm achieves a completion time which is equal to the completion time of the fastest single-source information propagation algorithm (i.e., flooding), and does so using significantly lower number of messages. The formal definition of the problem appears in Sect. 17.3, and the *TPP* collaborative algorithm and its performance analysis are discussed in Sect. 17.4.2. Experimental results are presented in Sect. 17.5, whereas the algorithm's robustness with regards to attacks by adversaries is discussed in Sect. 17.6. Conclusions and suggestions for future research appear in Sect. 17.7.

17.2 Related Work

Flooding a network with messages intended for a large number of nodes is arguably the simplest form of information dissemination in communication networks (specifically when previous knowledge about the network topology is limited or unavailable). Since the problem of finding the minimum energy transmission scheme for broadcasting a set of messages in a given network is known to be NP-Complete [12], flooding optimization often relies on approximation algorithms. For example, in [29, 57] messages are forwarded according to a set of predefined probabilistic rules, whereas [53] discusses a deterministic algorithm, approximating the connected dominating set of each node. Alternative information propagation techniques simulate various epidemic models [25, 55, 56, 63].

In this work, we apply a different approach – instead of a probabilistic forwarding of messages, we assign a TTL value for each message, in order to guide the flooding process. The analysis of the algorithm is done by modeling the messages as agents practicing *random walk* in a *random graph* overlay of the network. The optimal value of this TTL is shown, guaranteeing a fast completion of the task, while minimizing the overall number of messages sent. The use of a TTL dependent algorithm was discussed in [3] (demonstrating $O(\log^2 n)$ completion time) and [44] (where no analytic result concerning the completion time was demonstrated).

In addition, the algorithm's efficiency is further improved by selecting the TTL value in a way which focuses the collaborative efforts toward threats of high penetration probabilities. This approach is inspired in part by the observation made in [65] regarding a critical phase transition point of threats' penetration rates, defining the epidemic potential of the threat.

17.2.1 Flooding Algorithms

The simplest information propagation technique is of course the *flooding* algorithm. It is well known that the basic flooding algorithm, assuming a single source of information, guarantees completion in a worse case cost of $O(n^2)$ messages ($O(|E|)$

in expanders, and $O(n \times \log(n))$ in random graphs $G(n,p)$), and time equals to the graph's diameter, which in the case of a random graph $G(n,p)$ equals $O(\frac{\log n}{\log(n \times p)}) \approx O(\log n)$ [16, 22, 34]. We later show that the *TPP* algorithm discussed in this work achieves in many cases the same completion time, with a lower cost of $O(n \log n)$ messages (Corollaries 17.1 and 17.2).

There are various methods used to improve the efficiency of the basic flooding algorithm. One example is a probabilistic forwarding of the messages in order to reduce the overall cost of the propagation process, optionally combined with a mechanism for recognizing and reducing repetitions of packets transitions by the same device [50]. Other methods may include area based methods [50] or neighborhood knowledge methods [42, 51, 59, 60]. In many of these works, completion time is traded for a reduced overall cost, which results in a similar cost as the *TPP* algorithmproposed in this work (namely, $O(n \log n)$), but with a significantly higher completion time. Additional details on various flooding algorithms can be found in [66].

An extremely efficient flooding algorithm, in terms of completion time, is the network coded flooding algorithm, discussed in [17]. In this work, dedicated to $G(n,p)$ random graphs, a message is forwarded by any receiving vertex $\frac{k}{d(v)}$ times, where k is a parameter which depends on the network's topology [24]. Using this method, the algorithm achieves a completion time of approximately $O(\frac{n^3}{|E|^2})$. This algorithm, however,is still outperformed by the *TPP* algorithm. Specifically, the *TPP* algorithm would perform faster in graphs with average degree of less than $O\left(\sqrt{\frac{n}{\ln n}}\right)$.

17.2.2 Epidemic Algorithms

An alternative approach to be mentioned in this scope is the use of *epidemic algorithms* [9, 55, 63]. There exist a variety of epidemic algorithms, starting with the basic epidemic protocol [18, 28], through *neighborhood epidemics* [25] and up to *hierarchical epidemics* [56]. In general, all the various epidemic variants have a trade-off between number of messages sent, completion time, and previous knowledge required for the protocols (concerning the structure of the network). However, the most efficient results of such approaches are still outperformed by the *TPP* algorithm.

17.2.3 Distributed Coverage Algorithms

A different approach for a collaborative assimilation of an important piece of information throughout the network is the use of cooperative exploration algorithms (in either known or unknown environments), guaranteeing that all (or a large enough portion) of the graph is being "explored" by agents carrying the alerting messages. Planar networks can be sampled into $\mathbf{R}^2$, and then be collaboratively explored by a decentralized group of myopic agents (see additional swarm coverage examples in [5, 11, 38, 40, 52, 54, 61, 67]).

In [62], a swarm of ant-like robots is used for repeatedly covering an unknown area, using a real time search method called *node counting*. Using this method, the agents are analytically shown to cover the network efficiently. Another algorithm to be mentioned in this scope is the *LRTA** algorithm [41], that was shown in [39] to guarantee coverage time of $O(n^2)$ in degree bounded undirected planar graphs. Interestingly, in such graphs, the random walk algorithm is also known to require at most $O(n^2)$ time (and at least $\Omega(n(\log n)^2)$) [35].

In [64], a collaborative algorithm that guarantees coverage of regions in the $\mathbf{Z}^2$ grid, in $O(n^{1.5})$ time, using extremely simple and myopic "ant like" mobile agents and using no direct communication by the agents is discussed. In [4] this algorithm was later shown to guarantee (under some limitations) coverage of dynamically expanding domains (namely, domains in which explored vertices may become "unexplored" after a certain time of being exposed to an unexplored neighbor) in $O(n^{2.5}\log n)$ time.

17.2.4 Summary

Tables 17.1 and 17.2 compare the performance of the *TPP* algorithm (in terms of convergence time and overall message overhead) to the main leading works in this field. Each table examines a different set of assumptions concerning the network. In each table, the algorithm displaying the best result is marked using the "✓" sign.

As previously discussed, the efficiency of the *TPP* algorithm is derived from the fact that participating devices form a collaborative infrastructure, tuned to focus on threats of high penetration rates. As the *TPP* algorithm uses random elements, it also requires approximately $O(\ln^2 n)$ random bits by each device.

Another interesting approach with respect to decentralized information proliferation is the *population problem*, discussed for example in [6], in which a consensus among a decentralized group of n units is generated in $O(n\log n)$ time, with tolerance to the presence of $O(\sqrt{n})$ Byzantine agents. Additional information on population protocols can be found in [8].

17.3 The Collaborative Application Monitoring Problem

Given a mobile network of n devices, let us denote the network's devices by $V = \{v_1, v_2, \ldots, v_n\}$. Note that the network's topology may be dynamic.[2] Each device may occasionally visit the marketplace, having access to N new downloadable

[2]This will later come into effect when messages will be sent between the network's members, at which case the selection of "an arbitrary network member" can be assumed to be purely random.

Table 17.1 Performance comparison of *TPP* and available state of the art algorithms

	Time	Messages
TPP	In most cases $O(\ln n)$ and at most $O(\frac{n}{\ln n})$	$O(n \ln n)$
Flooding	O (*Graph's diameter*)	$O(\|E\|)$
Network coded flooding [17] using $G(n,p)$ overlay	$O\left(n^{-1} \cdot p^{-2}\right)$	$O(n)$
Neighborhood Epidemics [25] using $G(n,p)$ overlay	$O(n^c)$ for some constant c	$O(c \cdot n)$ for some constant c
Hierarchical Epidemics [56] using α-tree overlay	$O(\ln n)$	$O(\alpha \times n \ln n)$ for branching factor α
LRTA* [41] in planar degree bounded graphs	$O(n^2)$	$O(n^2)$
SWEEP [64] in the $\mathbf{Z}^2$ grid	$O(n^{1.5})$	$O(n^{1.5})$

Table 17.2 Performance comparison for random $G(n,p)$ graphs, with $p < O((n \ln n)^{-0.5})$

	Time	Messages
TPP	In most cases $O(\ln n)$ ✓ and at most $O(\frac{n}{\ln n})$	$O(n \ln n)$
Flooding	$O(\ln n)$ ✓	$O(n^2 p)$
Network coded flooding	$O\left(n^{-1} \cdot p^{-2}\right)$	$O(n)$ ✓
Neighborhood epidemics	$O(n^c)$ for some constant c	$O(c \cdot n)$ for some constant c ✓
Hierarchical epidemics using α-tree overlay	$O(\ln n)$ ✓	$O(\alpha \cdot n \ln n)$ for branching factor α

The "✓" sign marks the algorithm with the best performance

applications every month. We assume that downloading of applications is done independently, namely – that the probability that a user downloads application a_1 and the probability that the same user downloads application a_2 are uncorrelated.

For some malicious application a_i, let p_{a_i} denote the application's *penetration probability* – the probability that given some arbitrary device v, it is unaware of the maliciousness of a_i. The penetration probability of every new malicious application is 1. Our goal is to verify that at the end of the month, the penetration probability of all malicious applications released during this month are lower than a *penetration threshold* p_{MAX}, resulting in a "vaccination" of the network with regards to these applications. Formally, for some small ε we require that

$$\forall \text{Malicious application } a_i \quad \text{Prob}\left(p_{a_i} > p_{\text{MAX}}\right) < \varepsilon$$

The reason behind the use of the threshold p_{MAX} is increasing the efficiency of the collaborative system, defending against dominant threats. Given additional available resources, the parameter p_{MAX} can be decreased, resulting in a tighter defense grid (traded for increased convergence time and messages overhead).

We assume that any device v can send a message of some short content to any other device u. In addition, we assume that at the initialization phase of the algorithm each device is given a list containing the addresses of some X random network members. This can be implemented either by the network operator, or by distributively constructing and maintaining a random network overlay.

We assume that each device can locally monitor applications that are installed on it (as discussed for example in [7, 46]). However, this process is assumed to be expensive (in terms of the device's battery), and should therefore be executed as few times as possible. The result of an application monitoring process is a non-deterministic boolean value : $\{true, false\}$.

False-positive and false-negative error rates are denoted as

$$\begin{aligned} P(\text{Monitoring}(a_i) &= \text{ true } \mid A_i\text{is not malicious}) = E_+ \\ P(\text{Monitoring}(a_i) &= \text{ false } \mid \quad A_i\text{is malicious}) \quad = E_- \end{aligned}$$

We assume that the monitoring algorithm is calibrated in such way that $E_+ \approx 0$.

As we rely on the propagation of information concerning the maliciousness of applications, our system might be abused by injection of inaccurate data. This may be the result of a deliberate attack, aimed for "framing" a benign application (either as a direct attack against a competitive application, or as a more general attempt for undermining the system's reliability), or simply as a consequence of a false-positive result of the monitoring function. Therefore, in order for a device v to classify an application a_i as malicious, one of the following must hold:

- Device v had monitored a_i and found it to be malicious.
- Device v had received at least ρ alerts concerning a_i from different sources (for some *decision threshold* ρ).

In addition, note that the information passed between the devices concerns only applications discovered to be malicious. Namely, when an application is found to be benign, no message concerning this is generated. This is important not only for preserving a low message overhead of the algorithm but also to prevent malicious applications from displaying a normative behavior for a given period of time, after which they initiate their malicious properties. In such cases, soon after an application exits its "dormant" stage, it will be detected and subsequently reported, generating a "vaccination reaction" throughout the network.

17.4 Algorithm, Correctness, and Analysis

We shall now present the *TPP* algorithm. The algorithm is executed by each device separately and asynchronously, where no supervised or hierarchical allocation of tasks, as well as no shared memory are required. Table 17.3 presents the main notations used in the presentation and analysis of the proposed algorithm.

Table 17.3 Main notations used throughout this work

n	The number of devices participating in the *TPP* algorithm
X	The number of alert messages generated and sent upon the discovery of a malicious application
p_N	The ratio $\frac{X}{n}$
p_{MAX}	Penetration threshold – the maximal percentage of non-vaccinated devices allowed by the network operator
E_-	False negative error rate of the local monitoring mechanism
T	Delay time between each two consecutive local monitoring processes
N	Number of new applications introduced to the network each month
p_{a_i}	The penetration probability of application a_i
$1-\varepsilon$	Confidence level of the correctness of the convergence time estimation
α	The polynomial confidence level $\ln_n \varepsilon^{-1}$
ζ_{T,N,P_M,E_-}	The *vaccination factor*, defined as $\frac{T \times N}{p_{\text{MAX}}(1-E_-)}$
timeout	The Time-To-Live counter assigned to alert messages
ρ	Decision threshold – the number of alerts a device must receive in order to classify an application as malicious
C_S	Cost of sending a single message
C_M	Cost of locally monitoring a single application

17.4.1 Overview of the Results

The *TPP* algorithm is analytically shown to guarantee a successful completion of the monitoring mission (Theorem 17.2). The performance of the algorithm is then analyzed, and is shown to be superior compared to existing results in this domain (in terms of completion time and messages overhead). Upper bounds for the algorithm's completion time for the overall number of messages required are presented in Observation 1. Approximation of these bounds are given in Theorems 17.3 and 17.4. More explicit approximations of the bounds for *sparsely connected networks* (see Definition 17.1) are presented in Corollaries 17.1 and 17.2.

17.4.2 **TPP***: A Collaborative Monitoring Algorithm*

The *TPP* algorithm conceptually relies on the fact that in order to "vaccinate" a network with regards to malicious applications, it is enough that only a small number of devices will monitor this application. This way, although the application monitoring process is relatively expensive (in terms of battery and CPU resources), the amortized cost of monitoring each malicious application is kept to a minimum. A detailed implementation of the *TPP* algorithm appears in Algorithm 11.

At its initialization (lines 1 through 6), all the applications installed on the device are added to a list of *suspected applications*. In addition, an empty list of *known malicious applications* is created. Once an application is determined as *malicious*, it is added to the known malicious application list. In case this application was

Algorithm 11 *TTL Probabilistic Propagation*

```
1: Initialization :
2:    Let A(v) be the list of installed applications
3:    Let Ȧ(v) be the list of suspected applications
4:    Let Ã(v) be the list containing known malicious applications
5:    Initialize Ȧ(v) ← A(v)
6:    Initialize Ã(v) ← ∅

7: Interrupt upon encountering a new application a_i :
8:    Ȧ(v) ← Ȧ(v) ∪ {a_i}
9:    If a_i ∈ Ã(v) then
10:      Ȧ(v) ← Ȧ(v) \ {a_i}
11:      Issue an alert to the user concerning a_i
12:   End if

13: Interrupt receives malicious application a_i notice, for the j-th time :
14:   If j ≥ ρ then
15:      Ȧ(v) ← Ȧ(v) \ {a_i}
16:      Ã(v) ← Ã(v) ∪ {a_i}
17:      If a_i ∈ A(v) then
18:         Issue an alert to the user concerning a_i
19:      End if
20:   End if
21:   Decrease TTL of report by 1
22:   Forward report to a random network member

23: Execute every T time-steps :
24:   Select a random application a_i from Ȧ(v) for monitoring
25:   If a_i is found to be malicious then
26:      Issue an alert to the user concerning a_i
27:      Ã(v) ← Ã(v) ∪ {a_i}
28:      Ȧ(v) ← Ȧ(v) \ {a_i}
29:      Report a_i to X random network members
30:      Set TTL = timeout
31:   End if
```

also in the suspected application list (namely, it is installed on the device, but has not been monitored yet), it is deleted from that list. Once a new application is encountered it is compared to the known malicious application list, and if found, an alert is sent to the user (alternatively, the application can be chosen to be uninstalled automatically). This feature resembles the long-term memory of the immune system in living organisms. If the new application is not yet known to be malicious, the application is added to the suspected application list.

Once executed, a periodic selection of an arbitrary application from the list of suspected applications is done, once every T steps (lines 23 through 31). The selected application is then monitored for a given period of time, in order to discover whether it is of malicious properties (see details about such monitoring in Sect. 17.3). If the application is found to be malicious, it is removed from the list of suspected applications and added to the known malicious application list (lines 28 and 27). In addition, an appropriate alert is produced and later sent to X random devices. The alert message is also assigned a specific TTL value (lines 29 and 30). Once a network device receives such an alert message it automatically forward it to one arbitrary device, while decreasing the value of TTL by 1. Once TTL reaches zero, the forwarding process of this message stops (lines 21 and 22). In case a monitored application displays no malicious properties, it is still kept in the list of suspicious applications, for future arbitrary inspections.

A device may also classify an application as malicious as a result of receiving an alert message concerning this application (lines 14 through 20). In order to protect benign applications from being "framed" (reported as being malicious by adversaries abusing the vaccination system), a device classifies an application as malicious only after it receives at least ρ messages concerning it, from different sources (for a pre-defined *decision threshold* ρ). Note that when a device v receives an alert message concerning application a_i, it still forward this message (assuming that TTL > 0), even when v has not yet classified a_i as malicious (for example, when the number of alert messages received is still smaller than ρ). When the ρ-th alerting message concerning an application is received, the device adds the application to its known malicious application list and removes it from the suspected application list if needed. The selection of the optimal value of ρ is discussed in Sect. 17.6. The values of T, ρ, and TTL, as well as the number of generated alert messages can be determined by the network operators, or be changed from time to time according to the (known or estimated) values of n and N, and the penetration threshold p_{MAX}. Selecting an optimal value for TTL is discussed in Sect. 17.4.

17.4.3 Optimal Parameters for Guaranteeing Successful Monitoring

17.4.3.1 Outline of Analysis

In order to analyze the algorithm's behavior, we shall model the movements of the notification messages between the network's devices as random walking agents, traveling in a random graph $G(n,p)$. Taking into account the fact that the messages have limited lifespan (namely, TTL), a relation between the size of the graph and the lifespan of the agents is produced. Having the value of TTL that guarantees a coverage of the graph, the algorithm's completion time, as well as the overall number of messages sent, can then be calculated.

While analyzing the performance of the *TPP* algorithm we imagine a directed *Erdös-Renyi* random graph $G(V,E) \sim G(n, p_N)$, where $p_N = \frac{X}{n}$. The graph's vertices V denote the network's devices, and the graph's edges E represent message forwarding connections. Notice that as G is a random graph, it can be used for the analysis of the performance of the *TPP* algorithm, although the message forwarding connections of the *TPP* algorithm are dynamic. In addition, although the identity of the "neighbors" of a vertex v in the real network overlay may change from time to time (as the overlay graph can be dynamic), it can still be modeled by static selection of X random neighbors of v.

Observing some malicious application a_i, every once in a while some device which a_i is installed on randomly selects it for monitoring. With probability $(1-E_-)$ the device discovers that a_i is malicious and issues alerts to X network's members. We look at these reports as the system's "agents," and are interested in finding:

- The time it takes the graph to be explored by the agents. Namely, the time after which every device was visited by at least ρ agents (and is now immune to a_i).
- The total number of messages sent during this process.
- The minimal TTL which guarantees a successful vaccination of the network.

Note also that agents have a limited lifespan, equals to timeout. As the graph is a random graph, the location of the devices in which a_i is installed is also random. Therefore, as they are the sources of the agents, we can assume that the initial locations of the agents are uniformly and randomly distributed along the vertices of V. In compliance with the instruction of the *TPP* algorithm, the movement of the agents is done according to the random walk algorithm.

Application a_i is installed on $n \times p_{a_i}$ devices, each of which monitors a new application every T time steps. Upon selecting a new application to monitor, the probability that such a device will select a_i is $\frac{1}{N}$. The probability that upon monitoring a_i the device will find it to be malicious is $(1-E_-)$, in which case it will generate $n \times p_N$ alerting messages. The expected number of new agents created at time t, denoted as $\hat{k}(t)$, therefore, equals

$$\hat{k}(t) = \frac{n^2 \times p_{a_i} \times p_N}{T \times N}(1-E_-)$$

and the accumulated number of agents which have been generated in a period of t time-steps is therefore $k_t = \sum_{i \le t} \hat{k}(i)$.

The value of timeout (the assigned TTL) is selected in such a way that the complete coverage of the graph, and therefore its vaccination against a_i, is guaranteed (in probability greater than $1-\varepsilon$). We now artificially divide the mission to two phases, the first containing the generation of agents and the second discussing the coverage of the graph. Note that this division ignores the activity of the agents created in the second phase. Note also that the fact that the agents are working in different times (and in fact, some agents may already vanish while others have not even been generated yet) is immaterial. The purpose of this technique is to ease the analysis of the flow of the vaccination process.

Denoting the completion time by T_{Vac} we therefore have

$$T_{\text{Vac}} \le T_{\text{Generation}} + T_{\text{Propagation}}$$

It is easy to see that $T_{\text{Propagation}} \triangleq \mathit{timeout}$. We now artificially set

$$\begin{cases} \mathit{timeout} = \lambda \times (T_{\text{Generation}} + \text{timeout}) \\ T_{\text{Generation}} = (1-\lambda) \times (T_{\text{Generation}} + \text{timeout}) \end{cases}$$

From this we can see that

$$T_{\text{Generation}} = \frac{(1-\lambda)}{\lambda} \times \text{timeout}$$

We later demonstrate an upper bound for timeout, based on the number of agents created in $t \leq T_{\text{Generation}}$ (ignoring the activity of the agents created between $t = T_{\text{Generation}}$ and $t = T_{\text{Generation}} + \text{timeout}$).

We now examine the case of $\lambda = 0.5$ (which we show to be the optimal selection for λ, in the paper's Appendix). In this case, we can now write: $T_{\text{Vac}} \leq 2 \times timeout$.

Let us denote the number of agents created in the first $T_{\text{Generation}}$ time-steps by $k = k_{T_{\text{Generation}}}$. We now find the time it takes those k agents to completely cover the graph G, and from this, derive the value of timeout.

Recalling that upon the detection of a malicious application, devices are required to send notification messages to exactly X random network neighbors. We can still use a random graph $G(n,p)$ for analyzing the algorithm's behavior. Since our bounds are probabilistic, we can state that the following "bad event" occurs with very low probability (e.g., $2^{-\omega(n)}$). Event $E_{\text{low degree}}$, defined as the existence of some vertex $v \in V$ with $deg(v) < \frac{n \times p_N}{2}$. Using the *Chernoff* bound on G we get: $Prob[deg(v) < \frac{n \times p_N}{2}] < e^{-\frac{n \times p_N}{8}}$. Applying union bound on all vertices we get

$$Prob[E_{\text{low degree}}] < n \times e^{-\frac{n \times p_N}{8}} < 2^{-\omega(n)}$$

Similarly,

$$Prob[E_{\text{high degree}}] < 2^{-\omega(n)}$$

From now on we assume that $E_{\text{low degree}}$ and $E_{\text{high degree}}$ do not occur, and condition all probabilities over this assumption. In the private case of $\forall v \in V$, $deg(v) = p_N \times n$, every analysis that is based on the expected number of neighbors shall hold.

In order to continue analyzing the execution process of the *TPP* algorithm we note that as the initial placement of the agents is random, their movement is random and the graph G is random, we can see that the placement of the agents after every-step is purely random over the nodes. Using these observation, the number of agents residing in adjacent vertices from some vertex v can be produced:

Lemma 17.1. *Let $v \in V$ be an arbitrary vertex of G. Let $N_1(v,t)$ be the number of agents which reside on one of $Neighbor(v)$ (adjacent vertices to v) after step t:*

$$\forall t \geq 0 : E[N_1(v,t)] \geq \frac{p_N \times k}{2}$$

In other words, the expected number of agents who reside in distance 1 *from v after every step is at least $\frac{p_N \times k}{2}$.*

Proof. Upon our assumption, in $G(n, p_N)$ the number of incoming neighbors for some vertex v is at least $\frac{1}{2} p_N \times n$. In addition, for every $u \in V(G)$, $Prob[$some agent resides on $u] = \frac{k}{n}$. In addition, for every $u \in V$ such that $(u,v) \in E$ we also know that $deg(u) \leq \frac{3}{2} p_N \times n$. Combining the above together, we get $\forall t \geq 0 : E[N_1(v,t)] \geq \frac{p_N \times n \times k}{2n} \geq \frac{1}{2} p_N \times k$. □

Lemma 17.2. *For any vertex $v \in V$, the probability of v being notified at the next time-step that a_i is malicious is at least $1 - e^{-\frac{k}{2n}}$.*

Proof. The probability that an agent located on a vertex u such that $(u,v) \in E$ will move to v at the next time-step is $\frac{1}{p_N \times n}$. The number of agents that are located in adjacent vertices to v is $\frac{k}{2} p_N$. Therefore, the probability that v will not be reported about a_i at the next time-step is $(1 - \frac{1}{p_N \times n})^{\frac{1}{2} p_N \times k}$. Using the well known inequality $(1-x) < e^{-x}$ for $x < 1$, we can bound this probability from above by:

$$\left(e^{-\frac{1}{p_N \times n}}\right)^{\frac{1}{2} p_N \times k} \leq e^{-\frac{k}{2n}}$$

Therefore, the probability that v will be notified on the next time-step is at least $1 - e^{-\frac{k}{2n}}$. □

Interestingly, this fact holds for any positive p_N (the density parameter of G).

Let us denote by *ρ-coverage* of a graph the process the result of which is that every vertex in the graph was visited by some agent at least ρ times.

Theorem 17.1. *The time it takes k random walkers to complete a ρ-coverage of G in probability greater than $1 - \varepsilon$ (denoted as $T(n)$) can be bounded as follows:*

$$T(n) \leq \frac{2\left(\rho - \ln \frac{\varepsilon}{n}\right)}{1 - e^{-\frac{k}{2n}}}$$

Proof. Lemma 17.2 states the probability that some vertex $v \in V$ will be reported of a_i at the next time-step. This is in fact a Bernoulli trial with:

$$p_{\text{success}} = 1 - e^{-\frac{k}{2n}}$$

Now we bound the probability of failing this trial (not notifying vertex v enough times) after m steps. Let $X_v(m)$ denote the number of times that a notification message had arrived to v after m steps, and let $F_v(m)$ denote the event that v was not notified enough times after m steps (i.e., $X_v(m) < \rho$). We additionally denote by $F(m)$ the event that one of the vertices of G is not notified enough times after m steps (i.e., $\bigcup_{v \in V(G)} F_v(m)$). We use the *Chernoff* bound

$$P[X_v(m) < (1-\delta) p_{\text{success}} m] < e^{-\delta^2 \frac{m p_{\text{success}}}{2}}$$

in which we set $\delta = 1 - \frac{\rho}{mp_{\text{success}}}$. We can then see that

$$P[X_v(m) < \rho] < \epsilon^{-\left(1 - \frac{\rho}{mp_{\text{success}}}\right)^2 \frac{mp_{\text{success}}}{2}}$$

namely: $P[F_v(m)] < e^{\rho - \frac{mp_{\text{success}}}{2}}$. Applying the union bound we get

$$P[e_1 \cup e_2 \cup \ldots \cup e_n] \leq P[e_1] + P[e_2] + \ldots + P[e_n]$$

on all n vertices of G. Therefore we can bound the probability of failure on any vertex v (using Lemma 17.2) as follows:

$$Pr[F(m)] \leq ne^{\rho - \frac{mp_{\text{success}}}{2}} \leq ne^{\rho - \frac{m\left(1 - e^{-\frac{k}{2n}}\right)}{2}} \leq \varepsilon$$

and the rest is implied. □

We now show how to select a value of timeout that guarantees a successful vaccination process:

Theorem 17.2. *For every values of timeout that satisfies the following expression, the* TPP *algorithm is guaranteed to achieve successful vaccination for any penetration threshold* p_{MAX}, *in probability greater than* $1 - \varepsilon$*:*

$$\frac{2\left(\rho - \ln \frac{\varepsilon}{n}\right)}{\text{timeout}\left(1 - \text{e}^{-\text{timeout} \times \frac{\text{n} \times \text{p}_{\text{MAX}} \times \text{p}_{\text{N}}}{2\text{T} \times \text{N}}(1 - \text{E}_-)}\right)} = 1$$

Proof. Recalling the expected number of agents generated at each time step, the expected number of agents k that appears in Theorem 17.1 equals

$$E[k] = \sum_{i \leq T_{\text{Generation}}} \frac{n^2 \times p_{a_i} \times p_N}{T \times N}(1 - E_-)$$

A successful termination of *TPP* means that the penetration probability of (any) malicious application is decreased below the threshold p_{MAX}. Until this is achieved, we can therefore assume that this probability never decreases below p_{MAX}:

$$\forall t < T_{\text{Generation}} : p_{a_i} \geq p_{\text{MAX}}$$

Therefore, we can lower bound the number of agents as follows:

$$k \geq \text{timeout} \times \frac{n^2 \times p_{\text{MAX}} \times p_N}{T \times N}(1 - E_-)$$

Assigning timeout = m into Theorem 17.1, successful vaccination is guaranteed for

$$\text{timeout} = \frac{2(\rho - \ln\frac{\varepsilon}{n})}{1 - e^{-\frac{k}{2n}}} \leq \frac{2\left(\rho - \ln\frac{\varepsilon}{n}\right)}{1 - e^{-\frac{n \times p_{\text{MAX}} \times p_N}{2T \times N \times \text{timeout}^{-1}}(1-E_-)}}$$

and the rest is implied. □

17.4.4 Number of Messages and Time Required for Vaccination

From the value of timeout stated in Theorem 17.2, the vaccination time T_{Vac} as well as the overall cost of the *TPP* algorithm can now be extracted. The cost of the algorithm is measured as a combination of the overall number of messages sent during its execution and the total number of monitoring activities performed. Let us denote the cost of sending a single message by C_S and the cost of executing a single local monitoring process by C_M.

Observation 1. *For any* timeout $= \tau$ *which satisfies Theorem 17.2, the time and cost of the* TPP *algorithm can be expressed as:*

$$T_{\text{Vac}} = O(\tau), \quad M = O\left(k \times \tau \times C_S + \frac{k}{X} C_M\right)$$
$$= O\left(\frac{p_{\text{MAX}} \times p_N}{n^{-2} T \times N} \times (1 - E_-) \times \left(\tau^2 \times C_S + \frac{\tau}{n \times p_N} \times C_M\right)\right)$$

Let us assume that ε is polynomial in $\frac{1}{n}$, namely: $\varepsilon = n^{-\alpha} \quad s.t. \quad \alpha \in \mathbb{Z}^+$. Using the bound $(1 - x) < e^{-x}$ for $x < 1$ we can see that when assuming[3] that:

$$\text{timeout} \times \frac{\text{n} \times \text{p}_{\text{MAX}} \times \text{p}_\text{N}}{2\text{T} \times \text{N}} (1 - \text{E}_-) < 1$$

Theorem 17.2 can be written as:

$$\rho + (\alpha + 1)\ln n \geq \text{timeout}^2 \times \frac{\text{n} \times \text{p}_{\text{MAX}} \times \text{p}_\text{N} \times (1 - \text{E}_-)}{4\text{T} \times \text{N}}$$

[3]The intuition behind this assumption is as follows: we aspire that the number of messages each device is asked to send upon discovering a new malicious application is kept to a minimum. As the value of P_N is required to be greater than $\frac{\ln n}{n}$ in order to guarantee connectivity [23], it is safe to assume that $P_N = O\left(\frac{\ln n}{n}\right)$. Notice that under some assumptions, a connected pseudo-random graph can still be generated, such that $p_N = O(\frac{1}{n})$ (see for example [21]). However, as we are interested in demonstrating the result for any random graph $G(n,p)$, this lower bound of p_N is still mentioned. In addition, we later show that timeout $\approx \text{O}(\log \text{n})$. It is also safe to assume that $N \approx \Omega(\ln n)$ and that $P_{\text{MAX}} \approx O\left(\frac{1}{\ln n}\right)$. This assumption is later discussed in great details.

and therefore:

$$\text{timeout} \leq \sqrt{\frac{4T \times N(\rho + (\alpha+1)\ln n)}{n \times p_{\text{MAX}} \times p_N \times (1 - E_-)}}$$

Assigning this approximation of timeout into the above assumption yields:

$$\sqrt{\frac{2T \times N}{n \times p_{\text{MAX}} \times p_N(1 - E_-)} \times 2(\rho + (\alpha+1)\ln n)} < \frac{2T \times N}{n \times p_{\text{MAX}} \times p_N(1 - E_-)}$$

obtaining the following *sparse connectivity* assumption:

Definition 17.1. Let a network be called *sparsely connected* when:

$$p_N < \frac{T \times N}{n \times p_{\text{MAX}} \times (\rho + (\alpha+1)\ln n)(1 - E_-)}$$

We can now obtain the algorithm's completion time and cost:

Theorem 17.3. *Under the sparse connectivity assumption, the completion time of the* TPP *algorithm is:*

$$T_{\text{Vac}} \leq 4\sqrt{\frac{T \times N(\rho + (\alpha+1)\ln n)}{n \times p_{\text{MAX}} \times p_N \times (1 - E_-)}}$$

Theorem 17.4. *Under the sparse connectivity assumption, the overall cost of the* TPP *algorithm (messages sending and monitoring costs) is:*

$$M \leq k \times timeout \times C_S + \frac{k}{X} \times C_M \leq$$

$$\leq 4n(\rho + (\alpha+1)\ln n)C_S + 2C_M\sqrt{\frac{n(\rho + (\alpha+1)\ln n) \times p_{\text{MAX}} \times (1 - E_-)}{p_N \times T \times N}}$$

Proof. When the vaccination process is completed, no new messages concerning the malicious application are sent. The above is received by assigning the approximated value of timeout into Observation 1. □

Definition 17.2. Let the *vaccination factor* ζ_{T,N,P_M,E_-} be defined as:

$$\zeta_{T,N,P_M,E_-} \triangleq \frac{T \times N}{p_{\text{MAX}}(1 - E_-)}$$

Using the sparse connectivity assumption as an upper bound for p_N and $\frac{\ln n}{n}$ as a lower bound for p_N which guarantees connectivity [16], the following corollaries can now be produced:

Corollary 17.1. *The completion time of the* TPP *algorithm is*

$$T_{\text{Vac}} = O\left(\rho + \ln n + \frac{T \times N}{p_{\text{MAX}} \times (1 - E_{-}) \times \ln n}\right)$$

Proof. Assigning the upper bound for p_N into Theorem 17.3 immediately yields $O(\rho + \ln n)$. When assigning the lower bound of $p_N = \frac{\ln(n)}{n}$ the following expression is received:

$$T_{\text{Vac}} \leq 4\sqrt{\frac{T \times N \times (\rho + (\alpha + 1)\ln n)}{\ln(n) \times p_{\text{MAX}}(1 - E_{-})}}$$

However, using the sparse connectivity we see that

$$\frac{\ln n}{n} < \frac{T \times N}{n \times p_{\text{MAX}} \times (\rho + (\alpha + 1)\ln n)(1 - E_{-})}$$

which in turn implies that

$$\rho + (\alpha + 1)\ln n < \frac{T \times N}{\ln(n) \times p_{\text{MAX}}(1 - E_{-})}$$

Combining the two yields

$$T_{\text{Vac}} = O\left(\frac{T \times N}{\ln n \times p_{\text{MAX}} \times (1 - E_{-})}\right)$$

Note that although $O\left(\frac{T \times N}{p_{\text{MAX}}(1 - E_{-})}\right)$ is allegedly independent of n, by assigning the connectivity lower bound $P_N > \frac{\ln n}{n}$ into the sparse connectivity assumption, we can see nevertheless that

$$\zeta_{T,N,P_M,E_{-}} = \frac{T \times N}{p_{\text{MAX}}(1 - E_{-})} = \Omega(\rho \ln n + \ln^2 n)$$

□

For similar arguments, the vaccination's cost can be approximated as:

Corollary 17.2. *The overall cost of the* TPP *algorithm (messages sending and monitoring costs) is:*

$$\begin{aligned} M &= O\left(k \times timeout \times C_S + \frac{k}{X} \times C_M\right) \\ &= O\left((n\rho + n\ln n)C_S + \left(\frac{n}{\ln n} + n(\rho + \ln n)\zeta^{-1}_{T,N,P_M,E_{-}}\right)C_M\right) \end{aligned}$$

In networks of $E_{-} < 1 - o(1)$, provided that[4] $\rho = O(\ln n)$, and remembering that in this case $\zeta_{T,N,P_M,E_{-}} = \Omega(\ln^2 n)$ we can see that the dominant components of Corollary 17.2 become:

$$M = O\left(n \ln n C_S + \frac{n}{\ln n} C_M\right)$$

17.4.5 Messages Propagation Among Network Members

In order for a vaccination algorithm which relies on the propagation of valuable information concerning possible threats in the networks to be implemented, the network's devices must have a way of passing relevant information between them. Specifically, the proposed algorithm requires that any network member can perform the following two actions:

- Send a notification message concerning a malicious application to X random network members.
- Forward a received notification message to a single random network member.

Those two commands can be generalized to the requirement that any network member must be able to send upon request a message of some content to a list of (up to X) random network members. The trivial implementation of this feature of the algorithm would use a central server, which will be in charge of storing the network's devices who have registered into the vaccination service. Upon request, the server would be able to provide a list (of any required length) of random network's members. This could either be done one time during the registration of a new device to the service, or be refreshed on a periodic basic.

However, as we would like to decrease the dependency of the system on a centralized component to the minimum, we would prefer this service to be implemented in a decentralized fashion. The implementation of this aspect of the algorithm depends on the specific model of the network. For example, suppose the network is a mobile network, in which each member has a unique address which is its 7-digit phone number, and in which all of the network's node take part in the vaccination efforts. In this example, upon the request to forward a message to a random network member, a device can simply select a random 7-digit number as the message's destination. In case this random number is inactive, the sender will receive an appropriate notification from the network, and will select another number, until successfully guessing an active number.

When a previous knowledge of the addresses of the members of the vaccination service is unavailable, a virtual network overlay in the form of a random graph $G(n, p)$ (of neighbors probability $p = \frac{X}{n}$) can gradually be generated, to be used for messages forwarding (as new devices are joining the service, they are given a

[4]See Sect. 17.6 for more details.

list of X random network's members). One example for such an implementation is the use of "random scouts," to be sent by any device joining the service, and returning with a list of X random network members. In addition, each device will periodically execute an additional service, designed for estimating the current number of members in the service. When the value of this number is significantly higher than its number in the previous scouting process, the device will refresh its neighbors list by re-running the scouting algorithm. This can be implemented for example using an approach similar to the one described in [21], which enables the generation of an expander of requested properties from an unknown network, using only $O(n \log n)$ short ($\log^4 n$ length) random walks. Another implementation can be based on the approach of [20], that uses a *probabilistic group communication service*. Yet another possible implementation can be found at [15].

Note that one may suggest that entire vaccination mechanism can be implemented by a centralized server, receiving all the alerts that are generated by the network's units, and upon the detection of a malware, forward this information to all the members of the network. It is important to note that we would like to refrain from implementing the entire system in one central server for three main reasons. First, such an implementation would make the server a single point of failure, that when compromised, great amount of damage can be caused. Second, implementing the system in the form of a single server that collects all the alerts that are produced by the units significantly increases the chance of a "framing attack" (causing a benign application to deliberately be reported as a malicious one – see Sect. 17.6 for more details), creating total havoc in the system, and subsequently resulting in the abandonment of the system as a reliable solution for defending against network attacks. This problem is prevented in the proposed system architecture as the ability of a group of collaborating adversaries to cause damage is highly limited. Third, the purpose of this chapter is to propose a vaccination algorithm that can be implemented apart from in mobile networks also in other networks, including social based services, where the implementation of a centralized server may be hard, or even impossible.

17.5 Experimental Results

We have implemented the *TPP* algorithm in a simulated environment and tested its performance in various scenarios. Due to space considerations, we hereby describe one example, concerning a network of $n = 1{,}000$ devices, having access to $N = 100$ applications, one of which was malicious.[5] We assume that each device downloads 30 random applications, monitors one application every week, and can

[5]Note that the number of malicious applications does not influence the completion time of algorithm, as monitoring and notification is done in parallel. The number of message, however, grows linearly with the number of malicious applications.

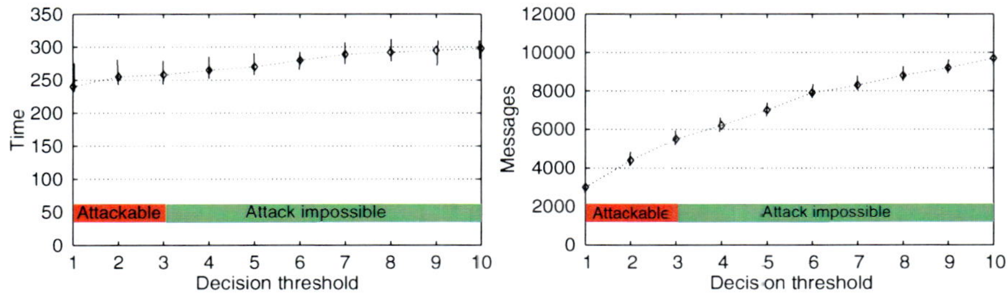

Fig. 17.1 An experimental result of a network of $n = 1{,}000$ members, with $N = 100$ applications, penetration threshold of $p_{\text{MAX}} = 0.01$, $p_N = 0.01$ and 100 adversaries that try to mislead at least 5% of the network into believing that some benign application is malicious. Notice how changes in the decision threshold ρ dramatically effect the adversaries' success probability, with almost no effect on the completion time

send notification messages to 10 random network members (namely, $p_N = 0.01$). We require that upon completion, at least 990 network members must be aware of the malicious application (namely, $p_{\text{MAX}} = 0.01$), with $\varepsilon = 0.001$. In addition, we assume that among the network members there are 100 adversaries, trying to deceive at least 50 network devices to believe that some benign application is malicious.

Figure 17.1 shows the time (in days) and messages required in order to complete this mission, as a function of the decision threshold ρ. We can see that whereas the adversaries succeed in probability 1 for $\rho < 3$, they fail in probability 1 for any $\rho \geq 3$. Note the extremely efficient performance of the algorithm, with completion time of $\sim$260 days using only five messages and at most 30 monitored applications per user. The same scenario would have resulted in 100 messages per user using the conventional *flooding* algorithm, or alternatively, in 700 days and 100 monitored applications per user using a non-collaborative scheme. Figure 17.2 demonstrates the decrease in completion time and message requirement as a result of increasing the penetration threshold p_{MAX}. Figure 17.3 demonstrates the evolution in the malicious application's penetration probability throughout the vaccination process. An interesting phenomenon is demonstrated in Fig. 17.4, where the number of adversarial devices is gradually increased, and their success in deceiving 5% of the network's members is studied. It can be seen that as the deception rate increases linearly, the success to generate a successful attack displays a phase transition – growing rapidly from "a very low attack success probability" to "very high attack success probability" with the increase of only 20% in the number of adversaries.

Many viruses, trojans or other malicious applications contain an integral part in charge of proliferating them throughout the network. We would therefore like to integrate this property into the analysis of our *TPP* vaccination algorithm. For this, we use a variation of the *SIR* (Susceptible – Infectious – Recovered) epidemic model. According to the SIR model, once becoming infected, a network device can infect other devices until it is "cured" (namely, receives the relevant

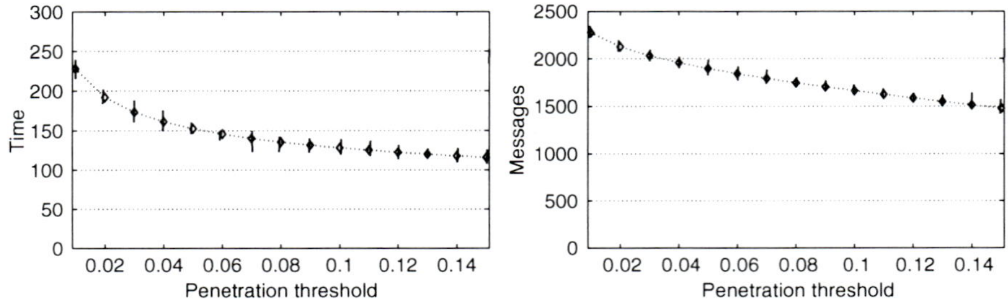

Fig. 17.2 The effect of decreasing the penetration threshold p_{MAX} on the algorithm's completion time and number of messages ($\rho = 1$)

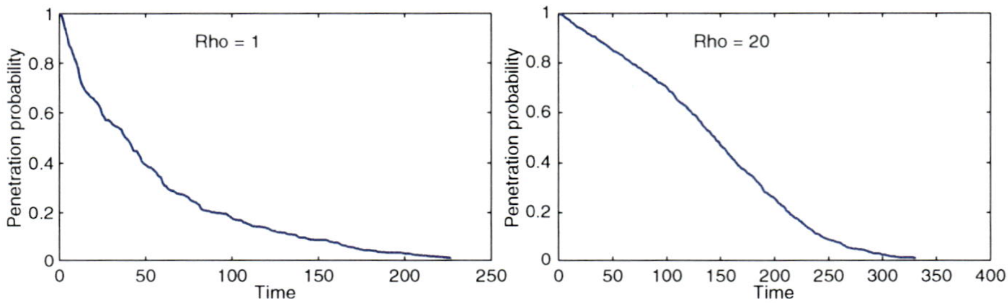

Fig. 17.3 The *penetration probability* of the malicious application, as a function of the time, with $\rho = 1$ (on the *left*) and $\rho = 20$ (on the *right*)

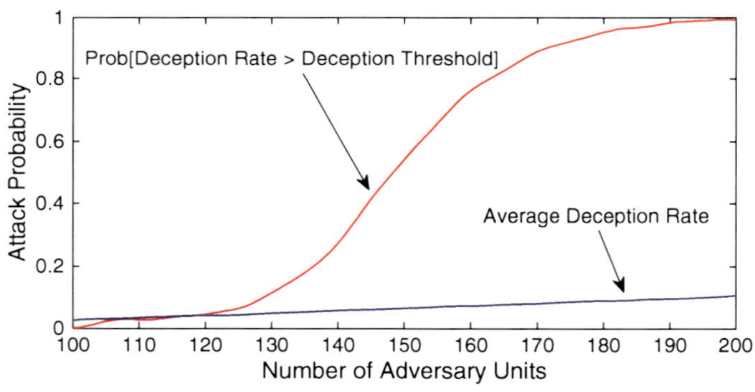

Fig. 17.4 An illustration of the phase transition that is displayed when observing the influence of the number of adversaries over their probability to successfully generate an attack. Note how an increase of 20% in the number of adversaries increases the probability to deceive a large enough portion of the network from less than 0.2 to approximately 0.8

vaccinated information). Once vaccinated, the device is immune to reinfections of this particular threat. This epidemic model can be represented as follows:

- $\text{Infectious}(t) \triangleq I(t) = I(t - dt) + (DI - DR) \times dt$
- $\text{Recovered}(t) \triangleq R(t) = R(t - dt) + DR \times dt$

- Susceptible$(t) \triangleq S(t) = S(t-dt) - DI \times dt$
- New_recoveries $\triangleq DR = \frac{1}{n} I \times p_R$
- New_infections $\triangleq DI = \frac{1}{n} I \times \frac{1}{n} S \times r_I$

In addition, at any given time, an infected device has a certain possibility of suffering from other symptoms of the infection (such as malicious OS crash, identity theft through key-loggers, etc.). As we would later be interested in the overall number of possible compromised devices, we would like to add this to our model, as follows:

- Infected_undmg$(t) \triangleq IuD(t) = IuD(t-dt) + (DI - DuR - DD) \times dt$
- Damaged$(t) \triangleq D(t) = D(t-dt) + DD \times dt$
- New_undmg_recoveries $\triangleq DuR = \frac{1}{n} IuD \times p_R$
- New_damages $\triangleq DD = \frac{1}{n} IuD \times p_D$

This model is initialized by the following values:

- $p_D \triangleq$ Damage probability
- $p_R \triangleq$ Recovery probability
- $r_I \triangleq$ Infection rate
- $S(0) = n - I_0$
- $D(0) = 0$
- $R(0) = 0$
- $IuD(0) = I(0) = I_0$

Let us denote the maximal level of infectious units throughout the evolution of the epidemic by $\widehat{I}$. A graphic illustration of this model is presented in Fig. 17.5.

Similar to many biological systems, devices infected with malicious applications can self-cure [20,46]. In addition, devices can be cured by receiving assistance from other network's devices (by receiving maliciousness notification messages from enough members of the network). Therefore, p_R is derived from the *TPP* algorithm's performance, and will be discussed in the following sections.

The infection rate r_I is a configurable parameter of the system, denoting the spreading speed of the threat. The value of r_I is provided by the operators or the monitors of the network, and can be changed in response to a detection of new threats. A larger value of r_I would cause the vaccination system to require a larger number of messages for its operation, while a lower value of r_I would allow the system to guarantee a proper vaccination using less messages. As to the initial number of infectious members, I_0, we can assume that it is equal to the initial penetration probability multiplied by the number of network's devices.

The effect of the *TPP* vaccination algorithm on an epidemically spreading network threat is illustrated in Fig. 17.5.

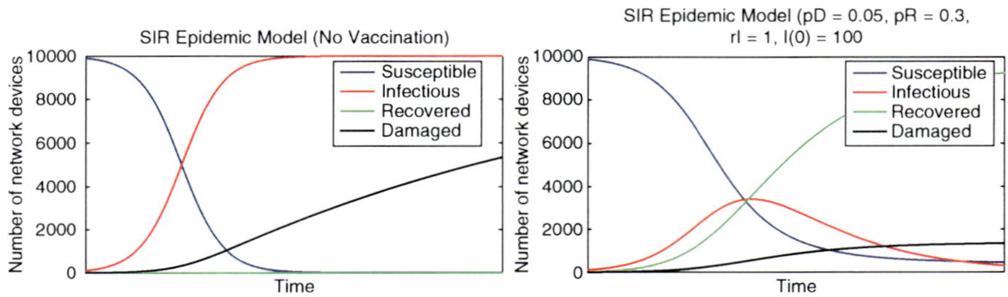

Fig. 17.5 An illustration of the effect of the *TPP* vaccination algorithm on an epidemically spreading network threat. The epidemic model is initialized with the following values: $n = 10{,}000$, $p_{\text{Damage}} = 0.05$, $p_{\text{Recovery}} = 0.1$ (due to the vaccination algorithm), $r_{\text{Infection}} = 1$, $I_0 = p_{\text{MAX}} = \frac{n}{100}$

17.6 Avoiding Benign Applications "Frame" Through Adversarial Use of the *TPP* Vaccination Algorithm

As mentioned in previous sections, the *TPP* algorithm is fault tolerant to the presence of adversarial devices which try to compromise the system's integrity by causing a large enough portion of the network to consider (one or more) benign applications as malicious. This aspect of the algorithm is crucial, as it is a collaborative algorithm which relies on the participation of as many network devices as possible – devices who should be guaranteed that the efficient collaborative monitoring will not be accompanied with erroneous results. In the *TPP* algorithm, this fault tolerance is being provided by the introduction of the ρ "decision threshold" into the decision system. Namely, the fact that an application is being marked as malicious only when at least ρ relevant messages (from different origins) are being received. Relying on the common agreement of several devices as a tool for overcoming system noise (which can be either coincidental or intentional) is often used in swarm based systems. For example, a similar mechanism called "threshold cryptography" is used for enabling the collaborative use of cryptographic tasks (see for example [26, 48].

Definition 17.3. Let us denote by $P_{\text{Attack}}(TTL, \rho, \frac{k}{n}, \varepsilon)$ the probability that a "framing attack" done by a group of k organized adversaries will successfully convince at least $\varepsilon \times n$ of the network's devices that some benign application a_i is malicious.

A trivial example is the use of very large values for TTL, which allow a group of k adversaries to convince the entire network that any given application is malicious, provided that $k > \rho$, namely

$$\forall k \geq 1, \quad \forall \rho \leq k, \quad \forall \varepsilon < 1, \quad \lim_{TTL \to \infty} P_{\text{Attack}}\left(TTL, \rho, \frac{k}{n}, \varepsilon\right) = 1$$

Theorem 17.5. *The probability that k attackers will be able to make at least an ε portion of the network's devices treat some benign application a_i as malicious, using the* TPP *algorithm with a decision threshold ρ is*

$$P_{Attack}\left(TTL,\rho,\frac{k}{n},\varepsilon\right) \leq 1-\Phi\left(\sqrt{n}\times\frac{\varepsilon-\tilde{P}}{\sqrt{\tilde{P}(1-\tilde{P})}}\right)$$

where

$$\tilde{P}=e^{\left(\rho-TTL\times\left(1-e^{-\frac{k\times p_N}{2}}\right)\right)}\times\left(\frac{TTL\times\left(1-e^{-\frac{k\times p_N}{2}}\right)}{\rho}\right)^{\rho}$$

and where $\Phi(x)$ is the cumulative normal distribution function, defined as

$$\Phi(x)=\frac{1}{\sqrt{2\pi}}\int_{-\infty}^{x}e^{-\frac{1}{2}t^2}dt$$

and also provided that

$$\rho > TTL\left(1-e^{-\frac{k\times p_N}{2}}\right)$$

Proof. We use Lemma 17.2 to calculate the probability that a device $v\in V$ will be reported of some malicious application a_i by a message sent by one of the k adversaries at the next time-step. This is yet again a Bernoulli trial with

$$p_s=1-e^{-\frac{(k\times n\times p_N)}{2n}}=1-e^{-\frac{k\times p_N}{2}}$$

Denoting as $X_v(t)$ the number of times a notification message had arrived to v after t steps, using *Chernoff* bound

$$P[X_v(t)>(1+\delta)t\times p_s]<\left(\frac{e^{\delta}}{(1+\delta)^{(1+\delta)}}\right)^{t\times p_s}$$

in which we set $\delta=\frac{\rho}{t\times p_s}-1$. We can therefore see that

$$\tilde{P}\triangleq P_{\text{Attack}}\left(TTL,\rho,\frac{k}{n},n^{-1}\right)=P[X_v(TTL)>\rho]<e^{(\rho-TTL\times p_s)}\times\left(\frac{TTL\times p_s}{\rho}\right)^{\rho}$$

It is important to note that the *Chernoff* bounds requires that $\delta>0$. This is immediately translated to the following requirement, necessary for the validity of this Theorem

$$\rho > TTL\left(1-e^{-\frac{k\times p_N}{2}}\right)$$

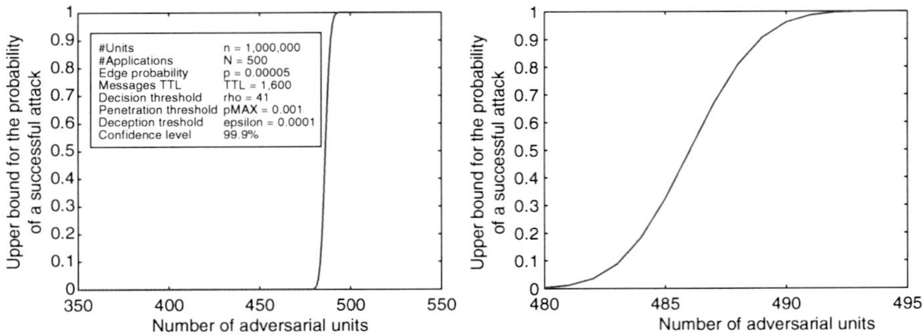

Fig. 17.6 An illustration of Theorem 17.5 – an upper bound for the success probability of a collaborative "framing attack" as a function of the number of adversarial devices. In this example, a changing number of adversaries are required to deceive at least 100 devices to think that some benign application is in fact malicious. Notice the phase transition point around 485 adversarial device

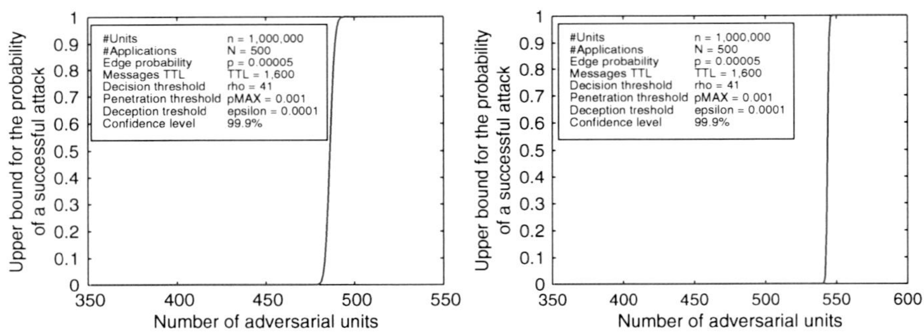

Fig. 17.7 An illustration of Theorem 17.5 – the influence of the deception threshold ε (namely, the size of the network portion which is needed to be deceived in order for an attack to be considered successful) on the number of adversarial devices required for a successful attack

As we want to bound the probability that at least εn of the devices are deceived, we shall use the above as a success probability of a second Bernoulli trial. As n is large, the number of deceived devices can be approximated using normal distribution, as follows:

$$P_{\text{Attack}}\left(TTL, \rho, \frac{k}{n}, \varepsilon\right) \leq 1 - \Phi\left(\frac{\varepsilon \times n - n \times \tilde{P}}{\sqrt{n \times \tilde{P}(1-\tilde{P})}}\right)$$

and the rest is implied. □

Theorem 17.5 is illustrated in Figs. 17.6–17.8.Figure 17.6 presents the execution of the vaccination algorithm in a network of 1,000,000 devices, where each device is connected to 50 neighbors. In this example, the network operators require that

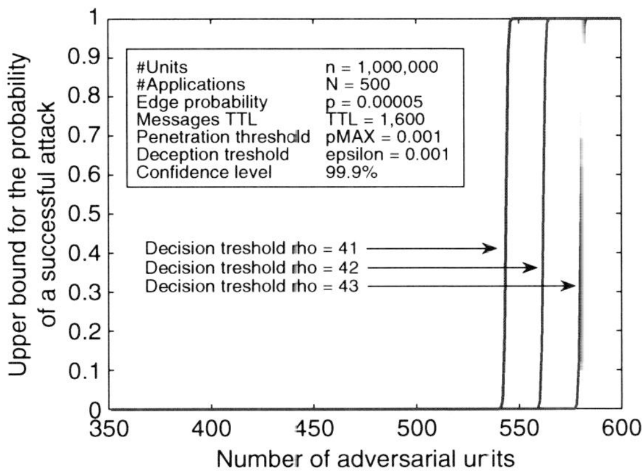

Fig. 17.8 An illustration of Theorem 17.5 – the influence of the value of decision threshold ρ on the number of adversarial devices required for a successful attack. Note how as the adversaries increase their numbers, all that has to be done in order to prevent them from launching successful attacks is simply to slightly increase the value of the decision threshold ρ. This feature of the *TPP* algorithm makes it economically favorable for networks operators. From experimental observations it can be seen that the resources required in order to defend a network against attacks of growing scales (both convergence time and number of messages) converge asymptotically, whereas the resources required in order to launch stronger attacks grow approximately linearly

the number of devices that may be deceived as a result of an adversarial attack would be at most 100. With the decision threshold ρ properly calibrated according to Theorem 17.5, it is shown that as long as the number of adversaries is below 480 the adversaries cannot launch a successful attack. However, as the number of adversaries increases, such an attack quickly becomes a feasible option. Specifically, by increasing the number of adversaries by 3% (from 480 to 494) the probability of a successful attack rises from 0 to 1. In this case, in order to compensate this change in adversaries numbers, all the devices operating the *TPP* algorithm have to do is simply increase the value of ρ by 1. Figure 17.8 shows how each such small increase in the value of the decision threshold ρ requires the adversaries to invest a lot of effort and increase their numbers by approximately 7%. Although severely preventing the adversaries attempts to launch successful attacks, the effect of such changes in the value of ρ on the "normative" devices of the network is very small, as can be observed in Fig. 17.1.

Note that in the proof of Theorem 17.5 we assumed that the adversarial devices may decide to send a false message concerning an application's maliciousness, but they must do so using the definitions of the *TPP* algorithm. Namely, each device may send at most $p_N \times n$ messages, and send these messages to random destinations. If adversarial devices had been allowed flexibility in those constraints as well, a small group of adversarial devices could have sent an infinitely large number of false messages, that would have been propagated throughout the network, resulting in a successful attack (similarly to the case where $TTL \rightarrow \infty$). Alternatively, had adversarial devices been allowed to chose the destination of the $p_N \times n$ messages they send, they could have all send them to the same destination v, thus guaranteeing that v would be deceived. More generically, a group of $k = \imath \times \rho$ of adversarial devices could have send their messages to $i \times p_N \times n$ different devices, guaranteeing their deception.

Definition 17.4. Let us denote by $P_{\text{Attack}-\text{Destination}}\left(TTL, \rho, \frac{k}{n}, \varepsilon\right)$ the attack success probability when adversarial devices are allowed to control the destination of the messages they produce.

The following Corollary can be drawn:

Corollary 17.3. *The probability that k attackers that can control the destination of the $p_N \times n$ messages they produce will be able to make at least an ε portion of the network's devices treat some benign application a_i as malicious, using the* TPP *algorithm with a decision threshold ρ is:*

$$P_{Attack-Destination}\left(TTL, \rho, \frac{k}{n}, \varepsilon\right) \leq P_{Attack}\left(TTL-1, \rho, \frac{k}{n}, \varepsilon - \frac{k}{\rho} \times p_N\right)$$

17.7 Conclusions and Future Work

In this work, we have presented the *TPP* TTL-based propagation algorithm, capable of guaranteeing the collaborative vaccination of mobile network users against malicious applications. The performance of the algorithm was analyzed and shown to be superior compared to state of the art in this field, guaranteeing fast collaborative detection of attack attempts, and do so using a lower network overhead.

The algorithm was also shown to be capable of overcoming the presence of adversarial devices who try to inject messages of false information into the network. The challenge of filtering out false information which is injected into a collaborative information propagation networks resembles the known "faulty processors" problem. For example, the following work discusses the challenge of synchronizing the clock of a communication network of size n when $\frac{n}{3}$ faulty processors are present [10]. Another interesting work in this scope is the work of [14] which discusses a collaborative fault tolerant "acknowledgment propagation" algorithm, for reporting on the receipt (or lack of) of sent messages.

It should be noted that during the analysis of the fault tolerance of the algorithm, we assumed that although an attacker can send reports of benign applications, or alternatively – refuse to forward messages passed through it, the basic parameters of the algorithm are still preserved. Namely, adversarial devices are not allowed to send more than X messages or generate messages with TTL values higher than the value allowed by the network operator. The exact implementation details of a cryptographic protocol of these properties, however, is out of the scope of this work,

Future versions of the work should investigate the generalization of the propagation mechanism, allowing the number of alert messages generated to be dynamically calibrated in order to further decrease the algorithm's overhead (for example, as a function of the number of alerts already received concerning the application). Alternatively, upon forwarding a message, devices might be requested to send more than a single copy of the message they received.

Another interesting topic to investigate is the use of the mechanism proposed in this work as an infrastructure for other security related problems. One example can be the problem of collaboratively coping with malicious beacons in hostile wireless environment, as discussed in [43]. Most of the existing localization protocols for sensor networks are vulnerable in hostile environments, requiring for the enhancement of the security of location discovery. The work of [43] presents voting-based methods to tolerate malicious attacks against range-based location discovery in sensor networks. This problem seems to benefit from the use a mechanism which is fault tolerant to the injection of false information, such as the algorithm we propose in this work.

References

1. Mcafee mobile security report 2008. Tech. rep. (2008). http://www.mcafee.com/us/resources/reports/rp-mobile-security-2008.pdf
2. Mcafee mobile security report 2009. Tech. rep. (2009). http://www.mcafee.com/us/resources/reports/rp-mobile-security-2009.pdf
3. Adamic, L.A., Lukose, R.M., Puniyani, A.R., Huberman, B.A.: Search in power-law networks. Phys. Rev. E **64**(4), 046,135 (2001). DOI 10.1103/PhysRevE.64.046135
4. Altshuler, Y., Bruckstein, A.M.: Static and expanding grid coverage with ant robots: Complexity results. Theor. Comput. Sci. **412**(35), 4661–4674 (2011)
5. Altshuler, Y., Yanovsky, V., Bruckstein, A., Wagner, I.: Efficient cooperative search of smart targets using uav swarms. ROBOTICA **26**, 551–557 (2008)
6. Angluin, D., Aspnes, J., Eisenstat, D.: A simple population protocol for fast robust approximate majority. Dist. Comp. **21**, 87–102 (2008)
7. Apap, F., Honig, A., Hershkop, S., Eskin, E., Stolfo, S.: Detecting Malicious Software by Monitoring Anomalous Windows Registry Accesses. Recent Advances in Intrusion Detection, pp. 36–53. Springer Berlin Heidelberg (2002)
8. Aspnes, J., Ruppert, E.: An Introduction to Population Protocols. Middleware for Network Eccentric and Mobile Applications, pp. 97–120. Springer Berlin Heidelberg (2009)
9. Bailey, N.: The Mathematical Theory of Infectious Diseases and its Applications (second edition). Hafner Press (1975)
10. Barak, B., Halevi, S., Herzberg, A., Naor, D.: Clock synchronization with faults and recoveries (extended abstract). In: PODC '00: Proceedings of the nineteenth annual ACM symposium on Principles of distributed computing, pp. 133–142. ACM, New York, NY, USA (2000). DOI http://doi.acm.org/10.1145/343477.343534
11. Batalin, M., Sukhatme, G.: Spreading out: A local approach to multi-robot coverage. In: 6th International IEEE Symposium on Distributed Autonomous Robotics Systems, (IEEE) (2002)
12. Cagalj, M., Hubaux, J., Enz, C.: Minimum-energy broadcast in all-wireless networks: Np-completness and distribution issues. In: The Annual International Conference on Mobile Computing and Networking (MOBICOM), ACM (Atlanta Georgia) (2002)

13. Mobile Broadband Growth report: HSPA/HSPA+ operator success stories worldwide, trends, forecasts, GSA (the Global mobile Suppliers Association) (2010)
14. Castelluccia, C., Jarecki, S., Kim, J., Tsudik, G.: Secure acknowledgement aggregation and multisignatures with limited robustness. Computer Networks **50**(10), 1639–1652 (2006)
15. Choy, M., Singh, A.K.: Efficient fault-tolerant algorithms for distributed resource allocation. ACM Trans. Program. Lang. Syst. **17**(3), 535–559 (1995). DOI http://doi.acm.org/10.1145/203095.203101
16. Chung, F., Lu, L.: The diameter of sparse random graphs. Advances in Applied Mathematics **26**, 257–279 (2001)
17. Crisostomo, S., Barros, J., Bettstetter, C.: Flooding the network: Multipoint relays versus network coding. In: 4th IEEE Intl. Conference on Circuits and Systems for Communications (ICCSC). IEEE (Shanghai, China), pp. 119–124 (2008)
18. Demers, A., Greene, D., Hauser, C., Irish, W., Larson, J., Shenker, S., Sturgis, H., Swinehart, D., Terry, D.: Epidemic algorithms for replicated database maintenance. In: In Proc. of the Sixth ACM Symposium on Principles of Distributed Computing, (Vancouver, Canada), pp. 1–12 (1987)
19. D.F. Zucker M. Uematsu, T.K.: Markup-based smartphone user interface using the web browser engine. In: Proceedings XTech. IDEAlliance, Amsterdam, pp. 25–27 (2005)
20. Dolev, S., Schiller, E., Welch, J.: Random walk for self-stabilizing group communication in ad hoc networks. IEEE Transactions on Mobile Computing, Switzerland **5**, 893–905 (2006)
21. Dolev, S., Tzachar, N.: Spanders: Distributed spanning expanders. In: Proceedings ACM SCS (ACM), Switzerland (2010)
22. Erdos, P., Renyi, A.: On random graphs. Publ. Math. Debrecen **6**, 290–291 (1959)
23. Erdos, P., Renyi, A.: On the evolution of random graphs. Publications of the Mathematical Institute of the Hungarian Academy of Sciences **5**, 17–61 (1960)
24. Fragouli, C., Widmer, J., Boudec, J.L.: A network coding approach to energy efficient broadcasting: from theory to practice. In: The 25th IEEE International Conference on Computer Communications (INFOCOM2006). IEEE, Barcelona, pp. 1–11 (2006)
25. Ganesa, D., Krishnamachari, B., Woo, A., Culler, D., Estrin, D., Wicker, S.: An empirical study of epidemic algorithms in large scale multihop wireless networks – technical report ucla/csd-tr 02-0013. Technical report, UCLA Computer Science (2002)
26. Gemmel, P.: An introduction to threshold cryptography. CryptoBytes, RSA Labs, pp. 7–12 (1997)
27. GetJar: Getjar statistics (2010)
28. Golding, R., Long, D., Wilkes, J.: The refdbms distributed bibliographic database system. In: In Proc. of USENIX. USENIX (the advanced computer systems association), Boston MA, pp. 47–62 (1994)
29. Haas, Z., Halpern, J., Li, L.: Gossip-based ad-hoc routing. IEEE/ACM Transactions of networks **14**(3), 479–491 (2006)
30. Hypponen, M.: Malware goes mobile. Sci. American **295**, 70 (2006)
31. Hypponen, M.: State of cell phone malware in 2007. Technical report FSECURE (2007)
32. J. Cheng S. Wong, H.Y.S.L.: Smartsiren: Virus detection and alert for smartphones. In: In Proceedings of the Fifth ACM International Conference on Mobile Systems Applications and Services (MOBISYS). ACM, Puerto Rico, pp. 258–271 (2007)
33. Jacoby, G., Davis, N.: Battery-based intrusion detection. In: Proceedings of the IEEE Global Telecommunications Conference, GLOBECOM04. IEEE, Dallas Texas, vol. 4, pp. 2250–2255 (2004)
34. Jacquet, P., Laouiti, A., Minet, P., Viennot, L.: Performance analysis of olsr multipoint relay flooding in two ad-hoc wireless network models. technical report 4260. Technical report, INRIA (2001)
35. Jonasson, J., Schramm, O.: On the cover time of planar graphs. Electron. Comm. Probab. **5**, 85–90 (electronic) (2000)
36. Kim, H., Smith, J., Shin, K.G.: Detecting energy-greedy anomalies and mobile malware variants. In: MobiSys '08: Proceeding of the 6th international conference on Mobile systems,

applications, and services, pp. 239–252. ACM, New York, NY, USA (2008). DOI http://doi.acm.org/10.1145/1378600.1378627
37. Kleinberg, J.: The wireless epidemic. Nature **449**, 287–288 (2007)
38. Koenig, S., Liu, Y.: Terrain coverage with ant robots: A simulation study. In: The International Conference on Autonomous Agents (AGENTS). (ACM), Montreol Canada 600–607 (2001)
39. Koenig, S., Szymanski, B., Liu, Y.: Efficient and inefficient ant coverage methods. Annals of Mathematics and Artificial Intelligence **31**, 41–76 (2001)
40. Kong, C., Peng, N., Rekleitis, I.: Distributed coverage with multi-robot system. In: IEEE International Conference on Robotics and Automation (ICRA). IEEE Orlando Florida, (2006)
41. Korf, R.: Real-time heuristic search. Artificial Intelligence **42**, 189–211 (1990)
42. Lim, H., Kim, C.: Multicast tree construction and flooding in wireless ad hoc networks. In: In Proceedings of the ACM International Workshop on Modeling, Analysis and Simulation of Wireless and Mobile Systems (MSWIM). ACM, Bodrum Turkey (2000)
43. Liu, D., Ning, P., Liu, A., Wang, C., Du, W.K.: Attack-resistant location estimation in wireless sensor networks. ACM Trans. Inf. Syst. Secur. **11**(4), 1–39 (2008). DOI http://doi.acm.org/10.1145/1380564.1380570
44. Lv, Q., Cao, P., Cohen, E., Li, K., Shenker, S.: Search and replication in unstructured peer-to-peer networks. In: ICS '02: Proceedings of the 16th international conference on Supercomputing, pp. 84–95. ACM, New York, NY, USA (2002). DOI http://doi.acm.org/10.1145/514191.514206
45. Moskovitch, R., Gus, I., Pluderman, S., Stopel, D., Glezer, C., Shahar, Y., Elovici, Y.: Detection of unknown computer worms activity based on computer behavior using data mining. In: CISDA 2007. IEEE Symposium on Computational Intelligence in Security and Defense Applications. IEEE, Honolulu, HI, USA, pp. 169–177 (2007)
46. Moskovitch, R., Pluderman, S., Gus, I., Stopel, D., Feher, C., Parmet, Y., Shahar, Y., Elovici, Y.: Host based intrusion detection using machine learning. In: 2007 IEEE Intelligence and Security Informatics. IEEE, ISI, New Jersey, pp. 107–114 (2007)
47. Mutz, D., Valeur, F., Vigna, G., Kruegel, C.: Anomalous system call detection. ACM Trans. Inf. Syst. Secur. **9**(1), 61–93 (2006). DOI http://doi.acm.org/10.1145/1127345.1127348
48. Narasimha, M., Tsudik, G., Yi, J.H.: On the utility of distributed cryptography in p2p and manets: the case of membership control. In: Proceedings of the 11th IEEE International Conference on Network Protocols. IEEE, Atlanta Georgia, pp. 336–345 (2003)
49. Nash, D.C., Martin, T.L., Ha, D.S., Hsiao, M.S.: Towards an intrusion detection system for battery exhaustion attacks on mobile computing devices. Pervasive Computing and Communications Workshops, IEEE International Conference on. IEEE, PERCOM, Hawaii, **0**, 141–145 (2005). DOI http://doi.ieeecomputersociety.org/10.1109/PERCOMW.2005.86
50. Ni, S., Tseng, Y., Chen, Y., Sheu, J.: The broadcast storm problem in a mobile ad hoc network. In: In Proceedings of the ACM/IEEE International Conference on Mobile Computing and Networking (MOBICOM), Seattle, Washington, pp. 151–162 (1999)
51. Peng, W., Lu, X.C.: On the reduction of broadcast redundancy in mobile ad hoc networks. In: MobiHoc '00: Proceedings of the 1st ACM international symposium on Mobile ad hoc networking & computing, pp. 129–130. IEEE Press, Piscataway, NJ, USA (2000)
52. Polycarpou, M., Yang, Y., Passino, K.: A cooperative search framework for distributed agents. In: IEEE International Symposium on Intelligent Control (ISIC), Mexico City, pp. 1–6 (2001)
53. Qayyum, L., Laouiti, A.: Multipoint relaying for flooding broadcast messages in mobile wireless networks. In: Proceedings of HICSS. IEEE Hawaii, (2002)
54. Rekleitis, I., Lee-Shuey, V., Newz, A.P., Choset, H.: Limited communication, multi-robot team based coverage. In: IEEE International Conference on Robotics and Automation. IEEE, Barcelona (2004)
55. van Renesse, R.: Power-aware epidemics. Reliable Distributed Systems, IEEE Symposium on. IEEE Computer Society, Osaka Japan, **0**, 358 (2002). DOI http://doi.ieeecomputersociety.org/10.1109/RELDIS.2002.1180210

56. van Renesse, R., Birman, K.: Scalable management and data mining using astrolabe. In: In Proc. of the First International Workshop on Peer-to-Peer Systems (IPTPS02). Springer, Cambridge MA, (2002)
57. Sasson, Y., Cavin, D., Schiper, A.: Probabilistic broadcas for flooding in wireless mobile ad-hoc networks. In: Proceedings of IEEE Wireless communication and networks (WCNC). IEEE, New Orleans, Louisiana (2003)
58. Shevchenko, A.: An overview of mobile device security. Kaspersky Labs, (available at www.viruslist.com/en/analysis?pubid$=$170773606) (2005)
59. Stojmenovic, I., Seddigh, M., Zunic, J.: Ahbp: An efficient broadcast protocol for mobile ad hoc networks. Journal of Computer Science and Technology **16**(2), 114–125 (2001)
60. Stojmenovic, I., Seddigh, M., Zunic, J.: Dominating sets and neighbor elimination-based broadcasting algorithms in wireless networks. IEEE Transactions on Parallel and Distributed Systems **13**(1), 14–25 (2002). DOI http://doi.ieeecomputersociety.org/10.1109/71.980024
61. Stone, L.: Theory of Optimal Search. Academic Press, New York (1975)
62. Svennebring, J., Koenig, S.: Building terrain-covering ant robots: A feasibility study. Autonomous Robots **16**(3), 313–332 (2004)
63. Vogels, W., van Renesse, R., Birman, K.: The power of epidemics: robust communication for large-scale distributed systems. SIGCOMM Comput. Commun. Rev. **33**(1), 131–135 (2003). DOI http://doi.acm.org/10.1145/774763.774784
64. Wagner, I., Altshuler, Y., Yanovski, V., Bruckstein, A.: Cooperative cleaners: A study in ant robotics. The International Journal of Robotics Research (IJRR) **27**(1), 127–151 (2008)
65. Wang, P., Gonzalez, M., Hidalgo, C., Barabasi, A.: Understanding the spreading patterns of mobile phone viruses. Science **324**, 1071–1075 (2009)
66. Williams, B., Camp, T.: Comparison of broadcasting techniques for mobile ad hoc networks. In: The Third ACM International Symposium on Mobile Ad Hoc Networking and Computing. Lausanne, Switzerland, pp. 9–11 (2002)
67. Zlot, R., Stentz, A., Dias, M., Thayer, S.: Multi-robot exploration controlled by a market economy. In: Proceedings of the IEEE International Conference on Robotics and Automation. IEEE, Washington DC (2002)

Index

M.T. Thai and P.M. Pardalos (eds.), *Handbook of Optimization in Complex Networks: Theory and Applications*, Springer Optimization and Its Applications 57, DOI 10.1007/978-1-4614-0754-6,